NARROW ROADS O

The Collected Papers of

W. D. HAMILTON

see National Geographic Feb 2008 for 'Basho + Narrow Roads.

T

Darv
ard
Gene,
most
20th
Ha
for h
the

how
ours s
ch a s
ally st
d direc
By the
ne *Jour*
Hamilto

NARROW ROADS OF GENE LAND

The Collected Papers of W. D. HAMILTON

VOLUME 1
Evolution of Social Behaviour

OXFORD · NEW YORK · HEIDELBERG

W. H. Freeman at Macmillan Press Limited
Houndmills, Basingstoke, RG21 6XS
W. H. Freeman and Company
41 Madison Avenue, New York, NY10010

British Library Cataloguing in Publication Data
A catalogue record for this book is available from the British Library.

Library of Congress Cataloging-in-Publication Data
Hamilton, W. D. (William Donald), 1936–
Narrow roads of gene land : the collected papers of W. D. Hamilton
p. cm. Includes bibliographical references and index.
Contents: Vol. 1, Evolution of social behaviour
ISBN 0–7167–4530–5 (pbk.)
ISBN 0–7167–4551–8 (hbk.)
1. Social behavior in animals. 2. Behavior evolution.
3. Sexual selection in animals.
I. Title
QL775.H33 1995 95–16280 575.01′62—dc20 CIP

Set by KEYWORD Publishing Services
Printed by The Maple-Vail Manufacturing Group

Contents

Preface

THIS book is the first of two volumes aiming to re-publish all my substantial scientific papers together. The introductions accompanying the papers explain why I studied a topic and where I was at the time. They note what has happened since the paper was written and provide a few recent references. Furthermore, they include a varying amount of anecdote and opinion. They are untechnical and are, I believe, easily readable by anyone interested in science.

In the first half of my life's work I studied mainly social aspects of the working of evolution; in the second I moved to a problem that is raised by and underlies this first interest: why is there sex? Roughly these halves of my working life define the two volumes. Volume 1 is about selection for social behaviour but a glimpse forward to the substance of Volume 2 comes near the end in Chapter 10. This is a review showing my growing interest in sexuality and some sign of how I think the debate will go. Obviously, a scientist's long-term interests seldom change abruptly so that similarly in Volume 2 there will be, besides much on 'sex itself', at least one paper that looks back to 'pure' social behaviour and also much thought on sexual social behaviour. If these distinctions sound a bit vaticanical, their oddness may perhaps prime the reader to await Volume 2.

Michael Rodgers, my editor at Spektrum, pursued me relentlessly for some 15 years to persuade me to write *some* book. For a long time—enough perhaps to have written several—I have told him I would love to write one soon but still there was a question I didn't sufficiently understand, that other research I was urgently engaged in . . . At length via a seemingly telepathic collaboration with Sarah Hrdy in California he has undermined my irresolute, procrastinating nature. First, Sarah sapped and breached my wall by making it sound easy (the 'briefest introductions' would do, the rest virtually was already done). Second, Michael Rodgers charged through the breach she had made.

Thus were started the far from short, indeed rambling, introductions that you will find.

What title could I give to a mixture of anecdotes, opinions, and serious science such as are included here? 'Rambling' seemed indeed a key word and at length there crossed my mind a book describing a tour that was made on foot in northern Japan in the seventeenth century. The tourist and writer Matsuo Basho was a poet and he interspersed lively descriptive and anecdotal prose concerning his journey with brief poems—*haiku*—that were also inspired by the people, events, and scenery he encountered. Even in their translation into English, in which most of their assonance must be lost, the poems had long appealed to me as including some of the deepest and most poignant thoughts and images ever recorded. It was therefore with obeisance to Basho and his 'Narrow Roads of Oku' that I chose my title 'Narrow Roads of Gene Land'.

My presumptuous and vague idea is that my introductions may simulate his diary and my papers, his poems. I realize that my title may seem as ridiculous to Japanese ears as it would be to English if I had called the book 'The Tragedies of William D. H. Shakespeare'. If I had, most literate Englishmen, of course, would mutter: 'Tragedies indeed'; correspondingly, I expect Japanese readers will mutter: 'Indeed, roads vanishingly narrow.' Never mind; the poet whose dreams will wander Japan's countryside for as long as it exists (as mere hours before his death at 50, his last marvellous indomitable *haiku* predicted) has not yet warned me to forebear. Great artists are omni-humans; we all think we are like them. Even so I will dare to lay my book direct at this master's feet, sensing in him a maverick fatalist who was indeed somewhat akin, believing that he may enjoy the slant of my papers and perhaps even see point and beauty in some images that I express.

But 'tragedies indeed', and 'roads vanishingly narrow'—both are fair and expectable comments! Believing in the explanatory power of evolution by natural selection is like migraine, or perhaps still more like being, as it was in the old days, a 'wise woman'. The majority of humanity seem to have difficulty in accepting that the 'oddness' of such a believer can be real—that is, simply an oddness and nothing else. As the migraine sufferer is suspected of malingering, and the woman who is merely literally wise, of witchcraft, so the evolutionist is always suspected of covert agendas unconnected with reality or the search for truth. In despair over the unending bemusement in friends and relatives and over the stream of articles and books that still pours forth stating Darwinism to be wrong, dead, right except for natural selection, superseded by this stale or ridiculous notion or that (all of which, evidently, the public eagerly buy and read, no matter what the competence of the writer or his knowledge of the evidence); puzzled, in short, by resistance to ideas that seem vastly more obvious and intuitive than, say, relativity or quantum mechanics, which every one accepts blithely with or without understanding, the evolution sufferer sometimes comes to believe it must be he who is mistaken. I describe this happening to me in this book. At other times the evolutionist may feel like one of the stranger 'genetic morphs' of his own theories—mutant carrier, say, of a fourth intellectual pigment of the retina capable of raising into clear sight patterns of nature and of the human future that are denied to the majority of his fellows, or perhaps just a person bewitched in babyhood to have revealed to him through blind sight, through such X-ray eyes, all the ravishing and foreboding beauty of the world that he now endures.

Whatever the nature of the ailment is, I certainly caught or inherited it early compared to most, and it burned me badly. The book can be read at two levels. One is for the contributions to evolutionary thinking that it contains—contributions that, I will proudly claim, have resisted all efforts for their derailment either by fact or further theory (apart from a few small and self-admitted errors) for periods that now vary from 15 to 30 years. The other is that it may be read, mainly via the Introductions that I have added, as a description of what it is like to live burdened and inspired by the evolution migraine—to be both subject to its piercing attacks and to the lucidity and enthusiasm that often follows them.

Under Stalin and Lysenko, brave men in Russia suffered and sometimes died for refusing to renounce genetics, evolution, or natural selection. Nikolai Vavilov, a pioneer of domestic plant origins, is one outstanding example. At quite an opposite pole from such martyrs, I have led a charmed life secure in the two countries that have been historically, perhaps, most favourable to evolutionary thought, and perhaps even more outstandingly favourable concerning its 'worst' aspect— the elucidation of natural selection. Mentioning here the seemingly unabating popular hostility to current (and especially Darwinian) evolution theory that continues worldwide, I am made the more deeply aware how there exist people who have made their way in the real markets of the world (and all over the world), who see the interest and value of right ideas for themselves, irrespective of their immediate profit-making or humanitarian applications. These in the end bow not to crowd opinions and have enough influence to give support to all such studies as lead to real understanding. Examples springing to mind are Caryl Haskins, Gordon Getty, Kasuo Inamori (not to forget, of course, such great rationalists of the farther past as Captain Cook, Benjamin Franklin, or even, say, the Medicis). My start may have been shaky and my position odd at times, but thanks to such more worldly people, no one, hardly even Darwin himself, can have enjoyed better opportunities for study than I have had. I very much hope that the papers that follow in the two volumes of this work, few as they are, will be considered to justify the help I have received.

It is in exactly this line that I am grateful for my salaried support and opportunities in Britain, in the United States, and in Brazil, to carry out the work represented in this book. My special thanks go to the Imperial College of London University, to the University of Michigan, and to Warwick Kerr for his invitations to Brazil; and most recently and overwhelmingly my thanks are due to the Fellows of the Royal Society of London who in 1984 appointed me a Royal Society Research Professor and have continued to support me generously up to the present (even before 1984, I should mention, I was in their debt for small grants while at IC).

Even though a full catalogue would speak strongly for the internationalism of interest I have already mentioned, to list all university departments, institutes, societies, academies, and congresses that have given me encouragement through their invitations to lecture would be tedious for the reader. Instead it must suffice to name the countries (in addition to the three above) from which such invitations have come: Australia, Brunei, Canada, Crimea, Denmark, Finland, France, Germany, Italy, India, Israel, Japan, Netherlands, New Zealand, Norway, Poland, Sweden, and Switzerland. To all these countries I express my thanks.

Numerous people played a more personal part in making these papers possible and especially helping to initiate them. My parents started me early into science, perhaps were surprised and dismayed (at times) by my maniacal response, but even in worst years, when it was unclear that their efforts on my education would lead to any career, never faltered in their support. My sister Mary and my other siblings have been crucial to that family base of strength that made me willing to step back to meditate. Other good friends in the years of Cambridge, London, and Ascot—Colin Hudson, Vahine Gibbings, Marion Luke, David Harris, Yura Ulehla—made the condition of my meditation, by then complexed by being a far-gone gene and altruism freak, more tolerable. Close after those came Christine Friess, soon to be my wife. She has given me her longstanding patience in respect of wasps in Brazil, whitefly on her tomatoes, sexless weevils felling her houseplants, integrals and Voronoi polygons far into the night, and eventually, in spite of all this, the joy, support, and distraction of a family.

Debts to my co-authors I hope appear sufficiently in the chapters concerned; others, of course, are to come in Volume 2. Robert Trivers, Richard Alexander, Edward Wilson, George Williams, Michael Orlove, and Richard Dawkins were never co-authors but have become known almost as if they had been through their endeavours parallel and supporting. Thanks, too, for early serious attention in their own papers to my work, must yet again fly the Atlantic to Gordon Orians, Jerram Brown, and Jon Seger. Regarding the production of this volume I am much indebted to five in particular. Starting from what now seems a long time ago there have been three who, standing largely apart from its chapters and themes, have helped the whole enterprise both directly and by good counsel: Sarah Hrdy, Naomi Pierce, and Katrina Mangin. More recently two have kept me afloat. First, for such even grain of sense and appearance as the book has, as well as for many sharp thoughts for improvement, I thank Sarah Bunney, my copy editor. Second, for an index that can at least help me, if no one else, to distinguish what I wrote from what I thought or wished that I wrote, I thank Jeremy John.

Having listed friends—with all too many still unmentioned since the net eventually spreads too wide—I am tempted to begin listing my enemies, those who have spurred my work on by being so outstandingly wrong, so critical, and so irritatingly indifferent. But this multitude of patrons, beyond those broad classes already hinted at, had better be left to be implied in the papers themselves.

CHAPTER 1

Willing sentinel

On a termite mound in the semi-desert of the southern Kalahari a meerkat risks predation while watching for hawks, jackals, and snakes. Around it in relative safety its companions forage over sand and leaf litter for their small prey. Groups are greatly benefited by the sentinel and, when too small to post one, rapidly die out.

CHAPTER 1

SHOULDERS OF GIANTS

The Evolution of Altruistic Behaviour

There be many shapes of mystery.
And many things God makes to be,
Past hope and fear.
And the end men looked for cometh not,
And a path is there where no man sought,
So hath it fallen here.

EURIPIDES[1]

UP to 1963 my only 'published' writings were a story written when a child and filling a half page in Sevenoaks Preparatory School Magazine and a review of a popular book on dolphins for a New Zealand literary journal.[2] The review was done when I was already a graduate student in London. The reasons why the editor of *Landfall* requested a review from a biology student so far away are, probably, first, that he did not know other biologists, second, I was his godson, and, third, that he knew from my rare letters that I was working on something to do with social behaviour. Even then the lives of dolphins had a reputation for being unusual and complex. My ideas about them, as about the book, were very lightweight and my review does not merit inclusion in this book. However, the topic of dolphins and my godfather's 'nepotism' in sending me the task are relevant in an odd way to the paper that follows. They hint at matters beyond biology, even a line of escape from the theory my paper describes. Dolphins indeed come near to defying Darwin himself by an act of a kind of which he said that, if a single instance could be convincingly documented, it would annihilate his theory completely.[3] It seems that sometimes a dolphin may push a drowning human—a mammal from a world almost as alien to theirs as a world of Martians would be to ours—to the sea surface where the human can breathe or, perhaps still more inexplicably appropriate, may push or carry him or her towards the shore, as first recorded in numerous Greek legends, such as the one concerning Arion.[4]

Just as *Homo*, the taxon of humans, is unrelated to *Tursiops truncatus*—that of the seemingly most foolishly anthropophilic of the dolphins—so I, within *Homo*, was unrelated to my New Zealand godfather Charles Brasch, who gave me my first chance of a publication. Yet my parents named him my god*father* and he acknowledged me as his god*son*: I thus received a kind of pseudo-nepotism from him—and in fact I received this on many occasions besides the one mentioned. Was this just a stupid mistake on his part, as biologists are inclined to interpret the dolphin's act to be? If not, what aberration or extension of the normal run of nepotism is here? That 30 years before Charles Brasch had hoped to marry my mother but since that time had stayed single and, living as he did in New Zealand, had hardly a hope even to see her, certainly cannot, via biology, help to resolve the issue.

Doubtless I was seeing these problems from a somewhat different perspective when I was twenty-seven, but one thing has not changed—this is my dislike for the idea that my own behaviour or behaviour of my friends illustrates my own theory of sociality or any other. I like always to imagine that I and we are above all that, subject to far more mysterious laws. In this prejudice, however, I seem, rather sadly, to have been losing more ground than I gain. The theory that I outline in the paper has turned out very successful. It certainly illuminates not only animal behaviour but, to some extent as yet unknown but now being actively researched, human behaviour as well. Clearly, the fact that I devoted several papers to expanding its implications and had proceeded initially unaware of most of the tentative approaches of various Neodarwinian forerunners means I was more hopeful of the idea's explanatory power than had been the forerunners—the three great pioneers of modern evolution theory, J. B. S. Haldane, R. A. Fisher, and Sewall Wright, who, as shown in the references in the paper, gave to altruism at most a few lines. Thus even if it is not quite true that mine was a path 'where no man sought', it does seem true that the pioneers had not tried very hard to apply their Mendelized Darwinism to social behaviour nor followed the signs of paths that they did find in this area. But, equally, I was surprised at the scope of my eventual success, and in particular did not anticipate the degree of relevance to humans that the findings eventually proved to have. Nor did I wish for this relevance: I am a child of the receding wave of the Romantic Movement and as such I still hanker for miracles.

Although first to be published, 'The evolution of altruistic behaviour' was written after the next two papers in this volume had reached almost their final form. A reviewer's criticism was forcing me to revise and split the first, longer manuscript concerning the same ideas and, with my hope of receiving my PhD at the time seeming to founder (see Chapter 2), I urgently needed something to represent the fruits of the 3 years I had spent doing research. I therefore quickly wrote this brief abstract of the ideas of the longer paper and submitted it to *Nature*. I received the editor's decision almost by return of post. In about three lines he regretted that he had no space for my manuscript and suggested that, given its specialized topic, it might be more appropriate to a 'psychological or sociological journal'. I was already well aware that there existed a prejudice against my topic and I now dimly perceived that there might have been a tactical mistake in giving my address as the Department of Sociology, London School of Economics. Later, after the paper had been accepted by *The American Naturalist*, John Hajnal, my supervisor at the LSE, agreed that the address might have been unhelpful, and he suggested that next time I use the equally justified address of Department of Statistics. Why not, for that matter, the Galton Laboratory of University College London in which I was also enrolled—or perhaps both?

The reason I had not given the Galton Laboratory as my address was that I was feeling a much greater debt to the LSE than to UCL. In 1960, at a dark time, LSE had arranged the studentship that still supported me and had encouraged my intention to study of altruism—or at least to pursue it as a sideline, along with the human demography for which for safety's sake they enrolled me (obviously it is not good for a college to have a student flop too completely in research). Earlier I had failed to interest any genetics department, including the Galton Laboratory, in my wish to take up the genetics and the evolution of altruism.

Norman Carrier, trained in mathematics and an experienced practical human demographer, was the first person I met at the LSE. A genial, very informal man he came as a breath of fresh air after my previous interviews. He listened to my confused ideas with an unfamiliar sympathy, speaking as though quite unaware of even a possibility that I might be a sinister new sucker budding from the roots of the recently felled tree of Fascism, a shoot that was once again so daring and absurd as to juxtapose words such as 'gene' and 'behaviour' into single sentences. It was on Carrier's encouragement that I applied for and won a Leverhulme Studentship I had not even known existed. Carrier was an enthusiast for any new idea or puzzle. Although little inclined to pure theory himself he told me he probably would not understand the genetics in what I did nor even, probably, my verbal arguments, but he expected that we could jump around together among my equations well enough, if there were to be any.

Meanwhile I could take his MSc course in human demography. I see Mr Carrier perched in my memory on a front desk before our class of about six (a scale I was accustomed to—the one I had been in just before in Fisher's Genetics Department at Cambridge had had four). His arms clasp his knees and his feet (on one occasion I remember them shod but sockless) rest on the desk seat. I recall a lively mixture of theory, demographic practicality, and anecdote. Consider an animal living 5 years at most, a sort of insect if you like, having one offspring the second, third, and fourth years of its life, then it dies . . . When talking of humans, he tells us, the life table you present to your public should have a radix of, say, 10 000: otherwise, as the numbers go down in the age classes you soon have fractional people, and the public and press complain (more than usual) that we statisticians have no respect for the individual. He explains how the quick putrescence and size of a human corpse is a boon to the actuary and demographer and is the reason why human death statistics are more accurate than those of births, for which it may not matter for years whether you report one or not. The essence is that people like keeping babies but not corpses.

After some months, perhaps a year, Carrier said my written sketches were getting too genetical for him and he passed me to John Hajnal (then a Reader in Demography, later Professor) to supervise. After about a year Hajnal decided exactly the same about my work and arranged for me to be part-supervised at the Galton Laboratory. By this time I was already of the status of a PhD student instead of MSc. Hajnal himself was not particularly sanguine that my ideas would find application to anything in nature (could genes really affect altruism, he wondered, was there evidence?), let alone could apply to humans, but he liked the sheer puzzle of how a self-destructive trait *might*, in a pure world of genetics, evolve—perhaps on Mars or somewhere like that—under the influence of natural selection. He was a stern but friendly critic of my logic. I owe a great deal to John Hajnal and Norman Carrier. I think of this paper as emerging as a somewhat unwanted love child sired by a certain dread 'other discipline' out of the crowded ant hill of nine o'clock students and leftward-inclined staff that the LSE was from 1960 to 1963, and doubtless still is. An awkward and unpredictable genie was steaming out from its overcrowded lifts and library, its stats rooms clattering with electric-powered but not yet electronic calculating machines (but many hand ones still); a half-forgotten chapter of a half-forgotten book was being dusted off and then—this presence. Was it dangerous? I think that even the head of my department, Professor D. V. Glass, did not like my production much, and perhaps the LSE as a whole is still slightly embarrassed, in so far as the address attached to this little-read first paper is ever noticed, by what I intended should express my allegiance and my thanks.

References and notes

1. Euripides (epilogue to several tragedies).
2. W. D. Hamilton, review of *A Book of Dolphins*, by Anthony Alpers, *Landfall* **15**, 94–6 (1961).
3. C. Darwin, *On the Origin of Species* (Harvard University Press, Cambridge, MA, 1859). Facsimile of the first edition with an introduction by E. Mayr, 1967; see p. 189 for Darwin's prediction.
4. A. Montagu, *The History of the Dolphin* (William Andrews Clark Memorial Library, University of California, Los Angeles, 1963).

THE EVOLUTION OF ALTRUISTIC BEHAVIOUR[†]

W. D. HAMILTON

It is generally accepted that the behaviour characteristic of a species is just as much the product of evolution as the morphology. Yet the kinds of behaviour which can be adequately explained by the classical mathematical theory of natural selection are limited. In particular this theory cannot account for any case where an animal behaves in such a way as to promote the advantages of other members of the species not its direct descendants at the expense of its own. The explanation usually given for such cases and for all others where selfish behaviour seems moderated by concern for the interests of a group is that they are evolved by natural selection favouring the most stable and co-operative groups. But in view of the inevitable slowness of any evolution based on group selection compared to the simultaneous trends that can occur by selection of the classical kind, based on individual advantage, this explanation must be treated with reserve so long as it remains unsupported by mathematical models. Fisher in the second edition of *The Genetical Theory of Natural Selection* (1958) rejects almost all explanations based on 'the benefit of the species'.[1] Wright (1948) in a summary of population genetics shows explicitly that a general advantage conferred on a group cannot alter the course of intragroup selection.[2] This point is very adverse to the following model of Haldane,[3] which seemed to offer a possibility for the evolution of altruism. Haldane supposed an increment to group fitness (and therefore to group rate of increase) proportional to its content of altruistic members and showed that there could be an initial numerical increase of a gene for altruism provided the starting gene frequency was high enough and the individual disadvantage low enough compared to the group advantage conferred. He concluded that genetical altruism could show some advance in populations split into 'tribes' small enough for a single mutant to approximate the critical frequency. He did not, however, sufficiently emphasize that ultimately the gene number must begin to do what the gene frequency tends to do, *ex hypothesi*, from the very first;

[†]*The American Naturalist* **97**, 354–6 (1963).

namely, to decrease to zero. The only escape from this conclusion (as Haldane hints) would be some kind of periodic reassortment of the tribes such that by chance or otherwise the altruists became re-concentrated in some of them.

There is, however, an extension of the classical theory, generalizing that which serves to cover parental care and still having the generation as the time-unit of progress, which does allow to a limited degree the evolution of kinds of altruism which are not connected with parental care.

As a simple but admittedly crude model we may imagine a pair of genes g and G such that G tends to cause some kind of altruistic behaviour while g is null. Despite the principle of 'survival of the fittest' the ultimate criterion which determines whether G will spread is not whether the behaviour is to the benefit of the behaver but whether it is to the benefit of the gene G; and this will be the case if the average net result of the behaviour is to add to the gene pool a handful of genes containing G in higher concentration than does the gene pool itself. With altruism this will happen only if the affected individual is a relative of the altruist, therefore having an increased chance of carrying the gene, and if the advantage conferred is large enough compared to the personal disadvantage to offset the regression, or 'dilution', of the altruist's genotype in the relative in question. The appropriate regression coefficient must be very near to Sewall Wright's coefficient of relationship r provided selection is slow. If the gain to a relative of degree r is k-times the loss to the altruist, the criterion for positive selection of the causative gene is

$$k > \frac{1}{r}.$$

Thus a gene causing altruistic behaviour towards brothers and sisters will be selected only if the behaviour and the circumstances are generally such that the gain is more than twice the loss; for half-brothers it must be more than four times the loss; and so on. To put the matter more vividly, an animal acting on this principle would sacrifice its life if it could thereby save more than two brothers, but not for less. Some similar illustrations were given by Haldane (1955).[4]

It follows that altruistic behaviour which benefits neighbours irrespective of relationship (such as the warning cries of birds) will only arise when (1) the risk or disadvantage involved is very slight, and (2) the average neighbour is not too distantly related.

An altruistic action which adds to a genotype's reproduction (inclusive of the reproduction of identical genes procured in the relative) by 1 per cent is not so strongly selected as a 1 per cent advantage in personal reproduction would be; for it involves also an addition of unrelated genes which are in the ratio of the existing gene pool—an addition which must be larger the more distant the relationship.

A multifactorial model of inheritance, which is doubtless more realistic, does not invalidate the above criterion, and provided fitness is reckoned in terms of 'inclusive' genotype-reproduction, and the dilution due to unrelated genes is allowed for, the classical treatment of dominance and epistasis can be followed closely.

Fisher in 1930 (1958) offered an explanation of the evolution of aposematic colouring based on the advantage to siblings of the self-sacrifice, by its conspicuousness, of a distasteful larva, and his discussion contains what is probably the earliest precise statement concerning a particular case of the principle presented above: 'The selective potency of the avoidance of brothers will of course be only half as great as if the individual itself were protected; against this is to be set the fact that it applies to the whole of a possibly numerous brood.'[5] It would appear that he did not credit the possibility that selection could operate through the advantage conferred on more distant relatives, even though these must in fact tend to be still more numerous in rough inverse proportion to the coefficient of relationship.

This discussion by Fisher is one of the few exceptions to his general insistence on individual advantage as the basis of natural selection; the other notable exception from the present point of view is his discussion of putative forces of selection in primitive human societies.[6]

References

1. R. A. Fisher, *The Genetical Theory of Natural Selection*, 2nd edn (Dover, New York, 1958); e.g. p. 49.
2. S. Wright, Genetics of populations, in *Encyclopaedia Britannica* (1948; 1961 printing), **10**: 111D–112.
3. J. B. S. Haldane, *The Causes of Evolution*, p. 208 (Longmans, London, 1932).
4. J. B. S. Haldane, Population genetics, *New Biology* **18**, 34–51 (1955).
5. Fisher 1958 (ref. 1), p. 177 *et seq.*
6. Fisher 1958 (ref. 1), p. 261 *et seq.*

CHAPTER 2

A cuckoo queen . . .

Under a rock in the Alps a wasp (right), *Polistes (Sulcopolistes) atrimandibularis*, guards white-capped coccoons containing her offspring, none of which will be workers. Her aggressive wing posture on the nest challenges (left) the workers of *Polistes binotatus* whom she has enslaved after evicting their mother. One worker replies to her but will become submissive if the *Sulcopolistes* approaches. The larger, more squared head of the parasite reflects powerful muscles: head blows and biting rather than the deadly sting are used in day-to-day domination. But the parasite's sting is also larger and is grimly curved in adaptation for lethal fighting when necessary.

. . . and a loser

At another site, a *Polistes binotatus* female (far left) crouches immobile and distant on the stone bearing the nest that she initiated (right) where her daughters are now enslaved. On the nest, the usurper (*Sulcopolistes*; far right, pale antennae) approaches a *binotatus* worker. Above the worker, a male (pale narrow face; first son of the *Sulcopolistes*) backs nervously from the camera. What 'emotion' has an evicted mother? Depression, it seems here; but perhaps there can also be a faint hope that the evictor will die, opening a chance for the loser to resume reproduction.

CHAPTER 2

HAMILTON'S RULE

The Genetical Evolution of Social Behaviour, I and II

It is not necessary that you leave the house. Remain at your table and listen. Do not even listen, only wait. Do not even wait, be wholly still and alone. The world will present itself to you for its unmasking, it can do no other, in ecstasy it will writhe at your feet.

FRANZ KAFKA[1]

IF Professor David Glass was cool towards my work at the London School of Economics, Professor Lionel S. Penrose, FRS, at University College was equally so. I believe he definitely saw in me the eugenical spectre I mentioned in the first introduction. Although supervised at the Galton Laboratory for 2 years I never had a desk there nor was ever invited to give any presentation to explain my work or my occasional presence to others. I think virtually no one I passed in the corridors or sat with in the library knew my name or what I was doing. Probably this was largely my own fault. I was a student who, if not offered or invited, wouldn't request, and I think I didn't ask for desk space either there or at LSE. In fact I had no idea at the time what was normal for graduate students and it is only after seeing arrangements made in other departments both in Britain and the US that my own situation as a graduate has come to seem odd. At the time I hardly knew what a PhD was. I just wanted access to libraries plus some pittance of support; these together would give me freedom to follow my puzzles. Apparently I needed affiliation to colleges to receive my grant and to be a part of London University; two colleges were kindly offering this and that was enough. The consequence of not having desk space was that most of my thinking, analyses, and writing, were done in my bedsitting room. For most of my graduate period this was in Chiswick, some 6 miles west from the West End locations (Kingsway and the Euston Road) where were the two colleges. This, too, may have been partly by my own choice. Certainly I well knew the aphorism of Kafka I have cited above and had begun to believe even as a schoolboy that the quickest way to understand what puzzled me was to spend more time thinking about what I already knew, not necessarily collecting more data. In practice, as is already evident, I made a big exception for going out to libraries for more information. I did start one experiment to look at the altruism or otherwise of seedling 'conkers' (*Aesculus hippocastanum*) while I was at London. But even before all this, still at Cambridge, I had made the decision that I would not even try to come abreast of the important work that was being done around me on the molecular side of genetics. This might well be marvellous in itself: I admitted the DNA story to concern life's most fundamental executive code. But, to me, this wasn't the same as reading life's real plan. I was convinced that none of the DNA stuff was going to help me understand the puzzles raised by my reading of Fisher and Haldane or to fill in the gaps they had left. Their Mendelian approach had certainly not been outdated by any of the new findings.

Thus my sorties to my colleges became mainly for computation and for reading and, rarely, to see my supervisors. Most other trips were to libraries all over London or to the London University's Senate House canteen to get a cheap meal. In so far as I needed a desk I carried mine with me in the form of a large canvas holdall whose depths held papers and books. The upper half had such things as a crumpled jersey, a raincoat, or a fresh loaf of bread to take home. I walked everywhere, rarely taking a bus or the Underground for distances less than 2 miles. My idea was, I think, first, to save money and get exercise and, second, to make London seem as far as possible an extension of the countryside of my boyhood Kent, the presence of which I could still feel a mere tantalizing 20 miles distant from where I walked. Always keen on physical experiences of topography I wanted to conjure and recreate somehow, below the bricks and mortar around me, the marshy fields that had once been there, the streams dividing them, and the heaths behind—to seek, in short, elemental forces underlying the city. Even a lonely child crying on the street did not tug my heart as hard as a bracken fern when I saw it, for example, in the valley of the stream once called the Fleet—a plant etiolated and pale, standing up from deep wells of basement windows, its fronds reaching to the level of the street.

From those same wells I recall also the soul-piercing yellow from the flat, star-like flowers of a ragwort. What plant was this, daring to flower under such hostile cliffs, where was it from? Was it a growth form of the more crinkled and robust plant I knew so well on the Kentish fields, there crawled upon and stripped by the stripy caterpillars of the cinnabar moth? But if the leaves in London might be laxer and greener because of a lack of light or to an excess of lead in these wells, why would the flowers be larger? Today, based on a theory of sex to be explained in the second volume of this work, I might give another conjecture in answer to this but in 1961 I think I simply imagined that, like me, the plants were longing to be away from these dark concrete canyons, back in their own countryside. Fixed in their cracks, as I thought, what else was there for them but to signal desperately with their bright petals to the rare and equally lost London bees crossing (again like me) the chasm of the Farringdon Road? Was it through such petals and then such an insect's aid that they hoped to create the seeds which an autumn storm could carry away? Such conjectures as I made about them, excepting perhaps this last one, turned out to be wrong. The ragworts may, after all, have been quite happy: it was in fact a rather triumphant species, a new conqueror of Britain, a railway-and-town ragwort from the start, not my ragwort of rural Badgers Mount in Kent. Only later did I come to know the full romance of the 'Oxford' ragwort, the inhabitant of those window wells and walls, and how it was by origin more an expatriate even than I and had come by a strange zig-zagging convergence to mix its faint scent with stronger urban ones, which, as it happened, were from just the same homeland as itself. That is to say, the flowers I saw were mixing scents with those of pizza, pasta, and oregano wafting from the Italian restaurants that served my fellow lonely pedestrians of the area (not me, I was too poor) around King's Cross and Euston stations. In short, with the help of Oxford botanists the species had immigrated from southern Italy some 150 years ago.[2]

In spite of his appointment as head of the institution Galton had created, Professor Penrose was strongly anti-eugenical and during his tenure realigned the institution's research to become more standard human genetics. Likewise, as editor of Galton's journal *Annals of Eugenics*, he changed the name to *Annals of Human Genetics*. An imaginative, speculative thinker in other very diverse fields—for example, artificial life, voting theory, and the pictorial representation of impossible structures—his principal work in genetics was elegant but a rather mainstream elucidation of human genes, chromosomes, and biometry. When, before going to LSE, I had come to him to ask if the Galton Lab could sponsor my work on the altruism problem, Penrose emphatically told me no and that he doubted that there was such a problem to be studied. At the same time, however, he acknowledged the worth of my Cambridge genetics degree; if I would like to join him to work out the nature of a new polysomy he had recently discovered, he could probably support me.

A joke of the Darwinian period, perhaps from *Punch* magazine, has two ladies conversing. One says: 'Have you heard that Mr Darwin says that we are all descended from an ape?' The other replies: 'Oh, my dear—that surely cannot be true! . . . But, if it should be true, let us pray that at least it will not become generally *known*!' All evolutionists are acquainted with this class of reaction if not with the exact situation or the words. It equally affects the extensions of Neodarwinism that are set out in the two papers of this chapter. In many people, perhaps even a majority, reading them leads to an instant automatic wish for both the evidence and the idea to go away. In the case of my ideas on social evolution I think the Victorian woman's reply may appropriately caricature reactions even of Penrose and some other good geneticists. What constitutes evolutionary biology's difficulty as a subject, and even more makes the difficulty of sociobiology, and more than that still, makes the special difficulty in accepting *human* sociobiology, is something quite different from what makes difficulty in, say, a physics topic. The difficulty over these social and evolutionary questions is not the rigour of chains of logic or maths, nor complexities of geometry, nor counterintuitive ideas. Most of the ideas of evolution are very intuitive if we could just set our minds free. The problem is rather that of thinking the socially unthinkable. Darwin put it well on the side of public expression when he said in a letter that writing about non-fixity of species was like 'confessing to a murder'. It is the challenge that such ideas cannot fail to direct at deeply entrenched and perhaps almost essential human myths. In other words, some genetical and evolutionary hypotheses, including those underlying the papers that follow (ideas that in any other field would seem quite unexceptional and often appear as just plain common sense), turn out to have, or are perceived to have, the unfortunate property of being solvents of a vital societal glue. It came as no surprise to me that E. O. Wilson (see Chapter 9), much more bold in controversy than I am, faced a furore over his writings about human sociobiology.

At this point, however, it will even the account a bit to admit that, as an avowed adversary of eugenics, Penrose may really have had good reasons to distrust me. When I first met him I was quite a strong believer in eugenics. More than most people I still am, as will be seen in some of the later chapters of this book. Francis Galton, Darwin's cousin and within the evolutionary era the initiator of the notion of eugenics, was much admired by me for his amazingly diverse contributions to biology, psychology, statistics, and genetics. Among his more practical contributions had been starting the department of which Penrose was now head. I had come to Galton's idea by my own parallel reasoning spurred by the common youthful wish to improve the world, and by reading Fisher. I much liked the notion that human-directed selection, whether to maintain standards or to speed the intellectual and physical progress of humanity, could be made both more effective and more merciful than the obviously inefficient and cruel natural process. 'What a book', wrote Darwin in a letter, 'a Devil's Chaplain might write on the clumsy, wasteful, blundering, low and horribly cruel works of nature!' Darwin probably meant that natural selection itself was one of those works. I think that at first it hardly crossed my mind that Galton and Darwin were considered by many to have laid the intellectual foundation for Nazi racial crimes. It is true that both writers from time to time revealed that they accepted the opinions about race differences that were nearly universal in their time, but the selection, natural or otherwise, that they were discussing in their works was normally at a level very far from that of human races. It was generally far even from the broad categories of behaviour such as criminality and feeblemindedness that shortly became so controversial. Thus, although I could see some faint connection between their thoughts (and, in Galton's case, advocated policies), and such attributes as criminality, I saw none with respect to a struggle between classes or races.

As I realized later there is good evidence that the notion of coercive racial policies came to Hitler's Germany, in practice as well as in theory, via Russian Communism and ultimately from Marx himself.[3] Marx had never really understood Darwin's natural selection theory and what he did understand he recast into terms of his own preoccupation with group struggles. The idea that in a fixed physical environment, leave alone in a (hypothetical) fixed social one, there could ever be a single ideal type of human being always seemed to me unnatural—such a uniform ideal just never looked like being Nature's way. On the contrary, Nature, for reasons which it was to become my life's work to try to understand, greatly values variation of all kinds within her units, does so at all levels, and seemingly goes to great lengths to protect it. The thought of a genetically or even merely racially uniform population of humans was both unnatural and appalling and I suspect, had they understood genes, would have been so equally to Darwin and Galton. However, the idea that wise governments ought to try to think more of the extreme long-term consequences of their policies, on the scale of many generations, and should adapt the laws they pass accordingly, or that education also should include discussion of such distant future issues, appealed to me strongly and still does. It also seems to me that scientific giants such as Louis Pasteur, who have applied themselves with spectacular success to improving the immediate human condition, may eventually sink a little in our estimation—not, for sure, as scientists but under their parallel esteem for the immediate pan-human benefactions—whereas others who worked more on the plane of population effects such as Adam Smith, Malthus, and Galton (perhaps here one should also include thinkers such as Macchiavelli who likewise dared touch more difficult 'realist' issues) may rise.

I never had a conversation about any of this with Professor Penrose but it is quite likely that in my interview with him when I first tried the Galton Lab I may have mentioned something about my interest, or at least may have let him know of my general admiration for Galton. Later, when I had successfully entered his lab by, as it were, a back door from the LSE, and sensed in him a kind of gloomy avoidance, I always thought that if we could discuss matters he might discover less difference in our ideas about humans than he believed. Then, as I thought wistfully, he might not look so dourly on me in corridors and courtyards, and his secretary, with whom I actually had more to do and whom I guessed reflected his attitude, might have been more friendly. All this, of course, was happening a mere 15 years after the war. If he did think he had a rabid eugenicist, and one seemingly not particularly bright at genetics as well (as suggested by my lack of interest in all the mainstream work his lab was doing), the situation that arose is not surprising.

Since that time I have changed my views considerably. Although I am now even more certain than then that if an omnipotent dictator chose to breed humans with exceptional abilities in any direction the job could be done, I am also certain that the products would not be very desirable—even leaving out any intervening ghastliness of the policies. Contrary to what is often fiercely argued, heritable variation in almost every human talent does seem to be abundant. New evidence on this is being added all the time. In short, the dictator's selection would work. My change in view results rather from my search for ultimate reasons for sex, a topic that will become prominent in Volume 2. I now believe that just about when the dictator was beginning to relax on her throne and to congratulate herself on her progress, she would find her various 'alpha' human stocks had become so poorly in general health that no one would like them, and no one would want to mate with them. Finally, she would be forced into sterilizing them along with all her other black sheep and starting over again. In doing so, she would find herself apologizing to many of her formerly rejected phenotypes. I believe that parallel problems and reactions to this have often occurred in the course of breeding domestic animals. The causes of dwindling profitability in breeds is not properly understood but is likely to be at least partly of this nature (as will be explained Volume 2).

More important even than my realization of the universality and subtlety of parasite and disease coevolution accompanying any large-bodied species, however, I have come to see that the above experiments could only ever be done under the supposed ruthless dictatorship. A majority in a democracy will never be persuaded to accept a positive eugenics policy and when the malthusian crunch comes, democracy (or the militocracy that is likely to have succeeded it in the last stages) will simply drift back, willy-nilly, to natural selection, as can be seen happening in many disaster-torn countries today. Perhaps that is indeed the best we can expect and is no more cruel than a controlled intervention aimed at joint population control and improvement (or simply maintenance). Moreover, perhaps trying to guess the human ideals of the future always has to be stultifying. The very concept of the good and the superhuman may be better left to be determined by the natural process unaided. What has to be realized, though, is that a utopia of 'right-to-life' large families and of unlimited medical fixing of all health issues from the womb onwards—for ever, so far as possible—is not really going to happen. Such utopia punctuated from time to time by social cataclysm, with the latter providing severe natural selection, is possible and perhaps is already being illustrated, as I have suggested. Possibly the general unexpectedness of cataclysms makes them in the long run more kindly than the constant worry about self and kin that would accompany continuous control of population and of genetic quality. But for any species even just maintaining its standards, there is no escaping control and selection of some kind: to this extent Malthus and Galton were both right.

I include this digression about my early interest in eugenics partly because no mention is made of it in the other primarily autobiographical sketches that will appear later in this volume. Having mentioned it, however, it is important to emphasize that my interest is not in fact particularly relevant to the interest in the evolution of altruism and selfishness expressed in the paper in this chapter. In one of the most common misinterpretations of the theory, it is supposed that because genes and altruism (or selfishness) are constantly being mentioned, the theme involves some individuals having more altruism and more altruistic genes than others. It is true that differences must sometimes exist in a small way as, one by one, new alleles are selected into the population in the manner the model describes. But the typical end result of such new alleles becoming universal is that all members of the population are exactly the same: individuals end behaving in a discriminating, nepotistic manner to one another as if differences in genes were still there and mattered, when in fact, at a particular time, there may be no differences at all, these being matters of the past. That 97 per cent or whatever of alleles are in common even between randomly selected members of a population or species, and unknown genes affecting social behaviour are likely to show similar overlap, does not bear on the validity of the argument in the least. The existence of such probable uniformity leads on, of course, to the thought that the final social situation that evolution creates is rather absurd. Although animals may have to suffer such absurdity, surely it is open to humans and other intelligent animals to realize that, if the genes are so alike, it really only makes sense to treat relatives and non-relatives as alike to the self to the degree of the measured similarity—in other words, to be extremely generous and loving towards everyone. This, however, is not what the theory predicts will happen and our intelligent animal ought to think it out more carefully. Even when it is near to complete monomorphism the evolved nepotistic programme of behaviour is still at work in maintaining its base. Simply by being there it is beating back all mutant 'cheat' genes that fail to perform altruism but still reap its fruits from the acts of others. Random aid to just anyone gives scope for unthriftiness to flourish and for simple non-altruism to arise. Since there is hardly an easier mutation to imagine than one that causes a constructive behaviour like altruism *not* to be performed, so creating a passive and successful receiver for the altruism of others, the point is serious. Thus the critics who cite the 97 per cent or whatever as meaning that nepotism is pointless are wrong even as regards any possible monomorphic end state.

All this is implied in the two-part paper in this chapter (in Part I as implied in the unrestricted set of allele frequencies allowed on page 35) but it is evident from the lack of emphasis I gave to this point in the later sections of the paper that I did not anticipate how the situation would be perceived. I do believe in the existence of considerable genetical differences in altruism, in selfishness, and in many other social attributes in most social animals, and this includes in humans, but this belief is certainly not predicated directly from kin-selection theory. Rather, it is underpinned more by a belief in a complexity of life sufficient to generate genetical variability in almost everything—the variability we actually see—and also, as will appear in Volume 2 of this compilation, by modern theoretical expectations connected with reciprocation and with disease selection. It certainly does not come from kin selection *per se*.

Once a paper has achieved a certain status of citability it sometimes takes off on a life of its own that may have little relation to what is in it. Citability can mean citation in a bad way, as in '. . . . as believed by Idiot (1973)', and I have just indicated one bad way, via a misinterpretation, my two-part 1964 paper is sometimes cited. But citability and the independent lives of citations just as often mean intentions by citers to be positive while they actually give misconstrued support. For example, many people have written as if it was mainly about social insects and in particular as if its overwhelmingly important point was how a peculiar genetical asymmetry of haplodiploid Hymenoptera (sawflies, ants, bees, wasps, and wasp-like parasites) may have predisposed the group of insects to social life. The justification for this seems to be that because eusocial insects (those with sterile workers and a queen) provide many of the most indisputable examples of altruism, discussion of situations in the social insects does take up quite a large section of Part II. However, anyone who really read Part II, leave alone read both parts together, could hardly fail to see that my aim was much broader: it was to present a theory applicable to social behaviour under relatedness for any group in the living world. Besides the special coverage of diploids, that of haploids was implied and even asexuals touched on; the lives in question could be multicellular or otherwise; and so on. Anyone reading Part II will see that I discuss some explicit cases for plants: why shouldn't plants be altruistic too?

In the early days I was continually being told that there was no such thing as altruism, or else that, why, of course!—everything is altruistic. I badly needed examples, therefore, where both self-sacrifice and the limits to it were indisputable. Social insects became one of my main examples, and warning cries exchanged by social animals apprising one another of danger became another. Perhaps because of the space I devoted to social insects, for a long time the combined paper was cited largely by people concerned with my arguments about them, whereas actually these arguments had mostly been afterthoughts hardly even started until the whole of what became Part I of the paper had been written. What came to make up Part II was done rather hurriedly. Having seemingly flunked my chance of a PhD I was feeling in need of something to represent the years I had used up. Through my haste, two rather bad errors concerning the relatedness coefficients and the selection of sex ratio under haplodiploidy, slipped into Part II. I took the opportunity of correcting these errors when the two parts were republished in the book that George C. Williams edited on group selection, published in 1971,[4] especially using an Addendum that I put in after the paper (and include again in this volume). In spite of these defects and many others that the paper has, the rise in citation of the combination, after a slow start, eventually became dramatic and when I wrote my *Current Contents* commentary the article had been for some years *Journal of Theoretical Biology*'s most cited paper ever: whether it retains that status today I don't know. In any case, well illustrating the theme of citations developing a life of their own, a brief note on memetic evolution (i.e. in this case an evolution of errors in the wording of the listed reference) by Seger and Harvey[5] proves that a very high proportion of the citing was being done (and probably still is) by people who had not read the work.

How did the ideas that led to this paper begin? In the autobiographical sketches that appear in later chapters, including the one just mentioned, I explain how as an undergraduate at Cambridge University I had come to see altruism in particular and social behaviour in general as a poorly understood part of evolution theory (see Volume 2). The most immediate stimulus was my lecturers largely endorsing a 'benefit of the species' interpretation that stood in sharp contrast to the writings of my then hero of twentieth-century evolution theory, R. A. Fisher. Fisher championed an almost complete predominance of selection at the level of individuals. His writings on evolution I had had to discover at Cambridge for myself because they appeared to be either unknown to or disapproved of by my lecturers. One told me that Fisher's work on statistics was admittedly very important but Fisher had no credentials even to be writing on biology. Many Cambridge biologists seemed hardly to believe in evolution or at least seemed to be sceptical of the efficacy of natural selection. The Cambridge view of evolutionary mechanism at the time is well illustrated in the following passage from Professor Sir Vincent Wigglesworth's introduction to a chapter on population aspects of entomology in his book *The Life of Insects*, published in 1964:

Insects do not live for themselves alone. Their lives are devoted to the survival of the species whose representatives they are. . . . We must now stand back and look at the insect as a member of the 'population' or 'species' to which it belongs. Indeed we have now reached the heart of the matter—the aim and purpose (so far as we can understand them) of the life of insects.[6]

I quote this as a criticism of no more than a tiny portion of Wigglesworth's own work, which was almost entirely concerned with mechanistic aspects of insect physiology. Within that ambit his work was diverse and excellent: it was on how the animal survived, metamorphosed, reproduced, and so forth, and he never interested himself beyond what evolutionists now refer to as the 'proximate'[7–9] aspects of adaptation—how things are engineered in the insect's body. He can have had little cause to reflect on 'aim and purpose' in wider contexts where the conflictual and strategic questions enter. In other words, in those contexts containing what we have come to call 'ultimate' aspects of adaptation—who adaptations are for, which genetic units benefit, and so on. The point of the citation is the Cambridge atmosphere of the time, an atmosphere that Wigglesworth breathed and influenced. His automatic endorsement of adaptation at the population or 'species' level (even so, slightly hesitant, as suggested by the parentheses) corresponds exactly to the 'alternative pseudoexplanations' I had in mind when discussing Fisher's idea about the evolution of distastefulness in insects (see page 49).

While I was still an undergraduate at Cambridge, and just beginning to try to devise models that would support altruism under natural selection, I did not have Fisher's remark on distastefulness consciously in my mind even though I had probably read it, nor did I have J. B. S. Haldane's equally brief discussion that he put into a popular paper in *New Biology* even though I know for certain I had read that. One reads and forgets. Hints not understood probably leave their traces, but one has to return to the topic in a better state of preparation and to re-read before such throwaway items become meaningful. Hence, my first algebraic attack on altruism as an undergraduate concerned Haldane's other more lengthy and obvious model, which I suspected to be, and proved to be, a failure within its stated conditions (the argument for this comes in the paper of Chapter

9). Perhaps the hints of a factor of two somehow lurking in the mist around full sibs existed by now within me, waiting for better times. But instead of any effort to clarify the idea I started looking for estimates of the importance of relatedness in quite other places—in the population genetics literature, within my own intuition based on my family and ordinary literature, and even through trying to read what social anthropology had had to say about what cultures around the world thought about blood relatedness. I was vaguely hoping to 'average' out of all that something that would be revealing for my purpose.

The naive last idea had occurred to me as an undergraduate long before my move to London. With it in mind, as well as through a general interest in human social behaviour, I had made an unsuccessful attempt at Cambridge to enrol for a course in social anthropology. The idea of social anthropology as a subsidiary subject to genetics was frowned on both by my 'home' Department of Genetics and by the Department of Social Anthropology. The then Reader in Social Anthropology, with whom I tried to discuss my intention, was Edmund Leach, destined by 1976 to be Professor Sir Edmund Leach and by that year one of the most bitter and outspoken critics of E. O. Wilson's *Sociobiology*. Whether it was he in person, on the anthropology side, who denied the relevance of genetics to his discipline (and presumably the reverse relevance also), or whether the attitude he later revealed concerning Wilson was simply universal and unquestioned in his department, in retrospect the refusal of Social Anthropology to consider the combination with genetics is not surprising. It was part of the requirement for the genetics degree that some subsidiary subject should be chosen and that the department in question should be willing to set an examination paper. My ideas about the connections of the two subjects at that time were undoubtedly naive and probably poorly expressed but I still think that my request to combine the two was reasonable and that at the very least someone should have tried to explain to me why the two disciplines could have nothing in common. Some one in Genetics could have told me, for example, and realistically enough: 'You have to realize that those people over there won't teach you science . . . social anthropology is done more like poetry.' At the time I would have found such a statement astonishing but at least it would have been a reason, and being told of such a view might have countered my naive supposition that reading in the two disciplines would be like seeing the same mountain range first from one side and then another and out of the two viewpoints somehow figuring a route to a pass. In any case, the blank refusal of both sides even to discuss the issue caused an immediate decision to leave Cambridge once I had finished my first degree and to seek opportunities for further development of my ideas elsewhere.

When my upper-second-class degree in genetics was attained and I began to look for a path to continue, I had two strategies in mind. One was to try heads of genetics departments to see if any would sponsor me and the other was to enrol in a course for a Diploma of Education so that I could become a schoolteacher. I believed that the pace of an education course would be slower than that of the Cambridge courses I had just completed and that during the new one I should have time to straighten out my thinking about altruism, and with further luck might even continue such straightening after I had become a schoolteacher, if that career proved really necessary. I had in mind certain precedents of excellent science done after a start in schoolteaching. One was Ronald Fisher himself, and the other David Lack, an important bird and population biologist. But the 'Dip.Ed.' scheme that I pursued simultaneously with my search for a sponsor for altruism only led to more frustration when I found that the education course I favoured, at Moray House in Edinburgh, would not accept my genetics degree as a basis for teaching biology at secondary-school level; they wrote that I would be qualified to teach in elementary schools only but I could be accepted for that. I had no particular objection to teaching young children, indeed imagined this might be very rewarding. All the same the odd restriction applied by Moray House was disheartening if it indicated the intellectual atmosphere of such a place: genetics as a subject seemed unknown to them. Or were they perhaps just sharper at reading between the lines of my application than I realized—or, yet another possibility, were they so 'nurture'-inclined that the word 'genetics' simply terrorized? After two further letters trying to explain that I had done a lot of both botany and zoology at Cambridge as well as the genetics, and that in any case it was the nature of genetics to cover all sides of biology, and finding these letters equally stonewalled, I tore up the correspondence. I began some not much more successful enquiries about the Dip.Ed. at the more local Institute of Education alongside Senate House in London. About the same time I had my negative interview with Professor Penrose and my self-esteem and optimism reached about their lowest points ever. Then two people, John Hudson, Professor of Horticulture at Leicester and father of a Cambridge friend, and C. O. Carter, a senior member of the Eugenics Society which I had also approached for advice, converged in directing me to Professor Glass and to the human demography course at LSE. I was not interviewed by Glass but by his colleague Mr Carrier, as described in the first introduction. The Leverhulme scholarship followed and I had found my niche.

That finally successful search, however, had been just one long, wild swing of an emotional pendulum and there were many to follow. Through most of the 3 years that followed my feeling of alienation in my work continued. At times I was sure I saw something that others had not seen—or, as I thought, if it had been noticed then they had certainly written about it very badly. At others I felt equally certain that I must be a crank. How could it be that respected academics around me, and many manifestly clever contemporary graduate students that I talked to, would not see the interest of studying altruism along my lines unless it were true that my enterprise were bogus in some way that was obvious to all of them but not to me? I could think of many bogus ideas that I had followed enthusiastically for quite long periods and then later had 'seen through.' I realized I had little talent in mathematics and even less training for it, so that my efforts to teach myself what was necessary to understand even the merely standard theoretical population genetics of the day were tedious in the extreme. The mistakes I was continually finding in my attempts to parallel published analyses in my field often drove me to despair. Most of the time I was extremely lonely. Sometimes I came to dislike my bedsitting room so much that, when even late libraries such as Senate House or Holborn Public closed and I was still in a mood to continue work, rather than return to my room I would go to Waterloo Station, where I continued reading or trying to write out a model sitting on the benches among waiting passengers in the main hall. From late-night returns to my army unit in Hampshire during my National Service, before university, I knew there would always be activity in the huge station until well past midnight.

Although I virtually never spoke to anyone in such places or in the libraries I frequented, or in my departments, having people around me who were yet not so crowding or bustling as to make work impossible seemed to soften loneliness a little. But often also, if the weather was at all tolerable, I would leave my room in Chiswick during the day and go and work on a park bench in the gardens of Chiswick House or at Kew. The former was only a few hundred yards from my room. But the beauty and the wild life of these gardens were at least as distracting as was the human pageant at Waterloo (the alcoholics there sheltering or craving company like me, the lovers parting, the fractious children herded by tired mothers). Out at Kew I remember the birds hopping up hoping to be fed; also, only too often, the sun shining too brightly on my pages, the air being too cold, or the wind scattering the reams of my wretched and erroneous algebra across the grass. No drunkards were at Kew but occasional unexpected conversations suggested, in retrospect, that I had been sized up by a homosexual. On the whole my room or else a London library were better for real work. As a result of my disillusion with Moray House I had now changed my reserve life plan from teacher to carpenter, in case my theory, as I was now calling it in more hopeful moments, proved in the end to be unpublishable. I was not particularly worried by the prospect of carpentry itself, much more by that of having to acknowledge defeat. I knew I was fairly good at woodwork as at other handicrafts, and in general had little doubt about my ability to find a job and keep myself alive when my student grant came to an end. I even thought how peaceful such work would be compared with my present continual struggle with the near impossible, or as it might be (if I was a crank) the actually impossible.

The analytical presentation of kin selection that I finally wrote up in the 1963 paper is not at all graceful; I was to discover a vastly better approach using the ideas of another solitary, George Price, some 6 years later (see Chapter 5). However, when at last the one I used came to me I recognized it at once as both more general and more appropriate than anything I had done before and it also had gained enough of a veneer of mathematically sophistry to begin to interest my

University College genetics supervisor, Cedric Smith. Previous to this it had been one plodding case after another for me—sibs, half-sibs, uncles, cousins (and these were fearsomely complicated), varying the competition assumptions, and so on. Now with enthusiasm and relief I plunged ahead with the new, more general ways of handling relatedness and multiple alleles while hurling to the bin stacks of my less general derivations. In essence what I had come to see was the simplification to be effected by attacking the problem from two new points of view. One was simple the 'gene's eye', as expressed in the 1963 paper and in the closing remarks of Part I of the 1964 paper. This view, of course, was later brilliantly developed by Richard Dawkins at Oxford. But, at least as we humans perceive the matter, it is not genes but *we*—whole diploid organisms—that make the decisions, so I had been delighted to find something approaching an individualistic view that I could justify for whole genotypes and which could serve as a guide to social adaptation. This was the idea of inclusive fitness. There were still many defects in the account. My long endeavour to generalize a maximizing property of the classical selection model was vitiated by my relatedness coefficients being only properly defined if there was no selection, which was obviously not true in my model. Indeed, selection was the whole point. So the 'proof' I came up with (imitating the approach of a recent real proof in a parallel case in the classical model) was really only suggestive of what would happen, not a watertight demonstration. Nevertheless, it was easy to see that the argument must apply with increasing accuracy as selection in the model was made weak. Another limitation was that my model, complicated as it seemed already, had only one locus variable at a time. This, however, was the same restriction as was required to get a maximization property in the classical model; in this respect the limitations of my results appeared to be no worse.

Later I came to feel that my struggles to include multiple alleles had been rather pointless. On the other hand my confidence that I had proved maximization of inclusive fitness, with or without multiple alleles under weak selection for the one-locus case, was important to me. I was and still am a Darwinian gradualist for most of the issues of evolutionary change. Most change comes, I believe, through selected alleles that make small modifications to existing structure and behaviour. If one could understand just this case in social situations, who cared much what might happen in the rare cases where the gene changes were great and happened not to be disastrous? Whether under social or classical selection, defeat and disappearance would, as always, be the usual outcome for genes that cause large changes. I think that a lot of the objection to so-called 'reductionism' and 'bean-bag reasoning' directed at Neodarwinist theory comes from people, who, whether through inscrutable private agendas or ignorance, are not gradualists, being instead inhabitants of some imagined world of super-fast progress. Big changes, strong interlocus interactions, hopeful monsters, mutations so abundant and so hopeful that several may be under selection at one time—these have to be the stuff of their dreams if their criticisms are to make sense. And, for sure, all of us, from a mathematical point of view, would like to know what will go on in such an exciting and turbulent world, and we would be interested also from a biological point of view if it can be proven to be at all realistic. I tried myself to put down some thoughts about genes with strong effects in Part I (pages 42–3), concluding that the outcomes seemed unlikely to be very different from the version of inclusive fitness theory that I described. However, in general it is certainly unfair to project the unrealistic difficulties of such a fast moving and major-gene world into that of normal Neodarwinism as if the difficulties of the former were the usual truth. What is in question here is the idea of a mutation 'causing' a whole complex behaviour pattern—the whole of the stinging behaviour of a honeybee, say—all at once. Such large changes became the common straw man set up by anti-sociobiologists but, to repeat myself, the strictures applying in the big-gene world are not valid against standard and gradualistic Neodarwinian change. What I most wanted my theory to cover is such a case as an allele making a small difference to how performing an action is conditional on the perception of, say, a familiar smell rather than a strange one, to how seizing a food item may go with a preliminary locomotor reflex in an animal instead of it being eaten where found, and so on (I am thinking here of the chicken running off with the worm, and the question of whether she runs more when foraging in company with her own chicks or with unrelated birds, and the like). Doubtless some misunderstandings could have been avoided if my generally gradualist position in the early papers had been made more plain.

By the end of 1962 my short paper (Chapter 1) had been accepted by *The American Naturalist*. This together with increasing confidence in the basic correctness of the analysis of the longer version decided me that I probably wasn't a crank. However, because the essence of the paper was still just the common sense about relatedness that I had seemed to be seeing all along, and because that common sense needed no algebra to support it, I was still wondering whether I had been making a mountain out of a molehill and whether the reaction to my definitely inelegant maths would be: 'True; but, of course, all fairly obvious.' As to the making of the mountain, if such it was, I was utterly tired of it and about the time my next manuscript was first submitted to *Journal of Theoretical Biology* I wrote to Professor Warwick Kerr in Brazil asking if I could come to his lab to test out on bees and wasps some ideas I had had about social insects and their evolution. I quickly received a warm invitation. I forget whether in 1963 I was in preparation to go to Brazil or had already left when the editor's reply arrived saying my long paper was generally acceptable to the journal but needed major revision and in particular must be split into two. As a result of learning Portuguese, settling into Kerr's lab in Rio Claro, and starting practical research, the combined revision and division of the paper went rather slowly. I sent the new version, which came to include a few new points of evidence derived from observations I had been making in Brazil and from my conversations with Kerr and his colleagues, early in 1964. It was accepted and a little later I corrected the proofs. The two-part paper came out in July while I was travelling up mainly overland from São Paulo towards Canada on my way home to Britain. Very probably the sun of the day that witnessed my paper going into the post from the offices of *JTB* (or Academic Press, or wherever the journal is finally posted from) would have seen me weaving my old American jeep between the corrugations, stones, and potholes of the Belém–Brasilia road (first of Brazil's transcontinentals, just 2 years old). At midday it would have blazed near vertically on the top of my head as I stopped at the roadside and collected wasps from some nest; later at sunset, if still able to pierce the haze, it would have seen me and my Brazilian companion, Sebastião Laroca (now at the University of Paraná, Curitiba), slinging our hammocks between low cerrado trees not far back from the stony or sandy piste where occasional lorries still groaned on into the night. For sure, both that day and that night I was blissfully untroubled about the finer points of measuring relatedness.

The paper is reprinted here as it first appeared in *JTB* except for corrections of a few typos and notes marking two errors. It includes as an appendix the Addendum discussing the errors taken from the republication in the book by G. C. Williams:[4] the text also is almost identical to that of the slightly altered 1971 version.

References

1. F. Kafka, *The Great Wall of China and Other Short Works* (Penguin, London, 1991; first published 1920).
2. M. Walters, *Wild and Garden Plants* (Collins, London, 1993).
3. G. Watson, *The Idea of Liberalism: Studies for a New Map of Politics* (Macmillan, London, 1985).
4. G. C. Williams, *Group Selection* (Aldine-Atherton, Chicago, 1971).
5. J. Seger and P. Harvey, The evolution of the genetical theory of social behaviour, *New Scientist* **87**, 50–1 (1980).
6. V. B. Wigglesworth, *The Life of Insects* (Wiedenfeld and Nicolson, London, 1964).
7. J. R. Baker, The evolution of breeding seasons; in G. R. de Beer (ed.) *Evolution: Essays presented to E. S. Goodrich* (Oxford University Press, Oxford, 1938).
8. G. C. Williams, *Adaptation and Natural Selection: A Critique of Some Current Evolutionary Thought* (Princeton University Press, Princeton, NJ, 1966).
9. R. D. Alexander, *The Biology of Moral Systems* (Aldine de Gruyter, New York, 1987).

THE GENETICAL EVOLUTION OF SOCIAL BEHAVIOUR, I[†]

W. D. HAMILTON

A genetical mathematical model is described which allows for interactions between relatives on one another's fitness. Making use of Wright's coefficient of relationship as the measure of the proportion of replica genes in a relative, a quantity is found the means of which incorporate the maximizing property of Darwinian fitness. This quantity is named 'inclusive fitness'. Species following the model should tend to evolve behaviour such that each organism appears to be attempting to maximize its inclusive fitness. This implies a limited restraint on selfish competitive behaviour and possibility of limited self-sacrifices.

Special cases of the model are used to show (1) that selection in the social situations newly covered tends to be slower than classical selection, (2) how in populations of rather non-dispersive organisms the model may apply to genes affecting dispersion, and (3) how it may apply approximately to competition between relatives, for example, within sibships. Some artificialities of the model are discussed.

1. INTRODUCTION

With very few exceptions, the only parts of the theory of natural selection which have been supported by mathematical models admit no possibility of the evolution of any characters which are on average to the disadvantage of the individuals possessing them. If natural selection followed the classical models exclusively, species would not show any behaviour more positively social than the coming together of the sexes and parental care.

Sacrifices involved in parental care are a possibility implicit in any model in which the definition of fitness is based, as it should be, on the number of adult offspring. In certain circumstances an individual may leave more adult offspring by expending care and materials on its offspring already born than by reserving them for its own survival and further fecundity. A gene causing its possessor to give parental care will then leave more replica genes in the next

[†]*Journal of Theoretical Biology* **7**, 1–16 (1964).

generation than an allele having the opposite tendency. The selective advantage may be seen to lie through benefits conferred indifferently on a set of relatives each of which has a half chance of carrying the gene in question.

From this point of view it is also seen, however, that there is nothing special about the parent–offspring relationship except its close degree and a certain fundamental asymmetry. The full-sib relationship is just as close. If an individual carries a certain gene the expectation that a random sib will carry a replica of it is again one-half. Similarly, the half-sib relationship is equivalent to that of grandparent and grandchild with the expectation of replica genes, or genes 'identical by descent' as they are usually called, standing at one quarter; and so on.

Although it does not seem to have received very detailed attention the possibility of the evolution of characters benefiting descendants more remote than immediate offspring has often been noticed. Opportunities for benefiting relatives, remote or not, in the same or an adjacent generation (i.e. relatives like cousins and nephews) must be much more common than opportunities for benefiting grandchildren and further descendants. As a first step towards a general theory that would take into account all kinds of relatives this paper will describe a model which is particularly adapted to deal with interactions between relatives of the same generation. The model includes the classical model for 'non-overlapping generations' as a special case. An excellent summary of the general properties of this classical model has been given by Kingman.[1] It is quite beyond the author's power to give an equally extensive survey of the properties of the present model but certain approximate deterministic implications of biological interest will be pointed out.

As is already evident the essential idea which the model is going to use is quite simple. Thus although the following account is necessarily somewhat mathematical it is not surprising that eventually, allowing certain lapses from mathematical rigour, we are able to arrive at approximate principles which can also be expressed quite simply and in non-mathematical form. The most important principle, as it arises directly from the model, is outlined in the last section of this paper, but a fuller discussion together with some attempt to evaluate the theory as a whole in the light of biological evidence will be given in the sequel.

2. THE MODEL

The model is restricted to the case of an organism which reproduces once and for all at the end of a fixed period. Survivorship and fertility can both vary but it is only the consequent variations in their product, net reproduction, that are of concern here. All genotypic effects are conceived as increments and decrements to a basic unit of reproduction which, if possessed by all the individuals alike, would render the population both stationary and non-evolutionary. Thus the fitness of $a^\bullet$ of an individual is treated as the sum of his basic unit, the effect δa of his personal genotype and the total e° of effects on him due to his neighbours which will depend on their genotypes:

$$a^\bullet = 1 + \delta a + e^\circ. \tag{1}$$

The index symbol $^\bullet$ in contrast to $^\circ$ will be used consistently to denote the inclusion of the personal effect δa in the aggregate in question. Thus equation (1) could be rewritten

$$a^\bullet = 1 + e^\bullet.$$

In equation (1), however, the symbol $^\bullet$ also serves to distinguish this neighbour-modulated kind of fitness from the part of it

$$a = 1 + \delta a,$$

which is equivalent to fitness in the classical sense of individual fitness.

The symbol δ preceding a letter will be used to indicate an effect or total of effects due to an individual treated as an addition to the basic unit, as typified in

$$a = 1 + \delta a.$$

The neighbours of an individual are considered to be affected differently according to their relationship with him.

Genetically two related persons differ from two unrelated members of the population in their tendency to carry replica genes which they have both inherited from the one or more ancestors they have in common. If we consider an autosomal locus, not subject to selection, in relative B with respect to the same locus in the other relative A, it is apparent that there are just three possible conditions of this locus in B; namely that both, one only, or neither of his genes are identical by descent with genes in A. We denote the respective probabilities of these conditions by c_2, c_1, and c_0. They are independent of the locus considered; and since

$$c_2 + c_1 + c_0 = 1,$$

the relationship is completely specified by giving any two of them. Li and Sacks[2] have described methods of calculating these probabilities adequate for most relationships that do not involve inbreeding. The mean number of genes per locus i.b.d. (as from now on I abbreviate the phrase 'identical by descent') with genes at the same locus in A for a hypothetical population of relatives like B is clearly $2c_2 + c_1$. One half of this number, $c_2 + \frac{1}{2}c_1$, may therefore be called the expected fraction of genes i.b.d. in a relative. It can be shown that it is equal to Sewall Wright's coefficient of relationship r (in a non-inbred population). The standard methods of calculating r without obtaining the complete distribution can be found in Kempthorne.[3] Tables of

$$f = \tfrac{1}{2}r = \tfrac{1}{2}(c_2 + \tfrac{1}{2}c_1) \quad \text{and} \quad F = c_2$$

for a large class of relationships can be found in Haldane and Jayakar.[4]

Strictly, a more complicated metric of relationship taking into account the parameters of selection is necessary for a locus undergoing selection, but the following account based on use of the above coefficients must give a good approximation to the truth when selection is slow and may be hoped to give some guidance even when it is not.

Consider now how the effects which an arbitrary individual distributes to the population can be summarized. For convenience and generality we will include at this stage certain effects (such as effects on parents' fitness) which must be zero under the restrictions of this particular model, and also others (such as effects on offspring) which although not necessarily zero we will not attempt to treat accurately in the subsequent analysis.

The effect of A on specified B can be a variate. In the present deterministic treatment, however, we are concerned only with the means of such variates. Thus the effect which we may write $(\delta a_{\text{father}})_{\text{A}}$ is really the expectation of the effect of A upon his father but for brevity we will refer to it as the effect on the father.

The full array of effects like $(\delta a_{\text{father}})_{\text{A}}$, $(\delta a_{\text{sister}})_{\text{A}}$, etc., we will denote

$$\{\delta a_{\text{rel.}}\}_{\text{A}}.$$

From this array we can construct the simpler array

$$\{\delta a_{r,c_2}\}_{\text{A}}$$

by adding together all effects to relatives who have the same values for the pair of coefficients (r, c_2). For example, the combined effect $\delta a_{\frac{1}{4},0}$ might contain effects actually occurring to grandparents, grandchildren, uncles, nephews, and half-brothers. From what has been said above it is clear that as regards changes in autosomal gene frequency by natural selection all the consequences of the full array are implied by this reduced array—at least, provided we ignore (1) the effect of previous generations of selection on the expected constitution of relatives, and (2) the one or more generations that must really occur before effects to children, nephews, grandchildren, etc., are manifested.

From this array we can construct a yet simpler array, or vector,

$$\{\delta a_r\}_{\text{A}},$$

by adding together all effects with common r. Thus $a_{\frac{1}{4}}$ would bring together effects to the above-mentioned set of relatives and effects to double-first cousins, for whom the pair of coefficients is $(\frac{1}{4}, \frac{1}{16})$.

Corresponding to the effect which A causes to B there will be an effect of similar type on A. This will either come from B himself or from a person who stands to A in the same relationship as A stands to B. Thus corresponding to an effect by A on his nephew there will be an effect on A by his uncle. The similarity between the effect which A dispenses and that which he receives is clearly an aspect of the problem of the correlation between relatives. Thus the term e° in equation (1) is not a constant for any given genotype of A since it will depend on the genotypes of neighbours and therefore on the gene frequencies and the mating system.

Consider a single locus. Let the series of allelomorphs be $G_1, G_2, G_3, \ldots,$ G_n, and their gene frequencies $p_1, p_2, p_3, \ldots, p_n$. With the genotype $G_i G_j$ associate the array $\{\delta a_{\text{rel.}}\}_{ij}$; within the limits of the above-mentioned approximations natural selection in the model is then defined.

If we were to follow the usual approach to the formulation of the progress due to natural selection in a generation, we should attempt to give formulas for the neighbour-modulated fitnesses $a^{\bullet}_{ij}$. In order to formulate the expectation of that element of e°_{ij} which was due to the return effect of a relative B we would need to know the distribution of possible genotypes of B, and to obtain this we must use the double measure of B's relationship and the gene frequencies just as in the problem of the correlation between relatives. Thus the formula for e°_{ij} will involve all the arrays $\{\delta a_{r,c_2}\}_{ij}$ and will be rather unwieldy (see section 4).

An alternative approach, however, shows that the arrays $\{\delta a_r\}_{ij}$ are sufficient to define the selective effects. Every effect on reproduction which is due to A can be thought of as made up of two parts: an effect on the reproduction of genes i.b.d. with genes in A, and an effect on the reproduction of unrelated genes. Since the coefficient r measures the expected fraction of genes i.b.d. in a relative, for any particular degree of relationship this breakdown may be written quantitatively:

$$(\delta a_{\text{rel.}})_{\text{A}} = r(\delta a_{\text{rel.}})_{\text{A}} + (1-r)(\delta a_{\text{rel.}})_{\text{A}}.$$

The total of effects on reproduction which are due to A may be treated similarly:

$$\sum_{\text{rel.}}(\delta a_{\text{rel.}})_{\text{A}} = \sum_{\text{rel.}} r(\delta a_{\text{rel.}})_{\text{A}} + \sum_{\text{rel.}}(1-r)(\delta a_{\text{rel.}})_{\text{A}},$$

or

$$\sum_{r}(\delta a_r)_{\text{A}} = \sum_{r} r(\delta a_r)_{\text{A}} + \sum_{r}(1-r)(\delta a_r)_{\text{A}},$$

which we rewrite briefly as

$$\delta T^{\bullet}_{\text{A}} = \delta R^{\bullet}_{\text{A}} + \delta S_{\text{A}},$$

where $\delta R^{\bullet}_{\text{A}}$ is accordingly the total effect on genes i.b.d. in relatives of A, and δS_{A} is the total effect on their other genes. The reason for the omission of an index symbol from the last term is that here there is, in effect, no question of whether or not the self-effect is to be in the summation, for if it is included it has to be multiplied by zero. If index symbols were used we should have

$\delta S^{\bullet}_{\mathrm{A}} = \delta S^{\circ}_{\mathrm{A}}$, whatever the subscript; it therefore seems more explicit to omit them throughout.

If, therefore, all effects are accounted to the individuals that cause them, of the total effect $\delta T^{\bullet}_{ij}$ due to an individual of genotype $G_i G_j$ a part $\delta R^{\bullet}_{ij}$ will involve a specific contribution to the gene pool by this genotype, while the remaining part δS_{ij} will involve an unspecific contribution consisting of genes in the ratio in which the gene pool already possesses them. It is clear that it is the matrix of effects $\delta R^{\bullet}_{ij}$ which determines the direction of selection progress in gene frequencies; δS_{ij} only influences its magnitude. In view of this importance of the $R^{\bullet}_{ij}$ it is convenient to give some name to the concept with which they are associated.

In accordance with our convention let

$$R^{\bullet}_{ij} = 1 + \delta R^{\bullet}_{ij};$$

then $R^{\bullet}_{ij}$ will be called the *inclusive fitness*, $\delta R^{\bullet}_{ij}$ the *inclusive fitness effect*, and δS_{ij} the *diluting effect*, of the genotype $G_i\, G_j$.

Let

$$T^{\bullet}_{ij} = 1 + \delta T^{\bullet}_{ij}.$$

So far our discussion is valid for non-random mating but from now on for simplicity we assume that it is random. Using a prime to distinguish the new gene frequencies after one generation of selection we have

$$p_i' = \frac{\sum\limits_{j} p_i p_j R^{\bullet}_{ij} + p_i \sum\limits_{j,k} p_j p_k \delta S_{jk}}{\sum\limits_{j,k} p_j p_k T^{\bullet}_{jk}} = p_i \frac{\sum\limits_{j} p_j R^{\bullet}_{ij} + \sum\limits_{j,k} p_j p_k \delta S_{jk}}{\sum\limits_{j,k} p_j p_k T^{\bullet}_{jk}}.$$

The terms of this expression are clearly of the nature of averages over a part (genotypes containing G_i, homozygotes $G_i\, G_i$ counted twice) and the whole of the existing set of genotypes in the population. Thus using a well-known subscript notation we may rewrite the equation term by term as

$$p_i' = p_i \frac{R^{\bullet}_{i.} + \delta S_{..}}{T^{\bullet}_{..}}$$

$$\therefore \quad p_i' - p_i = \Delta p_i = \frac{p_i}{T^{\bullet}_{..}} (R^{\bullet}_{i.} + \delta S_{..} - T^{\bullet}_{..})$$

or

$$\Delta p_i = \frac{p_i}{R^{\bullet}_{..} + \delta S_{..}} (R^{\bullet}_{i.} - R^{\bullet}_{..}). \qquad (2)$$

This form clearly differentiates the roles of the $R^{\bullet}_{ij}$ and δS_{ij} in selective progress and shows the appropriateness of calling the latter diluting effects.

For comparison with the account of the classical case given by Moran,[5] equation (2) may be put in the form

$$\Delta p_i = \frac{p_i}{T^\bullet_{..}}\left(\frac{1}{2}\frac{\partial R^\bullet_{..}}{\partial p_i} - R^\bullet_{..}\right),$$

where $\partial/\partial p_i$ denotes the usual partial derivative, written $\mathrm{d}/\mathrm{d}p_i$ by Moran.

Whether the selective effect is reckoned by means of the $a^\bullet_{ij}$ or according to the method above, the denominator expression must take in all effects occurring during the generation. Hence $a^\bullet_{..} = T^\bullet_{..}$.

As might be expected from the greater generality of the present model the extension of the theorem of the increase of mean fitness[6,7] presents certain difficulties. However, from the above equations it is clear that the quantity that will tend to maximize, if any, is $R^\bullet_{..}$, the mean inclusive fitness. The following brief discussion uses Kingman's approach.[8]

The mean inclusive fitness in the succeeding generation is given by

$$R^{\bullet\prime}_{..} = \sum_{i,j} p'_i p'_j R^\bullet_{ij} = \frac{1}{T^{\bullet 2}_{..}} \sum_{i,j} p_i p_j R^\bullet_{ij}(R^\bullet_{i.} + \delta S_{..})(R^\bullet_{.j} + \delta S_{..}).$$

$$\therefore \quad R^{\bullet\prime}_{..} - R^\bullet_{..} = \Delta R^\bullet_{..} = \frac{1}{T^{\bullet 2}_{..}}\left\{\sum_{i,j} p_i p_j R^\bullet_{ij} R^\bullet_{i.} R^\bullet_{.j} + 2\delta S_{..} \sum_{i,j} p_i p_j R^\bullet_{ij} R^\bullet_{i.} \right.$$

$$\left. + R^\bullet_{..}\delta S^2_{..} - R^\bullet_{..} T^{\bullet 2}_{..}\right\}.$$

Substituting $R^\bullet_{..} + \delta S_{..}$ for $T^\bullet_{..}$ in the numerator expression, expanding and rearranging:

$$\Delta R^\bullet = \frac{1}{T^{\bullet 2}_{..}}\left\{\left(\sum_{i,j} p_i p_j R^\bullet_{ij} R^\bullet_{i.} R^\bullet_{.j} - R^{\bullet 3}_{..}\right) + 2\delta S_{..}\left(\sum_{i,j} p_i p_j R^\bullet_{ij} R^\bullet_{i.} - R^{\bullet 2}_{..}\right)\right\}.$$

We have ( ) ≥ 0 in both cases. The first is the proven inequality of the classical model. The second follows from

$$\sum_{i,j} p_i p_j R^\bullet_{ij} R^\bullet_{i.} = \sum_i p_i R^{\bullet 2}_{i.} \geq \left(\sum_i p_i R^\bullet_{i.}\right)^2 = R^{\bullet 2}_{..}.$$

Thus a sufficient condition for $\Delta R^\bullet_{..} \geq 0$ is $\delta S_{..} \geq 0$. That $\Delta R^\bullet_{..} \geq 0$ for positive dilution is almost obvious if we compare the actual selective changes with those which would occur if $\{R^\bullet_{ij}\}$ were the fitness matrix in the classical model.

It follows that $R^\bullet_{..}$ certainly maximizes (in the sense of approaching a local maximum of $R^\bullet_{..}$) if it never occurs in the course of selective changes that $\delta S_{..} < 0$. Thus $R^\bullet_{..}$ certainly maximizes if all $\delta S_{ij} \geq 0$ and therefore also if all $(\delta a_{\mathrm{rel.}})_{ij} \geq 0$. It still does so even if some or all δa_{ij} are negative, for, as we have seen, δS_{ij} is independent of δa_{ij}.

Here then we have discovered a quantity, inclusive fitness, which under the conditions of the model tends to maximize in much the same way that fitness tends to maximize in the simpler classical model. For an important class of genetic effects where the individual is supposed to dispense benefits to his neighbours, we have formally proved that the average inclusive fitness in the population will always increase. For cases where individuals may dispense harm to their neighbours we merely know, roughly speaking, that the change in gene frequency in each generation is aimed somewhere in the direction of a local maximum of average inclusive fitness,[9] but may, for all the present analysis has told us, overshoot it in such a way as to produce a lower value.

As to the nature of inclusive fitness it may perhaps help to clarify the notion if we now give a slightly different verbal presentation. Inclusive fitness may be imagined as the personal fitness which an individual actually expresses in its production of adult offspring as it becomes after it has been first stripped and then augmented in a certain way. It is stripped of all components which can be considered as due to the individual's social environment, leaving the fitness which he would express if not exposed to any of the harms or benefits of that environment. This quantity is then augmented by certain fractions of the quantities of harm and benefit which the individual himself causes to the fitnesses of his neighbours. The fractions in question are simply the coefficients of relationship appropriate to the neighbours whom he affects; unity for clonal individuals, one-half for sibs, one-quarter for half-sibs, one-eighth for cousins, ... , and finally zero for all neighbours whose relationship can be considered negligibly small.

Actually, in the preceding mathematical account we were not concerned with the inclusive fitness of individuals as described here but rather with certain averages of them which we call the inclusive fitnesses of types. But the idea of the inclusive fitness of an individual is nevertheless a useful one. Just as in the sense of classical selection we may consider whether a given character expressed in an individual is adaptive in the sense of being in the interest of his personal fitness or not, so in the present sense of selection we may consider whether the character or trait of behaviour is or is not adaptive in the sense of being in the interest of his inclusive fitness.

3. THREE SPECIAL CASES

Equation (2) may be written

$$\Delta p_i = p_i \frac{\delta R^{\bullet}_{i.} - \delta R^{\bullet}_{..}}{1 + \delta T^{\bullet}_{..}}. \tag{3}$$

Now $\delta T^{\bullet}_{ij} = \sum_r (\delta a_r)_{ij}$ is the sum and $\delta R^{\bullet} = \sum_r (\delta a_r)_{ij}$ is the first moment about $r = 0$ of the array of effects $\{\delta a_{\text{rel.}}\}_{ij}$ caused by the genotype $G_i G_j$; it appears that these two parameters are sufficient to fix the progress of the system under natural selection within our general approximation.

Let

$$r^{\bullet}_{ij} = \frac{\delta R^{\bullet}_{ij}}{\delta T^{\bullet}_{ij}}, \quad (\delta T^{\bullet}_{ij} \neq 0); \tag{4}$$

and let

$$r^{\circ}_{ij} = \frac{\delta R^{\circ}_{ij}}{\delta T^{\circ}_{ij}}, \quad (\delta T^{\circ}_{ij} \neq 0). \tag{5}$$

These quantities can be regarded as average relationships or as the first moments of reduced arrays, similar to the first moments of probability distributions.

We now consider three special cases which serve to bring out certain important features of selection in the model.

(a) The sums $\delta T^{\bullet}_{ij}$ differ between genotypes, the reduced first moment $r^{\bullet}$ being common to all. If all higher moments are equal between genotypes, that is, if all arrays are of the same 'shape', this corresponds to the case where a stereotyped social action is performed with differing intensity or frequency according to genotype.

Whether or not this is so, we may, from equation (4), substitute $r^{\bullet}\delta T^{\bullet}_{ij}$ for $\delta R^{\bullet}_{ij}$ in equation (3) and have

$$\Delta p_i = p_i r^{\bullet} \frac{\delta T^{\bullet}_{i.} - \delta T^{\bullet}_{..}}{1 + \delta T^{\bullet}_{..}}.$$

Comparing this with the corresponding equation of the classical model,

$$\Delta p_i = p_i \frac{\delta a_{i.} - \delta a_{..}}{1 + \delta a_{..}}, \tag{6}$$

we see that placing genotypic effects on a relative of degree $r^{\bullet}$ instead of reserving them for personal fitness results in a slowing of selection progress according to the fractional factor $r^{\bullet}$.

If, for example, the advantages conferred by a 'classical' gene to its carriers are such that the gene spreads at a certain rate the present result tells us that in exactly similar circumstances another gene which conferred similar advantages to the sibs of the carriers would progress at exactly half this rate.

In trying to imagine a realistic situation to fit this sort of case some concern may be felt about the occasions where through the probabilistic nature of things the gene carrier happens not to have a sib, or not to have one suitably placed to receive the benefit. Such possibilities and their frequencies of realization must, however, all be taken into account as the effects $(\delta a_{\text{sibs}})_{\text{A}}$, etc., are being evaluated for the model, very much as if in a classical case allowance were being made for some degree of failure of penetrance of a gene.

(b) The reduced first moments $r^\bullet_{ij}$ differ between genotypes, the sum $\delta T^\bullet$ being common to all. From equation (4), substituting $r^\bullet_{ij}\delta T^\bullet$ for $\delta R^\bullet_{ij}$ in equation (3) we have

$$\Delta p_i = p_i \frac{\delta T^\bullet}{T^\bullet}(r^\bullet_{i.} - r^\bullet_{..}).$$

But it is more interesting to assume δa is also common to all genotypes. If so it follows that we can replace $^\bullet$ by $^\circ$ in the numerator expression of equation (3). Then, from equation (5), substituting $r^\circ_{ij}\delta T^\circ$ for δR°_{ij}, we have

$$\Delta p_i = p_i \frac{\delta T^\circ}{T^\bullet}(r^\circ_{i.} - r^\circ_{..}).$$

Hence, if a giving-trait is in question (δT° positive), genes which restrict giving to the nearest relative ($r^\circ_{i.}$ greatest) tend to be favoured; if a taking-trait (δT° negative), genes which cause taking from the most distant relatives tend to be favoured.

If all higher reduced moments about $r = r^\circ_{ij}$ are equal between genotypes it is implied that the genotype merely determines whereabouts in the field of relationship that centres on an individual a stereotyped array of effects is placed.

With many natural populations it must happen that an individual forms the centre of an actual local concentration of his relatives which is due to a general inability or disinclination of the organisms to move far from their places of birth. In such a population, which we may provisionally term 'viscous', the present form of selection may apply fairly well to genes which affect vagrancy. It follows from the statements of the last paragraph but one that over a range of different species we would expect to find giving-traits commonest and most highly developed in the species with the most viscous populations whereas uninhibited competition should characterize species with the most freely mixing populations.

In the viscous population, however, the assumption of random mating is very unlikely to hold perfectly, so that these indications are of a rough qualitative nature only.

(c) $\delta T^\bullet_{ij} = 0$ for all genotypes.

$$\therefore \quad \delta T^\circ_{ij} = \delta a_{ij}$$

for all genotypes, and from equation (5)

$$\delta R^\circ_{ij} = \delta a_{ij} r^\circ_{ij}.$$

Then, from equation (3), we have

$$\begin{aligned}\Delta p_i = p_i(\delta R^\bullet_{i.} - \delta R^\bullet_{..}) &= p_i\{(\delta a_{i.} + \delta R^\circ_{i.}) - (\delta a_{..} + \delta R^\circ_{..})\}\\ &= p_i\{\delta a_{i.}(1 - r^\circ_{i.}) - \delta a_{..}(1 - r^\circ_{..})\}.\end{aligned}$$

Such cases may be described as involving transfers of reproductive potential. They are especially relevant to competition, in which the individual can be considered as endeavouring to transfer prerequisites of survival and reproduction from his competitors to himself. In particular, if $r^{\circ}_{ij} = r^{\circ}$ for all genotypes we have

$$\Delta p_i = p_i(1 - r^{\circ})(\delta a_{i.} - \delta a_{..}).$$

Comparing this to the corresponding equation of the classical model (equation (6)) we see that there is a reduction in the rate of progress when transfers are from a relative.

It is relevant to note that Haldane[10] in his first paper on the mathematical theory of selection pointed out the special circumstances of competition in the cases of mammalian embryos in a single uterus and of seeds both while still being nourished by a single parent plant and after their germination if they were not very thoroughly dispersed. He gave a numerical example of competition between sibs showing that the progress of gene frequency would be slower than normal.

In such situations as this, however, where the population may be considered as subdivided into more or less standard-sized batches each of which is allotted a local standard-sized pool of reproductive potential (which in Haldane's case would consist almost entirely of prerequisites for pre-adult survival), there is, in addition to a small correcting term which we mention in the short general discussion of competition in the next section, an extra overall slowing in selection progress. This may be thought of as due to the wasting of the powers of the more fit and the protection of the less fit when these types chance to occur positively assorted (beyond any mere effect of relationship) in a locality; its importance may be judged from the fact that it ranges from zero when the batches are indefinitely large to a halving of the rate of progress for competition in pairs.

4. ARTIFICIALITIES OF THE MODEL

When any of the effects is negative the restrictions laid upon the model hitherto do not preclude certain situations which are clearly impossible from the biological point of view. It is clearly absurd if for any possible set of gene frequencies any $a^{\bullet}_{ij}$ turns out negative; and even if the magnitude of δa_{ij} is sufficient to make $a^{\bullet}_{ij}$ positive while $1 + e^{\circ}_{ij}$ is negative the situation is still highly artificial, since it implies the possibility of a sort of overdraft on the basic unit of an individual which has to be made good from his own takings. If we call this situation 'improbable' we may specify two restrictions: a weaker, $e^{\circ}_{ij} > -1$, which precludes 'improbable' situations; and a stronger, $e^{\bullet}_{ij} > -1$, which precludes even the impossible situations, both being required over the whole range of possible gene frequencies as well as the whole range of genotypes.

As has been pointed out, a formula for $e^{\bullet}_{ij}$ can only be given if we have the arrays of effects according to a double coefficient of relationship. Choosing the double coefficient (c_2, c_1) such a formula is

$$e^{\bullet}_{ij} = \sum_{c_2,c_1}{}^{\bullet}[c_2\mathrm{Dev}(\delta a_{c_2,c_1})_{ij} + \tfrac{1}{2}c_1\{\mathrm{Dev}(\delta a_{c_2,c_1})_{i.} + \mathrm{Dev}(\delta a_{c_2,c_1})_{.j}\}] + \delta T^{\bullet}_{..},$$

where

$$\mathrm{Dev}(\delta a_{c_2,c_1})_{ij} = (\delta a_{c_2,c_1})_{ij} - (\delta a_{c_2,c_1})_{..}, \mathrm{etc.}$$

Similarly

$$e^{\circ}_{ij} = \sum{}^{\circ}[''] + \delta T^{\circ}_{..},$$

the self-effect $(\delta a_{1,0})_{ij}$ being in this case omitted from the summations.

The following discussion is in terms of the stronger restriction but the argument holds also for the weaker; we need only replace $^{\bullet}$ by $^{\circ}$ throughout.

If there are no dominance deviations, i.e. if

$$(\delta a_{\mathrm{rel.}})_{ij} = \tfrac{1}{2}\{(\delta a_{\mathrm{rel.}})_{ii} + (\delta a_{\mathrm{rel.}})_{jj}\} \quad \text{for all } ij \text{ and rel.,}$$

it follows that each ij deviation is the sum of the i and the j deviations. In this case we have

$$e^{\bullet}_{ij} = \sum{}^{\bullet} r\mathrm{Dev}(\delta a_r)_{ij} + \delta T^{\bullet}_{..}.$$

Since we must have $e^{\bullet}_{..} = \delta T^{\bullet}_{..}$, it is obvious that some of the deviations must be negative.

Therefore $\delta T^{\bullet}_{..} > -1$ is a necessary condition for $e^{\bullet}_{ij} > -1$. This is, in fact, obvious when we consider that $\delta T^{\bullet}_{..} = -1$ would mean that the aggregate of individual takings was just sufficient to eat up all basic units exactly. Considering that the present use of the coefficients of relationship is only valid when selection is slow, there seems little point in attempting to derive mathematically sufficient conditions for the restriction to hold; intuitively, however, it would seem that if we exclude over- and underdominance it should be sufficient to have no homozygote with a net taking greater than unity.

Even if we could ignore the breakdown of our use of the coefficient of relationship it is clear enough that if $\delta T^{\bullet}_{..}$ approaches anywhere near -1 the model is highly artificial and implies a population in a state of catastrophic decline. This does not mean, of course, that mutations causing large selfish effects cannot receive positive selection; it means that their expression must moderate with increasing gene frequency in a way that is inconsistent with our model. The 'killer' trait of *Paramecium* might be regarded as an example of a selfish trait with potentially large effects, but with its only partially genetic mode of inheritance and inevitable density dependance it obviously requires a selection model tailored to the case, and the same is doubtless true of most 'social' traits which are as extreme as this.

Really the class of model situations with negative neighbour effects which are artificial according to a strict interpretation of the assumptions must be much wider than the class which we have chosen to call 'improbable'. The model assumes that the magnitude of an effect does not depend either on the genotype of the effectee or on his state with respect to the prerequisites of fitness at the time when the effect is caused. Where taking-traits are concerned it is just possible to imagine that this is true of some kinds of surreptitious theft but in general it is more reasonable to suppose that following some sort of an encounter the limited prerequisite is divided in the ratio of the competitive abilities. Provided competitive differentials are small, however, the model will not be far from the truth; the correcting term that should be added to the expression for Δp_i can be shown to be small to the third order. With giving-traits it is more reasonable to suppose that if it is the nature of the prerequisite to be transferable the individual can give away whatever fraction of his own property his instincts incline him to. The model was designed to illuminate altruistic behaviour; the classes of selfish and competitive behaviour which it can also usefully illuminate are more restricted, especially where selective differentials are potentially large.

For loci under selection the only relatives to which our metric of relationship is strictly applicable are ancestors. Thus the chance that an arbitrary parent carries a gene picked in an offspring is $\frac{1}{2}$, the chance that an arbitrary grandparent carries it is $\frac{1}{4}$, and so on. As regards descendants, it seems intuitively plausible that for a gene which is making steady progress in gene frequency the true expectation of genes i.b.d. in a n-th generation descendant will exceed $\frac{1}{2}^n$, and similarly that for a gene that is steadily declining in frequency the reverse will hold. Since the path of genetic connection with a simple same-generation relative like a half-sib includes an 'ascending part' and a 'descending part' it is tempting to imagine that the ascending part can be treated with multipliers of exactly $\frac{1}{2}$ and the descending part by multipliers consistently more or less than $\frac{1}{2}$ according to which type of selection is in progress. However, a more rigorous attack on the problem shows that it is more difficult than the corresponding one for simple descendants, where the formulation of the factor which actually replaces $\frac{1}{2}$ is quite easy, at least in the case of classical selection, and the author has so far failed to reach any definite general conclusions as to the nature and extent of the error in the foregoing account which his use of the ordinary coefficients of relationship has involved.

Finally, it must be pointed out that the model is not applicable to the selection of new mutations. Sibs might or might not carry the mutation depending on the point in the germ-line of the parent at which it had occurred, but for relatives in general a definite number of generations must pass before the coefficients give the true—or, under selection, the approximate—expectations of replicas. This point is favourable to the establishment of taking-traits and slightly against giving-traits. A mutation can, however, be expected to overcome any such slight initial barrier before it has recurred many times.

5. THE MODEL LIMITS TO THE EVOLUTION OF ALTRUISTIC AND SELFISH BEHAVIOUR

With classical selection a genotype may be regarded as positively selected if its fitness is above the average and as counterselected if it is below. The environment usually forces the average fitness $a_{..}$ towards unity; thus for an arbitrary genotype the sign of δa_{ij} is an indication of the kind of selection. In the present case although it is $T^{\bullet}_{..}$ and not $R^{\bullet}_{..}$ that is forced towards unity, the analogous indication is given by the inclusive fitness effect $\delta R^{\bullet}_{ij}$, for the remaining part, the diluting effect δS_{ij}, of the total genotypic effect $\delta T^{\bullet}_{ij}$ has no influence on the kind of selection. In other words the kind of selection may be considered determined by whether the inclusive fitness of a genotype is above or below average.

We proceed, therefore, to consider certain elementary criteria which determine the sign of the inclusive fitness effect. The argument applies to any genotype and subscripts can be left out.

Let

$$\delta T^{\circ} = k\delta a, \quad (\delta a \neq 0). \tag{7}$$

According to the signs of δa and δT° we have four types of behaviour as set out in the following diagram:

		Neighbours	
		gain; δT° +ve	lose; δT° −ve
Individual	gains; δa + ve	k +ve Selected	k −ve Selfish behaviour ?
	loses; δa − ve	k −ve Altruistic behaviour ?	k +ve Counter-selected

The classes for which k is negative are of the greatest interest, since for these it is less obvious what will happen under selection. Also, if we regard fitness as like a substance and tending to be conserved, which must be the case in so far as it depends on the possession of material prerequisites of survival and reproduction, k −ve is the more likely situation. Perfect conservation occurs if $k = -1$. Then $\delta T^{\bullet} = 0$ and $T^{\bullet} = 1$: the gene pool maintains constant 'volume' from generation to generation. This case has been discussed in Case

(c) of section 3. In general the value of k indicates the nature of the departure from conservation. For instance, in the case of an altruistic action $|k|$ might be called the ratio of gain involved in the action: if its value is two, two units of fitness are received by neighbours for every one lost by an altruist. In the case of a selfish action, $|k|$ might be called the ratio of diminution: if its value is again two, two units of fitness are lost by neighbours for one unit gained by the taker.

The alarm call of a bird probably involves a small extra risk to the individual making it by rendering it more noticeable to the approaching predator but the consequent reduction of risk to a nearby bird previously unaware of danger must be much greater. (The alarm call often warns more than one nearby bird of course—hundreds in the case of a flock—but since the predator would hardly succeed in surprising more than one in any case the total number warned must be comparatively unimportant.) We need not discuss here just how risks are to be reckoned in terms of fitness: for the present illustration it is reasonable to guess that for the generality of alarm calls k is negative but $|k| > 1$. How large must $|k|$ be for the benefit to others to outweigh the risk to self in terms of inclusive fitness?

$$\begin{aligned} \delta R^{\bullet} &= \delta R^{\circ} + \delta a \\ &= r^{\circ}\delta T^{\circ} + \delta a && \text{from (5)} \\ &= \delta a(kr^{\circ} + 1) && \text{from (7)} \end{aligned}$$

Thus of actions which are detrimental to individual fitness (δa – ve) only those for which $-k > \frac{1}{r^{\circ}}$ will be beneficial to inclusive fitness ($\delta R^{\bullet}$ + ve).

This means that for a hereditary tendency to perform an action of this kind to evolve the benefit to a sib must average at least twice the loss to the individual, the benefit to a half-sib must be at least four times the loss, to a cousin eight times, and so on. To express the matter more vividly, in the world of our model organisms, whose behaviour is determined strictly by genotype, we expect to find that no one is prepared to sacrifice his life for any single person but that everyone will sacrifice it when he can thereby save more than two brothers, or four half-brothers, or eight first cousins ... Although according to the model a tendency to simple altruistic transfers ($k = -1$) will never be evolved by natural selection, such a tendency would receive zero counterselection when it concerned transfers between clonal individuals. Conversely selfish transfers are always selected except when from clonal individuals.

As regards selfish traits in general (δa + ve, k – ve) the condition for a benefit to inclusive fitness is $-k < \frac{1}{r^{\circ}}$. Behaviour that involves taking too much from close relatives will not evolve. In the model world of genetically controlled behaviour we expect to find that sibs deprive one another of reproductive prerequisites provided they can themselves make use of at least one half of what they take; individuals deprive half-sibs of four units of reproductive potential if they can get personal use of at least one of them; and so on. Clearly from a gene's point of view it is worthwhile to deprive a large number of distant relatives in order to extract a small reproductive advantage.

References and notes

1. J. F. C. Kingman, *Proceedings of the Cambridge Philosophical Society* **57**, 574 (1961).
2. C. C. Li and L. Sacks, *Biometrics* **10**, 347 (1954).
3. O. Kempthorne, *An Introduction to Genetical Statistics* (Wiley, New York, 1957).
4. J. B. S. Haldane and S. D. Jayakar, *Journal of Genetics* **58**, 81 (1962).
5. P. A. P. Moran, *The Statistical Processes of Evolutionary Theory*, p. 54 (Clarendon Press, Oxford, 1962).
6. P. A. G. Scheuer and S. P. H. Mandel, *Heredity* **31**, 519 (1959).
7. H. P. Mulholland and C. A. B. Smith, *American Mathematical Monthly* **66**, 673 (1959).
8. J. F. C. Kingman, *Quarterly Journal of Mathematics* **12**, 78 (1961).
9. That is, it is 'aimed uphill'; that it need not be at all directly towards the local maximum is well shown in the classical example illustrated in Mulholland and Smith 1959 (ref. 7).
10. J. B. S. Haldane, *Transactions of the Cambridge Philosophical Society* **23**, 19 (1923).

THE GENETICAL EVOLUTION OF SOCIAL BEHAVIOUR, II†

W. D. HAMILTON

Grounds for thinking that the model described in the previous paper can be used to support general biological principles of social evolution are briefly discussed.

Two principles are presented, the first concerning the evolution of social behaviour in general and the second the evolution of social discrimination. Some tentative evidence is given.

More detailed application of the theory in biology is then discussed, particular attention being given to cases where the indicated interpretation differs from previous views and to cases which appear anomalous. A hypothesis is outlined concerning social evolution in the Hymenoptera; but the evidence that at present exists is found somewhat contrary on certain points. Other subjects considered include warning behaviour, the evolution of distasteful properties in insects, clones of cells, and clones of zooids as contrasted with other types of colonies, the confinement of parental care to true offspring in birds and insects, fights, the behaviour of parasitoid insect larvae within a host, parental care in connection with monogyny and monandry, and multi-ovulate ovaries in plants in connection with wind and insect pollination.

1. INTRODUCTION

In Part I of this paper,[1] a genetical mathematical model was used to deduce a principle concerning the evolution of social behaviour which, if true generally, may be of considerable importance in biology. It has now to be considered whether there is any logical justification for the extension of this principle beyond the model case of non-overlapping generations, and, if so, whether there is evidence that it does work effectively in nature.

In brief outline, the theory points out that for a gene to receive positive selection it is not necessarily enough that it should increase the fitness of its bearer above the average if this tends to be done at the heavy expense of related individuals, because relatives, on account of their common ancestry, tend to carry replicas of the same gene; and conversely that a gene may receive positive selection even though disadvantageous to its bearers if it causes them to confer

†*Journal of Theoretical Biology* 7, 17–52 (1964).

sufficiently large advantages on relatives. Relationship alone never gives grounds for *certainty* that a person carries a gene which a relative is known to carry except when the relationship is 'clonal' or 'mitotic' (e.g. the two are monozygotic twins)—and even then, strictly, the possibility of an intervening mutation should be admitted. In general, it has been shown that Wright's coefficient of relationship r approximates closely to the chance that a replica will be carried. Thus if an altruistic trait is in question more than $1/r$ units of reproductive potential or 'fitness' must be endowed on a relative of degree r for every one unit lost by the altruist if the population is to gain on average more replicas than it loses. Similarly, if a selfish trait is in question, the individual must receive and use at least a fraction r of the quantity of 'fitness' deprived from his relative if the causative gene is to be selected.

For a more critical explanation of these ideas and of the important concept of 'inclusive fitness', which will be freely referred to in what follows, the reader is referred to Part I of this paper.

2. THE GROUNDS FOR GENERALIZATION

It is clear that in outline this type of argument is not restricted to the case of non-overlapping generations nor to the state of panmixia on which we have been able to base a fairly precise analysis. The idea of the regression, or 'probabilistic dilution', of 'identical' genes in relatives further and further removed applies to all organisms performing sexual reproduction, whether or not their generations overlap and whether or not the relatives considered belong to the same generation.

However, perhaps we should not feel entirely confident about generalizing our principle until a more comprehensive mathematical argument, with inclusive fitness more widely defined, has been worked out. But even from this point of view there does seem to be good reason for thinking that it can be generalized—reason about as good, at least, as that which is supposed to give foundation to certain principles of the classical theory.

Roughly speaking the classical mathematical theory has developed two parallel branches which lie to either side of the great range of reproductive schedules which organisms actually do manifest. One is applicable to once-and-for-all reproduction (e.g. Kingman[2]); and this form is actually exhibited by many organisms, notably those with annual life cycles. The other is applicable to 'continuous' reproduction (e.g. Kimura[3]). This involves a type of reproductive process which is strictly impossible for any organism to practise, but which for analytical purposes should be approximated quite closely by certain species, for example, some perennial plants. Our model is a generalization in the former branch and there seems little reason to doubt that it can be matched by a similar model in the latter.

Even in the classical theory itself difficulties still face generalization between the two branches, and yet their continuance does not seem to cause much worry. For instance there does not seem to be any comprehensive definition of fitness. And, perhaps in consequence of this lack, it rather appears that Fisher's Fundamental Theorem of Natural Selection has yet to be put in a form which is really as general as Fisher's original statement purports to be.[4] On the other hand, the clarity of Fisher's statement must surely, for general usefulness, have far outweighed its defects in rigour.

3. VALUATION OF THE WELFARE OF RELATIVES

Altogether then it would seem that generalization would not be too foolhardy. In the hope that it may provide a useful summary we therefore hazard the following generalized unrigorous statement of the main principle that has emerged from the model.

The social behaviour of a species evolves in such a way that in each distinct behaviour-evoking situation the individual will seem to value his neighbours' fitness against his own according to the coefficients of relationship appropriate to that situation.

The aspect of this principle which concerns altruism seems to have been realized by Haldane[5] as is shown in some comments on whether a genetical trait causing a person to risk his life to save a drowning child could evolve or not. His argument, though not entirely explicit and apparently restricted to rare genes, is essentially the same as that which we have outlined for altruism in the Introduction.

Haldane does not discuss the question which his remarks raise of whether a gene lost in an adult is worth more or less than a gene lost in a child. However, this touches an aspect of the biological accounting of risks which together with the whole problem of the altruism involved in parental care is best reserved for separate discussion.

The principle was also foreshadowed much earlier in Fisher's (1930) discussion of the evolution of distastefulness in insects.[6] That this phenomenon presents a difficulty, namely an apparent absence of positive selection, is obvious as soon as we reject the pseudo-explanations based on the 'benefit to the species', and the problem is of considerable importance as distastefulness, construed in a wide sense, is the basis not only of warning coloration but of both Batesian and Mullerian mimicry. The difficulty of explaining the evolution of warning coloration itself is perhaps even more acute; here *a priori* we would expect that at every stage it would be the new ultra-conspicuous mutants that suffered the first attacks of inexperienced predators. Fisher suggested a benefit to the nearby siblings of the distasteful, or distasteful and conspicuous insect, and gave some suggestive evidence that these characters are correlated with gregariousness of the larvae. He remarked that 'the selective potency of the avoidance of brothers will of course be only half as great as if the individual itself were protected; against this is to be set the fact that it applies to the whole of a possibly numerous brood'. He doubtless realized that further selective benefit would occur through more distant relatives but probably considered it negligible. He realized the logical affinity of this problem with that of the evolution of altruistic behaviour, and he invokes the same kind of selection in his attempt to explain the evolution of the heroic ideal in barbaric human societies.

Another attempt to elucidate the genetical natural selection of altruistic behaviour occurring within a sibship was published by Williams and Williams in 1957.[7] Although their conclusions are doubtless correct the particular form of analysis they adopted seems to have failed to bring out the crucial role of the two-fold factor in this case.

A predator would have to taste the distasteful insect before it could learn to avoid the nearby relatives. Thus despite the toughness and resilience which is supposed to characterize such insects (qualities which the classical selectionists may have been tempted to exaggerate), the common detriment to the 'altruist' must be high and the ratio of gain to loss (k) correspondingly low. The risks involved in giving a warning signal, as between birds, must be much less so that in this case, as indicated in Part I, it is more credible that the condition

$$k > \frac{1}{\bar{r}}$$

is fulfilled even when cases of the parents warning their young and the young each other up to the time of their dispersal are left out of account. The average relationship within a rabbit-warren is probably quite sufficient to account for their 'thumping' habit.[8] Ringing experiments on birds indicate that even adult territorial neighbours must often be much closer relatives than their powers of flight would lead us to expect;[9] a fact that may be of significance for the interpretation of the wider comity of bird behaviour.

The phenomena of mutual preening and grooming may be explained similarly. The mild effort required must stand for a diminution of fitness quite minute compared to the advantage of being cleansed and cleared of ectoparasites on parts of the body which the individual cannot deal with himself. Thus the degree of relationship within the flocks of birds, troupes of monkeys, and so on where such mutual help occurs need not be very high before the condition for an advantage to inclusive fitness is fulfilled; and for grooming within actual families, of monkeys for instance, it is quite obviously fulfilled.

An animal whose reproduction is definitely finished cannot cause any further self-effects. Except for the continuing or pleiotropic effects of genes which are established through an advantage conferred earlier in the life history, the behaviour of a post-reproductive animal may be expected to be entirely altruistic, the smallest degree of relationship with the average neighbour being sufficient to favour the selection of a giving trait. Blest[10] has shown that the post-reproductive behaviour of certain saturnid moths is indeed adaptive in this way. His argument may be summarized in the present terminology as follows. With a species using cryptic resemblance for its protection the very existence of neighbours involves a danger to the individual since the discovery of one by a predator will be a step in teaching it to recognize the crypsis. With an aposematic species on the other hand, the existence of neighbours is an asset since they may well serve to teach an inexperienced predator the warning pattern. Thus with the cryptic moth it is altruistic to die immediately after reproduction, whereas with the warningly coloured moth it is altruistic to continue to live at least through the period during which other moths may not have finished mating and egg-laying. Blest finds that the post-reproductive lifespans of the moths he studied are modified in the expected manner, and that the cryptic species even show behaviour which might be interpreted as an attempt to destroy their cryptic pattern and to use up in random flight activity the remainder of their vital reserves. The selective forces operating on the post-reproductive lifespan are doubtless generally weak; they will be strongest when the average relationship of neighbours is highest, which will be in the most viscous populations. It would be interesting to know how behaviour affecting gene dispersion correlates with the degree of the effects which Blest has observed.

4. DISCRIMINATION IN SOCIAL SITUATIONS

Special case (b) of Part I of this paper has shown explicitly that a certain social action cannot in itself be described as harmful or beneficial to inclusive fitness; this depends on the relationship of the affected individuals. The selective advantage of genes which make behaviour conditional in the right sense on the discrimination of factors which correlate with the relationship of the individual concerned is therefore obvious. It may be, for instance, that in respect of a certain social action performed towards neighbours indiscriminately, an individual is only just breaking even in terms of inclusive fitness. If he could learn to recognize those of his neighbours who really were close relatives and could devote his beneficial actions to them alone an advantage to inclusive fitness would at once appear. Thus a mutation causing such discriminatory behaviour itself benefits inclusive fitness and would be selected. In fact, the individual may not need to perform any discrimination so sophisticated as we suggest here; a difference in the generosity of his behaviour according to whether the situations evoking it were encountered near to, or far from, his own home might occasion an advantage of a similar kind.

Although this type of advantage is itself restricted to social situations, it can be compared to the general advantage associated with making responses conditional on the factors which are the most reliable indicators of future events, an advantage which must, for instance, have been the basis for the evolution of the seed's ability to germinate only when conditions (warmth, moisture, previous freezing, etc.) give real promise for the future survival and growth of the seedling.

Whether the trend implied could ever spread very far may be doubted. All kinds of evolutionary changes in behaviour, especially those subject to the powerful forces of individual advantage, are liable to disrupt any *ad hoc* system of discrimination. This is most true, however, for discrimination in the range of distant relationships where the potential gains are least. The selective advantage when a benefit comes to be given to sibs only instead of to sibs and half-sibs indifferently is more than four times the advantage when a benefit of the same magnitude is given to cousins only instead of to cousins and half-cousins indifferently.

Nevertheless, if any correlate of relationship is very persistent, long-continued weak selection could lead to the evolution of a discrimination based on it even in the range of distant relationships. One possible factor of this kind in species with viscous populations, and one whose persistence depends only on the viscosity and therefore may well be considerably older than the species in question, is familiarity of appearance. For in a viscous population the organisms of a particular neighbourhood, being relatives, must tend to look alike and an individual which used the restrained symbolic forms of aggressive behaviour only towards familiar-looking rivals would be effecting a discrimination advantageous to inclusive fitness.

In accordance with the hypothesis that such discriminations exist it should turn out that in a species of resident bird, strongly territorial and minimally vagrant, the conflicts which proved least readily resolved by ritual behaviour and in which consequent fighting was fiercest were between the rivals that had the most noticeable differences in plumage and song. Whether much evidence of this nature exists I do not know. The rather uncommon cases of interspecific territory systems in birds, as recently reviewed by Wynne-Edwards,[11] seem to be contrary. If differences between interspecific and conspecific encounters were noticed by the original observers they are not mentioned by Wynne-Edwards; and in any case, the very existence of these situations, taken at face value and assumed to be stable and of long standing, is as contrary to the present theory as it is to Gause's principle. Likewise, the positive indications I can bring forward are rather few and feeble. Tinbergen[12] has observed a hostile reaction by herring gulls towards members of their colony forced to behave abnormally (caught in a net) and states that a similar phenomenon is sometimes observed with other social species. Personal observations on colonies of the wasps *Polistes canadensis* and *P. versicolor* have shown a very strong hostility when a wasp taken off a nest is returned to it in a wet and bedraggled condition. This type of reaction after a member of the colony has been much handled seems to be quite common in the social insects. It is perhaps specifically aroused by certain acquired odours, or these combined with the odour of venom. That bird-ringers, who would surely have noticed any social stigma that fell upon birds carrying their often very conspicuous rings, usually report that the rings were no apparent inconvenience to the birds is a counterindication whose force is slightly reduced by the fact that in passerines and most other common birds the legs are unimportant in social communication. It is similarly fortunate for the insect ethologist that spots of fresh oil-paint by themselves on bees and wasps seem to provoke very little reaction. Butterflies of the family Lycaenidae, especially males, are often to be seen jostling one another in the air, sometimes in groups of more than two. The function of this behaviour is obscure; the species do not seem to be at all strongly territorial. According to Ford[13] lepidopterists find that a bunch of jostling butterflies is rather apt to contain an unusual variety.

With the higher animals we may perhaps appeal to evidence of discrimination based on familiarity of a more intimate kind. Animals capable of forming a social hierarchy presumably have some ability to recognize one another as individuals, and with this present it is not necessary for the discrimination to be on the basis of 'racialistic' differences of appearance, voice, or smell. An individual might look extremely like certain members of a group and lie within the group's range of variation in every one of his perceptible characters and yet still be known for a stranger. Speaking from a wide knowledge of just such social animals Wynne-Edwards refers to 'the widespread practice of attacking and persecuting strangers and relegating newcomers to the lowest social rank' and gives several references.[14] The antagonistic nature of this discrimination is of course just what we expect.

As might be expected the evidence in the cases of closest relationship is much more impressive. Tinbergen[15] investigated the ability of herring gulls to recognize their own chicks by observing the reaction to strange chicks placed amongst them. He found that during the first 2 or 3 days after hatching strange chicks are accepted, but by the end of the first week they are driven away. Herring gulls will sometimes form the habit of feeding on the live chicks as well as on the eggs in their own breeding colony when they can catch them unattended, but Tinbergen records no case where an intruded chick was killed although this probably sometimes happens; the hostile behaviour he observed was half-hearted at first but became more definite as the age of the gull's own brood advanced. During the days which follow hatching, the chicks become progressively more mobile and the chance that they will wander into neighbouring nest-territories must increase. Therefore it seems a reasonable hypothesis that the ability to discriminate 'own young' advances in step with the chance that without such discrimination strange chicks would be fostered and the benefits of parental care wasted on unrelated genes. Supporting this hypothesis are the findings quoted by Tinbergen of Watson and Lashley on two tropical species of tern: 'The Noddies nesting in trees do not recognize their young at any age, whereas the ground-nesting Sooties are very similar to Herring Gulls in that they learn to recognize their own young in the course of four days.'[16] House sparrows will accept strange young of the right age placed in the nest but after the nestlings have flown 'they will not, in normal circumstances, feed any but their own young'.[17] Not all observations are as satisfactory for the theory as these however; we may mention the positive passion for fostering said to be shown by emperor penguins that have lost their own chick.[18] This and some other similar anomalies will be briefly discussed in the last section.

Tinbergen showed that herring gulls discriminate eggs even less than chicks, the crudest egg-substitutes being sufficient to release brooding behaviour providing certain attributes of shape and colour are present. This is what we would expect in view of the fact that eggs do not stray at all. It is in striking contrast with the degree of egg-discrimination which is shown by species of birds subject to cuckoo parasitism.

The theoretical principle which these observations seem largely to support is supplementary to the previous principle and we may summarize it in a similar statement.

> *The situations which a species discriminates in its social behaviour tend to evolve and multiply in such a way that the coefficients of relationship involved in each situation become more nearly determinate.*

In situations where relationship is not variable, as, for example, between the nestlings in an arboreal nest, there still remains a discrimination which, if it could be made could greatly benefit inclusive fitness. This is the discrimination of those individuals which do carry one or both of the behaviour-causing genes from those which do not. Such an ability lies outside the conditions postulated in Part I but the extended meaning of inclusive fitness is obvious enough. That genes could cause the perception of the presence of like genes in other individuals may sound improbable; at simplest we need to postulate something like a supergene affecting (1) some perceptible feature of the organism, (2) the perception of that feature, and (3) the social response consequent upon what was perceived. However, exactly the same *a priori* objections might be made to the evolution of assortative mating which manifestly has evolved, probably many times independently and despite its obscure advantages.

If some sort of attraction between likes for purposes of co-operation can occur the limits to the evolution of altruism expressed by our first principle would be very greatly extended, although it should still never happen that one individual would value another more highly than itself, fitness for fitness. And if an individual can be attracted towards likes when it has positive effects—

benefits—to dispense, it can presumably be attracted the other way, towards unlikes, when it has negative effects to dispense (i.e. when circumstances arise which demand combat, suggest robbery, and so on).

5. GENETICAL RELATIONSHIP IN COLONIES

In this section we discuss a small selection of the biological problems relating to life in colonies, choosing particularly those which the theory we have developed is able to illuminate in a simple and novel manner and those concerning which discussions in the existing literature are often unsatisfactory.

Clones

According to considerations advanced so far the coefficient of relationship between all members of a clone should be unity. If this is so our theory predicts for clones a complete absence of any form of competition which is not to the overall advantage and also the highest degree of mutual altruism. This is borne out well enough by the behaviour of the clones which make up the bodies of multicellular organisms. However, when we consider populations of free-living asexual organisms there appears to be a discrepancy in that competitive adaptation is hardly less conspicuous than it is for most wholly sexual populations and altruism if it exists, is not easily detected. To account for this discrepancy three points may be made.

In the first place it may be doubted how many apparently asexual populations are really as they seem. Repeated discoveries of sexual or recombinative processes in species formerly thought to possess none may cause a suspicion that pure clonal populations of any considerable size are uncommon; and taking into account the well-known generalization that asexual reproduction tends to give place to sexual with the onset of adverse conditions, it may be argued that fully competitive (i.e. stationary or declining) pure clonal populations must be less common still. In a mixed sexual–asexual population the levels of competition and altruism should, neglecting mutation, be appropriate to the average relationship.

Second, as regards the appearance of competitive adaptations, we may repeat what was noted in Part I, namely that to the new mutant all individuals have zero relationship (for the locus in question); any selfish mutation must therefore have an immediate advantage and its progress will be merely slowed down, not completely arrested, by the replica-destruction it comes to work in the later stages of its spread.

Third, as regards the absence of co-operation and altruism, we may note an adjustment to the metric of relationship which we have so far found it convenient to neglect but which will have a slight effect in reducing the relationship between individuals in a clonal population. This again involves mutation. Each step in the path of mitotic connection between two asexual organisms corresponds to a constant chance of mutation (m). The chance that a mutation does not occur $(1-m)$ can be multiplied along these paths just as is the factor $\frac{1}{2}$ along paths of meiotic connection in the ordinary calculation of r, and the grand product is likewise the expectation of replica genes in the relative. The number of generations for a given value of r to be reached is approximated by the formula

$$\frac{1}{2m}\log_e\frac{1}{r}.$$

This would apply to the minimum relationship but it is that borne to an individual by half the population and the average relationship is very close to it. With normal mutation rates the decrease in relationship will be slow. Thus if $m = 10^{-5}$ the number of generations for asexual descendants of a common ancestor to become as widely related as full-sibs or the gametes of a single sexual individual is about 39 660. A bacterium with continuously favourable growth conditions so that it divided once every 20 minutes would take $1\frac{1}{2}$ years to run through this many generations whilst a unicellular green alga such as *Chlorella*, dividing once every 15 hours, would take 68 years.

However, taking all three points together and especially considering the fact that a population will normally be started by many sexually produced spores, our apparent discrepancy is largely removed. Such obvious differences in co-operation and altruism as are apparent between a 'colony' of *Volvox* and a population of *Chlamydomonas*, or, to present the contrast another way, within and between colonies of *Volvox*, are at least plausibly accounted for. The co-operation of the cells in the *Volvox* colony, or coenobium as it is perhaps better called, can be regarded as due to the closeness of their relationship, a mere 14 cell generations being necessary to produce the 10 000 or so cells concerned (*V. globulina*).

Thus the classical 'evolutionary' series in the Chlorophyceae, starting with temporary cohesion of mitotic daughter-cells of a free-living unicellular form like *Chlamydomonas* and ending with forms with a large and highly differentiated soma, is well in accord with our theory.

Fusion of individuals or clones

If on the contrary such integrated colonies were found to be formed by the coming together of random members of the population or even by the cohesion of meiotic daughter-cells, there would be some cause for surprise, especially if a soma were formed without any sign of discord among the cells.

Something like this has in fact been noted by Jones in the Rhodophyceae.[19] The sporelings developed from either carpospores or tetraspores of *Gracillaria verrucosa* were found to fuse readily when they grew into contact. Jones suggested that the compound sporelings so formed might have an advantage over solitary ones in nature in being less likely to be smothered by sand in the littoral situations in which they grow since he had observed that they sent up fronds sooner and more strongly; but about four out of five of the component sporelings must nevertheless have been total losers by the arrangement to judge by the numbers of fronds sent up. Jones does not say whether the spores in question were from a single parent thallus, but he states that he has seen young plants resembling his compound sporelings in the wild.

Fusion of plasmodia is known in the Myxomycetes and Acrasiales.[20] But again, if the cultures in which this has been observed were made up from spores taken from a single sporulating plasmodium, as seems quite likely, the congregating cells or fusing plasmodia cannot be regarded as unrelated, and they could be segregants which happened to have received like combinations of the incompatibility genes normally effective in preventing fusion.

Knight-Jones and Moyse[21] give an interesting summary of the known facts concerning fusion in marine colonial animals (including reference to the above-mentioned case of *Gracillaria*). It seems that fusion of adjacent colonies does sometimes occur naturally in sponges and corals when contact is made in the early stages of growth; but old colonies tend to develop a line of demarcation where they meet and the same is true of the Bryozoa and the colonial ascidians, fusion even in the early stages being unknown in these groups.

The theoretical considerations which the present theory would apply to the cases of the three preceding paragraphs may be gathered from the discussions that will be given in the next section concerning fighting and co-operation. In general, it is fair to state as a matter of fact that the sexually produced individuals of a species do not, and usually will not, fuse with one another. Of course from such a statement, a large exception must be made for the fusion of haploids in the normal sexual cycle; but here it will be noted that except in respect of certain unusual types of chromosomes the discipline of the meiotic process must generally assure equal reproductive expectations for the two co-operating genomes.

Knight-Jones and Moyse emphasize the contrast between the mutual behaviour of zooids of a single colony and that occurring between the members of the dense clusters that arise from the gregarious settling of larvae: 'Such systems are strikingly more economical than is a barnacle population, in that the crowded and smothered barnacles die wastefully, but unsuccessful zooids are resorbed and their materials presumably transferred to help growth elsewhere.' According to the present view, clonal colonies of zooids are things of a very distinct kind from colonies of sexually produced organisms such as oysters or barnacles, and the co-operation of zooid individuals, which comes to reach such remarkable complexity in some of the pelagic Siphonophora, should in itself cause no surprise.

Colonies of social insects

The colonies of the social insects are remarkable in having true genetic diversity in the co-operating individuals.

Caution is necessary in applying the present theory to Hymenoptera because, of course, their system of sex determination gives their population genetics a peculiar pattern. But there seems to be no reason to doubt that the concept of inclusive fitness is still valid.

(a) *A hypothesis concerning the social tendencies of the Hymenoptera*
Using this concept it soon becomes evident that family relationships in Hymenoptera are potentially very favourable to the evolution of reproductive altruism.

If a female is fertilized by only one male all the sperm she receives is genetically identical. Thus, although the relationship of a mother to her daughters has the normal value of $\frac{1}{2}$, the relationship between daughters is $\frac{3}{4}$. Consider a species where the female consecutively provisions and oviposits in cell after cell so that she is still at work when the first of her female offspring ecloses, leaves the nest and mates. Our principle tells us that even if this new adult had a nest ready constructed and vacant for her use she would prefer, other things being equal, returning to her mother's and provisioning a cell for the rearing of an extra sister to provisioning a cell for a daughter of her own. From this point of view therefore it seems not surprising that social life appears to have had several independent origins in this group of insects or that certain divisions of it, represented mainly by solitary species which do more or less approximate the model situation (e.g. most halictine bees), do show sporadic tendencies towards the matrifilial colony.

It may seem that if worker instincts were so favoured colony reproduction could never be achieved at all. However, this problem is more apparent than real. As soon as either the architectural difficulties of further adding to the nest, or a local shortage of food, or some other cumulative hindrance, makes the adding of a further bio-unit to the colony $1\frac{1}{2}$ times more difficult than the creating of the first bio-unit of a new colony the females should tend to go off to found new colonies. Of course, in a more advanced state with differentiated workers, the existing workers would be expected to connive at the change-over to the production queens, which is, so to speak, the final object of their altruism. That in actual species the change-over anticipates the onset of adverse conditions is not surprising since they must be to a large extent predictable. In Britain where winter sets the natural termination the vespine wasps round off their colony growth at about the time one would expect but some bumblebees begin rather surprisingly early. If climatic termination were not in question and queen-production tended to come a little late so that the worker population had already risen above the number that could work efficiently on the nest workers might best serve their inclusive fitness by going off with the dispersing queens, despite the fact that in this case the special high relationship of workers to the progeny of the queen no longer holds. Descriptively this is roughly what happens in the meliponine bees[22] and, apart from the serious complication of the swarms having many queens each, it seems to be what happens in the polybiine wasps. In *Apis*, as is well known, it is the old queen who goes off with some of her daughters, leaving a young queen together with sister workers. This oddity cannot be so easily derived in the imagination from semi-social antecedents in colony reproduction (it could come more readily from the habit of the whole colony absconding under adverse conditions) and like other peculiar features in honeybees it hints at a long and complicated background of social evolution. As attempts to represent the actual course of evolution and its forms of selection the above outlines are in any case thoroughly naive; they are merely intended to illustrate certain possible courses which would accord with our principles.

The idea that the male-haploid system of sex determination contributes to the peculiar tendency of the Hymenoptera towards social evolution is somewhat strengthened by considering other relationships which may be relevant.

Figure 2.1 shows a hymenopteran pedigree and Fig. 2.2 shows the coefficients of relationship between the individuals lettered on the pedigree.

The relationships concerning males are worked out by assuming each male to carry a 'cipher' gene to make up his diploid pair, one 'cipher' never being considered identical by descent with another.[23] For all male relationships we then have

$$r = \tfrac{1}{2}c_1,$$

where c_1 is the chance that the two have a replica each. The convenience of this procedure, which is arbitrary in the sense that some other value for the fundamental mother–son and father–daughter link would have given an equally coherent system, is that it results in male and female offspring having equal relationships to their mother which matches with the fact that when the sex ratio is in its equilibrium condition individuals of opposite sex have equal reproductive values.[24]

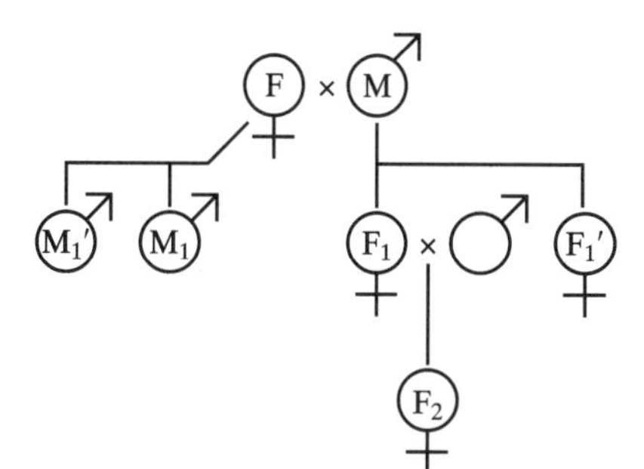

Figure 2.1 A hymenopteran pedigree.

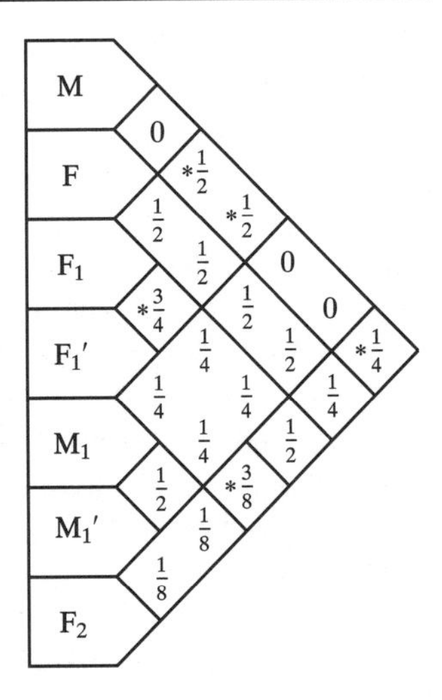

Figure 2.2 Coefficients of relationship for the pedigree of Fig. 2.1. Asterisks indicate the coefficients that would diminish in cases of polyandrous insemination assuming the fathership of particular offspring to be unknown.

The relationships whose values are affected by polyandrous insemination of the female are indicated in Fig. 2.2 by asterisks. It will be seen that among those unaffected, because fertilization is not involved, are the relationships of a female to her son, $r = \frac{1}{2}$, and to her brother, $r = \frac{1}{4}$. According to our theory these values indicate that workers should be much less inclined to give up their male-producing in favour of the queen's than they are to give up their female-producing in favour of a singly-mated queen. Laying by workers is known to occur in each of the main social groups, bees, wasps, and ants. The extent to which the practice occurs in normal colonies remains largely obscure; but in some species it is so prevalent that observers have been led to suggest that all the male members of the population are produced in this way.[25,26] In fairness, however, rather than emphasize this apparently detailed fit of our hypothesis, it should be pointed out that male-egg production by workers is in any case the simplest possible manifestation of an incipient selfish tendency since it does not require the complicated preliminary of mating.

Males are related to their brothers as well as to their sisters with $r = \frac{1}{4}$; their relationship to their daughters is $\frac{1}{2}$. Hence the favourable situation for the evolution of worker-like instincts cannot ever apply to males, and in conformity with this, working by males seems to be unknown in the group. Again, however, it must be admitted that another explanation of the fact could be advanced: except for the faintest ambiguous suggestion in one genus (*Trypoxylon*)[27] there is not even any parental care by males even in the solitary nesting species, so that the evolution of worker behaviour would have difficulties of initiation in this sex.

While this point must be fairly taken, nevertheless it may be that the male-haploid system is still the prime cause of the very different behaviour of males.

It can be shown that it causes a selection pressure towards a sex ratio which is markedly female-biased.[28] This may be seen as due to the fact that in the replacement of the gene pool in each generation the females have a bigger contribution to make than the males, so that, so long as the numerical deficiency of males has not gone too far, it is more profitable to produce females than males. And if a chronic deficiency of males does occur it is clear that the male sex will tend to evolve adaptations for polygamous mating which must be almost completely incompatible with the evolution of male parental care. The argument concerning the sex ratio must properly take into account the relative expensiveness of producing the two sexes. Thus if individuals all incur the same expenditure irrespective of sex, which must be the case, for instance, with a bee which provisions a series of cells with equal amounts of food, the ratio is the well-known 1:1.618; only when a male is merely half as expensive as a female does the ratio sink to the usual 1:1. The argument does not apply, however, if there is thelytoky, polyembryony, etc. and it does not apply once a worker caste has come into existence. If worker laying takes place a more male-biased ratio should prevail.

(b) *Multiple-mating and multiple-insemination in Hymenoptera*

Following these considerations of sex ratio, however, it is not surprising to find in most solitary and even moderately social Hymenoptera that the male carries more sperm than is necessary to fill the spermatheca of a single female. Generally it seems that he carries far more than enough. Possibly only in some very highly social species is multiple-insemination necessary to fill the spermatheca. This is an important point in favour of our hypothesis since it predisposes to the production of the very highly intrarelated families which the male-haploid system makes possible. But to what extent, over the range of groups and species, the females actually produce such families remains a large question. The literature contains many references to multiple matings by female Hymenoptera, spread over many of the major groups of the order. How frequently such multiple mating is accompanied by a significant degree of multiple insemination, and how the phenomena are distributed with respect to incipient, advanced, or retrogressing social life are matters too wide and complex to be reviewed here. For the present it must suffice to quote the very small amount of work known to the author which bears directly on multiple insemination.

Concerning female wild bees in general, Michener *et al.* state that 'Spermathecas with only a few sperms have not been found, in spite of some search, although specimens with the spermatheca only half-full are known'.[29] But in a survey of some Australian halictines, Michener[30] found that on the whole the number of sperms in spermathecas was small in comparison with his experience of American halictines. Without knowledge of the quantity of sperms which the male can provide or of mating behaviour, one cannot be sure what this argues about multiple-insemination, but it suggests that it may be uncommon. Taken together with Michener's notable failure to find any small-ovaried worker-type bees, which according to him are a feature of most common halictines of other continents, this observation seems, therefore, against our hypothesis. But Michener notes as another general feature the short adult lifespan of the Australian bees and concludes that, 'There is no evidence that any female lives long enough to encounter her adult progeny', which at least offers another possible reason why worker behaviour has failed to appear. Plateaux-Quénu[31] thought queens of the quite highly social *Halictus marginatus* were probably multiply inseminated because she found some queens towards the beginning of the period of fertilization with only partially filled spermathecae. Michener and Lange[32] present evidence that a female of the solitary (though gregarious-nesting) anthophorine bee, *Paratetrapedia oligotricha*, in Brazil, taken in copula, was engaged in receiving her second insemination, this apparently being the only direct evidence of such a thing in a primitive bee known to them at the time.

Multiple insemination of a high order effectively producing a progeny of multiple paternity seems to be firmly established for the honeybee.[33,34] On the other hand, it would seem not to occur in the Meliponinae.[34] It occurs in the socially very advanced fungus-growing ants (e.g. *Atta sexdens*).[35] But in another myrmicine of a different tribe I found no evidence but of single inseminations, using Kerr's sperm-counting methods.

Suppose a female is mated by n males and they are respectively responsible for proportions

$$f_1, f_2, \ldots f_s, \ldots f_n, \left(\sum_s f_s = 1\right)$$

of her female progeny. The average relationship between daughters is then

$$\tfrac{1}{2}\left(\tfrac{1}{2} + \sum_s f_s^2\right).$$

In particular, if all males contribute equally we have

$$\bar{r} = \tfrac{1}{2}\left(\tfrac{1}{2} + \frac{1}{n}\right),$$

which is the lowest average relationship for a given value of n. If two males contribute equally we have $\bar{r} = \frac{1}{2}$ as for normal full-sibs. Clearly multiple-insemination will greatly weaken the tendency to evolve worker-like altruism and $n > 2$ in the model situation described above should prevent its incipience altogether. Using Taber and Wendel's estimate of $\bar{n} = 8$,[33] which Kerr's different method roughly confirmed, we get $\bar{r} = \frac{5}{16}$, which doubtless should be raised a little to allow for inequality in the contributions of the drones. It does seem at first rather surprising that altruism towards sisters so much less related than full sisters can be maintained at its observed pitch of perfection. But even the limiting value of $\bar{r}$ is no lower than $\frac{1}{4}$ and we may well imagine that once established the biological advantage of the social mode of reproduction, which is evidenced by the success of the social insects in general, proves sufficient to outweigh even a twofold higher value in personal offspring. It may also be argued that the firm establishment of highly differentiated trophogenic worker castes creates a gulf which a sexualized mutant is unlikely to cross successfully, especially when as in honeybees so much depends on the behaviour of the other workers. For example, a mutant sexualized worker of the honeybee will eclose from a worker cell and will therefore be small. Even if it secretes queen-substance it is unlikely to be as attractive as a proper queen and is likely to be killed. If it escapes it cannot found a colony on its own. Thus, if the trend to multiple insemination occurs after the firm establishment of the worker caste, its threat to colonial discipline is a rather remote one.

In species of social Hymenoptera which found their colonies through single fertilized females the difficulties and dangers of seeming to be royal are less important; but the mutant will still suffer handicaps from its probable small size and lack of food reserves. In ants it will be further handicapped by its lack of wings. Nevertheless with ants there are strong indications that trends of worker sexualization have occurred in the evolution of the group.[36] It may be remarked that the sexualized worker is likely to have a smaller spermatheca and so to restore single insemination, which will, according to our theory, restore the basis for re-evolving strong worker altruism.

An ability of females to lay unfertilized eggs which develop into females would open another possible avenue for selfish selection. Again, the menace will be greatest when multiple insemination of queens occurs, for then when a worker had inherited the causative gene from its father there would be a better chance, especially when the gene frequency was low, that it would have some normal worker sisters to help rear its offspring. In general, whether we are concerned with parthenogenetic production of males or females we need only follow Sturtevant's argument[37] and visualize the drastic or fatal overproduction of sexual or egg-laying forms which would occur in the 'son' or 'daughter' colonies due to an egg-laying worker to see the potent counterselection to which a fully penetrant causative gene will become exposed. Clearly the situation is worse for the gene when it is common than when it is rare so that an equilibrium is possible.

Female-to-female parthenogenesis by workers does occur sporadically in honeybees and shows geographical variation in its incidence. In the South African race, *Apis mellifera capensis* Esch., it seems that worker eggs always develop into females.[38,39] But whether this is explicable as a selfish trait is rather doubtful. To be such the laying-workers would have to try to get their eggs cared for in queen cells. Despite what Flanders[40] seems to quote

Onions as having stated—that in queenless hives 'Uniparental workers do not construct either queen cells or drone cells', and that 'a queenless colony gradually disintegrates'—Dr Kerr informs me that these bees do eventually construct queen cells in an emergency and can thereby secure the perpetuation of their colony; but he found that they did so somewhat tardily compared to queenless colonies of the familiar honeybees. Of course for them, possessing this unusual ability, the need to initiate queen-rearing at once is not so urgent. Also in some other races diploid eggs laid in queen cells by workers in hopelessly queenless hives may sometimes be reared and so save their colonies from extinction.[41]

Female-to-female parthenogenesis is also present in various species of ants. For example, in the ant *Oecophylla longinoda*, parthenogenesis of a clonal type seems to have become a normal mode in the reproduction of the colony.[42] Here the workers and not the mother queen produce the new generation of queens, which is suggestive at least that the situation had its origin through the selection of a selfish trait.

(c) *Termites*

The special considerations which apply to the Hymenoptera do not seem to have been noticed by Williams and Williams.[7] The discussion which they base on their analysis of the full-sib relationship would, however, be applicable to the termites where this relationship is ensured in the colony by having the queen attended by a single 'king'. Termites of both sexes have an equal relationship ($r = \frac{1}{2}$) to their siblings and their potential offspring. Thus the fact that both sexes 'work' is just what we expect; we need only a bio-economic argument to explain why restriction of fertility to a few members has proved advantageous to the sibship as a whole. On this point the present theory can add little to previous discussions.

When either king or queen dies the worker castes rear a substitute or 'neotene' from the eggs or young nymphs already present. The neotene mates with the surviving parent. The progeny which come from such a mating will still be related to the old workers with $r = \frac{1}{2}$. They will be related among themselves by $r = \frac{5}{8}$. They will also tend to be highly homozygous and such matings are in fact said to be somewhat infertile.

It is surprising, however, if increasing the tendencies to social cohesion by such close inbreeding can ever pay off as a long-term policy against the disadvantages of decreasing adaptive flexibility. That it may be a successful short-term policy for a species is perhaps indicated by the frequency of mention of brother–sister mating in the literature on social insects; but these statements are not always based on very firm evidence.

(d) *Pleometrosis and association; population viscosity in the social insects*

However, it does seem necessary to invoke at least a mild inbreeding if we are to explain some of the phenomena of the social insects—and indeed of animal sociability in general—by means of this theory. The type of inbreeding which we have in mind is that which results from a high viscosity of population or from its actual subdivision into small quasi-endogamous groups.

In some ants (e.g. *Iridomyrmex humilis*), at least one species of stingless bee (*Melipona schencki*)[43] and apparently most species of wasps of the subfamily Polybiinae it is normal to have at least several 'queens' engaged in egg-laying in each nest. This phenomenon is known as pleometrosis. Colony reproduction is by swarming with several or many fertilized females—potential queens—in each swarm. Clearly this social mode presents a problem to our theory. Continuing cycle after cycle colonies can come into existence in which some individuals are almost unrelated to one another. Such situations should be commoner the higher the number of founding queens, but less common in so far as there is any positive assortment of true sisters in the swarms. They would be very favourable to the selection of genes causing selfish behaviour and this in turn would be expected to lower the efficiency of social life and to reduce the species. Yet though selfish behaviour is certainly not absent—witness the large proportion of unfertilized wasps in egg-laying condition,[44] and the common occurrence of dominance behaviour—it does not seem to do the colonies much harm and the species concerned are highly successful in many cases. For example, the genus *Polybia* includes several very abundant species in the Neotropics and has obviously undergone considerable speciation with the whole system in working order.

Wasps of the widespread genus *Polistes*, doubtfully placed in a separate subfamily from the Polybiinae, present a rather similar problem. In this case it seems that there is usually or always only a single principal egg-layer on the nest; she dominates the others and they succeed in laying only a few eggs if any.[45] But with many of the species and races that inhabit warmer lands it is common for the initial building of the nest to be the work of two or more fertilized queen-sized wasps. This phenomenon has been called 'association'.[44] Even at this stage the dominant wasp does least work and probably all the egg-laying, and, probably due to their more arduous and dangerous lives, the auxiliaries (as the subordinate queen-like wasps are called) tend to disappear in the course of time so that a queen assisted by her daughter workers becomes the normal situation later on. Here it is the ready acceptance of non-reproductive roles by the auxiliaries that we have difficulty in explaining. There is good reason to believe that the initial nest-founding company is usually composed of sisters,[46,47] which brings the phenomenon closely into line with the pleometrosis of the polybiines. But it is doubtful if the wasps have any personal recognition of their sisters and if a wasp did arrive from far away it is probable that it would be accepted by the company provided it showed submission to the one or two highest ranking wasps. Dominance order does sometimes change and an accepted stranger has before it the prospect of rising in rank and ultimately subduing or driving off the queen. Thus an innocent rendering of assistance is not always easy to distinguish from an attempt at usurpation as

Rau has pointed out, and the readiness to accept 'help' is really just as puzzling as the disinterested assistance which some of the auxiliaries undoubtedly do render.

The geographic distribution of the association phenomenon in *Polistes* is striking.[47] We may state it as a general, though by no means unbroken, rule that northern species approximate to the vespine mode of colony foundation and tropical species to the polybiine to the extent above described. The single species *Polistes gallicus* illustrates the tendency well. At the northern edge of its range in Europe its females usually found nests alone. In Italy and southern France the females found nests in companies; while in North Africa the species is said to found colonies by swarming with workers.[44] We here suggest two hypotheses which could bring these facts into conformity with our general theory.

The first posits a general higher viscosity of the tropical populations. This will cause, through inbreeding, all coefficients of relationship to have higher actual values than we would get taking into account only connections through the past one or two generations. And it will also increase the tendency for casual neighbours to be related, which is clearly of potential importance for the association phenomenon.

Populations of *Polistes* certainly are very viscous. Generally the wasps have a strong attachment to their place of birth,[46] and like to found nests near the parental nest. They are weak flyers. And they do show a very pronounced tendency to local variation. But whether these remarks apply any more strongly to tropical than to temperate populations I do not know. Polybiine wasps seem to be weaker flyers than vespines and also have indications of a tendency for swarms to build not far from the parent nest. Polybiines also show much geographical variation.

By its very nature the so-called temperate climate may tend to force a greater degree of vagrancy on the insects inhabiting it, both through its pronounced seasons and its seasonal irregularities. A discussion of this idea from a similar biological point of view can be found in Wynne-Edwards.[48] As one further factor relevant at least to *Polistes* we suggest that if, as seems probable, the genus is of tropical origin the northern species will be derived from former races, which themselves tended to be made up of vagrant colonist wasps that had flown north. Thus there would have been selection for wasps willing and able to found nests alone; and in general, in the course of such a spreading colonization, a species would be expected to shed some of its co-operative adaptations. But if the spread was very slow, as it may well have been, these factors would hardly apply.

The second hypothesis appeals to the lack of marked seasons in the tropics causing a lack of synchronism in the breeding activity of insects. This will tend to cause inbreeding because it rarifies the mating population. Thus a polybiine nest may be in active production of sexual wasps when its nearest neighbours are not and its progeny may therefore be more inclined to mate among themselves. The same doubtless applies to *Polistes* in a really equable tropic environment and with *Polistes* we again have an important correlative effect that when a nest-founding wasp accepts an adventive helper the chance that she is a sister is also increased. However, with *Polistes*, multiple-queen nest founding does occur even where the wasps are constrained by the climate to follow a definite seasonal cycle. Queens may come together in the spring after hibernation to found their colonies. Rau[46] records some interesting observations on *P. annularis* in the United States showing the variability of its nesting behaviour and he mentions his general experience that the hibernated queens return to the old nest for a short time before going off to found nests. Such behaviour should help to ensure that in cases of associative founding the co-foundresses are sisters. The cases of hibernating yet associating *Polistes* would seem to dismiss any hypothesis that the differences we have noted between northern and tropical wasps is due solely to factors following from the necessity for hibernation. In *Vespula*, queens do often hibernate in the parental nest and yet do not show association in nest-founding.

To the extent that they are valid, the above hypotheses would also help to extenuate previously discussed difficulties concerning the maintenance of reproductive altruism despite multiple insemination of queens. It may be remarked that although modern work rather indicates that its breeding system is far from viscous the honeybee does seem to maintain local races quite readily. With *Atta sexdens* I have noticed that males and females come to earth from their nuptial flight in local concentrations, but whether these are associated each with an established colony or represent some wider nuptial gathering is not clear.

(e) *Aggressiveness*

The aggressiveness of the workers of social insects towards disturbers of their nests is one of the most conspicuous features of their altruism. The barbed sting and the function of sting autotomy are physical parallels of the traits of temperament. The correlation of these characters with sterilization does seem to hold very well throughout the social Hymenoptera. Queens are always timid and reluctant to use their stings compared to workers. In *Polistes*, workers, unless very young, are more aggressive than auxiliaries, and auxiliaries more than the reigning queen. Races of honeybees in which laying workers occur more frequently or appear more readily when the hive becomes queenless are generally milder than the races where they are less prevalent.[38,49] Polybiine wasps, pleometrotic and lacking pronounced caste differences, are generally somewhat less fierce than vespines.

However, aggressiveness is also clearly a function of the size of the colony, or perhaps even more of the worker:queen ratio. This applies not only to particular colonies as they grow larger but also in a general way to variation in mature colony size between species. This effect, too, is not very surprising, for, to take the extreme case, we can see that it is only when its nest is overpopulated and its services in other directions superfluous that the worker can afford to throw its life away. Typically the vespines have the higher worker:queen ratio, so that from this point of view as well, it is not surprising that the polybiines are generally speaking milder wasps. It is interesting to learn that even in the limited north–south range covered by the islands of Japan, *Polistes* shows in this respect also its previously noted tendency to bridge the two types. Yoshikawa[50] gives an interesting comparison of northern and southern Japanese species and it is seen that northern species are both fiercer and have the larger colonies. Iwata (quoted by Yoshikawa) believes that the fierceness is a function of the colony size. Although no properly associative *Polistes* occur in Japan, Yoshikawa[47] has found a case of temporary association in a southern species, suggesting a slight or vestigial tendency. Perhaps this factor may play a part in the difference in fierceness. Existence of auxiliaries would seem incompatible with a high degree of worker differentiation and will therefore tend to counter the development of high worker altruism. But just why it appears to be also incompatible with higher worker:queen ratios is not entirely clear.

(f) *Usurpation*

Its made or half-made nest is obviously a valuable property to a queen bee, wasp, or ant. If it is ready provisioned or staffed by workers and set for the rearing of sexual brood it is even more valuable. It is therefore not surprising that usurpation has become a major evolutionary and behavioural issue with the nesting Hymenoptera.

On the one hand we have the great array of parasites. Often, especially in bees and wasps, the host and parasite species seem to be closely related, suggesting that the habit arose out of petty intraspecific usurpation. But the present theory indicates considerable difficulties for the sympatric emergence of a parasitic race. Unless the evolving complex of characters could include a strong tendency to vagrancy the usurper would in too many cases destroy the genes on which its own behaviour was founded. One allopatric race invading the territory of another with at least partial reproductive barriers already present should create a more promising situation for progress in usurper-instincts. A situation like this, involving occasional parasitism, is suggested for two species of *Bombus* in Britain.[51] Plateaux-Quénu[52] has observed a half-provisioned nest of *Halictus marginatus* being used by a female of *H. malachurus*. Both these species are social on about the same level as *Bombus*.

On the other hand, we have the sensitivity about adventive females which is so widespread in the nesting Hymenoptera, including the parasites themselves. According to Plateaux-Quénu, conspecific usurpation is frequently attempted, albeit before the appearance of the workers, in the nest aggregations of

Halictus malachurus and sometimes succeeds. A successful conspecific usurpation, strongly resisted, has actually been observed in *Polistes fadwigae* by Yoshikawa[53] and I have observed what was probably an attempt, persistent but unsuccessful, in *P. versicolor*. Something similar seems to have been seen by Kirkpatrick[54] with *P. canadensis*. (In the light of observations of Sakagami and Fukushima[55] an alternative interpretation that this was an attempt to thieve larvae for food in these cases should be borne in mind. But I have not seen thieving in either *P. versicolor* or *P. canadensis* even in artificial situations that should encourage it. It would in any case be normally very difficult to perform in associative species.) And with the same species I have found that if a dominant wasp is transferred from one nest to another a mortal fight, usually with the reigning dominant, begins immediately; whereas a young worker similarly transferred may sometimes be accepted and, perhaps because of its submissiveness, seldom receives so severe an attack. Extreme suspicion concerning wasps which approach the nest in a wavering uncertain manner sometimes prevents a genuine member of the colony from rejoining it, at least for some time, in *P. canadensis*. This is especially apt to happen with young wasps, perhaps returning from their first flight; and it may be a rather paradoxical result of such a reception that they sometimes end up working on a nearby nest not their own. Possibly it is the danger of usurpation, joint with that of parasitoids, that keeps so large a proportion of a *Polistes* colony idle on the nest when one would have thought they could be much more usefully employed out foraging.

As the very existence of association necessitates, antagonistic behaviour is not so marked in the very early stages of nest-founding: then, with *Polistes versicolor*, a considerable amount of swapping of wasps may take place from week to week within a local group of initiated nests—for example, all those located around the buildings of a household and usually not far from a last year's abandoned nest from which very likely all or most of the wasps are derived. The same sort of thing has been noted by Ferton[56] for *P. gallicus* and by Rau[46] for *P. annularis*. But even at this stage fights are sometimes seen severe enough for the combatants to fall off the nest.

In these associative *Polistes* the great variation in the degree of association—from lone nest-founding to companies of 12 or more crowded on and about a tiny nest-initial—the frequent abandonment of young nests, the quarrels, the manifest concern about adventive wasps, combine to create an impression which is very reminiscent of the breeding affairs of the South American cuckoos *Crotophaga ani* and *Guira guira* as described by Davis.[57] In their broad features the situations are indeed so similar as to suggest similar trends of selection must be at work in populations similarly patterned with respect to relationship. In these birds, much as in *Polistes*, we have a basic ability to rear young independently complicated by a tendency of some birds to assist altruistically (perhaps most marked in *Crotophaga*) and of others to play the cuckoo (most marked in *Guira*, which also sometimes parasitizes other birds). A striking difference from *Polistes* of course is the presence of males, playing parts in close parity with those of females. And the systems also differ in that usually several birds succeed in laying in the communal nest, which is more like what is found with certain primitively social xylocopine bees (see, for example, Michener[58]) than like *Polistes*. When the clutch becomes very large through this cause a large proportion of eggs may fail to hatch. Eggs are sometimes taken out and dropped. Such action by a particular bird might serve to increase the proportion of its own eggs in the clutch. For all the seeming confusion and inefficiency these birds are, like *Polistes versicolor* and *P. canadensis* in the same area, widespread and apparently successful.

(g) *Pleometrosis in Halictinae*

The social halictine bees closely parallel the systems found in *Polistes*. Worker populations are of comparable size. The state of affairs found in *Augochloropsis sparsilis*[59] and in *Lasioglossum inconspicuum*[60] shows that these species have a class closely corresponding to the auxiliaries of warm-climate *Polistes*. But since at least some of the halictine nests are pleometrotic it seems more probable that some of their auxiliaries become layers later on rather than dying young as workers as they tend to do in *Polistes*. Probably only a minority of the species of Halictinae have any trace of a worker caste and the group also differs in the wide range of types of sociability which their tunnel-nesting encourages. For instance, quite a common situation with burrowing bees, both Halictinae and others, is for several females to be using a common entrance tunnel while each owns a separate branch tunnel further back.

Michener[58] has recently suggested that the road to sociability and the development of a worker caste has lain in this group through a stage like this followed by a stage like that found in *Augochloropsis sparsilis*. This we are inclined to doubt since even if the nest system users are for some reason always sisters the genetic relation of sister eggs will always be twice that of niece eggs irrespective of multiple insemination, so that on the present theory social evolution via the matrifilial colony always offers the easier route to worker altruism. Hymenopteran societies in which the queen (or queens) have auxiliaries but not, later, filial workers, seem in fact to be unknown. The classical theory concerning the evolution of the social insects has always posited a wide overlap of generations allowing mother and daughters to co-exist in the imaginal state as one of the preconditions for the evolution of this kind of sociability, and it is surely significant that it is never observed where this condition is lacking, as it might well be if genetic interest in nieces were sufficient to encourage reproductive altruism.[61] That such altruism could arise through genetic interest in the offspring of unrelated bees sharing the same excavation, as Michener actually suggests, seems to me incredible.

(h) *Tunnel-guarding by bees*

There is, however, another important type of social behaviour to which Michener has re-drawn attention which might well arise on the basis of much lower relationships. One of the potential advantages when two or more females share a common entrance tunnel is that the entrance can be defended against parasites by a single bee, leaving the others free to forage. Instincts for guarding a narrow entrance seem to be widespread in the nest-excavating bees and also occur in Meliponinae.[58] Michener has seen females of *Pseudogapostemon divaricatus*, a workerless but entrance-sharing halictine, apparently taking turns at the duty and he and other observers have seen guard bees of this and other species repulse mutillids[58,60] and parasite bees.[60] The menace of intruding parasites may give such co-operation a very high advantage. But it would seem that once established the system should give an even higher advantage to the sporadic 'shirker', so that it is a little difficult to see how guarding could become perfect. Perhaps it is not. One may, however, construct a simple imaginary system that would render it so: the bees could evolve an instinct which allowed them to leave duty at the nest entrance only on the stimulus of another recognized tenant coming in, or better, of another bee coming up from behind; this would ensure that there was always a bee on duty or at least somewhere in the nest system. By going out when supposed to be on duty, a bee would jeopardize her own brood as much as, if not more than, the broods of the others, so that selection would tend to stabilize the instinct. Interestingly Claude-Joseph and Rayment both have claimed to have observed guarding on this system, but Michener[30] is inclined to doubt these claims because his careful observations on *P. divaricatus* in Brazil had revealed that the behaviour was more irregular than might appear at first sight, bees remaining on guard for some time and allowing others to go out past them. In a highly pleometrotic nest system, shirking might be relatively easier and safer for the isolated social deviant but the spells of guard-duty demanded would also be much shorter and therefore the selective incentive to shirking much less. Nevertheless, even if it is possible to account for the evolution of guard-instincts without a basis of relationship between the bees, it is hard to see how other socially disruptive practices, such as robbing within the nest-system, could fail to evolve unless a bee's co-tenants are also usually the carriers of some part of its inclusive fitness.

6. EQUAL-STATUS SITUATIONS

By an equal-status situation we simply mean a social situation where there is no obvious and regular difference in age, caste, or sex between the individuals concerned. Several apparently of this nature have already been mentioned, including the nest system of independently working solitary bees discussed just previously. Now, using other examples, we indicate some other kinds of argument which it may be useful to apply to these situations.

Co-operation

In certain ants, notably *Lasius flavus* and *L. niger*,[62] it is known that companies of several queens will co-operate in excavating the initial nest. Since these have just come to earth from a vast mating-flight they are unlikely to be close relatives. According to Waloff, the queens of *L. flavus* usually cohabit peacefully in the nest chamber and even keep their eggs in a common pile, but about the time cocoons are first formed they tend to separate, some taking a portion of the brood (not necessarily a very fair one it seems) to a particular corner of the nest. There is evidence that the queens so separated tend to control distinct sectors in the developing nest, each having its own worker population; and whether by death of queens—by fighting or otherwise—or by migration of a 'sector', most nests of *L. flavus* end up haplometrotic. In *L. niger* fighting between the queens is regular and generally only one survives in the initial nest chamber.

If we imagine a situation where, of the queens which succeed in co-operatively establishing an initial nest, only one is allowed to survive and use it, rather as happens with *L. niger* except that the survivor is chosen at random and not according to fighting prowess, we see that unrelated queens will evolve instincts to co-operate as a group of n if the chance that they succeed in establishing the chamber is more than n-times the chance that one would succeed if alone. When engaged in digging the queens are very helpless and it is not difficult to imagine that a team gets itself underground so much more quickly than an individual that this criterion is met. As to the continued amity once the chamber is made Waloff's observations[62] on experimental multi-queened initial nests showed that for some reason the queens survive better and rear their first workers sooner when in a group than when alone; if sufficiently marked in a state of nature such an effect could explain the continued amity. With the appearance of the first workers the queen and her brood tend to become more independent and we expect behaviour to change accordingly.

It will be seen that in essentials this situation has much in common with that previously described concerning the fusion of sporelings of *Gracillaria verrucosa*.

In both cases we have a strong presumption that a stage in which selection very strongly favours the united group over the lone individual gives place to conditions where the individual would be better off in the absence of its close companions. According to our theory whether these new conditions will bring on an overt struggle or fighting will depend very large on the degree of relationship in the group in question, or rather on the degree of relationship that has held on average in the multitude of similar situations which have occurred during the evolutionary development of the behaviour.

Fights

The argument to be applied to fights is merely another form of the argument applied above to co-operation. If two evenly matched unrelated animals holding one unit of reproductive potential each are in a typical situation which holds out the prospect of a fight, and if their instincts have been nicely adjusted by natural selection to suit the average outcome, then they will fight only if the expectation of reproductive potential for the winner is more than one unit. If they are sibs they will fight only if the expectation of 'winner's r.p. $+\frac{1}{2}$ loser's r.p.' is greater than $1\frac{1}{2}$. Thus if one inevitably dies in the fight the winner must normally gain by more than 50 per cent or the two will prefer to co-exist. In the case of a 'hymenopteran full-sistership' they will not fight to the death unless the expected gain to the winner is more than 75 per cent. But with the honeybee, with the amount of multiple-insemination discussed previously, about a 40 per cent increase will be enough. In this case we may put it that unless the presence of an extra young queen can increase the growth of the colony by more than 40 per cent the reigning young queen will prefer to do without her. Thus the mutual animosity of young queens is not very surprising. The 'piping' by a still imprisoned queen incidentally would seem to have the characteristics of altruism. The females of various species of Hymenoptera (in *Vespula, Halictus*, etc.) are said to fight in spring for the possession of the maternal nest in which they hibernated together. But in these cases we do not know about multiple-mating and anyway there is probably no question of a fight to the death, beaten females are usually expelled and presumably go off to discover or excavate other nest sites.

It may be noted that the larvae of 'gregarious' parasitoid Hymenoptera in whose case there is normally no question of 'going off' do not fight even if overcrowded.[63] 'Gregarious' refers to species where the adult normally lays several eggs per host. In 'solitary' species, which lay only one egg per host, the first instar larva is adapted for fighting and always attempts to kill any other larvae in the same host; normally there is only one survivor. The gregarious larvae in a host are not necessarily 'hymenopteran full-sisters', however, even apart from the question of polyandrous insemination. In cases of polyembrony they will be clonal. The same would be true if the mother reproduces by thelytoky, and, as Dr G. Salt has reminded me, this certainly occurs in some ichneumon-flies; and female-to-female parthenogenesis of one kind or another is widespread throughout the parasitoidal Hymenoptera. In such cases the comparison to the batch of females competing for the nest is still less valid, although in itself the difference in social behaviour between these two types of parasitoid according to relationship remains very striking. In the parasitoidal

Diptera with normal sexual reproduction our theory predicts that the competition between gregarious larvae should be fiercer; whether this is observed I have been unable to ascertain, but gregarious cases are certainly much less common in Diptera and this at least is what we expect.

Parental behaviour to minimize sibling competition

Of course, the above argument is of potentially much wider application. It may be applied to broods of insect larvae feeding under circumstances where the exigencies of competition are not so inflexible as with parasitoids—for example, to broods feeding on plants. Competition within such broods should according to our theory be fiercest in species where the female is inseminated by many males. Fierce competition will waste the energies of the brood and the ovipositing behaviour of adult females should tend to evolve so as to minimize this wastage which spells a lowering of total surviving progeny. Hence over a range of species the habit of laying eggs in batches should correlate with monandrous insemination of females. The correlation should be stronger for cases where the larvae are also gregarious.

This reminds us of Fisher's suggestions concerning the evolution of distastefulness and warning coloration and we note that he appears to have tacitly assumed that the broods he discusses would be of full sibs. Probably this is fair for his cases of Lepidoptera. But as regards the sawfly larvae which he also cites, we have all the diverse hymenopteran possibilities already mentioned, both those dependent on multiple mating and, for some species, those due to female-to-female parthenogenesis. Since we know polyandrous insemination to be a distinct possibility for the Hymenoptera it is of interest to note that D'Rozario[64] found evidence for the gooseberry sawfly, *Nematus ribesii*, that though the males are readily polygynous the female ceases to be attractive after one mating. *N. ribesii* is a good Dzierzon-rule species and concerning the sex ratio it is said that 'females predominate'.[65] If we assume equilibrium sex ratio and equally costly males and females the average relationship is actually a little under one-half.[66] In accordance with Fisher's suggested correlation the eggs are laid close together and the larvae are aposematic and fairly gregarious. However, to the counter-instance admitted by Fisher—the butterfly *Anosia plexippus* which 'scatters her eggs although she has solitary, inedible, conspicuous, larvae'—it will be fair to add another; the case of the moth *Panaxia dominula* which also scatters her eggs although her larvae are conspicuous and presumably distasteful. In the vegetation they tend to be found concentrated on the preferred food plants but are probably not truly gregarious. I have noticed in the wild that the female moth ceases to be attractive as soon as she enters copulation; thus females are probably only once mated, and the case is contrary to my suggested correlation as well as to Fisher's. But of course, though these few instances help to outline the situation, they carry little weight for or against the hypothesis.

When a brood is still under parental care the parent or parents involved will be concerned to minimize the wasteful effects of sibling competition. Their disciplinary task will be easiest if the brood is of full-sibs. In the vast majority of cases it is so, either due to monogamy or to polygamy combined with parental care by the female alone. In the unusual cases of birds where polyandry is combined with male parental care it seems that the male is always monogynous and broods a clutch given him by a single female.[67] But in some Ratites male parental care for polymaternal broods does seem to occur,[68] and in lekking birds there would seem to be a distinct possibility of polypaternal clutches. Doubtless many more exceptions could be found. The notable case of the polyandrous social insects has already been discussed; we merely note here that the method of rearing larvae in cells is ideal for preventing direct competition and where this method is not adopted, as in *Bombus* and the social xylocopine bees, we have added reason for expecting that the queens are effectively monandrous. Nevertheless larval competition seems to be severe in some species of *Bombus*.[69] Although the cases where full-sibships are not the rule cannot amount to much numerically compared to the vast array of cases where they are, we do not intend to suggest that diminishment of sibling competition is the sole evolutionary *raison d'être* of permanent mating ties and bi-parental broods. The cases where the tie continues, as in many birds, from brood to brood and even sometimes until one of the mates dies are sufficient to show that other factors must be operating as well.

There are some rather puzzling cases where the parent seems deliberately to provoke competition in the brood—for example, by associating more eggs with a food supply than it could ever fully support. As just one example we have the case of *Bombus* just mentioned: in *B. agrorum* it has been found that only 30 to 40 per cent of the eggs laid become eventual adults. Mortality is greatest in the late egg and early larval stages and cannibalism among the larvae is suggested.[70] However, the habit of many hawks of having one more nestling than it is normally possible to rear is fairly obviously a special strategy allowing for the chance that the breeding season will turn out a good one; and explanations of a like nature may appear for the other cases eventually.

The strong tendency of plants to produce seeds of standard sizes irrespective of the size of the plant shows that how available food reserves are apportioned between seeds is not a matter of indifference to the fitness of a plant. This is indeed just what we would expect provided the situations into which the seeds disperse are not too varied. Thus for one seed to expand selfishly at the expense of its neighbours may or may not be advantageous to the inclusive fitness of its genotype but it is almost certainly not in the interest of that of the parent plant. Wind pollination will tend more to produce half-sibships among the seeds in an ovary than will insect pollination. Hence according to our theory if seeds in general have genotypic control of their own growth, as they surely must to some extent, wind-pollinated plants will tend to have the more pressing difficulties in respect of uniform seed production. Hence it is rather to be expected that the situation which most lays itself open to this type of competition, the ovary with numerous closely placed ovules, will be uncommon in wind-pollinated plants. By comparison with entomophilous plants this is certainly the case, although there do seem to be a few anomalous genera (e.g. *Populus* and *Juncus*). In a great many anemophilus genera carpels or gynoecia originally with two, three, or four ovules end up, through more or less regular abortions, as one-seeded 'fruits'. But sometimes the seeds may nevertheless be quite closely placed, as in the pine cone, the birch 'catkin', the maize cob, etc.

The remarks at the end of section 3 of Part I of this paper apply to this case as well as to the above problems of animal parental care. We note again that the selfish genes for seed growth tend to waste their powers a little not only because of the assortation due to relationship but also because of the purely chance occurrence of extreme situations where gene replicas are largely in competition with one another. But this extra effect can only be of importance when the number of seeds in the ovary is very small. A much more important contrary factor must be the tendency of wind-borne pollen grains to arrive one by one rather than all at once as with insect pollination, so exacerbating the disciplinary problem of the wind-users. But on this point, even more than on others in the above discussion, our ideas are as yet rather unclear.

7. ANOMALIES

Here and there in the literature are found records of behaviour where relationship is conspicuously disregarded, or harms or benefits are dispensed apparently in contravention of our principles.

However, in every case known to me it seems possible to claim either that the situation has been misinterpreted or that the observation concerns a biological error; that is, a rare occurrence in an unusual situation or something of the kind. The latter would seem to be the case for instance with the unusual cases of adult birds feeding the young of other species.[71]

Where apparently gratuitous interspecies assistance is recorded more regularly, misinterpretation must be suspected. A non-apparent return benefit signifying a symbiosis, or some degree of positive deception signifying some sort of cuckoo-parasitism, are possibilities that should be borne in mind. For instance, it has been reported that different species of xylocopine bees of the genus *Exoneura* in Australia will sometimes pool their broods in a common nest (Rayment, quoted by Sakagami[72]). The finding by Michener[73] of a seemingly very similar situation in the related genus *Allodapula*, together with signs of adaptation to parasitism by one of the species, strongly suggests that the situations Rayment has observed contain at least some mild element of parasitism. And Michener's[74] further finding of two species parasitic on *Exoneura*, clearly derived from the genus itself and hardly separable from it taxonomically, point the same way. These two species are not adapted for pollen-collecting and hence must be fully dependent, but at least one of the supposed parasites in *Allodapula* does collect pollen and so presumably does contribute something to the nest.

Among birds the Cuculidae are a thoroughly anomalous family as regards parental care. We have already mentioned *Guira* and *Crotophaga*. Kendiegh[68] gives a summary of knowledge of reproductive behaviour in other genera. *Geococcyx californicus* also seems to have many females laying in each nest. The two North American species of *Coccyzus* show a situation rather like that which Rayment has found in *Exoneura*.[72] The species are reported sometimes to lay in each other's nests. But both have brooding instincts and a case has been recorded where both species incubated on the same nest.

At the level of single species we may instance the occasional exceptions to the rule that nesting Hymenoptera know their own nests and do not, even if they safely could, transfer to others. As regards the transference of workers, which seems to be not uncommon in some wasps, some cases are perhaps errors due to the powers of visual recognition not being equal to the situation. A strong basis of relationship between neighbour nests, which I believe is usual with the species of *Polistes* in which I have observed worker transference (*canadensis* and *versicolor*), would greatly reduce the selection against such errors. Then there may be situations in which transferral is really in the interests of inclusive fitness—for example if a colony is dying out,[75] or happens to find itself with more workers than can usefully be employed on it, or if a wasp brings in food when all the larvae on its own nest are completely sated. This last explanation may perhaps apply to the cases of cross-provisioning by sphecids in a dense nest aggregation observed by Tsuneki (quoted by Sakagami[72]), and to the cases which Deleurance[76] has observed in *Polistes* in the wild where a worker pays visits to two nests. In birds there is a parallel of a sort in a practice of nesting guillemots and razorbills.[77] It seems that parent birds will sometimes feed the hungriest chicks in the dense nest aggregation rather than their own.

As regards the already mentioned fostering passion shown by emperor penguins that have lost their chicks, some doubt as to whether the observations have been correctly interpreted would seem to remain.[68] But taking the statements at their face value we might suggest, for instance, that it has something to do with heat conservation. Perhaps the parent penguin is so closely adapted to living with its offspring that it is, at the stage in question, at a positive disadvantage without a chick nestling in the brood-pouch. But such a situation would hardly come into being unless there were strong general relationship within the flock. We seem to need to postulate this in any case to explain some other social behaviour of penguins, for example, the way adelie penguins parents are said to leave their young in the care of only a few adults while they go off on long fishing expeditions. On the other hand, some apparently social behaviour such as the formation of the crêche in severe weather is easily interpretable as being almost entirely selfish.

Acknowledgements

The work presented in both parts of this paper was carried out during tenure of a Leverhulme Research Studentship at the London School of Economics and a Medical Research Council Scholarship.

The author has to thank Mr J. Hajnal and Dr C. A. B. Smith for much helpful discussion and advice concerning the analytical parts, and also Professor O. W. Richards for reading the first draft of Part II. It is also a pleasure to thank Professor W. E. Kerr for his helpful comments on the present version of Part II and for the kind hospitality of his laboratory in Brazil.

References and notes

1. See page 31.
2. J. F. C. Kingman, *Proceedings of the Cambridge Philosophical Society* **57**, 574 (1961).
3. M. Kimura, *Heredity* **12**, 145 (1958).
4. R. A. Fisher, *The Genetical Theory of Natural Selection*, p. 37 (Dover, New York, 1958; first published 1930). See points raised by P. A. P. Moran, *The Statistical Processes of Evolutionary Theory* (Clarendon Press, Oxford, 1962), pp. 60 and 66.
5. J. B. S. Haldane, *New Biology* **18**, 34 (1955).
6. Fisher 1930 (ref. 4), p. 177 *et seq.*
7. G. C. Williams and D. C. Williams, *Evolution* **11**, 32 (1957).
8. For relevant observations see H. V. Thompson and A. N. Worden, *The Rabbit*, pp. 104 and 217 (Collins, London, 1956).
9. For example, D. Lack, *The Life of the Robin*, pp. 114–16 (London, Pelican, 1953).
10. A. D. Blest, *Nature* **197**, 1183 (1963).
11. V. C. Wynne-Edwards, *Animal Dispersion in Relation to Social Behaviour*, p. 391 (Oliver & Boyd, Edinburgh, 1962).
12. N. Tinbergen, *The Herring Gull's World*, p. 49 (Collins, London, 1953).
13. E. B. Ford, *Butterflies*, p. 256 (Collins, London, 1945).
14. Wynne-Edwards 1962 (ref. 11), p. 136.
15. Tinbergen 1953 (ref. 12), p. 224 *et seq.*
16. Ibid., p. 228.
17. D. Summers-Smith, *The House Sparrow*, p. 50 (Collins, London, 1963).
18. J. Prevost, *New Scientist* **16**, 444 (1962).
19. W. E. Jones, *Nature* **178**, 426 (1956).

20. For example, G. W. Martin, *Botanical Review* **6**, 356 (1940). A detailed description for *Dictostelium discoideum* is given in J. T. Bonner, *Scientific American* **180** (6), 44 (1949).

21. E. W. Knight-Jones and J. Moyse, *Symposia of the Society for Experimental Biology* **15**, 72 (1961).

22. J. S. Moure, P. Noguiera-Neto, and W. E. Kerr, *Proceedings of the Tenth International Congress of Entomology, Montreal* **2**, 481 (1962).

23. (1995) The rule given here for coefficients involving males is vacuous and must be discarded—see Addendum of the 1971 republication that follows this paper; for a more satisfactory and complete account, see Chapter 8 of this volume. The values here given in Fig. 2.2 are correct as gamete-for-gamete coefficients in the sense of Chapter 8. Points made from here to the end of subsection (a) need to be treated with caution; points of the general discussion of the paper, however, are hardly affected by these errors and changes.

24. W. F. Bodmer and A. W. F. Edwards, *Annals of Human Genetics* **24**, 239 (1960).

25. W. M. Wheeler, *The Social Insects*, p. 220 (Harcourt, Brace, New York, 1928).

26. O. W. Richards, *The Social Insects*, p. 81 (Macdonald, London, 1953).

27. T. W. Kirkpatrick, *Insect Life in the Tropics*, p. 254 (Longmans, London, 1957).

28. (1995) My error of Note 23 is involved again here; again, see the Addendum following, and also Chapters 4 and 8.

29. C. D. Michener, E. A. Cross, H. V. Daly, C. W. Rettenmeyer, and A. Wille, *Insectes Sociaux* **2**, 237 (1955).

30. C. D. Michener, *Journal of the Kansas Entomological Society* **33**, 85 (1960).

31. C. Plateaux-Quénu, *Annales de Biologie* **35**, 327 (1959).

32. C. D. Michener and R. B. Lange, *Kansas University Science Bulletin* **35**, 69 (1958).

33. S. Taber and J. Wendel, *Journal of Economic Entomology* **51**, 786 (1958).

34. W. E. Kerr, R. Zucchi, J. T. Nakadaira, and J. E. Butolo, *Journal of the New York Entomological Society* **70**, 265 (1962).

35. W. E. Kerr, *Revista Brasileira de Biologia* **21**, 45 (1961)

36. C. P. Haskins and E. F. Haskins, *Insectes Sociaux* **2**, 115 (1955); C. D. Michener and M. H. Michener, *American Social Insects*, p. 126 (Van Nostrand, New York, 1951); W. M. Wheeler and J. W. Chapman, *Psyche* **29**, 203 (1922); Wheeler 1928 (ref. 25), pp. 173–4.

37. A. H. Sturtevant, *Quarterly Review of Biology* **13**, 74 (1938).

38. W. E. Kerr and V. de P. Araujo, *Garcia de Orta* **6**, 53 (1958).

39. Other interesting peculiarities reported for this race are its mild temperament and the presence in workers of large queen-like spermathecae. However, Dr Kerr tells me that he has checked that laying-workers do not have sperm in their spermathecae.

40. S. E. Flanders, *Insectes Sociaux* **9**, 375 (1962).

41. G. C. Butler, *The World of the Honeybee*, p. 58 (Collins, London, 1954).

42. A. Ledoux, see E. O. Wilson, *Annual Review of Entomology* **8**, 345 (1963).

43. W. E. Kerr, *O Solo* **41**, 39 (1949); and personal communication.

44. O. W. Richards and M. J. Richards, *Transactions of the Royal Entomological Society of London* **102**, 1 (1951).
45. For example, J. Gervet, *Insectes Sociaux* **9**, 231 (1962).
46. P. Rau, *Annals of the Entomological Society of America* **33**, 617 (1940); and personal observations of *Polistes versicolor*.
47. K. Yoshikawa, *Mushi* **30**, 37 (1957).
48. Wynne-Edwards 1962 (ref. 11), p. 463.
49. S. F. Sakagami and Y. Akahira, *Kontyu* **26**, 103 (1958) and *Evolution* **14**, 29 (1959).
50. K. Yoshikawa, *Journal of Biology, Osaka City University* **13**, 19 (1962).
51. J. B. Free and C. G. Butler, *Bumblebees*, p. 77 (London, Collins, 1959).
52. C. Plateaux-Quénu, *Insectes Sociaux* **7**, 349 (1960).
53. K. Yoshikawa, *Insectes Sociaux* **2**, 255 (1955).
54. Kirkpatrick 1957 (ref. 27), p. 277.
55. S. F. Sakagami and K. Fukushima, *Journal of the Kansas Entomological Society* **30** (1957).
56. C. Ferton, *Annales de la Société Entomologique de France* **70** (1), 83 (1901).
57. D. E. Davis, *Auk* **57**, 179 and 472 (1940).
58. C. D. Michener, *Proceedings of the Tenth International Congress of Entomology, Montreal* **2**, 441 (1958).
59. C. D. Michener and R. B. Lange, *Science* **127**, 1046 (1958).
60. C. D. Michener and A. Wille, *Kansas University Science Bulletin* **42**, 1123 (1961).
61. As regards the traces of similar sociability that exist in birds, with *Crotophaga* and *Guira*, recent evidence suggests the possibility of both aunt-like and sister-like altruism, although just how widely genetical relationship may range within groups is not known. Other recorded cases in birds suggest *immature* progeny helping the mother to rear subsequent broods (A. Skutch, *Auk* **52**, 257 (1935)). These immatures would doubtless reproduce normally later on.
62. N. Waloff, *Insectes Sociaux* **4**, 391 (1957).
63. G. Salt, *Symposia of the Society for Experimental Biology* **15**, 96 (1961).
64. A. M. D'Rozario, *Proceedings of the Royal Entomological Society of London* A. **15**, 69 (1940).
65. A. D. Imms, O. W. Richards, and R. G. Davies, *A General Textbook of Entomology* (Methuen, London, 1957).
66. (1995) The original footnote on sex ratio was clearly wrong. The facts given now suggest that relatedness in gregarious sawfly larvae might well be above one-half.
67. Wynne-Edwards 1962 (ref. 11), pp. 237–8.
68. S. C. Kendiegh, *Illinois Biological Monographs* **22**, 1 (1952).
69. Free and Butler 1959 (ref. 51), pp. 16, 19–21.
70. Ibid., p. 16.
71. Summers-Smith 1963 (ref. 17), p. 50.
72. S. F. Sakagami, *Insectes Sociaux* **7**, 231 (1960).
73. C. D. Michener, *Annals of the Entomological Society of America* **54**, 532 (1961).

74. C. D. Michener, *Journal of the Kansas Entomological Society* **34**, 178 (1961).
75. Ed.-Ph. Deleurance, *Insectes Sociaux* **2**, 285 (1955).
76. Ed.-Ph. Deleurance, *Colloque Internationale de la Centre National de Recherche Scientifique* **34**, 142 (1955).
77. J. Fisher and R. M. Lockley, *Sea Birds* (Collins, London, 1954).

ADDENDUM†

Sex ratio and social coefficients of relationship under male haploidy

Under the male-haploid system a male expects to contribute only half as much as a female to the gene pool of distant future generations. I previously thought this would cause females to value daughters more than sons until a fairly female-biased sex ratio had been produced. But gametes from sons carry the mother's genotype in full concentration whereas gametes from daughters carry it diluted by half. This exactly offsets the difference in numbers of progeny. Thus a 1:1 population sex ratio is as basic to male haploidy as to normal reproduction. Indeed, as shown in a later publication (Chapter 4; for example, Fig. 4.3), it is actually more strongly stabilized in the male-haploid case. In the amended discussion of the later paper the 'golden ratio', $\frac{1}{2}(3-\sqrt{5}) : \frac{1}{2}(\sqrt{5}-1)$ or 1:1.618, appears only as the unstable equilibrium sex ratio of a rather unrealistic special case, and other reasons are adduced to explain why male-haploid sex ratios do so very often show excess of females.

In the context of social interactions between males and females it is now clear, first, that a distinction must be made between measurement of fitness effects in terms of expected numbers of successful gametes and measurement in terms of the total fitnesses of individuals, and, second, that coefficients will not be reflexive. Neither distinction is necessary when the interacting individuals are of the same sex.

The coefficients given in Fig. 2.2 are in general probabilities that random gametes from the individuals in question are identical by descent. They correspond to Haldane and Jayakar's coefficient ϕ_{AB}.[1] My own rules for the calculation of this coefficient are certainly wrong. The rules of Haldane and Jayakar also fail in certain cases, so I give below a new set based on their rules:

1. List all paths by which A and B are connected through latest common ancestors.

†Reprinted from G. C. Williams (ed.), *Group Selection*, pp. 87–9 (Aldine-Atherton, Chicago, 1971).

2. Reject any path having two males in succession.
3. Count the steps in each remaining path, treating any double steps 'female–male–female' (whether apical or not) as single steps, and missing out steps to terminal males if any.
4. If n_i is the number of steps counted in the i-th path from A to B, and ϕ_i is the coefficient of inbreeding of the latest common ancestor (zero if male) on this path, then the coefficient of relationship is given by

$$\phi_{AB} = \sum_i 2^{-(n_i+1)}(1 + \phi_i).$$

The coefficients apply directly on a basis of gamete-for-gamete measurement for actions of males towards females, and on a fitness-for-fitness basis they apply for actions of females towards males. In the remaining contexts, gamete-for-gamete female toward male and fitness-for-fitness male towards female, the values of the coefficients must be doubled.

The kinds of sacrifices which have to be considered as possible among social insects, such as costs of rearing larvae and the sacrifice of one life in the defence of others, are most reasonably viewed in terms of fitness-for-fitness measurement of effects. Thus the possibility of alteration of the argument brought in by the present correction mainly concerns actions of males towards females. The unit coefficients appearing for the son-to-mother and the son-to-daughter relationships are the most striking changes. At first sight they may even seem paradoxical. In the realm of theory, however, it is clearly true that if a son were to sacrifice his life to rejuvenate completely a dying mother he could expect the genes that he loses to be replaced. In practice he has no possibility of doing this. On the other hand there are plenty of ways in which the male can attempt to increase his production of daughters. The theoretical implication of the unit coefficient with daughters is also plain, and if the male does not exactly sacrifice his life for a daughter he does normally use it up in efforts which, in a stable population, gain him the expectation of one daughter. This unit coefficient, it must be noted, does not imply any special likelihood for the evolution of parental care. On the contrary, that the males' efforts are always towards polygamy rather than parental care is as expected. In parental care haploid males would be liable to waste some effort completely unless they were able to discriminate diploid from haploid larvae among the progeny of their mates. With normal males, at least in respect of autosomal genotype, discrimination is unnecessary.

Reference

1. J. B. S. Haldane and S. D. Jayakar, *Journal of Genetics* **58**, 81–107 (1962).

CHAPTER 3

Granny chimpanzee

Thinking about the Baldwin Effect — or the advantage of evolving a menopause?

C H A P T E R 3

LIVE NOW, PAY LATER

The Moulding of Senescence by Natural Selection

—Are you drunk or what are you trying to say? asked Cranly facing round on him with an expression of wonder.
—The most profound sentence ever written, Temple said with enthusiasm, is the sentence at the end of the zoology. Reproduction is the beginning of death.
He touched Stephen timidly at the elbow and said eagerly:
—Do you feel how profound that is because you are a poet?

JAMES JOYCE[1]

It seems that Stephen Daedalus was either a poet too much or too little: he did not reply and presently watched as his friend Cranly chased 'gipsy' Temple away. For my part I sympathized with Temple most among all Joyce's students in *A Portrait of the Artist as a Young Man*; he seemed more scientific than the rest. From the time I first read it I suspected his idea could well be profound. Later, combining some thoughts from R. A. Fisher, P. B. Medawar, and G. C. Williams along with new ones of my own, I became convinced of it as the paper in this chapter shows.

I began writing it from the notes and tables I had already accumulated soon after I returned from Brazil in 1964. I was then working at the Imperial College Field Station. I had applied for a lectureship advertised in genetics at Imperial College, University of London, and, by standards of academic jobs today, obtained the post with ridiculous ease. All my appointment had needed was an application letter and an interview by Professor O. W. Richards, the then head of the Department of Zoology and Applied Entomology. Richards introduced me to two or three of his colleagues, who I suppose were asked their opinions about me, but there was no committee and nothing remotely resembling a job seminar. I had no PhD and no thesis; I did have two published papers and Richards had read the long one. Universities were expanding and jobs for academics almost fell from the trees. I learned much later that for my job there had been only one other applicant, that he was offered it and he declined.

Besides various unfinished theoretical projects I wanted to continue the work on bees, wasps, and their allies that I had begun at Kerr's lab in Brazil. Although by now fond of London I greatly preferred living in the country. Thus instead of accepting space in the main department in South Kensington, London, I chose to have an office-cum-lab at the Field Station at Sunninghill in Berkshire about 30 miles west from where I had been living in Chiswick. I rented from a very old lady a bedsitting room in her house about a mile away at the edge of Windsor Great Park. The room was not much different from that which I had occupied in Chiswick although one plus was a surprise species sharing the room with me: a small shiny black insect that I later discovered was the bacon beetle. I should explain that I am generally fond of indoor insects, think of them like pets, and generally see an advantage over pets in that they don't need to be fed. These particular beetles probably bred out from the mummified mice that reposed in dark corners of the house due to the activities of my landlady's cat: that they weren't out of any bacon of mine is quite certain because I hadn't any, the room's cooking facilities being even more reduced than those I had had in Chiswick.

From the house I walked every day to work, usually across the small but beautiful and wild Silwood Park owned by the Field Station. From my lab window I looked back on the same park. Silwood House, the building housing the whole of the entomological and plant-pathological side of the Field Station including my lab, was a Victorian mansion and the park its estate. In crossing it was difficult, as I will describe later, to avoid distraction by the insects and other wildlife. Besides this I had by now other demands on my research time in the form of studies on insect sex ratios that I had started. But unfinished projects from my previous period in London had to have priority and this paper was the first result.

Themes of both this paper and the next, on sex ratios, had been closely interwoven with my thinking on kin selection. They represented the most difficult and most interesting side issues of my main theme, those issues that needed special care and that had philosophical interest in their own right. By this time the theme of senescence no longer excited me much because I had decided that most of my findings on it were not new and that in any case, regarded as adaptation, it was rather a rag-bag phenomenon to boot; however, there were a few ideas that might be novel in my work and I felt I ought to try to publish them before moving on. Perhaps when I came to write up the paper my general lack of enthusiasm showed because, after submission, one of the referees to whom the editor of *JTB* sent the manuscript described it in his report as 'verbal diarrhoea'. Probably from this referee's point of view it still is, because all I can recall altering in the required rewriting is the Introduction and the way I show what the puzzle is.

The argument about how natural selection works on organisms that reproduce over a period I had worked out in fair detail, almost as far as I have taken the matter in this paper, by the time I left the London School of Economics. My thinking had been helped on by my course with Norman Carrier and by discussions with my subsequent supervisor at LSE, John Hajnal. Suggestions of where to find the vital statistics for humans who had no modern birth control or medicine to aid their infant survival came from the demographers at LSE, and the computing of the curves for Figs 3.1 to 3.3 was done on the electrical hand calculators in the LSE stats rooms. As I write I almost hear the rhythmic clatter of that room with, as its background, the dancing, clunking rhythm of Scandinavian Facits executing their favourite step—the calculations of sums of squares.

This paper almost of necessity is about organisms that bear offspring sequentially or are iteroparous as biologists say, but organisms that reproduce just once (semelparous) are covered as a special case of the model. Humans, birds, earthworms, and perennial plants are good examples of iteroparity. Most of us know by personal experience of our ages with respect to those of our cousins how human generations can overlap and merge. Agaves, salmon and annual insects and plants, on the other hand, show semelparity, and, given the additional assumption that their lifespans are the same, their generations do not overlap. I had avoided confronting the extra mathematical difficulties of overlapping generations and iteroparity in Part I of the 1964 paper by making the organisms of my model semelparous with constant lifespans. With the sequential reproducers, quite apart from any special problems that social selection might raise, there were more basic ones about how effectively selection would work creating adaptations at one age versus another, and this is the crux of the 1966 paper. These difficulties had puzzled me since I first read Fisher's book *The Genetical Theory of Natural Selection* (1930) as an undergraduate. Fisher had explained an important quantity, reproductive value, which could be associated with iteroparous reproducers as a function of age. The idea seemed at first to be Fisher's own for he gave no citation. I later realized, however, that his idea had been implicit if put to less use, in works of the continental pioneer mathematical biologists Volterra and Lotka,[2] while the fundamental equation of population growth, out of which reproductive value ultimately springs, dates as far back as the great eighteenth-century Swiss mathematician Leonhard Euler. At the time, with difficulty and wasting time with various false starts, I grasped Fisher's concept. He gave a diagram for humans that is basically similar to the one in the paper that follows and he described but didn't illustrate the curve of human mortality rate that is an ingredient. He followed his description of the course of this mortality with the comment that: 'It is probably not without significance in this connexion that the death rate in man takes a form generally inverse to the curve of reproductive value'. Just one more sentence followed expanding his thought and revealing that he was thinking of the 'incidence of natural death' being 'to a large extent moulded by the effects of differential survival'.

The paper summarizes my attempt to find out what those brief comments by Fisher could mean. I had struggled similarly with many other passages in his work and had always concluded that he had not been writing vaguely, understood his topic well, and was making a suggestion that was deep and valid. Hence my work for this paper became a psychological milestone for me eventually, in the odd sense that in the end I concluded that Fisher for once was not making a deep point: the inverse relation of the two curves is largely 'without significance' because no U-shaped curve of mortality can fail to be associated with a humped curve of reproductive value. Moreover, although evolutionary 'moulding' by differential survival does work to produce the steeply rising late tail of intrinsic mortality, it certainly does not directly produce high infant mortality: for that, other selective factors—possibly the social ones I mention in Chapter 2 (page 75)—had to be operating. Thus the only correspondence in the curves that may be non-trivial is the point that mortality starts to rise at the age where reproduction approximately begins, and even this may be considered the less original in being roughly the notion, I believe, that Temple was trying to express to Stephen and Cranly in Joyce's book.

In spite of its representing a unique case where I thought I had seen farther, or at least more clearly, on one issue than the giant on whose shoulders I had for so long been trying to balance, I ended less proud of this paper than I was of my other two London themes. This was because I had found that I had forerunners in the theory of senescence in P. B. Medawar and G. C. Williams, and that what I had been doing was little more than to put a mathematical gloss to their ideas. Fortunately for the paper's survival after I realized this I also saw that Medawar and Williams had not done a very complete job. Medawar had started out towards the correct theory but then had inexplicably bowed to Fisher's cryptic comment cited above, and had tried to write something in keeping in spite of the comment being quite out of accord with his own logic. G. C. Williams, on the other hand, seemed to have understood the matter right through and had merely sacrificed a little of the generality that I attained by his assuming the populations under selection to be static. But, when all work for my paper had been done, I had to admit that even this defect was trivial because most populations had to be almost static in the long term. In short, it needs a very peculiar species and a very unstable ecology to make the elegant Eulerian weightings in the general version of the theory really necessary.

My main reservation about Williams's paper was that I couldn't see why he should bring in senescence of leaves and other discardable organs: these had to be selected, it seemed to me, in quite different ways, towards whatever adaptation is most efficient for the life of the plant as a whole. The outcome of my more general analyses that I did feel to be original, and which also linked nicely with my kin-selection interest, is the suggestion to explain the rise of infant mortality as we move back towards the moment of birth. Right-to-lifers, it seems to me, ought seriously to consider this backward ascent of mortality and the explanation for it I offer in the paper. They would also do well to look carefully at what is known about how mortality of foetuses is scheduled within the gestation period: all certainly seems arranged as if Mother Nature never remotely intended all of her embryos conceived to be born, nor even intended (at least during the palaeolithic, when humankind was in the main flurry of construction) that all human babies born should be raised. It seems almost certain that direct human suffering is minimized the earlier a foetus or neonate dies and it is surely totally certain that this is the case regarding the suffering that parents experience as a result of such a death. Furthermore, an important increase in suffering for all concerned comes when a child is old enough to think about life and to begin to see a possible death ahead of it as a dread and a threat: the child feels the dread and the parents likewise, the latter both through sympathy with the child and through the loss and waste of their life that will be entailed to themselves. In yet another way of looking at this, if happiness and suffering are linked to the impending increases and decreases of inclusive fitness (as we should expect them to be; Chapter 2), a foetus or baby that has detected in itself some fatal physiological flaw is expected to decide for itself to die at the earliest opportunity, and, in executing this decision, it should die relatively calmly and happily, aware perhaps in a subconscious way that it is doing the 'right thing'. This is because for itself it has nothing to lose. By bringing the event forwards, it is helping its sibs and parents who are likely carriers of the same thanatic gene it has found occasion to express.

The part of my paper that I enjoyed working out most is the submodel of a non-senescing organism that I used to show that no life schedule, even under the most benign ecology imaginable, could escape my spectrum of forces of senescence: I felt myself to be proving Temple to be right not only in Dublin's Trinity College Square but in the farthest reaches of almost any bizarre universe.

Returning from such phantasies to more ordinary possibilities and to reproductive schedules similar to our own, I suspect that it is the starkness of Temple's theme that makes it a lead balloon not only with Stephen and Cranly but also, more seriously, with the ageing voters and rich university alumni today. Indeed, considering the idea's almost obvious validity, it has to be its inherent pessimism that accounts for its extremely slow penetration into the funded mainstream of research on ageing. Slowness has been apparent ever since the decade 30 years ago when the evolutionary approach to the topic started. Most donors are middle-aged at the least and the whole of the voting population is getting older too. All such people concur, I suspect, in wanting 'something to be done about ageing'; moreover, they want that it should be done if not yesterday at least before they themselves are many years older. They are not at all content to be told that there already exists a theory admirably outlining the whole subject and showing that after a few hundred years of draconian eugenic measures, which no one could ever endure, the human lifespan might be stretched out just a little—stretched, to say, a three-score years and fifteen instead of the biblical three score and ten. In return for their cash, universities therefore probably humour their old alumni by promising to waft money towards departments of gerontology that they have staffed with congenital optimists. These staff have, in effect, a brief to study non-evolutionary aspects of sensescence, and to ensure that there are breakthroughs simmering and around the corner all the time. Much of the work done in this way, at least as regards finding a practicable programme for extension of active life, seems to me comparable with the alchemists' search in past ages for that elixir that was aimed to solve exactly the same problem. The pity is that although the investigation may be by no means worthless in itself, furious study of particular aspects (as if they had a chance to be the whole) distracts both from unavoidable truth and from realistic social programmes. Meanwhile, fear of the truth sees to it that the only effective theory is hardly ever cited. Recently, however, there seems to have been something of an outbreak of realism and an excellent book of Neodarwinist slant (not from a department of gerontology) may be helping this along.[3]

A story about research and development undertaken to 'improve' car parts by a Detroit manufacturer illustrates an aspect of the senescence theory remarkably well. The company is said to have employed a man to visit car breakers' yards near Detroit and to record all the parts that were still found to be in working order when the car was scrapped. If a part was found to be usually working, research and development to improve that part was discontinued at the factory. Obviously, such policy slowly creates cars that tend to have several systems failing within a short time (applying this thought to cars I have owned leads me to guess that a cousin of the Detroit car pathologist was working for British car companies at much the same time). This story bears more on the senescence of semelparous organisms than on that of the sequential reproducers treated in my paper, but the tendency to have multiple systems becoming senescent at once, leading to multiple failures, must apply to some extent to all. It is the variety of failing systems that is fatal to any notion of an elixir. Elixir treatment will need to be magical indeed if it can fix or prevent all of such various conditions as worn and carious teeth, hardened and misting eye lenses, cartilage proteins ever more cross-linked with sulphur bonds (like vulcanized rubber), while it is also working its still more radical wonders in keeping DNA repair mechanisms in perfect order—all this within dozens of types of cell.

There is not much more to say. My new environment of entomologists at Silwood Park proved very congenial and even in the unlikely topic of this paper it began to provide me with examples I otherwise would never have thought of. Where else but in the entomological Mecca of Britain that Silwood was in those days would I have heard, for example, about greenfly great-grandmothers having a menopause? Finally, let me transcribe a pencilled note I made in the margin of the oldest original reprint of the paper that I still keep. It is written near where, on page 119, I say that an 'irregular downward movement is occurring fastest at the left-hand end [of the mortality schedule]'. My pencilled note says: 'Thus ageing animal should climb *down* his evolutionary tree: young man's youthful features in trends which made *old* gorilla—therefore Dad's comment, how we "revert to type."' This remark of my engineer father, which was uttered probably about the time I wrote the paper (he died in 1972), took me by surprise and is the only evolutionary thing I ever recall him saying. I also recall that the phrase emerged suddenly rather than reflectively as if he had heard or read it, but if he did I don't know where. However, our expression 'going ape' is, I suppose, just another caricature of the same idea.

Therefore one last confession. I, too, am probably coward enough to give funds for 'elixir' gerontology if anyone could persuade me that there is hope: at the same time I want there to be none so that I will not be tempted. Elixirs seem to me an anti-eugenical aspiration of the worst kind and to be no way to create a world our descendants can enjoy.[4] Thus thinking, I grimace, rub two unrequestedly bushy eyebrows with the ball of a happily still-opposable thumb, snort through nostrils that each day more resemble the horse-hair bursts of an old Edwardian sofa, and, with my knuckles not yet touching the ground, though nearly, galumph onwards to my next paper.

References and notes

1. J. Joyce, *A Portrait of the Artist as a Young Man*; in H. Levin (ed.) *The Essential James Joyce* (Triad/Panther, St Albans, 1977; first published 1916).
2. F. Scudo and J. Ziegler, *The Golden Age of Theoretical Ecology* (Springer, Berlin, 1978).
3. M. R. Rose, *Evolutionary Biology of Aging* (Oxford University Press, Oxford, 1991).
4. The idea is, of course, extremely old. Epictetus, the Stoic philosopher, wrote: 'If we were useful alive, should we not be still more useful to mankind by dying when we ought, and as we ought? And now the remembrance of Socrates is not less, but even more useful to the world than were the things he did and said when alive.' Book IV, Chapter 1 (translated by Elizabeth Carter).

THE MOULDING OF SENESCENCE BY NATURAL SELECTION†

W. D. HAMILTON

The consequences to fitness of several types of small age-specific effects on mortality are formulated mathematically. An effect of given form always has a larger consequence, or at least one as large, when it occurs earlier. By reference to a model in which mortality is constant it is shown that this implication cannot be avoided by any conceivable organism. A basis for the theory that senescence is an inevitable outcome of evolution is thus established.

The simple theory cannot explain specially high infant mortalities. Fisher's 'reproductive value', the form of which gave rise to an erroneous opinion on this point, is shown to be not directly relevant to the situation. Infant mortality may evolve when the early death of one infant makes more likely the creation or survival of a close relative. Similarly, post-reproductive lifespans may evolve when the old animal still benefits its younger relatives.

The model shows that higher fertility will be a primary factor leading to the evolution of higher rates of senescence unless the resulting extra mortality is confined to the immature period. Some more general analytical notes on the consequences of modifications to the reproductive schedule are given.

Applications to species with populations in continual fluctuation are briefly discussed. Such species apart, it is argued that general stationarity of population can be assumed, in which case the measurement of consequences to fitness in terms of consequences to numerical expectation of offspring is justified.

All the age functions discussed are illustrated by graphs derived from the life table of the Taiwanese about 1906, and the method of computation is shown.

1. INTRODUCTION

Consider four hypothetical human genes. Suppose all are limited in their expression to the female sex and also age-limited in the following way: each gives complete immunity against some lethal disease but only for one particular year of life. Suppose the first gives immunity for the first year, the second for

†*Journal of Theoretical Biology* **12**, 12–45 (1966).

the fifteenth, the third for the thirtieth, and the fourth for the forty-fifth. What are the relative selective advantages of these genes?

If for further simplicity parental care is ignored and it is assumed that the menopause always comes before age 45, it is at once obvious that the fourth gene is null, whereas all the others do confer some advantage. It is also fairly obvious that the third gives less than the second. But how much less? Does the second give a maximum because it occurs at the age of puberty? Does the first give less than the second?

The importance of questions of this kind for an evolutionary theory of senescence has been realized for some time. Most of the answers that will be given in this paper agree with the theory of Williams.[1] Although perhaps not obvious, they are so simple that it is surprising to find almost no indication that they had been realized earlier. Several writers have in effect answered the last two questions in the affirmative, which is for the one inexact and for the other wrong.

Even Williams's discussion failed to clear up completely the previous confusion of thought on the subject. Thus he regarded his theory as consistent with the views of Medawar.[2,3] But Medawar in his 1952 lecture combined the development of a model which did lead him to the outlines of what we believe to be the correct theory with tentative adherence to a logically inconsistent opinion about the forces operating in the immature period. This latter seems to have been taken over uncritically from Fisher who had written that he thought it 'probably not without significance ... that the death rate in man takes a course generally inverse to the curve of reproductive value'.[4] As may be seen from the diagrams given in this paper a human curve of reproductive value (see Fig. 3.3) rises to a maximum shortly after the attainment of reproductive maturity, while the curve of force of mortality (see Fig. 3.2a) has a minimum at or slightly before it. Fisher argued that with an earlier age at marriage, such as is very probable for our remote human and semi-human ancestry, the peak of reproductive value would have been earlier. Indeed it does seem quite likely that under primitive ancestral conditions the two turning points would have closely approximated the age of puberty. Hence apparently came Medawar's idea that with the onset of reproduction there is reversal of selection for age-to-onset modifiers such that for deleterious effects before this age selection tends to make them occur earlier instead of postponing them.

I hope to make it clear that the correspondence to which Fisher draws attention in the above statement is really largely trivial and that in the context to which they were restricting themselves the idea which he tacitly and Medawar explicitly assumed is without foundation. It must be admitted, however, considering the peculiar form of pre-adult mortality in humans, the hypothesis was attractive, and I shall show reason to think that with a bio-economic basis something like it could be revived (section 9(b)).

This paper outlines a fairly general approach to the problem of assessing how the age at which a gene acts affects its influence on fitness. Every effect on fitness must be capable of analysis into components manifesting in the schedules of mortality and fertility separately. Thus as regards the individual's own fitness the problem has two sides. The argument will be developed in terms of hypothetical effects on mortality and most of the discussion will also be confined to this side. The other has fundamental interest but it will merely be shown at the end (section 9(c)) that effects on fertility are tractable, mathematically at least, by the same approach: it will be seen at the same time that the simplest implications are here rather obvious while more detailed ones are biologically doubtful.

2. MEASURES OF FITNESS AND MORTALITY

For an organism with 'non-overlapping' generations the obvious measure of Darwinian fitness is the expectation of offspring as measured at birth. If the organism practises parental care 'birth' should be considered to occur, for the purposes of this definition, at the age at which the offspring becomes independent. Although difficulties may arise, due to the participation of mates of different types in the parental care, the continuance of care right up to the period when the offspring themselves reproduce, and so on, the concept is essentially straightforward. The logarithm of the expectation is a parallel measure of fitness which may have advantages for some purposes.

For an organism which reproduces repeatedly the concept of fitness is not so easily defined. The expectation of offspring suffers from the objection that early births are worth more than late in an increasing population, and vice versa in a decreasing one, and that there is no single measure of generation time which will serve as the unit for progress under natural selection. If the fitness differentials are all small and the population almost stationary these objections are of little importance, but in general the measure of fitness known as the Malthusian parameter[4] has the advantage that it does take into account the relative values of early and late offspring by giving them appropriate weightings in the expectation.

If l_x is the fraction of a birth cohort of the type which survives to age x and f_x the age specific fertility rate at age x, the Malthusian parameter m is defined to be the one real positive root for m of the equation

$$\int_0^\infty e^{-mx} l_x f_x \, dx = 1. \tag{1}$$

Clearly m is a logarithmic measure of fitness, the corresponding multiplicative measure being $\lambda = e^m$.

To make these measures correspond to the 'per generation' measures of the 'non-overlapping' organisms we have to pick a generation time T and use mT and λ^T as the logarithmic and multiplicative measures, respectively. But a single unchanging T applicable to all types in an evolving population cannot be found; the present case is in this respect more complex than the non-overlapping one. It is best to be content with λ or m based on some convenient unit of time. This would still have difficulties for exact Mendelian-demographic models but is good enough for the limited aims of the present study.

The remaining preliminary is to decide on the measure in which the changes in mortality are to be specified.

Given statistics of deaths by age and census population by age the life table may be constructed in the actuarial manner. At an early stage q_x is tabulated, estimating the proportion of people who have reached their xth birthday who will die before their $(x+1)$th birthday. This leads to the corresponding proportion of survivors $p_x = 1 - q_x$ and then to

$$l_x = p_0 \cdot p_1 \cdot p_2 \cdots p_{x-1}. \tag{2}$$

There are, of course, other possible procedures in life-table construction but this is the one preferred when the data are as stated and it is also the one most convenient for reference in the present discussion.

q_x is an average death rate for a period during which ability to resist death is not necessarily constant. We now define an instantaneous death rate, representing in negative the concept of ability to resist death, by a limiting process. Attention is transferred from q_x, which applies to the unit age interval following age x, to ${}_{\delta x}q_x$, which is defined to apply to the part of this interval extending from x to $x + \delta x$. We then have

$$\frac{{}_{\delta x}q_x}{\delta x} = \frac{1 - {}_{\delta x}p_x}{\delta x} = \frac{l_x - l_{x+\delta x}}{l_x \delta x} = \frac{l}{l_x} \cdot \frac{-\delta l_x}{\delta x} \xrightarrow[\delta x \to 0]{} -\frac{1}{l_x} \cdot \frac{\mathrm{d}l_x}{\mathrm{d}x} = \mu_x. \tag{3}$$

The instantaneous death rate μ_x is usually known as the 'force of mortality'.

The cologarithm of ${}_{\delta x}p_x$, the natural logarithm taken with changed sign, is also relevant to this limit, for

$$\begin{aligned} \frac{\operatorname{colog}_{\delta x}p_x}{\delta x} = \frac{\log l_x - \log l_{x+\delta x}}{\delta x} &= \frac{-\delta \log l_x}{\delta x} \xrightarrow[\delta x \to 0]{} -\frac{\mathrm{d} \log l_x}{\mathrm{d}x} \\ &= -\frac{1}{l_x} \cdot \frac{\mathrm{d}l_x}{\mathrm{d}x} = \mu_x. \end{aligned} \tag{4}$$

And for estimating $\mu_{x+\frac{1}{2}}$ from the ordinary life table

$$\operatorname{colog} p_x = \log l_x - \log l_{x+1} \tag{5}$$

is preferable to q_x in being free from the particular slight bias apparent in

$$q_x = \frac{l_x - l_{x+1}}{l_x}. \tag{6}$$

3. THE EFFECT OF A BRIEF MORTALITY CHANGE

In practice, to obtain an estimate of the parameter m defined in equation (1) an analogous equation is used:

$$\sum_{1}^{\infty} \lambda^{-x} l_x F_x = 1, \tag{7}$$

in which F_x is the mean number of same-sex births in the age interval $x - \frac{1}{2}$ to $x + \frac{1}{2}$ to a parent living through it.

Exactly, supposing the l_x and F_x are exact, equation (7) gives the Malthusian parameter as it would be for a parthenogenetic organism if the births in each age interval ($x + \frac{1}{2}$ to $x + \frac{1}{2}$) were actually concentrated on the birthday. This might very nearly happen if breeding was strictly regulated by the seasons, but in other cases the inaccuracy will not be great so long as the age intervals are not too long. In the case of the human life schedules, for example, so far from it mattering that reproduction is hardly seasonal at all we may take a quinquennial age interval as giving a result quite as accurate as the data, the restriction to one-sex reproduction and the other artificialities, will justify.

We require to differentiate λ in the above equation with respect to colog p_a, where a is the age at which the effect is supposed to occur. From equations (4) and (5) it is evident that this is analogous to differentiating equation (1) w.r.t. $\mu_x \delta_x$ and in value the derivative will correspond to the effect of a constant change of force of mortality between ages x and $x + 1$. Clearly a finite instantaneous effect on μ_x can have no effect on m and in the continuous treatment we can only arrive at the fitness differential due to a very localized effect by finding the effect of $\delta\mu_x \delta x = \Delta_x$ and then letting $\delta x \to 0$ while Δ_x is held constant. Discussion in terms of the discrete treatment is here simpler on the whole: it not only steps off from the potentially available forms of schedule, but should be easier to follow. We shall therefore concentrate on this treatment and merely write the analogous expressions for the more ideal continuous one as we go along.

We now proceed by differentiating λ in each term of the sum represented in equation (7) w.r.t. p_a. From the whole, noting that p_a is only a factor in the terms for which $x < a$, we arrive at

$$-\frac{\mathrm{d}\lambda}{\mathrm{d}p_a} \sum_{1}^{\infty} x \lambda^{-(x+1)} l_x F_x + \frac{1}{p_a} \sum_{a+1}^{\infty} \lambda^{-x} l_x F_x = 0.$$

Therefore

$$\frac{\mathrm{d}\lambda}{\mathrm{d}p_a} = \frac{\dfrac{1}{p_a} \displaystyle\sum_{a+1}^{\infty} \lambda^{-x} l_x F_x}{\dfrac{1}{\lambda} \displaystyle\sum_{1}^{\infty} x \lambda^{-x} l_x F_x}.$$

Since

$$\frac{\mathrm{d}m}{\mathrm{d}\lambda}=\frac{1}{\lambda} \quad \text{and} \quad \frac{\mathrm{d}p_a}{\mathrm{d}\log p_a}=p_a,$$

it follows that

$$\frac{\mathrm{d}m}{\mathrm{d}\log p_a}=\frac{\sum_{a+1}^{\infty}\lambda^{-x}l_xF_x}{\sum_{1}^{\infty}x\lambda^{-x}l_xF_x}. \tag{8}$$

This is the derivative we require. The corresponding formula for the continuous treatment is

$$\frac{\mathrm{d}m}{\mathrm{d}\Delta_a}=\frac{\int_a^{\infty}\lambda^{-x}l_xf_x\mathrm{d}x}{\int_0^{\infty}x\lambda^{-x}l_xf_x\mathrm{d}x}. \tag{9}$$

The standard iterative calculation for solving equation (7) (e.g. Coale[5]) to obtain the Malthusian parameter can be made to yield both the numerator and denominator expressions in the above formula. The value of the former for each successive value of a is given by cumulating the column which sets out $\lambda^{-x}l_xF_x$ from the bottom; the total will equal unity provided λ has been accurately found. The actual procedure is illustrated in Table 3.1, columns 7 and 8. The denominator expression is also familiar in demographic mathematics and may be denoted by W, with W used for the ideal quantity as it appears in equation (9). This is one of the parameters which can be considered as measuring the length of a generation, being the mean age of mothers at childbirth for all births occurring in the stable population.

4. THE EFFECT OF A PROLONGED MORTALITY CHANGE

Rather as l_x can be regarded as compounded of the chances of surviving each successive year of life up to age x (equation (2)), so p_x itself can be regarded as compounded of the chances of evading the various independent factors which threaten life during the xth year. Among those of such factors which persist from year to year a few remain constant or nearly so, but most must tend to change with the intrinsic changes of habits and physiology that accompany ageing. Despite such actual inconstancies, the illumination of which is the object of our study, we will proceed to discover the effect on m of multiplying every p_x for which $x > a$ by a *constant* factor slightly greater than unity. This is equivalent to adding a constant to $\log p_x$ (which is necessarily negative), and to subtracting a constant from μ_x, throughout the same age range. In other

Table 3.1 Age functions of mortality and fertility for Chinese women in Taiwan about 1906

1	2	3	4	5	6	7	8	9	10
Age	**Survivorship**	**Fertility**		**Weights for exponential growth**	**Weighted net reproduction**			**Stable age distribution**	**Basic reproductive value**
		Gross	**Net**						
x	l_z	F_x	$l_x F_x$	λ^{-x}	$g_0(x)$	$g_1(x)$	$g_2(x)$	$\lambda^{-x} l_x$	$\widehat{v_x/v_0}$
	(Barclay) (Interpolation)	(Tuan)	(Interpolation) (2 × 3)		(4 × 5)	($\sum 6 \uparrow$)	($\sum 7 \uparrow$)	(2 × 5)	(7 ÷ 9)
0	*1.0000*			1.00000		*1.00000*	5.60479	*1.00000*	*1.00000*
	0.7300			0.95852		1.00000		0.69972	1.42914
1	*0.6498*			0.91876			4.60479		
	0.6100			0.88065		1.00000		0.54790	1.82515
2	*0.5945*			0.84412			3.60479		
	0.5810			0.80910		1.00000		0.47029	2.12635
3	*0.5680*		0.03246	0.77554	0.02518		2.60479		
	0.5520	*0.25600*	*0.14122*	0.74337		0.97482		0.41008	2.37715
4	*0.5353*		0.29987	0.71253	0.21366		1.62997		
	0.5151	*0.80235*	*0.41325*	0.68298		0.76116		0.35177	2.16380
5	*0.4948*		0.42876	0.65465	0.28068		0.86881		
	0.4731	*0.82955*	*0.39246*	0.62749		0.48048		0.29687	1.61849
6	*0.4514*		0.35621	0.60146	0.21426		0.38833		
	0.4287	*0.75690*	*0.32452*	0.57651		0.26623		0.24718	1.07707
7	*0.4061*		0.29088	0.55260	0.16074		0.12210		
	0.3838	*0.64001*	*0.24564*	0.52968		0.10549		0.20329	0.51891

8	*0.3615*		0.17568	0.50770	0.08919		0.01661		
	0.3404	*0.27713*	*0.09433*	0.48664		0.01630		0.16565	0.09840
9	*0.3193*		0.03428	0.46646	0.01599		0.00031		
	0.2973	*0.02500*	*0.00743*	0.44711		0.00031		0.13293	0.00232
10	*0.2753*		0.00073	0.42856	0.00031				
	0.2519			0.41078				0.10347	
11	*0.2285*			0.39374				0.07744	
	0.2052			0.37741					
12	*0.1819*			0.36176					
	0.1581			0.34675				0.05484	
13	*0.1344*			0.33237					
	0.1118			0.31858				0.03562	
14	*0.0892*			0.30536					
	0.0697			0.29270				0.02040	
15	*0.0502*			0.28056					
	0.0367			0.26892				0.00987	
16	*0.0232*			0.25776					
	0.0167			0.24707				0.00414	
17	*0.0103*			0.23682					
	0.0067			0.22700				0.00152	
18	*0.0031*			0.21758					
Totals			**1.61887**		**1.00000**				
		3.58694	***1.61885***			**5.60479**			

Except in column 4 all figures italicized are basic data or values known *a priori*. Ages are in units of 5 years.

words, the effect we are considering can be thought of as the total or partial elimination, from a certain age onwards, of a certain constant risk of death.

Every l_x for which $x \leqslant a$ is unchanged by this effect.

For $x > a$, every l_x now contains a factor k^{x-a}. Hence with k as defined

$$\frac{\mathrm{d}\left(\sum_{1}^{\infty} \lambda^{-x} l_x F_x\right)}{\mathrm{d}k} = \frac{1}{k} \sum_{a+1}^{\infty} \lambda^{-x(x-a)} l_x F_x.$$

Writing the differential d log k as d log $p_{(a\ldots\infty)}$ so as to be more explicit we obtain corresponding to equation (8) by similar reasoning

$$\frac{\mathrm{d}m}{\mathrm{d}\log p_{(a\ldots\infty)}} = \frac{\sum_{a+1}^{\infty} (x-a)\lambda^{-x} l_x F_x}{\sum^{\infty} x\lambda^{-x} l_x F_x}. \tag{10}$$

In the continuous treatment the corresponding formula is

$$\frac{\mathrm{d}m}{\mathrm{d}\mu_{(a\ldots\infty)}} = -\frac{\int_{a}^{\infty} (x-a)\lambda^{-x} l_x f_x \mathrm{d}x}{\int_{0}^{\infty} x\lambda^{-x} l_x f_x \mathrm{d}x}. \tag{11}$$

The denominator expressions are W and W as before. The value of the numerator in equation (10) is very easily obtained from the standard calculation: we simply cumulate the previous column which gave

$$\sum_{x+1}^{\infty} \lambda^{-x} l_x F_x,$$

again from the bottom, as illustrated in column 8 of Table 3.1. Like the denominator the numerator now has the form of a moment but differs in referring to only a part of the total unit of area under the function $\lambda^{-x} l_x F_x$ and in being the moment about the abscissa a of the partial area instead of about the origin. When $a = 0$ the derivative is numerically unity. We then have

$$\delta m \simeq -\delta\mu_{(0\ldots\infty)},$$

which means that a mutant giving a life-long mortality reduction of 0.01 would confer a selective advantage of about 0.01. This fact has no particular novelty: it emerges equally well in the treatment of stable population theory in which the rate of increase or Malthusian parameter is conceived in terms of the imbalance of births and deaths. If b is the birth rate per head and d the death rate per head, for a population in its stable age distribution, the rate of natural increase[6] is given by

$$r = b - d,$$

whence the stated result. Of course, it must be remembered that the change in mortality by affecting the rate of increase alters the stable age distribution with consequent effects on b and d: hence such statements are only approximate.

If the mortality effect terminates at age b we would have

$$\frac{\mathrm{d}m}{\mathrm{d}\mu_{(a\ldots b)}} = \frac{\int_a^\infty (x-a)\lambda^{-x} l_x f_x \mathrm{d}x + (b-a)\int_b^\infty \lambda^{-x} l_x f_x \mathrm{d}x}{W} \tag{12}$$

This gives equation (9) as $b \to \infty$ and tends to zero as $b \to a$. By defining $\Delta_{(a\ldots b)}$ analogous to Δ_a, we can find the substitution

$$\mathrm{d}\mu_{(a\ldots b)} = \frac{1}{b-a}\mathrm{d}\Delta_{(a\ldots b)}.$$

Hence, if equation (12) is correct, we have

$$\frac{\mathrm{d}m}{\mathrm{d}\Delta_{(a\ldots b)}} = \frac{\int_a^b (x-a)\lambda^{-x} l_x f_x \mathrm{d}x}{(b-a)W} + \frac{\int_b^\infty \lambda^{-x} l_x f_x \mathrm{d}x}{W} \tag{13}$$

in which the left-hand term vanishes as $b \to a$, leaving equation (11).

5. THE EFFECTS OF AGE-OF-ONSET MODIFIERS

By integrating by parts we find that

$$\int_a^\infty \left(\int_x^\infty \lambda^{-t} l_t f_t \mathrm{d}t \right) \mathrm{d}x = \int_a^\infty (x-a)\lambda^{-x} l_x f_x \mathrm{d}x. \tag{14}$$

Thus adopting the notation

$$g_0(a) = \lambda^{-a} l_a f_a, \tag{15}$$

$$g_1(a) = \int_a^{\infty} \lambda^{-x} l_x f_x \mathrm{d}x, \tag{16}$$

$$g_2(a) = \int_a^{\infty} (x-a)\lambda^{-x} l_x f_x \mathrm{d}x, \tag{17}$$

we have

$$g_0(a) = -D(g_1(a)) = D^2(g_2(a)).$$

This fact, that the age functions appearing in equations (9) and (11) are successive indefinite integrals of a function, is implicit in the cumulation method by which, as we have indicated, the corresponding discrete age functions can be obtained. With a corresponding notation for these functions

$$g_0(a) = \lambda^{-a} l_a f_a, \tag{18}$$

$$g_1(a) = \sum_{a+\frac{1}{2}}^{\infty} \lambda^{-x} l_x F_x, \tag{19}$$

$$g_2(a) = \sum_{a+1}^{\infty} (x-a)\lambda^{-x} l_x F_x, \tag{20}$$

we see that each $g_i(a)$ will give an estimate of the corresponding $g_i(a)$ value. The cumulation indicated in (19) will give a column of values which are, reading from the top, $g_1(\frac{1}{2}), g_1(1\frac{1}{2}), \ldots$, whereas the cumulation in (20) gives $g_2(0), g_2(1), \ldots$. This is conveniently shown in the tabulation by setting the numbers a half-row up as has been done in Table 3.1.

It is easily seen that $g_2(0) = \mathrm{W}$, the estimate of W.

As derivatives with respect to age, $g_0(a)$ and $g_1(a)$ can evidently be regarded as measuring the effect of fitness of changes in age of action and age of onset, respectively, of mortality changes of the two kinds we have discussed.

The occurrence of *age-of-onset* modifying genes is a plausible genetical hypothesis. For animals which grow continuously and reproduce repeatedly they would seem at least as plausible as the genes whose effects they are supposed to modify. In effect, of course, a gene which brings forward the age of onset of, say, a specific immunity is the same as a gene which directly confers an equal immunity for the limited period in question.

The occurrence of *age-of-action* modifiers seems somewhat less plausible. This is partly because it is less easy to envisage the effects which they are supposed to modify. From the relation of $g_0(a)$ to the reproductive schedule, it is evident that even if suitable subject effects did occur, only those manifesting within the fertile period would be subject to this kind of evolutionary improvement.

6. THE INDIRECT RELEVANCE OF REPRODUCTIVE VALUE

The quantities

$$\int_a^\infty l_x f_x \mathrm{d}x \quad \text{and} \quad \frac{1}{l_a}\int_a^\infty l_x f_x \mathrm{d}x$$

are the expectations of births occurring after age a to persons chosen at age 0 and at age a, respectively.

The quantities

$$w_a = g_1(a) = \int_a^\infty \lambda^{-x} l_x f_x \mathrm{d}x \tag{21}$$

and

$$\frac{v_a}{v_0} = \frac{\lambda^a}{l_a}\int_a^\infty \lambda^{-x} l_x f_x \mathrm{d}x \tag{22}$$

are clearly similar except that in them the births have been weighted in a particular way. As mentioned before, the weights are those necessary to correct for the different values of early and late births in a non-stationary population from the point of view of contribution to the population of the distant future. Roughly, each is inversely proportional to the size of the population into which the birth occurs relative to the size at the respective reference ages 0 and a, although exactly this statement is only true if the population is in its stable age distribution and therefore growing exponentially.

All our other results are likewise reducible to simpler statements about the effects of mortality changes on expectations of offspring, as can be seen when the weights are removed from expressions as they stand. However, although it may be justifiable actually to ignore the weights in a wide range of cases, we retain generality for the present discussion.

Thus w_a can be considered to measure more exactly what Williams[1] meant by 'reproductive probability'. Unfortunately, it seems impossible to have a phrase which combines this brevity with greater precision, but 'expected reproduction beyond age a' is at least more explicit.

It should be clear from the analysis in preceding sections that v_a, which is Fisher's reproductive value, is not directly relevant to the rate of selection of a genotype whose special effect manifests at age a. Through various causes a proportion of people, $(1 - l_a)$, fail to survive to age a and in them the genotype is never expressed. Thus although v_a indicates the magnitude of the effect of a set age-localized genotypic effect on mortality *when this is allowed to express itself*, in rating the consequent natural selection of the genotype this effect on fitness must be considered diluted owing to a kind of 'poor penetrance' of the genotype: w_a takes account of this dilution, whereas v_a is so constructed as to ignore it.

It may be argued that reproductive value does have at least an indirect relevance to senescence through its relevance to parental care. Acts of parental care necessarily involve living individuals, and it should now be evident that the ratio of the reproductive values is just as important as the coefficient of relationship in determining ideally adaptive social behaviour: the coefficient gives the chance that the offspring carries a replica of a behaviour-causing gene of the parent (see Parts I and II of the paper in Chapter 2), while the ratio gives the relative conditional expectation of its reproduction. The inclusive fitness of an individual is maximized by its continually acting in ways that cause increases in its inclusive reproductive value. It is implied that the self-sacrificing tendencies of parental care should follow a course inverse to that of reproductive value.

But against this, it must be remembered that our main thesis concerns the necessary failure of ideal adaptation, and the reason why this should tend to increase with age. As Williams emphasized, ideally with non-social species natural selection improves viability for all ages for which there is prospect of any future reproduction. And if it is plausible to postulate for the working of his theory of senescence the existence and promotion of genes which reduce viability in late life and raise it in early life, then surely the natural selection of genes which are not physically pleiotropic, and cause unchanging patterns of parental care, is even more plausible. If so, the fact that the intensity of parental devotion does not increase very noticeably with age is not surprising.

7. THE MODEL OF A NON-SENESCING ORGANISM

The absurdity of the idea that reproductive value outlines the forces of selection tending to prevent senescence may be shown by reference to the concept of a non-senescing organism.

Such an organism is supposed to have a mortality which does not change with age. As regards reproduction it can be assumed that so far from showing any senescent decline, fertility actually increases exponentially as the organism gets older. No real organism could increase its fertility indefinitely, of course (although some probably do keep up a gradual increase for a long time, for example some fish), but this assumption prejudices the case against the evolution of senescence as strongly as possible.

To give an idea of how such reproductive expansion could be supposed to occur, a *Volvox*-like organism may be imagined with all its cells undergoing synchronous division. After two divisions each cell has given rise to a tetrad; suppose one cell of each tetrad is expelled as a spore which starts a new colony, while the other three separate to take up as far as possible equipotential positions on the sphere ready for the next round of growth. Such an organism both grows and expands its fertility according to a geometric progression (in this case the growth ratio is three per two cell generations). Its population dynamics ought strictly to be treated by means of series.[7] But approximating fertility by a continuous exponential function would be generally very nearly correct and leads to simpler-looking mathematics.

Suppose, therefore, the following schedules of survivorship and fertility:

Age range	Survivorship	Fertility
$x < \alpha$	Not defined	0
$x \geq \alpha$	$l_\alpha e^{-\mu(x-\alpha)}$	$f_\alpha e^{c(x-\alpha)}$

where c is the logarithmic growth rate of fertility, f_α the fertility at the beginning of reproduction, l_α the survivorship up to age α, and μ the constant force of mortality that supervenes at least from age α onwards. For the present discussion the course of mortality from age 0 to the age α need not be defined.

From equation (1) the fundamental equation for this case is

$$\int_\alpha^\infty e^{-mx} f_\alpha l_\alpha e^{(c-\mu)(x-\alpha)}\, dx = 1$$

or

$$e^{-m\alpha} l_\alpha f_\alpha \int_0^\infty e^{(c-\mu-m)z}\, dz = 1.$$

It is clear that it is possible to choose m sufficiently large to make the integral converge, and therefore that a value exists which is the solution of the equation. Thus integrating and evaluating at the limits:

$$e^{-m\alpha} l_\alpha f_\alpha = m + \mu - c. \tag{23}$$

By some numerical method this may be made to yield the real root for m to any desired approximation.

Using equations (21) and (22) we may now write down for this case formulae for the 'expected reproduction beyond age a' and for reproductive value at age a, valid in each case for the age range $a \geq \alpha$.

$$w_a = e^{-m\alpha} l_\alpha f_\alpha \int_a^\infty e^{-(m+\mu-c)(x-\alpha)} \, dx$$

$$= \frac{e^{-m\alpha} l_\alpha f_\alpha}{m+\mu-c} e^{-(m+\mu-c)(a-\alpha)}$$

$$\left.\begin{aligned} &= e^{-(m+\mu-c)(a-\alpha)} \\ &= e^{-e^{-m\alpha} l_\alpha f_\alpha (a-\alpha)} \end{aligned}\right\} \text{ from (23)}$$

$$\frac{v_a}{v_0} = \frac{e^{m(a-\alpha)} l_\alpha f_\alpha}{l_a} \int_a^\infty e^{-(m+\mu-c)(x-\alpha)} \, dx. \tag{24}$$

Substituting for l_a, evaluating the integral as before and performing cancellations:

$$\frac{v_a}{v_0} = \frac{f_\alpha}{m+\mu-c} e^{c(a-\alpha)}$$

$$= e^{c(a-\alpha)+m\alpha}, \text{ from (23).}$$

It is at once seen that w_a has the form of a descending exponential while v_a has the form of an ascending one provided c is positive. The expansion of v_a in this case simply reflects the fact that as we examine organisms from later and later age groups we find them not only still 'as good as new' in viability but at the same time displaying more and more enlarged reproductive capacity, so that the expectation of offspring to a still living organism is continually increasing. On the other hand, w_a must always have a descending form if it exists at all for otherwise the integral of the fundamental equation (1) would not converge, and the stable population theory would be inapplicable. From the point of view of biology the range of cases where it is possible to find a root m to satisfy the equation is quite wide enough. (Very hypothetical organisms are easily defined whose characteristics take them outside the range of the mathematics of the present discussion. An example would be one whose fertility increased indefinitely like e^{x^2}.) One of the most extreme may be illustrated from the above model: mortality is zero at all ages and c is positive. (It will be observed that while increases of c always reduce senescence, the same is not always true for increases in m. If the increase in m is caused by increases in l_α or f_α or by decrease in α, senescence is actually enhanced.) We have

$$e^{-m\alpha} l_\alpha f_\alpha = m - c.$$

It is easily seen that there is still a real positive roof for m, and that w_a retains its descending form.

The circumstances of this organism have to be imaged to be such that although individuals increase their fecundity exponentially, and continue to do so indefinitely, the population has still met with no checks to increase whatsoever, so that all its members are immortal. It is striking to find that even under these utopian conditions selection is still so orientated that, given genetical variation, phenomena of senescence will tend to creep in. The form of w_a, which is the same as the $g_1(a)$ discussed earlier, shows unequivocally that any mutation causing an improvement in early fecundity at the expense of an equal detriment later will give a raised Malthusian parameter, so that the mutant form will gradually come to numerical preponderance in the population; and if we allow any incipient incidence of mortality we likewise see that selection will favour resistance to it at early ages to a certain extent at the expense of greater vulnerability at later ages.

Thus it may be stated that *for organisms that reproduce repeatedly, senescence is to be expected as an inevitable consequence of the working of natural selection.*

There is no need to emphasize that senescence, or rather a tendency to complete exhaustion by the reproductive effort, and consequent death, is implied *a fortiori* for all organisms with reproductive schedules more limited than the one we have been discussing.

In the last example we were concerned with an extreme and impossible case. Returning to a condition of the model which is biologically possible, indeed the only one which is permanently possible, consider the case $m = 0$. Equation (24) becomes

$$w_a = e^{-l_\alpha f_\alpha (a-\alpha)}.$$

Hence tendency to senesce should be weakest if matters are arranged such that $l_\alpha f_\alpha$ is as small as possible: immature survival should be low and starting fertility low. In other words, immature mortality should be high and reproduction should not begin suddenly. From (23)

$$l_\alpha f_\alpha = \mu - c;$$

therefore the growth rate of fertility would be unimportant if it were balanced wholly by a general increase in adult mortality, but if the rise in c causes any rise in immature mortality as well, as is very likely, it will tend to reduce the rapidity of senescence.

The preceding equations support some of the points made by Williams[1] about apparent connections between the phenomena of survival and reproduction and the rapidity of senescence. In saying this we are applying the present model outside its strict field, to organisms that already do show senescence. But arguments such as were used by Williams seem to show that conclusions of similar general nature would be obtained if the present model could be reworked with μ replaced by some increasing function of age. Thus it seems justified to use the model to emphasize certain necessary connections which Williams says little about.

If $c = 0$ we have Williams's point that high rates of adult mortality should lead to high rates of senescence. The illustration which he gives in this connection, some evidence that birds and bats have low rates of senescence compared to flightless birds and mammals, is certainly striking. It is likely, however, that high levels of adult mortality are usually more a consequence of high fertilities than of degree of adaptation in adult life. Bats are very much less fecund than rodents. Among species that do not have a special infant mortality (e.g. as implied in section 9(b), dispersive animals which lay their eggs separately) it should be found that the highest rates of senescence accompany the highest fertilities.

Regarding Williams's third point we have already shown that an increasing schedule of fertility does not necessarily retard senescence since it can be wholly balanced by increase of μ (in the model at least); but if it is even partly balanced by increase of immature mortality his point holds, and this is very likely to be the case. It does seem likely at least with the molluscs, crustaceans, and poikilothermic vertebrates which he cites; we shall shortly refer to indications of high 'infant mortalities' in the first and last. It may seem rather contrary to the previous paragraph that these long-lived animals are so fecund. But it may be that the state of gradually increasing fertility could only be established if the progeny are relatively small at birth and therefore numerous.

8. NOTES ON THE COMPUTED EXAMPLES

As a further preliminary to discussing the fit of our *a priori* concepts to the reality, graphs illustrating the various functions for certain real cases are given. The notes of this section are necessary to explain the choice of data for these graphs and the computational procedures, and also to give some cautions as to generality. They may be skipped by the reader only interested to follow the main theme of the paper.

Much the best data are available for the human species. The physiological phenomena of the human lifespan must have become established very nearly in their present-day form long before the advent of any civilization, and during a period when cultural advance was on the whole slow. The statistical manifestations of these phenomena, however, both as regards senescence and expressed fertility, are being changed very rapidly by modern cultural advances, as is well known. Therefore in attempting to reconstruct the forces of selection that were operative in the stem of hominid evolution it is appropriate to start from a present-day population whose economy is as backward as possible.

Unfortunately there are no very good data for contemporary peoples in a pre-agricultural phase, nor even for those with the most primitive forms of agriculture.

The data which were collected by the Japanese for the Chinese farming population of Taiwan about the end of the last century seem to be about the best available for the present purpose. The general level of mortality was extremely high, one-half of those born being dead before the age of 26. The fertility was also high with an average of about seven children per woman living right through the reproductive period. This fertility was sufficient to give a positive rate of increase over 1 per cent per annum. At the time the influence of modern contraceptive methods must have been quite negligible.

Specifically, the basic data used in our 'male' and 'female' computations were obtained as follows:

(a) Male and female survivorship schedules (l_x)

Life tables for 1906 are given in Barclay.[8]

(b) Female fertility rates (F_x for $x = \frac{1}{2}, 1\frac{1}{2}, \ldots$)

The data collected by Tuan[9] have been used. They refer to a local sample of women from a farming population whose fertile period centred about the same time as the life table was made. When combined with Barclay's l_x these fertilities lead to a per annum rate of increase of 1.7 per cent, whereas Barclay gives 1.1 per cent for the general population. The interpolation to obtain the quinquennial F_x for $x = 1, 2, \ldots$ was performed by applying Lotka's[10] algorithm to the net-reproduction schedule. Though somewhat tedious to compute and perhaps unnecessarily elaborate in view of the accuracy of the basic data in this case, this method does ensure a net reproduction rate (NRR) and rate of increase (m) unchanged from those computed from the more conventional formulae based on the half-birthday pivots.

(c) Male fertility rates (F_x)

These do not seem to be available for the population in question and so a hypothetical schedule was made up as follows. Its basis is the set of male fertilities given for 1950–53 in Taiwan in the *United Nations Demographic Yearbook, 1954*. This lumps all births occurring to men over 45. Therefore from the age 45 onwards we have grafted the curve obtained from some data for men in Poona City, India.[11] By comparing the female fertilities for 1950–53 with those indicated by Tuan's data for women about 1890 (the data used) we see that an important change is that the early rise in fertility has been postponed by about 1.5 years, presumably due to an overall latening of marriage. We have therefore carried the early rise for the males forwards through an approximately equal age interval. Fertilities for quinquennial age-points 3, 4, 5, ... were measured from the curve so obtained. When combined with the appropriate l_x schedule the net reproduction rate was found to be fairly close to that of the females. The final adjustment was to multiply through the F_x by a factor, found by trial and error, which makes the Malthusian parameter equal to that of the females ($m = 0.017$). There is in fact no good reason to suppose that the 'male' and 'female' rates were exactly equal;[12] but this assumption is convenient for the present elementary discussion.

Some of the most important parameters summarizing the data obtained in this way are given in the following table (where GRR is gross reproductive rate).

	GRR	NRR	W (years)	*w* (years)
Females	3.59	1.62	28.0	28.8
Males	4.09	1.69	30.8	31.9

Besides the gross and net reproduction rates, two measures relevant to the length of a generation are recorded: W has already been defined and w is the average age at childbirth,

$$w = \frac{\sum_1^\infty x l_x F_x}{\sum_1^\infty l_x F_x}.$$

Graphs of the basic data are given in Fig. 3.1a (survivorship) and 3.1b (fertility, or 'gross reproduction').

Figure 3.2a shows forces of mortality as calculated from Barclay's life tables. The values plotted were $\log l_x - \log l_{x+1} \simeq \mu_{x+\frac{1}{2}}$ for the quinquennial survivorships, but a smooth curve has been drawn and the scale shows rates per annum.

The way in which the basic l_x and F_x are used to obtain the $g_i(x)(i = 0, 1, 2)$ is shown for the Taiwanese females in Table 3.1. Figure 3.2b shows the schedules of the $g_i'(x)$, which are the $g_i(x)$ of the table divided by W, in this case 5.6048 (quinquennial age units).

The average age at marriage for the Taiwanese women in question was about 19.[9] No doubt this is rather too late to be typical of a pre-agricultural people. Probably throughout most of our ancestry in the palaeolithic marriage has been, at least in the scarcer sex, a fairly close sequel to puberty just as it continues to be for girls in most present-day food-gathering peoples. In order to produce a hardly-growing population with a fertility beginning even earlier than that shown in Fig. 3.1b we should have to postulate a yet higher level of infant mortality, although the possibility that with more savage conditions

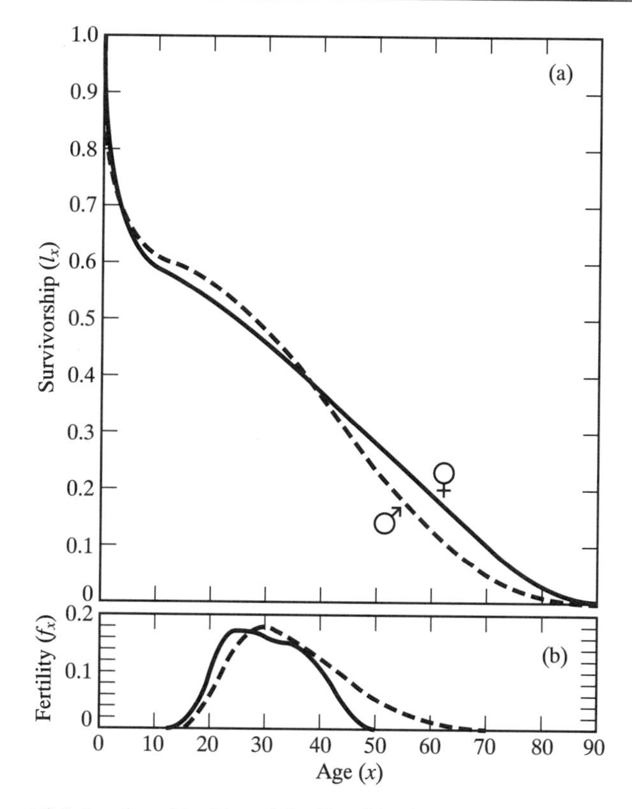

Figure 3.1 Survivorship (a) and fertility (b) of the Taiwanese about 1906.

fertility might actually be lower overall, either intrinsically or due to malnutrition or some such factor, is one that should be borne in mind. Such adjustments will in any case produce only minor changes in the picture of the selective forces which is given in Fig. 3.2b.

Table 3.1 also shows the computation of Fisher's reproductive value, v_x, and apart from their multiplication by a standardizing scale factor it is the values from this table, and the corresponding one for males, which are graphed in Fig. 3.3a. It would be digressing too far from the present theme to describe the standardizing procedure in detail, but the basic principle may be stated. It is to select v_0 *(male)* and v_0 *(female)* so that reproductive value has unit mean when the average is taken over all individuals, by age and sex, as they are distributed in the stable population. Thus for a stable population a scoring of value will give the same total as a count of heads; any particular value taken from our graph is, therefore, in effect, a comparison of a person of that age and sex with the 'average person', or population average, which is valued at unity. Included in the comparison, of course, are all persons who are past reproduction and

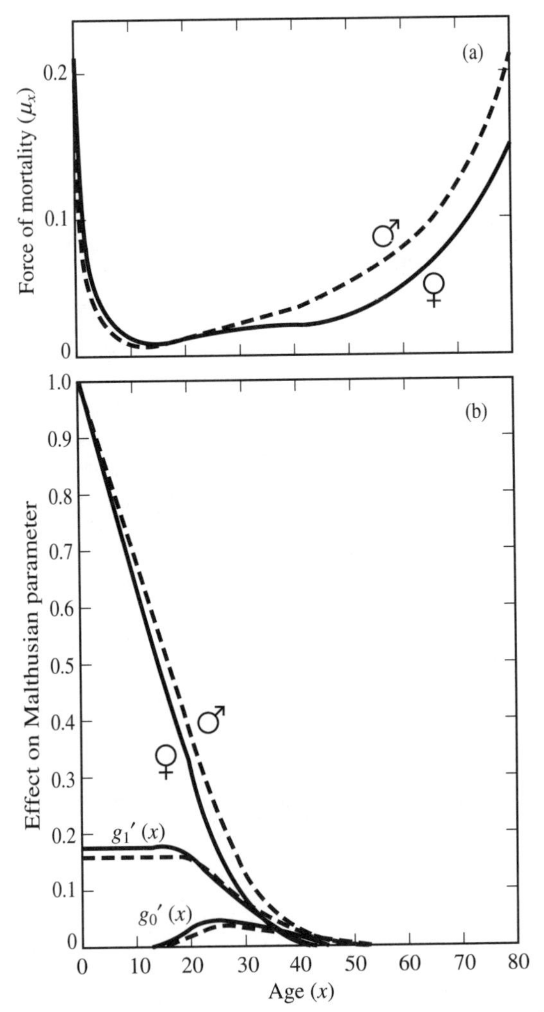

Figure 3.2 The age pattern of mortality (a) and of certain forces of selection acting on mortality (b) for the Taiwanese about 1906.

therefore have zero values; improving post-reproductive survival will therefore inflate the standardized curves without altering their shape. This may seem undesirable for a standardization, but considering how little even present-day medicine has been able to improve survival in old age the prospect of much inflation from this cause is rather remote.

The standardization dictates that the ratio v_0 (*male*): v_0 (*female*) shall be the inverse of the sex ratio at birth.[13] Wanting better information a sex ratio at birth of 105 males: 100 females has been used. There are some hints that a higher ratio would have been justified for the population in question.[14,15]

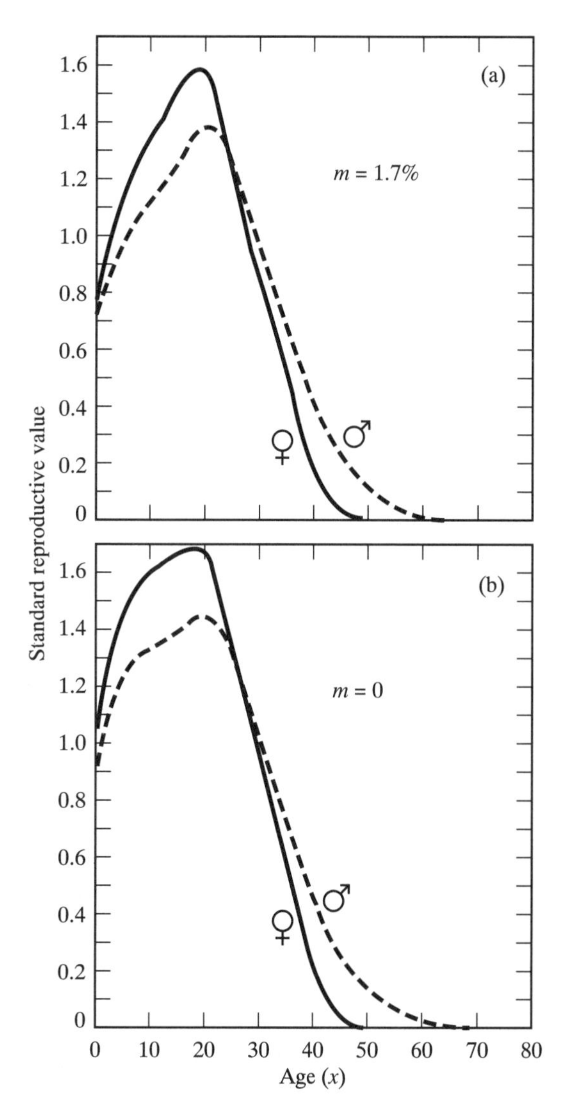

Figure 3.3 Reproductive values for the Taiwanese about 1906 with the rate of increase as observed (a) and hypothetically zero (b).

Figure 3.3b shows the curves as they would be if the fertilities were all reduced in the same proportion and just enough to give a stationary population. Figure 3.3a shows them as they are for the calculated rate of increase of 1.7 per cent per annum.

The age at which the male and female curves cross depends very largely on the differential mortalities of the sexes. It may be mentioned that the age of crossing shown is considerably later than it would be for most other demographically known populations. Before the beginning of reproduction by either sex the curves must proportionally follow reciprocals of the relative frequencies of the sexes and therefore if they crossed in this age range they would do so at the age when numerical equality was attained. If this age is after reproduction has begun, as it usually is, the crossing tends to occur earlier, owing to the tendency of females to run through their reproduction more quickly. This is the case with the data under study: reproductive values equalize round about 25, but males remain commoner than females until almost 40. This is unusually late. On comparisons available to him Barclay thought it anomalous that males were surviving much better in childhood (ages 1 to 10) than females, but more recent studies have shown Taiwan was not so unusual in this respect among countries of high mortality.[16] In our data the age of life-table equality seems to be extra late due to a famine mortality in 1906 in which females again fared worse; thus for 1910 the age is about 5 years earlier.

It is fair to say that as a whole the data given by Barclay, relevant as it is to a long-established social economy, cannot be reconciled with Fisher's theory of the sex ratio. It seems possible that we have here a case where a human cultural factor, the emphasis on maintaining a male line of descent, with its resultant preferential treatment for male children, has balanced the sex-ratio selection at some distance from its natural equilibrium. A Taiwanese couple could reasonably expect a greater total number of grandchildren if they concentrated on supplying the existing deficiency of females; but they show no inclination to do so.

Because curves of reproductive value have very seldom been illustrated a set applying to an animal very different from man, the crustacean 'water flea', *Daphnia pulex*, are given in Fig. 3.4. The data used for these was published by Frank, Boll, and Kelly.[17] It has the unusual advantage of comprising a series of tables for more and more unfavourable conditions of crowding, so that it can be used to show the effect of changing rate of population increase on the schedules under discussion without having to resort to the sort of *a priori* assumption about changing quantity with unchanging pattern that was used to produce Fig. 3.3b. Some idea of how justified such procedures are likely to be can perhaps be gained by comparing the shapes of the basic schedules as illustrated by the above authors; but at the same time even these life tables are not to be considered very natural; the authors disclaim that they should be and one likely source of artificiality will be mentioned later (section 9(b)).

These data refer to asexual reproduction of *Daphnia*. The selection on the form of the life schedules supposed to be taking place will operate mainly between different clones and to some extent between different branches of a clone when these differ by a mutation. Thus, while progress per generation may be slower, it should still be of the same kind, and in fact the overall similarity of such life curves to those of humans is an indication that it is so.

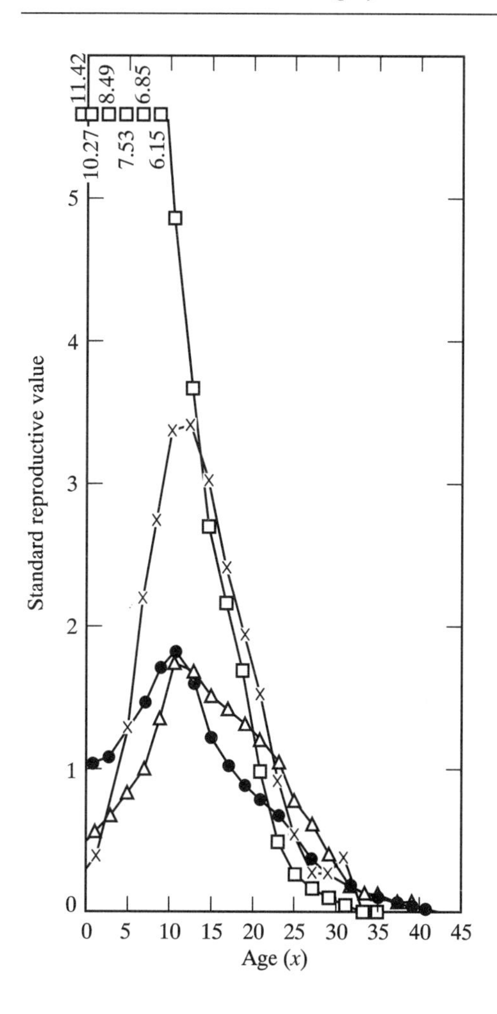

Figure 3.4 Reproductive values of *Daphnia pulex* in similar cultures with various degrees of crowding: ×, 2 *Daphnia*/cm³; △, 16/cm³; ●, '23'/cm³; □, 32/cm³.

The data cover a wide range of conditions, from one which gave a very high rate of increase (GRR = 64.3 and doubling time 2.4 days) to one which gave a high rate of decline (GRR = 0.23 and halving time 6.1 days). We illustrate the following cases:

1. 2 *Daphnia*/cm³. This was the second lowest density and gave the second highest rate of numerical increase but actually the highest rate of increase in biomass. The irregularity in the tail of the curve is probably not a biological phenomenon.[18]
2. 16/cm³. This density gave the most moderate positive increase.
3. '23'/cm³. In order to obtain a curve for a strictly stationary condition of population an artificial device was used. It is evident from the remarkably straight regression of m on numerical density that the density that would have given zero increase lies between 16 and 24/cm³, but closer to the latter.

Weights were therefore calculated by which the NRR's for these densities must be multiplied so as to give an arithmetic mean of unity. The survivorships and net fertilities ($l_x f_x$) were then 'mixed' in proportions corresponding to these weights so as to give a joint table. Naturally it turns out to differ only slightly from that for 24/cm^3.

4. 32/cm^3. This was the highest density used and gave the most rapid decline. Because of the preponderance of old organisms in the stable age distribution in this case and of the fact that their reproductive value is low both by reason of the small amount of reproduction left to them to run, and of its immediacy which in the declining population disvalues it, the standardization factor greatly inflates the curves as a whole, so that only a part can be shown. The part not shown is described by writing the values at their correct ages along the top of the figure.

9. THE MOULDING OF SENESCENCE

To what extent and in exactly what way life schedules will be moulded by natural selection depends on what sort of genetical variation is available. This was strongly emphasized by Williams, who was careful to outline characteristics of the genes which he felt necessary for the working of his ensenescing process, to show that some genes of the right type certainly do exist, and to argue that their general abundance was reasonably certain. The genes imagined were supposed to be pleiotropic in their effects on fitness with the partial effects occurring at different ages. In particular, Williams postulated a type of gene which has its good effects in early life offset by bad effects later on. The present author agrees as to the importance of the part which such genes must have played in the evolution of senescence, and has already found it convenient to refer to such possible pairs of effects in section 7. He feels, nevertheless, that Williams may have been unnecessarily restricting the scope of his theory by making it depend upon genes which are pleiotropic in the ordinary sense of the word. In a certain sense all genes which are age-specific in their action are pleiotropic, simply by the fact that they cause positive effects at some ages and null ones at others. If the net positive is a benefit, the following events may be expected: the mutant will spread, the spread will be accompanied by a rise in population, and this rise will in time be checked by density-dependant adverse factors of the environment which bear on the life schedules, raising mortalities or reducing fertilities in a pattern that has nothing to do with the particular pattern of the positive effect which initiated the sequence. When population has been made stationary again, the overall result will tend to be that fertility has been down-graded and mortality up-graded all along their length, so that the beneficial effect is itself slightly down-graded and all the null effects now appear as slight disadvantages.

This suggests the following picture of the evolution of the curve of force of mortality. It is continually being 'nibbled' from above, the nibbles representing the spreading of more or less age-specific advantageous mutations through the population. They may be closely age-specific or they may involve a lowering of the curve along a strip of considerable length. Following each nibble the whole curve, after more or less delay, makes a small ascent. The delay corresponds to the period of increasing population: the ascent to the coming into operation of the Malthusian checks to increase. The nibbling takes place fastest at the left-hand end of the curve and towards the right-hand end finally with infinite slowness at the age where reproduction ends. The greater the speed with which nibbles occur the sooner they can be succeeded by others. Thus the irregular downward movement is occurring fastest at the left-hand end, and the compensatory general upward movement results in a kind of dynamic equilibrium in which the curve trails upward indefinitely at the right. In the absence of complications due to parental care or other altruistic contributions due to post-reproductives, the curve should be roughly asymptotic to the age of the ending of reproduction.

The implication of this view that rising late mortality is due to severity of the environment in conditions of overpopulation rather than to intrinsic senescent changes does not square well with the fact that when the environment is made as favourable as possible deterioration is still observed. Therefore it seems evident that Williams's pleiotropic effects certainly do exist, even if the late effects—the intrinsic changes of senescence—are, as many writers have suggested, of negligible importance to the wild organism because hardly ever expressed.

Whichever view deserves most emphasis the general hypothesis accounts well for some features of the μ_a curve; for example, the fact that its minimum does come near the age where reproduction begins and the fact that it does rise indefinitely with increasing age, but still fails conspicuously over certain important features. We consider the most striking discrepancies under the next two subheadings. Under the third a few points will be made on the implications of the theory for the moulding of fertility. Under the fourth, finally, it will be considered whether the covering of non-zero Malthusian parameters in our analysis has any practical value for interpretations of patterns of senescence.

(a) Post-reproductive life

It is evident that the rise in mortality in the later reproductive ages of humans is by no means asymptotic to the age at which reproduction ends; the indefinite rise comes too gradually and too late. This is particularly evident from the curves given for the Taiwanese women, where the rather definite age of the menopause seems to be conspicuously ignored by the as yet gently rising curve of force of mortality. It is, moreover, a matter of common knowledge that the post-menopausal woman normally remains a useful and healthy member of the community for some time. A woman does sometimes live to twice the age of her menopause. The theoretical predictions come much closer to being realized with the men. But again it seems that the curve is not rising fast enough in ages where the chances of fatherhood have become quite negligible.

As remarked by Williams, an obvious excuse for this discrepancy is to be found in the factor of parental care. The vigour of the post-reproductive adult can be attributed to beneficial effects of continued survival on the survival and reproduction of descendants. In fact, the 15 or so years of comparatively healthy life of the post-reproductive woman is so long in itself and so conspicuously better than the performance of the male that it inevitably suggests a special value of the old woman as mother or grandmother during a long ancestral period, a value which was for some reason comparatively little shared by the old male. In view of the weakness of selection on post-reproductives, however, no great weight is attached to this indication; it seems quite possible that the effect is non-adaptive and a mere side effect of differences in sexual physiology established by the more powerful selection on younger age groups.

It would be interesting to be able to make similar comparisons for other mammals in which parental care is realtively shorter. Krohn[19] mentions slight evidence that post-menopausal life may be relatively shorter in female monkeys than in women. The data of Leslie and Ransom[20] for the vole, *Microtus agrestis*, indicate less post-reproductive life. Figures of unstated source in a table by Bodenheimer[21] show the rat, *Rattus norvegicus*, as having a longer post-reproductive life than humans, but a contrary statement that it is shorter by Weisner and Sheard[22] who were actually working on the life table of the rat is doubtless more reliable.

Aphis fabae equals the human female in the potential length of its post-reproductive lifespan (G. Murdie, personal communication; Banks and Macaulay[23]). Although aphids do not have any obvious post-natal care for their offspring it could well be that the presence of the old parent in the group does have some beneficial effect. It remains in contact with the food supply and does develop in a characteristic way. The data of Barlow[24], consistent although not very naturally obtained, seems to show that *Myzus persicae* has a post-reproductive period while *Macrosiphum euphorbiae* does not. Mr W. O. Steel has pointed out to me the fact that *M. persicae* is one of a group of species, which also includes *A. fabae*, which adhere tightly to their food plants and are slow to leave them when disturbed, whereas *M. euphorbiae* belongs to a group in which there is a rapid escape reaction. If aphids fall off a plant in a shower and then climb back on it, clones will tend to be mixed in the aggregations that reform, and the suggested basis for the evolution of post-reproductive life will have vanished. At the same time there certainly are exceptions to the rule, if it is one at all, that readily falling aphids do not have a post-reproductive life. Dr G. Murdie has drawn my attention to the very interesting situation with

Acyrthosiphum pisum. In general behaviour this aphid tends to the 'escaping' more than to the 'adhering' group, so that the fact that it does have a post-reproductive period seems rather against the hypothesis that the existence of such a period has to do with parental benefits. But information given by Lowe and Taylor[25] shows that within this single species there exist two strains, probably differing only by a single dominant gene, which themselves differ in tending to the 'adhering' and 'escaping' conditions respectively, and that between these strains the predicted correlation holds very well. Both strains finish reproducing at about the age of 17 days whereas total lengths of life average 27 and 20 days, respectively.

With *Daphnia pulex* comparison of the life-schedule figures given in Frank *et al.*[17] with those given in this paper for humans shows that though there seems to be some post-reproductive life it is relatively shorter.

(b) Infant mortality

None of the schedules of forces of selection shown in Fig. 3.3 can account for the rise in mortality in younger and younger pre-reproductive ages. There is no justification for the idea that the simple forces of age-dependant selection would drive early deleterious effects to occur earlier. Only our 'prolonged mortality changes' are driven at all in the immature period and these, if deleterious, are driven to occur later.

It is again suggested that in the human case parental care brings the necessary trend. The whole set of circumstances in which it is envisaged such a trend could work is, however, by no means exactly coextensive with parental care; it is better described as involving in all cases some possibility of *sibling replacement*. It is clear that human parents do possess some ability to replace offspring which die while still dependant. To mention the least subtle of the means which can contribute to this we have the inhibition of ovulation during lactation. For later deaths, even if there is no conscious or custom-based birth control by the parents (which cannot be taken for granted even in the most primitive societies), there is the possibility of replacement of a more probabilistic kind, through the other siblings being better cared for and surviving better in consequence. (It must be admitted that there is no evidence on the effectiveness of this or the preceding factor. Evidence that immediate replacement occurs in a *modern* society seems to be slight.[26]) It is obvious that there is better chance of nearly full replacement if an offspring dies of a congenital disorder in infancy than at age 10, and better again if it dies in the earliest stages of pregnancy. Thus modifiers of recurrent lethal mutations of any kind which cause them to express themselves earlier will be selected.

This is an easily understood case, but evidently the field of sibling replacement must be shown to be much wider if it is to account for anything like the whole of the phenomena of infant mortality.

Suppose that the catching of a disease in the immature period is inevitable and that the first infection has only two possible outcomes: death or survival with perfect immunity against second infection; and suppose that the probabilities do not change according to age of first infection. Then if there is any degree of sibling replacement at all, a gene bringing forward the expected age of first infection will be selected, for it can easily be seen that the more commonly the gene is appearing in a progeny the larger its expected completed size will be, while at the same time the expected frequency of the gene within the progeny is unchanged.

If the bringing forward of the susceptibility involves a disadvantage, for instance by a slight increase in the chance that death is the outcome of infection, the situation is more complex and will require mathematical analysis to delimit the possibilities. This is because it is then no longer true that the proportion of the gene is unchanged by the amount of replacement that goes on; positive selection could fail even if it was guaranteed that the amount of replacement could more than compensate for the extra mortality. This is in effect a problem of a more 'altruistic' versus a more 'selfish' trait (see the paper in Chapter 2, part I).

Analytical treatment also seems necessary before much can be said about the age at which the minimum of mortality is expected. The actual age of this minimum in humans was a point which raised difficulties for the erroneous theory that it was connected in some way with the reproductive schedule (see section 1, and Comfort[27]). Comfort quotes 10 to 12 in the male and Barclay's curves show about 12 to 13 for both sexes of the Taiwanese. Even with the Taiwanese the great majority have not begun to reproduce by 13, and probably most of them could not do so. On the present view this point creates no absolute difficulty, although the lateness of the minimum does seem surprising.

The hypothesis that infant mortality in humans has been evolved through sibling replacement would be strengthened if traces of a deep-rooted motivation to assist the process could be demonstrated in human parents. If the process works biologically, or has worked, to find this would be no more surprising than to find parents motivated, as they are, to care for their more promising offspring. In modern societies the effect can be expected to be masked to a very large extent by learned ideas, but even so there is evidence that it exists. The birth of a malformed baby produces distress and sometimes actual repugnance in its mother. From what is known of the near universality of infanticide in more primitive societies, it seems likely that in former times repugnance was the natural prelude to infanticide or abandonment.

The sort of replacement discussed so far, by the parent, would not be possible in all cases of parental care. For example, it could not occur in an insect which lays only one batch of eggs in its lifetime even though it has a post-reproductive life during which it guards the eggs. Even if such an insect could proceed to a second batch of eggs it is not likely that the second could be any larger or occur any sooner if the first had a high mortality. However, the possibility of another kind of replacement would be present in most of such cases and also in many others where parental care is absent: it only requires some sort of batching of the eggs which results in the early competition being mainly between siblings. A few references to cases where severe sibling competition occurs were given in a previous paper (see Chapter 2, Part II); to these may be added the striking embryonic cannibalism that occurs in the egg-cocoons of certain prosobranch molluscs, for example the whelk, *Buccinum undatum*.[28] Parsons[29] has suggested that competition causes the elimination of some congenitally defective foetuses in the mouse. Cases of milder competition must be extremely widespread.

It is often assumed that a high fecundity in an animal or plant implies a high infant mortality. It is reasonable to expect that infant mortality will be high in an absolute sense. But whether it is high relative to what comes later, which is the phenomenon we are concerned with, is very seldom known. Searching among published life tables it is difficult to find any that extend back reliably to age zero, and when they do they usually apply to animals reared carefully with a view to the study of the lifespan. However, whereas most of the invertebrate survivorship curves illustrated by Comfort[27] show no or little infant mortality, the curve he reproduces for voles, equally reared in captivity, shows the same sort of high infant mortality as is found in humans.[30] The only invertebrate curves given that are at all similar to this are those for a spider and a snail and these have just the sort of egg-laying habits that might bring the factor of sibling competition into force. Most aphid life tables show negligible infant mortality, but rearing conditions have usually been highly protected and hygienic; it would be interesting to know more about mortalities in the field, since apart from the fact that early deaths of recessive segregants can play no part in this case, conditions would seem generally right for the development of infant mortality. The only life tables known which have any consistent trace of it are Barlow's of 1962[24] for *Macrosiphum euphorbiae*. The infant mortality expected for intracompetitive groups of siblings might not appear as such in the life table and yet might be present in the sense that some members of each group have become stunted and are slowly dying, or at least have lost all prospect of reproduction apart from the remote contingency that they escape from the crowding. Situations of this kind seem to be quite widespread among aquatic organisms, ranging from coelenterates to fish.[31] The *Daphnia* data[17] we have been using may present a case in the 'late childhood' mortality which increasingly appears at the three highest densities. But the fact that the data refer to artificial isolated cohorts may also partly account for this: under the constant food regime the young *Daphnia* may have 'grown into' competition as their needs and capacities for feeding increased in a way that would not happen in a natural population of the same numerical density and culture condition. In any case it is unlikely that self-sacrifice by the smaller individuals is involved in all the situations cited by Rose, for unless the early competition is wholly between sibs the phenomenon could be simply an adaptation for the inter-sibship struggle, whereby the earliest batches of young successfully inhibit those hatching later.

Of the guppy, *Lebistes*, a fish which lacks parental care, Comfort[32] says the survival curves are not very different from those for a small mammal under laboratory conditions; but since the curves do not extend back to age zero and as far as they go show no special infant mortality it may be doubted whether his remark is intended to apply to this part of the curve. If there is any infant mortality it is probably restricted to high densities and is largely due to cannibalism by the adults.[31] However, fish in general seem to be rather against the hypothesis, since it is hard to believe that the very prolific fish mortality can be anything like as high in adults as it is in the vast crowds of the very young; and there are statements that infant mortality is relatively very high.[33] Clearly in this respect it will be of great interest to be able to compare infant mortalities in related species of fish that either cluster or disperse their eggs, or mix them in a common spawning ground.

(c) The moulding of the fertility schedule

Although changes of fertility with age are an aspect of senescence, and as has been shown a most important one, it is not intended to discuss this side of the process in any detail. From its dependence on the organism's ability to accumulate the reserves necessary for reproduction, which cannot be done at unlimited speed even in the best circumstances, the problem seems to be biologically more complex than that concerning the mortalities: in other words the question of what genetic effects selection can be working on is more stringent. For example, it is not so plausible that a gene could simply add an element of fertility at a given age without affecting the rest of the schedule as it is that a gene might cause the elimination of a single element of mortality. Materials that went into the extra fertility would have to represent either materials 'saved' from potential earlier fertility or a subtraction from growth capital which would tend to cause all subsequent fertilities to be lowered.

Nevertheless, it is felt that obtaining the schedule of forces moulding the age-specific fertilities by arguments analogous to those of section 3 may give a useful preliminary view of the problem. They are found to have the form of the stable age distribution, being

$$\frac{\mathrm{d}m}{\mathrm{d}F_a} = \frac{\lambda^{-a} l_a}{W}. \tag{25}$$

For a stationary population this reduces to

$$l_a/w.$$

Accepting that reproduction cannot be indefinitely brought forward without affecting its quality, this schedule suggests why the highest fertility occurs shortly after the onset of reproductive ability. But with a population that is decreasing sufficiently rapidly the stable age distribution becomes negatively skewed, or 'undercut'. Under such circumstances it is not advantageous to have offspring as early as possible; they are best scheduled for the age which is modal for the stable age distribution. With the *Daphnia* data only '32/cm^3' culture implies an undercut stable age distribution; but it is the '8/cm^3' culture which gave both the highest mean age, and the highest standard deviation, of fertility for the whole series, despite the fact that increase was still very fast at this density (GRR = 33.2 and doubling time 3.6 days); with the higher densities the mean scheduled age of reproduction declines steadily. Thus there is virtually no fit to our expectations.

The estimates of skewness and leptocurtosis for the same data show even less consistency than the estimates for the standard deviations; the data are clearly inadequate for indicating any trend in these parameters. Regarding the theoretical tendencies of the higher-order characteristics of the fertility distributions, however, it is worth noting an easily obtained mathematical result which shows that we should beware of supposing that positive skewness of fertility is advantageous under conditions of increase and negative skewness under decrease. With the net reproduction schedule $l_a F_a$ the opposite is true, for if we differentiate m in the cumulant expansion of the fundamental equation[34] with respect to the nth cumulant κ_n, of this schedule while all other cumulants are held constant we obtain

$$\frac{\partial m}{\partial \kappa_n} = \frac{(-m)^n}{n!} \cdot \frac{1}{W}.$$

For all odd cumulants the sign of this expression is the opposite of that of m. Thus if there existed genes capable of changing the shape of the net reproduction schedule without altering its quantity or mean position, it would be only during periods of population *decrease* that positive skewness would be selected. This is easily understandable when we remember that positive skewness involves the relative lengthening of the late tail of the distribution and that in the decreasing population late actual reproduction is advantageous. Such effects are, of course, extremely hypothetical. Equivalent distortions of the fertility schedule itself would be just slightly less so, but their effects are less easy to analyse. If we care to assume constant mortality μ so that $l_a = \mathrm{e}^{-\mu a}$, a similar approach via the expansion of the fundamental equation in terms of the cumulants κ_n' of the f_a gives

$$\frac{\partial m}{\partial \kappa_n'} = \frac{\{-(m+\mu)\}^n}{n!} \cdot \frac{1}{W}.$$

Thus tendencies to negative skewness of the f_a are favoured even when $m = 0$. Presumably genetic effects on skewness of the kind supposed are impossible. The bringing-forward tendency on the distribution as a whole (i.e. on the mean, κ_1') even when $m = 0$ is, of course, consistent with the effect indicated by equation (25). The fact that positively skewed fertility schedules are so commonly observed must mean that for developmental reasons extending the tail of the schedule in this direction is generally impracticable.

(d) Permanently fluctuating populations

Aphids and *Cladocera* are examples of a type of organism for which the generality of the mathematical treatment using the Malthusian parameters might be of direct value. Their numbers in any population range very widely. Mr M. J. Way tells me that with certain species of aphids the seasonal range is often of the order of 1000-fold; but that neither the build-up of population to the peak nor the subsequent decline is at all simple or even monotonic. The same sort of situation has been inferred from studies on laboratory populations of *Daphnia*. Slobodkin[35] concludes that an intrinsic tendency to population fluctuation means that 'It is extremely unlikely that *Daphnia* ever exist in an equilibrium in nature', although given time in an unchanging environment equilibrium could occur. Such species presumably have a long evolutionary experience of fluctuations, seasonal or otherwise, and however irregular they may be, clones whose individuals were capable of reacting by appropriate modifications of their life schedules to clues of the environment which showed that a lush period had begun, or that competition for food was about to become severe, should be favoured over clones which maintained the intrinsic forms of their schedules invariant.

It is obvious in fact that this type of species is specially adapted for the rapid colonization of transitory habitats. With aphids not only are the optimum growth rates very high compared with those of most other insects, but there has been apparent sacrifice of all other kinds of protective adaptation which might impede immediate fecundity. The phenoplastic changes found at the extreme of overcrowding are also clearly adaptive; typically special emigrant or dormant forms are produced. The present quantitative analysis with its basic assumption of like producing like cannot cover the abrupt changeover to production of a different form. Nor has it been found to give any special points about the periods of expansion which existing data could be expected to confirm. Speeding up of the life cycle to secure early reproduction is one thing certainly anticipated, and it does occur: some evidence has already been seen in the *Daphnia* data and Comfort[36] mentions several other cases. But this follows also from general growth conditions on the purely mechanistic theory that senescence is some cumulative effect of somatic metabolism; we cannot say more than that it also fits in with the evolutionary view.

In general and until more comprehensive life tables are available, it seems that for most species it will be quite adequate to discuss the forces of selection moulding senescence in terms of simple expectations of offspring, as was done by Williams.

Acknowledgements

The work presented in this paper was mainly carried out during tenure of a Medical Research Council Scholarship at the London School of Economics and at University College London. The author thanks Mr J. Hajnal for much helpful discussion and advice.

References and notes

1. G. C. Williams, *Evolution* **11**, 398 (1957).
2. P. B. Medawar, *An Unsolved Problem in Biology* (H. K. Lewis, London, 1952).
3. P. B. Medawar, *Ciba Foundation Colloquia on Ageing* **1**, 4 (1955).
4. R. A. Fisher, *The Genetical Theory of Natural Selection* (Dover, New York, 1958; first published 1930); here p. 29.
5. A. J. Coale, *Population Studies* **2**, 92 (1957).
6. A. J. Lotka, *Elements of Mathematical Biology*, p. 111 (Dover, New York, 1956; first published 1924).
7. L. C. Cole, *Quarterly Review of Biology* **29**, 103 (1954).
8. G. W. Barclay, *Colonial Development and Population in Taiwan* (Princeton, Princeton University Press, 1954); here p. 172.
9. C. Tuan, *Population Studies* **12**, 40 (1958).
10. A. J. Lotka, *Annals of Mathematical Statistics* **19**, 190 (1948).
11. V. M. Dandekar and K. Dandekar, *Survey of Fertility and Mortality in Poona District* (Gokhale Institute of Politics and Economics Publication No. 27, 1953).
12. P. H. Karmel, *Population Studies* **1**, 249; **2**, 240 (1948).
13. Fisher 1930 (ref. 4), p. 159.
14. Barclay 1954 (ref. 8), p. 159.
15. Y. S. Kang and W. K. Cho, *Human Biology* **34**, 38 (1962).
16. J. Hajnal quoted in D. V. Glass and D. E. C. Eversley, *Population in History* (Edward Arnold, London, 1965).
17. P. W. Frank, C. D. Boll, and R. W. Kelly, *Physiological Zoology* **30**, 287 (1957).
18. Ibid., p. 296.
19. P. L. Krohn, *Ciba Foundation Colloquia on Ageing* **1**, 141 (1955).
20. P. H. Leslie and R. M. Ranson, *Journal of Animal Ecology* **9**, 27 (1940).
21. F. S. Bodenheimer, *Problems of Animal Ecology*, p. 27 (Oxford University Press, Oxford, 1938).
22. B. P. Wiesner and N. M. Sheard, *Proceedings of the Royal Society of Edinburgh* **55**, 1 (1936).
23. C. J. Banks and E. D. M. Macaulay, *Annals of Applied Biology* **53**, 229 (1963).
24. C. A. Barlow, *Canadian Journal of Zoology* **40**, 145 (1962).

25. H. J. B. Lowe and L. R. Taylor, *Entomologia experimentalis et applicata* **7**, 287 (1964).
26. H. B. Newcombe, *Eugenics Review* **57**, 109 (1965).
27. A. Comfort, *Ageing: The Biology of Senescence* (Routledge & Kegan Paul, 1964); here p. 52.
28. V. Fretter and A. Graham, *British Prosobranch Molluscs*, Ch. 16 (Ray Society, London, 1962).
29. P. A. Parsons, *Nature 198*, 316 (1963).
30. The data of Weisner and Sheard 1936 (ref. 22) on the rat show an infant mortality but its nature is surprising. An estimated 40 per cent of the litters suffered total mortality, apparently through kronism. These averaged four young per litter; the remainder averaged eight and among these mortality was only 6 per cent. Thus under the favourable breeding conditions of this study sibling replacement, within litters at least, seems not to occur.
31. S. M. Rose, *Ecology* **41**, 188 (1960).
32. Comfort 1964 (ref. 27), p. 108.
33. Ibid., p. 154.
34. For example, Lotka 1948 (ref. 10).
35. L. B. Slobodkin, *Ecological Monographs* **24**, 69 (1954).
36. Comfort 1964 (ref. 27), p. 19 *et seq.*

CHAPTER 4

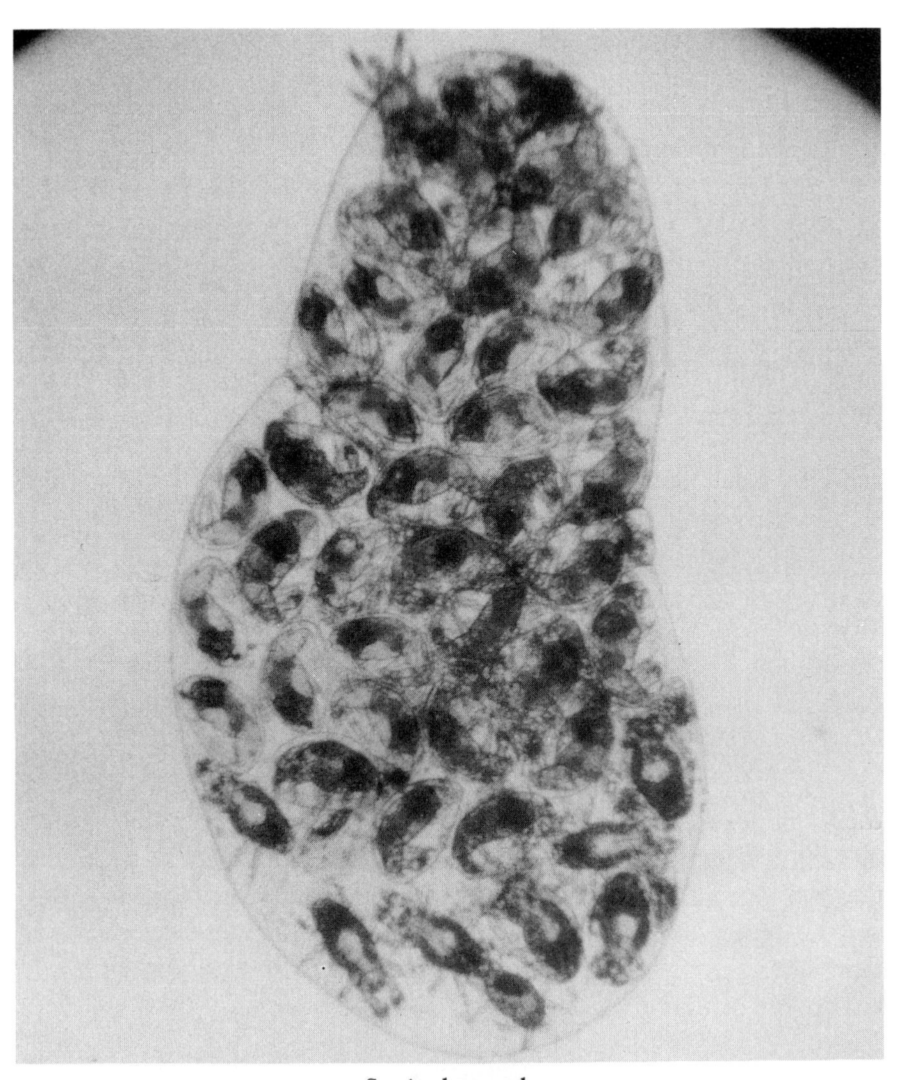

Sex in the womb

Eggs within this inflated mother mite are hatching to give fully formed adults. By far the most will be female but one male is always among the first. In the lowest part of the photograph this first male (the smaller, more oval body) is seen copulating with a sister while another waits nearby. The forelimbs and 'head' of the mother are at the top: prior to transferral into water on a microscope slide for photography, this female imbibed sap from stereum fungus threads in a rotting oak branch. The mother is alive but shortly, in natural course, will burst, liberating her mated female offspring who wander off. Males remain with any eggs that, at this stage, are still unhatched. The form of the first pair of legs of the mother differs from that of her hatched daughters: the latter are a special dispersal morph, their lobster-like fore claws enabling them to cling to the hairs on the legs of insects emigrating from the branch. Varied ratios of disperser and normal females are found in broods, but males are never dispersers.

CHAPTER 4

GENDER AND GENOME

Extraordinary Sex Ratios

None will part us, none undo
The knot that makes one flesh of two,
A. E. HOUSMAN[1]

IN his *The Genetical Theory of Natural Selection* R. A. Fisher included a certain passage about sex-ratio selection that after his death became famous among evolutionary biologists. It is a good example of what may be called his slit-trench style—short and deep. Many have fallen into this trench and been slow either to escape or to feel sheltered there, and I was one. Nowadays the trench catches fewer because there are many other good accounts of how selection works on the sex ratio; indeed, I believe that, in a purely qualitative way, the opening passage of my 1967 paper presents Fisher's idea more clearly than he did. However, by the time I was able to write such a stepwise account, my struggles had revealed to me that there was a lot that Fisher's argument had not covered. I saw that there were at least two kinds of influence that might affect selection of tendency to produce one sex or the other. First, there is the question of what genetic element is imagined to be controlling the sex ratio, and in which sex and at what stage the element acts. Second, there is the question of whether the population might be grouped such that the all-with-all (panmictic) competition assumed in Fisher's argument would not apply. If it does not apply then we particularly need to know about the dispersal movements typically performed by each sex. It turns out that all such issues, either alone or in combination, can change the evolved sex ratio dramatically away from the expectation of equal investment that Fisher had argued for the most straightforward situation.

The following paper was a start in considering such cases from the selectionist (and hence still basically Fisherian) point of view. As usual, forerunners had considered particular cases and had pointed out some of their curious expectations, but again I like to think that my article brought all these and more into a single system—or, re-using my first image, expanded Fisher's central trench to become part of a wide system of others.

I am more proud of this paper than of almost any other I have written. Perhaps this is most of all because I believe it helped sex-ratio theory well on its way to being the section of evolutionary theory that best proves the power and accuracy of the Neodarwinian paradigm as a whole. Brilliant work by others has gone on to show for sex-ratio theory a predictive power almost comparable to what is standard in the 'hard' sciences of physiology, chemistry, and physics. But as to the supposed 'hardness' of the mentioned disciplines, however, I must pause to note here that there is an important sense in which they are actually less hard than ecology and evolution. In population or evolutionary biology it is extremely difficult both to quantify and to understand all of the numerous factors in play to an accuracy that allows quantitative prediction. In addition, the theory making the predictions really has to be more subtle. It may even in some cases, following recent ideas of mathematical 'chaos', indicate accurate prediction to be impossible even in principle. I think a supreme theorist such as Von Neumann, who worked in the hard sciences and at least delved around the fringes of biology and economics (founding, for example, game theory whose concepts appear abundantly in their reshaped biological versions in this and later papers), would readily acknowledge the greater difficulty of these softer sciences and would admit to ending his forays in them with less-defined and testable ideas than he reached, say, with physics. However, with sex ratios we can be more hopeful. Sex ratio based on enough data quantifies through its very nature and often comes ready-averaged over many of its irrelevant stochastic influences as well as over the ordinary incidence of chance variation. We can thus expect that, if the theory is right, it can lead to convincing tests. Such tests have been done.[2–5]

Other recent topics in evolutionary theory to which I believe this paper contributed are:

1. The levels-of-selection debate.
2. The idea of conflict within the genome.
3. The 'evolutionarily stable strategy' or ESS (referred to in this paper as the 'unbeatable strategy').
4. The initiation of game theoretic ideas in evolutionary biology.
5. Finally and more indirectly, by emphasizing the costliness of male production for females and for population growth, as well as the ever-ready 'option' (among small insects, for example) of parthenogenesis, it helped to initiate debate over the adaptive function of sex.

As usual, these contributions were mostly not wholly new. The least definite is probably (1). As I see it now this was because I was still myself somewhat confused on how to handle multiple simultaneous levels of selection. I see nothing I have to retract from my remarks about group and species selection in the section on 'sex ratios and polygyny', but others I give in Note 42 of the paper seem rather simplistic to me now: within a few years, thanks to a singular American about now arriving and becoming known to me in London (and in these accounts to appear in more detail in the next chapter), I was restating what I say in that section much more quantitatively and precisely.

By the time I had written the paper the idea that the group-selection issue might be open to some form of algebraic treatment was developing fast. However, dimly lit by the same light that was now breaking over my nearby foothills, there began to loom a new cloudy massif behind—a problem about sex itself. To the scaling of such a mountain, all problems of mere sex ratio, even most of those of sociality which in the organisms we know are so obviously interwoven with sex and genetics, had to be subordinate. But glimpses of the massif came intermittently at first and when I was writing this paper what was giving me the most persistent musings and headaches was simpler and closer to home—just a pair of loosely connected thoughts. First, there had come the realization that the genome wasn't the monolithic data bank plus executive team devoted to one project—keeping oneself alive, having babies—that I had hitherto imagined it to be. Instead, it was beginning to seem more a company boardroom, a theatre for a power struggle of egotists and factions. Emergent from the potential strife I was having to imagine, in parallel with others,[6] a kind of parliament of the genes, and the signs suggested a rowdy parliament at that. Second, there was the realization that a certain teasing problem of game theory that I had first encountered in idle reading at Cambridge, and had continued to think about in London almost as leisure distraction, was with me again and beginning to loom with a stony insistence right in the centre of my chosen field. The idea of fitness being treated as equivalent to a 'payoff' in a biological version of Von Neumann's game theory had crossed my mind as soon as I read his earliest account but it was a complete surprise to find an exact evolutionary analogue to one of that theory's most famous puzzles coming to light right in the midst of my own work on population genetical dynamics. I discuss the famous so-called Prisoner's Dilemma of game theory, now realized to have multifarious applications in biology just as in human life, in more detail in Chapter 6, and then in various places and one major paper in Volume 2. Here, I merely note the issue to be not unconnected with my first topic—my realization of the existence of intragenome conflict.

Seemingly inescapable conflict within diploid organisms came to me both as a new agonizing challenge and at the same time a release from a personal problem I had had all my life. In life, what was it I really wanted? My own conscious and seemingly indivisible self was turning out far from what I had imagined and I need not be so ashamed of my self-pity! I was an ambassador ordered abroad by some fragile coalition, a bearer of conflicting orders from the uneasy masters of a divided empire. Still baffled about the very nature of the policies I was supposed to support, I was being asked to act, and to act at once—to analyse, report on, influence the world about me. Given my realization of an eternal disquiet within, couldn't I feel better about my own inability to be consistent in what I was doing, about my indecision in matters ranging from daily trivialities up to the very nature of right and wrong? In another metaphor, I was coming to see that I simply am the two or the many quarrelling kids who are pretending to false unity for a few minutes just so that their father will withdraw his threats and take them to the beach. As I write these words, even so as to be able to write them, I am pretending to a unity that, deep inside myself, I now know does not exist. I am fundamentally mixed, male with female, parent with offspring, warring segments of chromosomes that interlocked in strife millions of years before the River Severn ever saw the Celts and Saxons of Housman's poem—before Europe as a whole existed or saw any of the human violence that became later, for sure, embedded in my ancestry.

As to the putting together of materials for the paper, anyone reading it will realize that it contains a miscellany of themes, subtopics, and data whose sole uniting thread is their connection to sex ratio. The argument of the main model for 'local mate competition', as it later came to be called, was well under way when I left for Brazil in 1963, although at that stage my notes were less general and constituted a mere collection of cases. I had taught myself computer programming through a course offered by the University of London and conducted at its computer centre in Gordon Square, which was a mere block away from University College. Never having much confidence in maths or logic, or at least not in the lines of these that I generated for myself, I was beginning to use simulation repeatedly to check some of the more unexpected predictions from my thoughts or my algebra. Motivated at first by some demographic mathematical problems, I had learned a now long-forgotten 'high-level' language invented by the London University unit that ran the course I attended. It was called CHLF or, later, EXCHLF. When even the latter minor dinosaur of a language headed for extinction I traded it reluctantly for another larger dinosaur, FORTRAN, much disliking at first the fact that the new one wouldn't (then) allow me zero indices in my arrays, besides having various other obviously awkward features. At the same time, in the things it did well, FORTRAN had a undeniable power and in general had an aptness for quantitative scientific tasks that still keeps its particular ancient frame ever renascent, paying even for preservation of its nineteenth-century Hollerith punched-card six-space markings, those once-useful claws that are now diminished to useless toe nails. But in the real world, of course, the dinosaurs did survive—they became birds—and, similarly, the language I was forced to choose and have come rather to love also survives phoenix-like: even today FORTRAN has no need to be listed in any Red Data Book.

The university computer itself, first a 'Mercury' and later an 'Atlas', was in the basement of the building in Gordon Square. To run my sex-ratio programme at a certain morning hour I would queue with others along the old kitchen staff's route to the basement, my roll of five-hole punched tape at the ready. In just such a patrician building only a few blocks away Darwin had lived. Waiting on the stairs I could muse on this rather solemn region of London with the British Museum, centres of medicine, University College, H. K. Lewis's venerable bookshop (now no longer there), and so on. What had been his—Darwin's—relationship, if any, with his servants and with his basement, across there in Gower Street? Would his basement have been where he stored his collections, fresh from the *Beagle*? Would he have run down to check things—or had a butler to bring them to him? Or would his basement be for the coal and for the servants only? . . .

A silence in the chemical, crystallographic, and other more ordinary chatter around me makes me look down and over the heads of the queuers and I see a technician passing a long reel of tape through a reader: that is the compiler going in and my thoughts jump back to my program and what today it may reveal or mangle. Soon, presto—what dazzling automation. So fast my tape flies that the paper ribbon stands level in the air a yard beyond the reader's head. What lightening digestion of my holes and all by a machine—how different this is from the piecemeal calculations and clatter of the LSE stats rooms. Toggles are set by the technician to guide the magic monster right before our eyes; but for the most part there It is, attempting, by Itself, to the accompaniment of various musical notes and flashing lights, to digest what I have given it. Compilation failure, the usual fate, brings on whistles or even a little derisive tune. If my program survives, more tape spews like spaghetti from a chattering punch on the far side.

This tape was then rolled and rushed by me to another room to see the results on paper via a teletype. In my case this generally signalled the second level of failure; it would be a glimpse of some terse, uncomplimentary remarks, upon which I would race to yet another basement room where I could punch alterations to my code on to yet more tape and then, with black Sellotape, splice the new piece in. Haste was necessary because of the chance that, once ready, I might be back in the queue again on the stairs for a second run before the 'development' session ended, thus debugging my program one step further. It would be more garbage in and more out in one day but always a little advance. If you knew the hole code for tapes and had not much wrong, then using a hand punch to add holes and sticking on tiny scissored circles of the black tape to obscure others, you could correct small mistakes by altering a few letters. Such short cuts could sometimes get you back in the queue even faster. The latter practice—changing things by sticking on flecks of tape—was forbidden because sometimes the bits came off and adhered to the electronic eyes of the reader, messing them for everyone; however, as I mentioned, the tape went so fast that the technician usually wouldn't spot flecks and nine times out of ten the alteration passed through unnoticed. The tedium of having to wait 24 hours to get one more run was, of course, the incentive in all this; the need to read my way rapidly through hole codes to the points needing changing passed on to me a desire to keep programs short and this in turn had a permanent effect in inducing (I am told) the terrible programming style I still have.

In short, it was calvinist, ascetic computing in those days, no lounging in front of a monitor, one's keyboard on one's knees: now, in contrast, how sybaritic, how degenerate, is the life of a programmer! As for capacity, even the university's sole whole mainframe—the mighty Atlas that I attended so humbly, reading and digesting and spitting back my programs so astonishingly fast—must have been less capacious and vastly less speedy than the machine I rest on my knees as I write this.

My first programs, as I said, came out of my LSE demography phase, the very first being written to calculate all the eigenvalues of a population growth ('Leslie') matrix (to solve equation (7) of the paper of Chapter 3). But before long, realizing that sex ratio was itself a crucial determinant of population growth rate and also having noticed that there were gaps in Fisher's theory that needed filling, I was running population simulation programs that traced the course of sex ratio under varied genetical and ecological assumptions. Those programs, laid aside while I was in Brazil, were revived and rewritten in FORTRAN and on to punched cards in my new job at Silwood; there their results continued to provide checks on the formula solutions that I increasingly found myself able to obtain analytically.

'Increasingly', however, was not to be for long and the results set forth in this paper are near to the high point of my small success in analytical theoretical biology. Since then there have been occasional brief inspirational re-ascents, but on the whole my course has been downhill with respect to proper mathematics. I have come to rely more and more on computer simulation for guiding me through difficult issues. Less and less do I attempt the corresponding analyses. This is regrettable, of course, because analytical results—solved equations—are so much more powerful for understanding. However, putting to one side for the moment my father's principle of 'reversion to type' as well as the dreary issue of my own dwindling stock of neurones as causes, I can argue that my difficult problems now are actually much more difficult than those I had when I wrote this paper. It is a fact that not even youthful brains, packed to their barely sealed fontanelles with the most logical and maths-loving of neurones, have been able to reach the point of study where I now stand in respect of, say, the modelling of the selection of sex. Perhaps no one analyses my problems simply because no one sees where I stand as a place worth being; but, for sure, no mathematical analyst is here.

For these much simpler problems of sex ratio, at the time of this paper, there was a little more in the way of successful analysis still to come. As some of the later papers show, small discrepancies between the exact simulations and the approximate derivations I mention in this paper have come to be understood. For example, the value I mention as 'between 0.070 and 0.072' on page 151 turns out to be simply 1/14 and that 'between 0.215 and 0.205' on page 159 is likewise simply 3/14. As will be seen later (Chapter 13) the cause of my missing the exact target in this first attempt is ironic: I was neglecting to use proper measures of gene propagation via relatedness in my derivation of the 'unbeatable' (or ESS) formulas—in short, was neglecting inclusive fitness! As for the 3/14 when it was found, I am charmed by this even more than by my approximation of 1/4 as given in this paper (page 152), exciting as that was, and I see these two values as being perhaps the neatest results relevant to nature that I have ever obtained. The approximate unbeatable sex ratio of 1/4 for paired wasps had been nice but perhaps someone will sometime point out to me how this is in someway obvious. Yet who would claim that an integer factor involving

7 was obvious, and that some inkling of this hovered in the tiny mind (or at least behaviour) of wasps that are paired to share a fly pupa and have evolved to do the best thing?

Already at LSE I had searched enough in the entomological literature for facts about social insects and their origins to have noticed that there was a real puzzle about departures from Fisher's equalizing principle of the sexes when it came to small and usually habitually inbreeding insects. I well remember reading with delight as well as with near incredulity F. Balfour Browne's account of the strange life and sex ratio of *Melittobia acasta*, a tiny hymenopterous parasitoid of the cells of solitary bees and wasps. I think it may have been reading W. M. Wheeler, the ant authority, that led me on to another equally strange species of this genus, *M. chalybii*. Wheeler was interested in the alternative morphs of the female of this species, one of these having characters slightly suggestive of a worker caste. 'Slightly' was the word, however, and I am now certain that the morph difference has more to do with fast breeding than dispersal (see Chapter 11). However, co-operation of a kind is probably occurring in the colonies of this beewaspandflyophagous wasplet and a further more explicit example of this within the genus was to be noticed later.[7]

For a long time, therefore, I thought of sex ratio under population subdivision plus incest as my 'Melittobia problem'. Both in Brazil and after my return my file bearing this title expanded rapidly. In parallel with my steps towards a solution, I began to find an ever-increasing set of similar examples in the field and in the entomological literature. My situation at Imperial College was ideal for such work. Not only was there the knowledge of the staff and graduate students, but, at South Kensington, where I had to give my lectures, all the best entomological libraries in the country were clustered nearby, to say nothing of the good general entomological library back in Silwood House itself run by the widow of a Dutch insect systematist. Mrs. Van Emden was the most dedicated librarian I have known and fortunately she was kindly disposed to me. With excellent knowledge of the entomological journals she would sometimes point out to me papers relevant to my interests. Above all, where necessary, she would freely translate parts of them out of German. I might digress here to note that the ways librarians in this period helped me were actually varied indeed. Some like Mrs. Van E. were kindly ferrets for me among dusty stacks (old information was quite as useful to me as new on my particular track). But some helped me simply by being muses, flowers for me to watch in tired moments in the dark caves where we worked. Young and beautiful, absorbed in the rites of their profession, often they would illuminate for me by the half hour a distant desk or corner among the shelves, would pass close by, or even bring the books they had found for me or else those of my yellow index cards that had defeated them. Even a few words might be exchanged. Many of the librarians I admired, of course, I never spoke to, and the more beautiful they were the more likely this outcome was. I had an odd pride on my ability to use any library of any institution without help and without gaining any permission to work there, even though I knew that if I asked for it such permission would virtually always be given. The key of my strategy was the card catalogue: this had to be spotted quickly upon first entry and headed for as if I had used it not merely an hour ago or even only yesterday but for years—which, of course, often came to be the case. The only library I recall failing to penetrate successfully in this way was the library of the very private and patrician Zoological Society of London, which is near to and connected with the London Zoo. Why did I eventually raise my siege on the Zoological's library, I wonder—was its librarian more sharp-eyed than the rest, was she more dragonly unbeautiful, more coldly aristocratic? I am not sure.

Not only in the libraries but everywhere around me in the fields and the woods—out of the leaf sheaths of Silwood grasses, under bark of its trees, in my trap nests among parasitoids (*Melittobia*-like, though not of that genus) that were killing my bees, in the basement insect-rearing rooms of the Field Station—poured examples of my strange sex ratios. There and out in the park (and if I needed, also in those larger and more ancient and royal woods around Windsor) I found one by one the small insects whose gametocytes and tiny nervous systems set stage for the genomic intrigues I came to dramatize and simplify in this paper. In the varying and truly extraordinary sex ratios I recorded (see, for example, Chapter 13), the struggles of my various factions could often be seen; and through them I built my confidence that my reasoning about the selective forces was correct.

Silwood was peaceful, perhaps too peaceful for the farthest flights of thought. We were an institution eminent in the study of insect ecology and population change but no one in my time shared my interest in evolution. Hence I was not involved in any collaboration. Few questioned me or passed me their manuscripts to check and I remained on the whole blessedly unknown to journal editors. I worked in my lab, in the computer room of the experimental nuclear reactor that came to be built at the field station, in libraries there or in London, or in the field. All was just as I chose. Whether I was swinging my canvas bag and still footing it between the libraries in the great groves of stone and brick to the East, or in the West or North busy in the real woods and fields around Silwood, how well I could now think with A. E. Housman of one faithful, undemanding companion how 'Bound for the same bourne as I/On every road I wandered by/Trod beside me, close and dear/The beautiful and death struck year'. This paper marks barely my third year of life and death with this mistress; 10 more were to come.

Other companions were no less constant. Whether in the guarded pages of the libraries (not so well guarded to judge by my forays, but then I never harmed any book) or out wild in woods and fields where I wandered, the strange crew of tiny arthropods that I was discovering were my companions too and a never-ending delight. I like to think that this paper, in such parts especially as my listings in Table 4.1, begins the repaying of a debt to those animals that other papers (Chapters 12 and 13) will later continue.

References and notes

1. A. E. Housman, *The Collected Poems of A. E. Housman* (Jonathan Cape, London, 1960).
2. E. A. Herre, Sex ratio adjustment in fig wasps, *Science* **288**, 896–8 (1985).
3. E. A. Herre, Optimality, plasticity, and selective regime in fig wasp sex ratios, *Nature* **329**, 627–9 (1987).
4. J. J. Boomsma and A. Grafen, Intra-specific variation in any sex ratios and the Trivers–Hare hypothesis, *Evolution* **44**, 1026–34 (1990).
5. J. J. Boomsma and A. Grafen, Colony-level sex-ratio selection in the eusocial Hymenoptera, *Journal of Evolutionary Biology* **3**, 383–407 (1991).
6. E. Leigh, *Adaptation and Diversity* (Freeman Cooper, San Francisco, 1971).
7. B. J. Donovan, Co-operative material penetration by M. *hawaiiensis* (Hymenoptera, Eulophidae) and its adaptive significance, *New Zealand Entomologist* **6**, 192–3 (1976). This

paper came to my notice about 1977 and documents another striking achievement of melittobian sociality, which, however, remains well short of eusociality. Females were observed to take turns as excavators as they bite escape tunnels through the walls of the polystyrene containers used for their rearing. In nature, of course, these co-operative exits would be cut through other media such as clay, wax, or tough coccoon material, the typical building materials of various prey species.

EXTRAORDINARY SEX RATIOS†

W. D. HAMILTON

A sex-ratio theory for sex linkage and inbreeding has new implications in cytogenetics and entomology.

The two sexes are usually produced in approximately equal numbers. Fisher[1] was the first to explain why, under natural selection, this should be so, irrespective of the particular mechanism of sex determination. His rather tersely expressed argument has been clarified by subsequent writers[2] and seems to be widely accepted. In bare outline, the factor of parental care being ignored, it may be given as follows:

1. Suppose male births are less common than female.
2. A newborn male then has better mating prospects than a newborn female, and therefore can expect to have more offspring.
3. Therefore parents genetically disposed to produce males tend to have more than average numbers of grandchildren born to them.
4. Therefore the genes for male-producing tendencies spread, and male births become commoner.
5. As the 1:1 sex ratio is approached, the advantage associated with producing males dies away.
6. The same reasoning holds if females are substituted for males throughout. Therefore 1:1 is the equilibrium ratio.

The argument is not affected by the occurrence or non-occurrence of polygamy, or by any differential mortality of the sexes, provided this is uncorrelated with the sex-ratio genotypes.

More precisely, what has been called 'Fisher's principle' of the sex ratio states that the sex ratio is in equilibrium when, in the population as a whole, the totals of effort spent producing the two sexes are equal. If the totals are not equal, producers of the sex corresponding to the lesser total have an advantage.

†*Science* **156**, 477–88 (1967).

This article is concerned with situations where certain underlying assumptions of Fisher's argument do not hold. It will be seen that such situations must be quite widespread in nature. As regards ecological assumptions, for example, Fisher's argument is restricted to the actually unusual case of population-wide competition for mates. A contrary case wherein the competition is local is discussed in some detail. In some features it has an unexpectedly close similarity to certain types of situations considered in the 'theory of games'. Already the above outline seems to show that an individual supposedly able to choose the sexes of offspring would do best under natural selection by selecting the extreme opposite to the current sex ratio of the population—that is, by producing a unisexual progeny of whichever sex was currently in the minority. This game-like feature, which has already led one writer[3] to refer to genetically determined sex ratios as 'strategies', in the sense of a play by the individual against the population, becomes accentuated as we proceed into circumstances of local competition.

Before considering local competition, however, it is convenient to discuss the consequences of failure of some of the genetic assumptions latent in Fisher's argument.

SEX-LINKED DRIVE UNDER RANDOM MATING

Fisher's argument does apply to all cases where sex-ratio control is by genes acting in the homogametic sex, or in the female under the male-haploid system (contrary to some earlier statements of mine—see Part II of the paper in Chapter 2 of this volume), or by genes on the autosomes acting in the heterogametic sex. In all these cases the total number of the gene-bearer's grandchildren is a true measure of the propagation of the gene. This is not so in the case of sex-linked genes acting in the heterogametic sex.

For simplicity of argument, suppose the male is heterogametic. Then grandchildren through daughters are obviously irrelevant to the fitness of a gene on the Y chromosome. This fitness is measurable entirely by the number of sons. Sex-ratio control in the male is effectively the same as genetic control over the relative success of the X-bearing and Y-bearing sperm in fertilization. Suppose the Y chromosome has mutated in a way which causes it always to win in the race to fertilize. A male with the Y mutant then produces nothing but sons. Provided these sons, who also carry the mutant, cannot be in any way discriminated against in the unrestricted competition for mates (a situation which is implied if mating is random for the whole population), the Y mutant will have a constant selective advantage. As the mutant spreads, the population sex ratio will become more and more male-biased and the population itself will become smaller and smaller; finally the population will be extinguished, after the last female has chanced to mate with a male carrying the mutant.

Figure 4.1a represents such an episode, in which the starting condition is one mutant-bearing male in 1000 males. On the basis of the admittedly severe assumption that females produce only two offspring each, it was found that in the fifteenth generation the expected number of females is less than one.

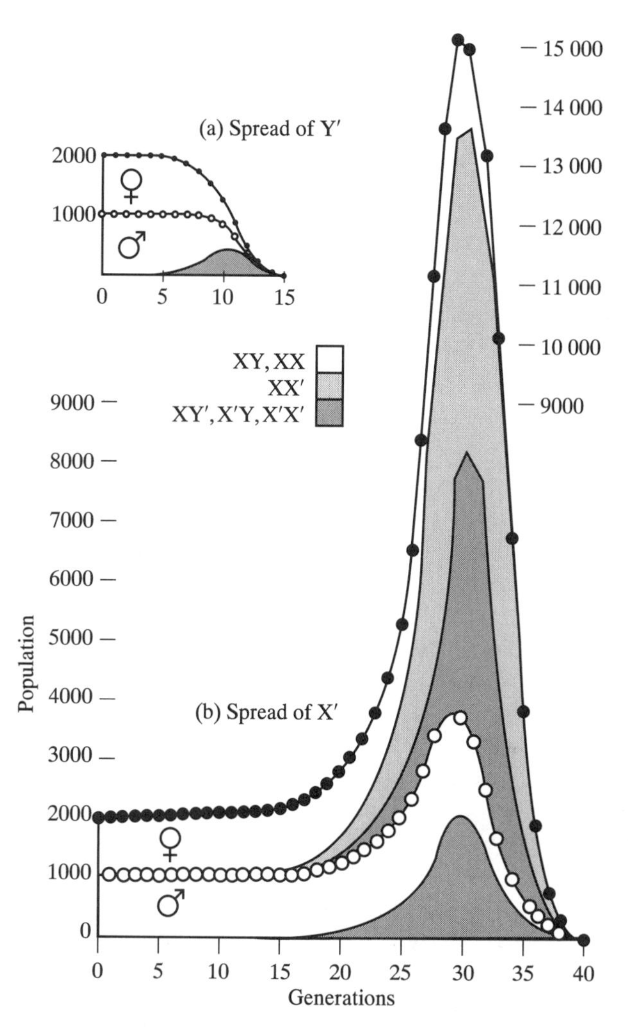

Figure 4.1 Population and its distribution by sex and genotype in the course of natural selection of (a) a Y chromosome and (b) an X chromosome, having complete drive in spermatogenesis. Mating is random, and normal males give a sex ratio of $\frac{1}{2}$. It is assumed (1) that mated females have two offspring each, so that, before mutation produces the driving chromosome, the population is stationary, and (2) that males can fertilize only two females each, so that, in (b), from the first generation in which the sex ratio is less than $\frac{1}{3}$ (generation 27), some females have no offspring because they are unmated. Both populations start with one chromosome of the driving type in 1000 chromosomes. Extinction is considered to occur after the first generation in which the expected number of females is less than one.

A similar extreme mutation on the differential X chromosome occasions a similar theoretical disaster but brings it about much more slowly. This is partly because in this case selection is intrinsically slower. Unlike the Y, the mutated X chromosome is not exposed to selection in every generation, and, overall, as might be expected from the proportion of its generations that an ordinary X chromosome spends in males, its spread is about one-third as fast. In Fig. 4.2 the two types of transience are compared, and the transience of an autosomal gene causing fully effective meiotic or gametic drive is also shown. As might be expected, the speed of the autosomal gene is intermediate. The progress towards extinction due to a driving X mutant is delayed by the occurrence of polygamy. Indeed, so long as the males remain sufficiently numerous to fertilize all the females, the population should show accelerating expansion (Fig. 4.1b).

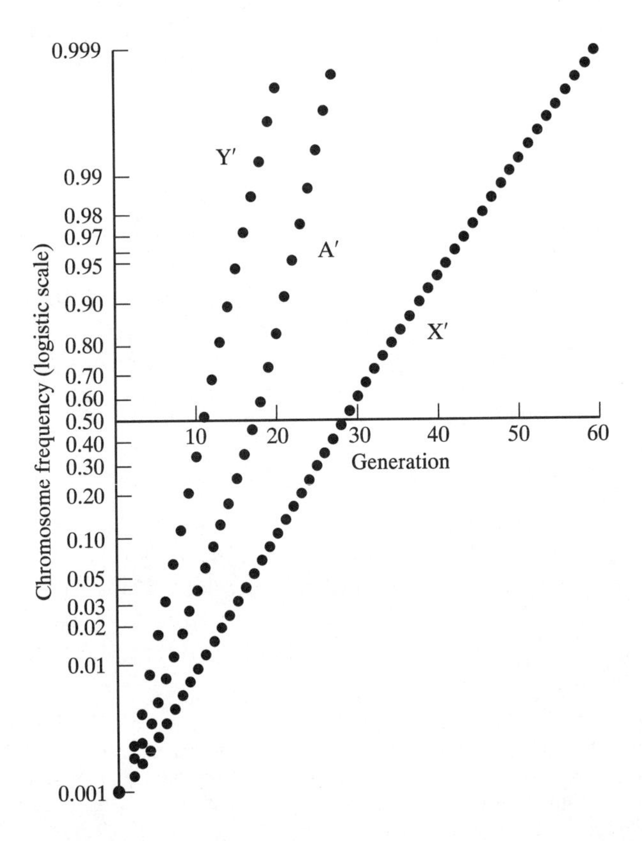

Figure 4.2 Logistic plots showing the progress in frequency of occurrence of sex chromosomes (X′, Y′) and an autosome (A′) having complete drive in spermatogenesis. Each course starts with one chromosome of the mutated type in 1000 chromosomes.

Cases of X-linked 'drive' are not uncommon in wild populations of some species of *Drosophila*, and their potential threat to the species has been recognized.[4] Cases of autosomal drive have been under study and discussion for some time.[5] With perhaps one exception, mentioned below, no equally striking cases of Y-linked drive have been reported. Perhaps, in view of the general inertness of the Y chromosome, this is not surprising. It is surprising, however, that the exceptional latent danger to the species presented by this form of drive has received so little comment; I suggest (and to the best of my knowledge this is the first time the suggestion has been made) that it may help to explain why the Y chromosome is so often inert. A population in which a driving Y mutant was spreading could be saved by another mutation, on an autosome or on the X chromosome, which was capable of inactivating the relevant region of the Y mutant. That such a mutation would spread follows from Fisher's principle. In doing so it would slow down (although it would never arrest) the spread of the Y mutant and, at the same time, would cover up its effect.

It seems probable that something like this has happened in the mosquito *Aedes aegypti*, if the genetic situation inferred by Hickey and Craig[6] is correct. In this case the 'Y' is apparently a male-determining gene, certainly not a whole chromosome, and it seems that, in the strains where the driving Y mutant exists, there may be several different 'X' alleles which restrain the action of the Y mutant in differing degrees. It is hybrid males from outcrosses of such strains that often give extreme male-biased sex ratios. All-male progenies were sometimes recorded.

Hickey and Craig have pointed out the potentiality of their effect for biological control. They did not attempt to demonstrate the capability for spread of their driving Y gene but did show in some experiments that it could maintain itself at high frequency, with consequent detriment to the experimental population. Some other experiments showed, however, a rapid masking of the sex-ratio effect. This may have been due to reconstitution of the restraining mechanism which operated in the paternal population. Clearly, if the view presented here is valid, a principle of application in biological control will involve repeated backcrossing of the hybrid male-producing males to females of the susceptible population. Unless this is done prior to liberation of male-producing males, some part of the genetic mechanism on other chromosomes which previously masked the sex-ratio effect (e.g. a recessive gene) is likely to be introduced at the same time. It should be worthwhile to look for male-producing effects in racially hybrid males in other outbreeding species with male heterogamety, and also perhaps for genetical devices by which the Y chromosome could be freed from the inhibitory action of the rest of the genome. The implied method of biological control is in theory very powerful, since the mere seeding of a population with a few prepared males could cause its extermination or at least its reduction to a density where mating was no longer effectively random.

The fact that no other case like that of *Aedes aegypti* has yet been reported, in contrast to several cases of X-linked drive, accords with the drastic rapidity of spread of the driving Y mutant and its immediately adverse effect on reproductive potential. Presumably an outbreeding species cannot long continue to exist if it has a Y chromosome likely to mutate in this way. The same does not apply if the female is heterogametic. An extreme sex-ratio swing which is very suggestive of the spread of a driving Y chromosome has recently been reported in a butterfly population.[7] As with the driving X chromosome under male heterogamety, such an event would be advantageous to the population at first. The frequent discovery of driving X chromosomes in wild populations of *Drosophila* is not very surprising, although, in order properly to understand their existence in a permanent polymorphism, there has to be postulated either a severe disadvantage to females homozygous for the driving chromosome[8] or some basis of interpopulation competition, as Novitski[4] suggested and as is shown in the model given below.

Altogether, the evidence for the workability of the suggested differential evolution of the Y chromosome is good. There is evidence that genes and other chromosomes can suppress the 'sex ratio' trait in *Drosophila*,[9] that heteropycnosis may be a cytological manifestation of such inactivation,[10] and that heteropycnotic regions sometimes do cause disturbed segregations.[11] In general, it is gametogenesis rather than fertilization that is the likely arena of the dysgenic effects that have been considered. Hickey and Craig show that their effect is probably due to events that either suppress production of X-bearing sperm or cause such sperm to degenerate, and McCloskey[12] has recently emphasized the extreme scarcity of evidence of gene activity in animal gametes in general. However, McCloskey admitted that his experiments with *Drosophila* could not exclude the possibility, significant for the present hypothesis, that genes in the main heterochromatic regions might be active in the sperm, and he also noted some hints that mammalian sperm may not be so inert genetically as the sperm of *Drosophila*. Therefore, just possibly the indications of high primary sex ratios in some mammals,[13] and of higher activity of Y-bearing sperm in rabbits and cattle,[14] are relevant evidence of a stronger tendency of the Y chromosome to evolve spermatic drive. Somewhat against this possibility, however, is the evidence, for plants (in which gene activity in the penetrating pollen tubes is well known), that it is the X-bearing pollen that tends to win under competitive conditions.[15] This is understandable on the basis of Fisherian theory only if it is the genotype of the style that controls the race, for, as Lewis has pointed out,[15] abundant pollination will usually imply that males are common in the neighbourhood, so that to favour the production of females is the correct policy. With occasional pollination the sex ratio produced tends to be 1:1. This is probably the sex ratio of the pollen grains, but that a male-biased ratio is not in some way brought about from sparse pollination may be partly due to the factor of local competition, now to be considered.

SEX-LINKED DRIVE WITH LOCAL COMPETITION FOR MATES

In a 'viscous' population—that is to say, one where the individual can mate only with a rather permanent set of neighbours who tend also to be his relatives—it is clear that the initial spread of a driving Y mutant from the point where the mutation occurred will bring about a local collapse of population due to the lack of females, and vacant space will tend to be filled by normal immigrants from surrounding areas. Thus some sort of equilibrium could occur.

Another biological situation in which the spread of the aggressive Y mutant is eventually checked is as follows. Free-moving females search for isolated food objects, or 'hosts'. Each host is colonized by a certain number and is eventually exhausted through feeding of the progenies. The subpopulation of adults reared on a host mates randomly *within itself*; no males successfully mate outside their own group. Inseminated females emigrate to take part in population-wide competition to discover new hosts. In this system, matings will be a mixture of sibmatings and outcrosses, the former becoming common when the number of females settling on a host is small. It is at once obvious that the driving Y mutant cannot supervene completely in such a population, provided the population is large and the females are fecund, for the grouping sets a limit to the extent to which a female's brothers can be outnumbered by the mutants among her suitors, and there always remains a finite chance that she will mate with a brother.

To facilitate further analysis of such a situation, let us suppose that always exactly n females settle on a host and that the progenies reared are equal. Suppose that there exist two types of Y chromosome, Y^a and Y^b.

Females accept only one insemination, so the host-seeking females will be of two types, according to the contents of their spermathecae: XY^a-bearing and XY^b-bearing, or, more briefly, type a and type b. Suppose these two types are constrained to give progenies in which males form fractions x_a and x_b of the totals. Such fractions are convenient measures of the sex ratio, and hereafter in this article all sex ratios are given in this form.

Consider the set of hosts which happen to be shared by r type a females and by $n - r$ type b females. The sex ratio, X_r in the progeny of this set will be

$$X_r = [rx_a + (n - r)x_b]/n.$$

Through fair competition for mates, a particular type-a female may be expected to be responsible for the spermathecal contents of a number of emigrant females proportional to

$$\frac{x_a}{X_r}(1 - X_r).$$

This may be called the Y-chromosome fitness in such a female in such a group. It is required to know whether the average for type a, taken over all such sets, is above or below the whole-population average for Y-chromosome fitnesses. This will indicate whether the Y^a chromosome is gaining or losing in frequency.

Consider therefore the $n-1$ associates of a type-a female selected at random from the whole set of those that have succeeded in finding hosts. Let F_r represent the probability distribution of the number $(r-1)$ of type-a females among these associates. The average Y^a-chromosome fitness, W_a, is then seen to be

$$W_a = \sum_1^n F_r \frac{x_a}{X_r}(1 - X_r).$$

The general average for Y-chromosome fitness must be

$$W = 1 - X,$$

where X is the sex ratio of the whole population. Explicitly,

$$X = px_a + (1-p)x_b,$$

where p is the frequency of Y^a.

The selective advantage, E_a, of the Y^a chromosome with respect to the population average is

$$E_a = W_a - W = \sum_1^n F_r \frac{x_a}{X_r} - x_a - (1 - X) = \sum_1^n F_r \frac{x_a}{X_r} + (1-p)(x_b - x_a) - 1. \quad (1)$$

Differentiating, we obtain

$$\frac{dE_a}{dx_a} = \frac{1}{n}\sum_1^n F_r(n-r)\frac{x_b}{X_r^2} - (1-p).$$

When $x_a = x_b = c$, this reduces to

$$\frac{1}{nc}\sum_1^n F_r(n-r) - (1-p),$$

which is zero only when

$$c = \frac{n - \sum_1^n F_r r}{n(1-p)}. \quad (2)$$

Now consider the simple and plausible case where the settling on hosts is random. The distribution of r will be binomial and so will be the distribution of the number of a-type associates of the randomly selected a-type female:

Then

$$F_r = \binom{n-1}{r-1}(1-p)^{n-r}p^{r-1}.$$

$$\sum_{1}^{n} F_r r = (n-1)p + 1$$

and equation (2) becomes

$$c = (n-1)/n. \tag{3}$$

I am satisfied, although I have not strictly proved, that this is indeed the only 'unbeatable' sex ratio for the situation: the Y chromosome that produces it has a selective advantage over any other, whatever the current gene frequency.

The above model equilibria do not seem very realistic. Species in which the sex ratio comes to be controlled by Y-linked drive probably become quickly extinct even if they are protected to some extent by grouping of the sort described. Unfortunately no general analysis of the less drastic situation in which there is gametic drive by the X chromosome has been achieved. However, deterministic computer simulation of the population genetics for $n = 2$ shows that the 'unbeatable' ratio is between 0.070 and 0.072. This result must also hold if control of the sex ratio depends on the genotype of the male in male haploid species. Such control is certainly unusual, but it seems to occur in the eulophid *Dahlbominus fuscipennis*.[16] This is an inbreeding species, and its sex ratio, about 0.09, is of the order expected. But the case is peculiar, and the underlying theory still has a major gap in that it is not known whether the 'unbeatable' sex ratio for $n = 3$ is above or below 0.07. In view of the possible relevance to the X-linked sex-ratio effects in *Drosophila*, this point deserves further study.

Although male-haploid inheritance is identical to the sex-linked type, from the point of view of sex ratios a difference enters if the males fail to fertilize all the females. This difference is due to the ability of the females, in cases of male haploidy, to produce all-male broods parthenogenetically. With sex ratios as extreme as those just mentioned, such failure of polygamy would not be surprising. In the case of random mating it is known to introduce a tendency toward unstable alternation of sex ratios which is quickly disastrous if the mated females are being forced to produce only female offspring.[17] From this aspect, also, the breeding structure of the model should give a measure of security.

NON-SEX-LINKED CONTROL WITH LOCAL COMPETITION FOR MATES

Apart from the supposition of sex-linked control, the above model contravenes Fisher's tacit assumption of unrestricted competition for mates. It is interesting, therefore, to consider what this same model situation implies when this is the only assumption broken—that is, when control of sex ratio is of one of the kinds which has been noted to give the relationships of Fisher's principle under the condition of panmixia. The model breeding structure forces sons into competition with one another, so that although there may be, in general, a shortage of males, production of males does not necessarily pay off in terms of grandchildren: whether it does so depends on the sex ratios of offspring produced by other females of the group.

An approximate analysis can follow the lines of the foregoing analysis very closely, provided it is supposed that there are only two types of female, A and B, and that

$$\text{fitness} \propto \text{N of inseminations by sons} + N \text{ of daughters.} \tag{4}$$

Obviously such a simple scheme cannot give a proper genetical representation of a situation, but, as mentioned below, it has been found to give a fair correspondence to the genetical model for $n = 2$, while, for the panmictic case, it has already been accepted in the literature.[18]

Using these assumptions and obvious developments of our notation, we have

$$W_{\mathrm{A}} = \sum_{1}^{n} F_r \frac{x_{\mathrm{A}}}{X_r}(1 - X_r) + (1 - x_{\mathrm{A}}).$$

Since we are now concerned with the production of daughters as well as of sons, the mean fitness in the population must be taken as $2(1 - X)$.

Therefore

$$E_{\mathrm{A}} = \sum_{1}^{n} F_r \frac{x_{\mathrm{A}}}{X_r} + 2(1 - p)(x_{\mathrm{B}} - x_{\mathrm{A}}) - 1.$$

The foregoing equation is the same as equation (1) except for the factor 2 in the second term. Hence, by just the same argument we can arrive at the unbeatable sex ratio $\frac{1}{2}$

$$(n - 1)/2n \tag{5}$$

for the case of random association.

As n increases, the situation undergoes a particular kind of approach to the Fisherian case and, correspondingly, the formula shows the unbeatable ratio tending to approach $\frac{1}{2}$.

At the other extreme, the case $n = 1$ obviously corresponds to a system of completely sibmated lines. That the theoretical equilibrium ratio is then zero merely implies that a female's advantage depends wholly on the number of fertile emigrant females she can produce from her host. It is not in her interest to produce more males than are necessary to ensure the fertilization of all her daughters.

A BIOFACIES OF EXTREME INBREEDING AND ARRHENOTOKY

Among small arthropods, wherever reproduction is quite regularly by brother–sister mating there seems to be extreme economy in the production of males, as predicted. Most of the cases which have been noted are listed in Table 4.1.

It is probable that all of them reproduce by arrhenotoky—that is, by the process in which males are derived always, and only, from unfertilized eggs. This process has been found to be associated with haploidy of the males in all cases that have been investigated cytologically; therefore, male haploidy is a strong supposition for most of the cases in Table 4.1. However, with a few species not only is there no direct evidence that males are usually impaternate but also there is none that female-to-female parthenogenesis (thelytoky) does not occur. But in view of the general conformity of such cases (for example, the fig insect and the moth ear mite) with the others, and of what is known of related species, it will be surprising if they turn out to be not normally arrhenotokous.

Taken together, the data suggest the outline of an ideal extreme biofacies which may be described as follows:

1. The primary sex ratio is spanandrous—that is, females greatly preponderate.
2. Reproduction is arrhenotokous.
3. There is at least one male in every batch of offspring.
4. There is gregarious development, as a group of siblings, from egg to adult.
5. Adult males eclose first and can mate many times.
6. Mating takes place immediately after (or even before) eclosure of adult females.
7. Males are disinclined, or unable, to emigrate from the batch.
8. Females can store sperm; one insemination serves to fertilize the whole egg production.

Arrhenotoky is a mode of reproduction that readily permits the production of biased sex ratios, with control of the sex ratio normally dependent on the phenotype of the mother. Therefore it seems either that male-haploid organisms have found themselves pre-adapted for life in niches of the sort characterized by the model or that the evolution of male haploidy has actually accompanied, in several independent lines, an evolutionary trend to occupy such niches. The latter alternative seems not unlikely for the ancestors of the male-haploid groups included in Table 4.1, although undoubtedly many of the

Table 4.1 Insects and mites having usual sibmating combined with arrhenotoky and spanandry (blanks are left where clear data are not available; in some cases equivocal evidence is mentioned in the notes)

Family, genus, and species	Host	Number in typical batch of progeny		Tendency towards inclusion of one male per batch	Usual site of mating	Functional wings present		Males pugnacious	Thelytoky known in species (s), genus (g), family (f)	References and notes
		♂	♀			♂	♀			
Agaontidae										
Blastophaga psenes	Wild fig	22	235		In fig	–	+		–	44
Torymidae										
Monodontomerus spp.	Bee larva	1	12	+	In host cell	+	+			45
Pteromalidae										
Nasonia vitripennis	Fly pupa	2	19		Just outside host	–	+		s	35,46
Encyrtidae										
Dusmetia sangwani	Scale insect	1	5	+	In host	+	–		f	47
Thysanidae										
Thysanus elongatus	Parasitoid larvae	1	5			+	+		g	48
Eulophidae										
Melittobia acasta	Pupa or fly or aculeate	1	46	+	In host cocoon	–	+	+	f	35,49
Melittobia chalybii	Aculeate pupa	2	50	+	In host cocoon	–	+	+	f	50
Pleurotropis parvulus	Leaf-mining beetle larva	4	13	+	Just outside leaf-mine				f	51
Trichogrammatidae										
Trichogramma semblidis	Alder fly egg mass	10	60		On host		+		g	52
Prestwichia aquatica	Aquatic insect egg	1	8	+	In host	–			s?	53
Elasmidae										
Elasmus hispidarum	Leaf-mining beetle larva	4	8	+	Just outside leaf-mine			+		51,54

Mymaridae										
Caraphractus cinctus	Water beetle egg	5* / 1†	25* / 2†	+	Underwater near host	+	+		f	41
Anaphoidea calendrae	Weevil egg	1	6	+		+	+		f	55
Anaphoidea nitens	Weevil ootheca	1	3	+	Just outside host		+		f	56
Scelionidae										
Telenomus fariai	Bug egg	1	6	+	In host				g	57
Asolcus spp.	Bug egg mass	8	40	+	On host	+	+	+	f	36,58
Bethylidae										
Cephalonomia quadridentata	Beetle larva or pupa	1	3	+	In group of cocoons		–		f?	59
Sclerodermus immigrans	Beetle larva	4	20		On mass of cocoons	+	–		s?	60
	Moth larva	2	8							
Perisierola emigrata					In group of cocoons				f?	61
Scolytidae										
Xyleborus compactus	Twig tissues	1	9		In parent gallery	–	+		–	28
Thripidae										
Limothrips denticornis	Grass plant	3	20	+	Grass, leaf sheath	–	+		f	62
Laelaptidae										
Myrmonyssus phalaenodectes	Moth				Moth ear	–	–		–	63
Pyemotidae										
Pyemotes ventricosus	Mother	4	86	+	On mother	–	–		–	51
Siteroptes graminum	Mother	7	140	+	In mother	–	–		–	64
Acarophenax tribolii	Mother	1	14	+	In mother	–	–		–	34
Tarsonemidae										
Tarsonemoides spp.	Mother	5	65		In scolytid's egg niche	–	–		–	65

*In *Dytiscus* egg; †in *Agabus* egg.

cases in the Hymenoptera must be derived secondarily from relatively outbreeding species.

No clear example of the biofacies has been found in the literature on arrhenotokous aleyrodids and coccids. For the coccids, an independent and detailed theory of the evolution of male haploidy has already been proposed, by S. W. Brown.[19] This is based on an 'island' breeding structure, with random mating within the almost-isolated subpopulations, and the fact that all female coccids are wingless while males are often winged certainly suggests random mating. The very complex chromosome cycles known in various fungus gnats of the genus *Sciara* all have a common resemblance to the typical situation of malc haploidy in that (1) virtually no chromosomes of paternal origin are passed on by a male, and (2) the sex of an embryo depends on some influence transmitted by its mother.[20]

Although all four combinations of the winged and wingless condition according to sex are known in the Sciaridae, cases in which the female alone is wingless are more numerous than the converse cases;[21] in this respect the group tends slightly towards the condition of the coccids. Furthermore, almost entirely unisexual broods are characteristic of some species, and it is clear that, in these, outbreeding must result. It fits well with Fisher's principle that the two types of female, arrhenogenous and thelygenous, which exist in these cases, are thought to be themselves produced according to a simple backcross mechanism which ensures their numerical equality and, consequently fixes the sex ratio at $\frac{1}{2}$.[22]

Thus, the Sciaridae, like the Coccidae, tend to support the view that the evolution of male haploidy can take place under relatively panmictic conditions. Nevertheless, with the Sciaridae, the decaying plant bodies and fungi on which the larvae feed are hosts of the kind specified in our model, and there is one report,[23] concerning *Sciara semialata*, of a combination of characters very suggestive of the biofacies, although not in its extreme form. These characters were a sex ratio of about $\frac{1}{4}$, winglessness of males only, and the *Sciara* habit of migrating in a compact column prior to pupation. Furthermore, facts which similarly suggest the biofacies, although somewhat less strongly, are available for gall midges of the related family Cecidomyiidae,[24] and this family has unusual chromosome cycles resembling those of the Sciaridae.

Although it is clear from Table 4.1 that winglessness in the male sex alone is rather characteristic of the biofacies, the opposite combination is not so rare as might have been expected. Three species are listed in Table 4.1 which fit quite well in most respects but have winged males and wingless females. I see no simple 'explanation' of this anomaly, but in any case it hardly affects the general correspondence, since, from the accounts on which these listings are based, it is clear that sibmating usually does occur.

SEX RATIOS WITH POLYGYNY

It is frequently suggested that biased sex ratios are adaptations of the populations that manifest them. In particular, the evolutionary ability to economize in the production of males to the point that gives maximum capacity for increase (an ability which, according to the model, is restricted to sibmating lines) is sometimes inputed to any population, whatever its breeding structure. In one sense the idea of adaptation for the benefit of the population is even less secure than that of adaptation for the benefit of the species. It is true that a population consisting of numerous subgroups has many potential sites for 'mutation', and that 'mutation' by genetic drift (for example, the complete transience of an altruistic gene) is distinctly possible if the subgroups are small.[25] Species, on the other hand, can at least reproduce in isolation from their competitors. The extent to which populations can do so without losing identity through hybridization is not at all clear. As regards sex ratio, even occasional outbreeding should cause the breakdown of any population adaptation. Vagrant males arriving in the predominantly female groups can so effectively propagate the genes which caused their own production that male-producing tendencies must spread. This is the essence of Fisher's principle. Accordingly it is not surprising to find that highly polygynous species with outbreeding habits do not have female-biased primary sex ratios. In polygynous mammals and birds a nonbreeding surplus of males is commonly observed during the breeding season, and, in mammals at least, the sex ratio at birth may be actually male-biased.

With insects, evidence on this point might be expected to come from the scolytid bark beetles. Chararas[26] gives a classification that is useful by biological criteria. First, he divides the group according to whether mating takes place prior to dispersion from the larval host or after arrival at the new host; second, he subdivides the latter group according to whether the males are usually polygynous or monogynous in the new borings.

All three systems occur in the scolytid subfamily Ipinae. The widest survey of representatives of the first system is given by Browne,[27] who calls it the system of 'extreme polygamy'. Polygamy *is* extreme, but, since this follows from the fact that males are produced in much smaller numbers than females, *spanandry* seems a better term, and is used here. In all species of the tribes Xyleborini, Eccopterini, and Webbini for which Browne gives data, the males are not only relatively few from birth but are also flightless, short-lived, and often blind. Obviously this system closely corresponds to the 'extreme biofacies' discussed above; some of the facts for *Xyleborus compactus*, the only species of these tribes for which arrhenotoky has been established,[28] are given in Table 4.1.

With the other two groups considerable outbreeding is expected. Therefore, according to Fisher's principle, a sex ratio of $\frac{1}{2}$ should occur, irrespective of polygamy. This seems to be generally the case. According to Browne, in the Crypturgini, while *Poecilips gedeanus* clearly has the spanandrous system (only 21 males were observed among 196 beetles reared in 18 broods), *Carposinus perakensis* has typical 'monogamy':[27] 'The young adult bores directly out from its pupal cell'; the male is not degenerate and probably mates with the female in her incipient new gallery, where he subsequently takes part in caring for the nest; and 'among 58 young adults examined, 26 were male'. It is implied[27] that the sex ratio of $\frac{1}{2}$ holds generally with monogamous species, and other authors confirm this.[26]

With the Ipinae of the third group the male flies first to the new host, where he cuts a nuptial chamber. In response to the male's attractant scent, several females (but not more than eight) colonize the nuptial chamber of a successful male, are mated by him, and construct radiating egg galleries. Sex counts show a primary sex ratio of $\frac{1}{2}$;[26,29] I know of only one possible exception.[30] What happens to the surplus males is in no case entirely settled. Probably their numbers are somewhat reduced by the greater hazards involved in their pioneer role, but, at the same time, there is evidence that many males fail to attract any mate and die as bachelors.[26,29]

With bark beetles of the hylesine genus *Dendroctonus* the situation is rather less clear. *Dendroctonus frontalis*, which is monogamous, has a primary sex ratio of $\frac{1}{2}$.[31] So does *D. pseudotsugae*, which is slightly polygamous.[32] For *D. monticolae* a sex ratio of about $\frac{1}{3}$, both before eclosure and after attack, has been given, on the basis of numerous counts.[33] The system of *D. monticolae* suggested that some mating might take place before dispersal, since the 'polygamy' amounted to the observation of some monogamous pairs and some lone but inseminated females. But Reid found that, in fact, less than 1 per cent of the females had mated before dispersal. Such a proportion of inbreeding certainly could not explain the bias in the sex ratio, but perhaps it was not typical. Another species, *D. micans*, is quite clearly on the road to inbreeding and spanandry. In this case the social biology seems closely similar to that of *Xyleborus*, except that the sex ratio is less extreme and males are not degenerate. At least very occasionally the males of *D. micans* fly to join the female in the new gallery.[26] Evidently, as a whole the Scolytidae support the view that polygamy is not associated with primary sex-ratio bias unless there is also inbreeding.

THELYTOKY

Browne remarked[27] that 'the evolutionary trend in the extremely polygamous Scolytidae appears to be towards the elimination of the male'. He considered, however, that the occurrence of thelytoky in the group was by no means proved. There is no doubt that in the Hymenoptera similar trends have ended in thelytoky in many different evolutionary lines. From an evolutionary point of view, when sibmating is invariable, the sexual breeding system, arrhenotokous or otherwise, gives none of the usually cited advantages over asexual reproduction; it is in effect already a tree whose branches all are, or tend quickly to become after each mutation, completely homozygous. Therefore, in a case of the extreme biofacies, nothing of immediate importance is lost by a changeover to thelytoky, while convenience and perfect economy can be gained. Table 4.1 shows roughly the extent to which thelytoky occurs in taxa related to the species listed.

MULTIPLE SETTLING

As noted above, the argument leading to $(n-1)/2n$ as the 'unbeatable' sex ratio when genetic control is by females cannot be expected to apply exactly even to the simplest genetical system of inheritance. Therefore, to gain some idea of how reliable the formula may be expected to be, the case $n = 2$ has been investigated by simulation on a computer. The model is deterministic, the sexual system is male-haploid, and only two alleles are present. The behaviour of the model was found to be complex, and here only the results with obvious bearing on the unbeatable ratio are mentioned. When sex ratios 0.23, 0.22, and 0.21 were ascribed to the three female genotypes *GG*, *Gg*, and *gg*, respectively, from any starting frequency the population went slowly homozygous for *g*. When the ascribed ratios were 0.22, 0.21, and 0.20, it went slowly homozygous for *G*. When the ratios were 0.225, 0.215, and 0.205, it was apparent that the sex ratio would equilibrate between the two latter values. Thus, the unbeatable sex ratio seems to lie between 0.215 and 0.205. Numerous other runs of the model, with a wide variety of trios of sex ratios, gave no results contradictory to this view.

It is not understood why the unbeatable ratio differs from $\frac{1}{4}$ in this direction. There seems no reason, however, to think that the discrepancies in cases $n > 2$ are any worse than those in the case $n = 2$, for it is known that when n is infinite, corresponding to panmixia, the formula correctly gives the ratio as $\frac{1}{2}$.

It would be interesting to see how some other characteristics of the model, besides the unbeatable ratio, alter as n is increased. In the case of panmixia, if a population sex ratio of $\frac{1}{2}$ cannot be attained homozygously, it is yet established as an equilibrium, provided of course that the range of genotypic sex ratios covers this value. This is shown in the runs of the appropriate deterministic model which are graphed in Fig. 4.3. The outcome is different with the model for random settling in pairs. Thus, in a run in which the female genotypes were assigned sex ratios 0.4, 0.25, and 0.1, the equilibrium sex ratio was 0.175 $[p(G) = 0.25]$. With ratios 0.3, 0.2, and 0.1, there was a stable equilibrium sex ratio at 0.267 $[p(G) = 0.835]$.

Although the theoretical position is even less clear for $n > 2$, the ratio now lies at least in some respects, within known bounds. Since the great majority of recorded sex ratios of Hymenoptera are in the range $\frac{1}{4}$ to $\frac{1}{2}$, it is thought likely that the model does at least exemplify the forces that are operating.

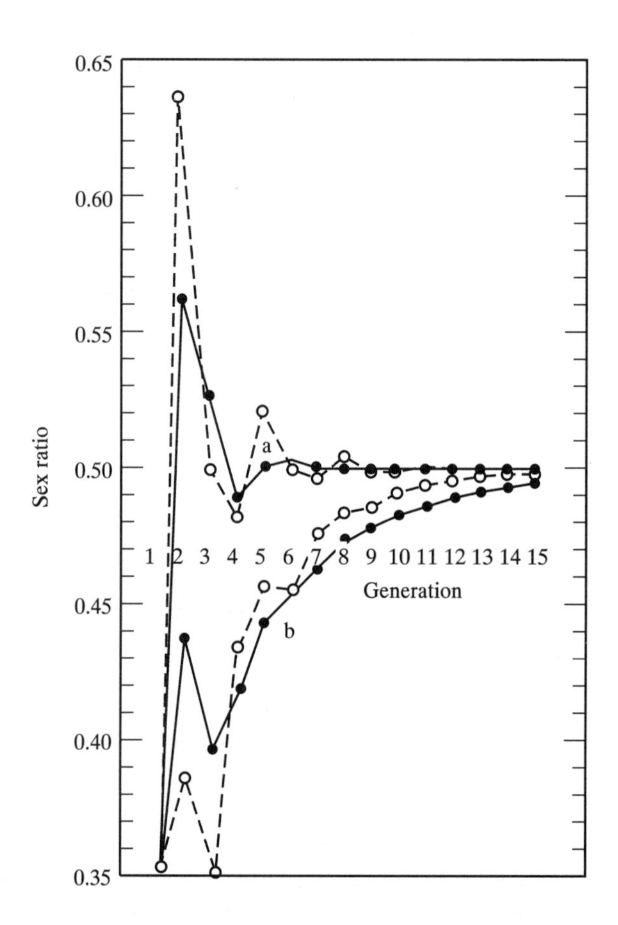

Figure 4.3 Equilibration of population sex ratio at $\frac{1}{2}$, under conditions of random mating. In all cases, control of the sex ratio is by the female, and it is assumed that the three genotypes *GG*, *Gg*, and *gg* give sex ratios 1, $\frac{1}{2}$, and 0, respectively. Dashed lines represent male haploids. Equilibrium frequencies of the genotypes *GG*, *Gg*, and *gg* are known, by analysis, to be as follows: in females, $\frac{1}{2}[(2)^{\frac{1}{2}} - 1], 2 - (2)^{\frac{1}{2}}$, and $\frac{1}{2}[(2)^{\frac{1}{2}} - 1]$; in diploid males, $\frac{1}{2}$, $(2)^{\frac{1}{2}} - 1$, and $1\frac{1}{2} - (2)^{\frac{1}{2}}$. For females, the equilibrium under male haploidy is the same; for males, it is *G*, $1/(2)^{\frac{1}{2}}$, and *g*, $1 - 1/(2)^{\frac{1}{2}}$. Runs were started by disturbing equilibrium populations as follows: males were left undisturbed; for (a), all female heterozygotes were removed; for (b), all *GG* and most *Gg* females were removed, leaving 1 *Gg* for every 999 *gg*. Under male haploidy the approaches to equilibrium are faster and less even; but in case (a) the final approach to the equilibrium sex ratio is from above—it is not oscillatory.[66]

OCCASIONAL OUTBREEDING: SEX-RATIO GAMES

Among the species listed in Table 4.1, *Acarophenax tribolii*[34] is the one most likely to conform, in respect of sibmating, to the ideal biofacies. The males of this species usually complete their life cycle, and die, before they are born. In general the lives outlined in Table 4.1 are not so secluded that outbreeding could not occur. For instance, not all mating of *Siteroptes graminum* takes place before the rupture of the maternal hysterosoma; galleries of *Xyleborus* sometimes coalesce; and so on. Pugnacity of males is a clear independent indication that outbreeding occurs, either through migration of males or through multiple settling as in the model situation. The males could hardly have evolved fighting instincts if their rivals were always of identical genotype, as brothers would be in a long sibmated line. The extreme pugnacity of *Melittobia acasta* males[35] must indicate that it is quite common for two females to attack the same host. With *Asolcus* both sexes are pugnacious. The possessive fighting of the females when parasitizing new hosts[36] would tend to prevent multiple settling—although not very effectively, to judge from the behaviour described. But the pugnacity of the males must, on the contrary, tend to cause outbreeding, since all but one of the males released from a batch of host eggs are driven away and must attempt to acquire sexual ownership of other batches. It is doubtless quite possible for them to do so if they find batches where emergence has not yet begun. Thus the pugnacious behaviour must both reflect and aggravate the biological need for it.

The somewhat anomalous cases, already mentioned, wherein the male is winged and the female is wingless may be interpreted rather similarly. With *Dusmetia sangwani* and *Cephalonomia quadridentata*, occasional outbreeding is encouraged not only by the males' ability to fly but also by the not uncommon occurrence of all-female groups. In *C. quadridentata*, Van Emden[37] recorded, among 108 groups of progeny from females known to have mated, 30 which were wholly female and, as if to balance this, nine which were wholly male. In the same record there were 59 groups which had just one male in an otherwise female group; statistically, this pattern, the ideal extreme spanandry, is still very distinctive. But from this and other cases in Table 4.1 it is evident that the typical occurrence of such extreme ratios does not necessarily imply perfect conformity with the biofacies in other respects.

The model analysis is not applicable to the sex-ratio problems arising from the intrusion of occasional migrant males. In essential features it is applicable to the problem of occasional double parasitism by females, and some recently described actual cases perhaps provide the best evidence yet obtained that, in an evolutionary sense, parasitoid species do really play the sort of sex-ratio game that is suggested by the model.

In a number of parasitoids it has been noted that more males are produced in laboratory culture than in the wild.[38] The theory shows that, if crowded conditions of mass culture lead to an increase in outbreeding and if the parent sample contained sufficient genetical variation in sex-ratio tendencies, there could be rapid selection of the more-male-producing genes and shift of the general sex ratio towards $\frac{1}{2}$. However this may be, several comparative studies have shown a more distinct phenomenon—that the sex ratio rises immediately when females are kept in crowded cultures. In some such cases (see, for example, ref. 39) experiments indicate that this rise is due, at least in part, to the greater survival of male larvae under conditions of superparasitism. But recently cases have been reported in which a female, on being crowded or on detecting previous parasitism, seems to alter her oviposition behaviour so as to produce a higher proportion of unfertilized eggs. Wylie[40] has shown that this happens in the case of *Nasonia vitripennis*. The normal sex ratio is variously given as 0.10 to 0.30, and this is one of the cases where, under conditions of superparasitism, more male larvae survive. In spite of this usually obscurative factor, by manipulating eggs so as to prevent overparasitization, Wylie showed that females laid more male eggs in response both to the detection of previous parasitism and to the simultaneous presence of other individuals walking over the host, especially if these others were females. When both factors operated together, a sex ratio of 0.59 was obtained, as against 0.20 in the controls. More tentatively, Jackson[41] reached similar conclusions with *Caraphractus cinctus*. The factor of differential mortality could not be controlled in this case, and it must be admitted that, on the basis of the results published, the effect in this case does not seem to have been established absolutely. The response appeared to be connected with crowding of the adults and not with perception of previous attack, even though an experienced female certainly could detect such attack, as Jackson showed.

In these species, double parasitism, at least, must certainly occur occasionally in the wild. From the analysis given above it should be evident that double parasitism confronts the female parasitoid with a difficult sex-ratio problem. In the way in which the success of a chosen sex ratio depends on choices made by the co-parasitizing females, this problem resembles certain problems discussed in the 'theory of games'. In the foregoing analysis a game-like element, of a kind, was present and made necessary the use of the word *unbeatable* to describe the ratio finally established. This word was applied in just the same sense in which it could be applied to the 'minimax' strategy of a zero-sum two-person game. Such a strategy should not, without qualification, be called optimum because it is not optimum against—although unbeaten by—any strategy differing from itself. This exactly is the case with the 'unbeatable' sex ratios referred to. But whereas in the foregoing cases the 'game' could be construed only rather artificially, as occurring between successive mutations acting rigidly in the statistical structure of the population as a whole, we are now concerned with a refined version which is very realistically game-like. Ability to adjust the sex ratio according to clues given by the immediate situation is undoubtedly potentially advantageous, and it leads to individual females behaving as players in a literal sense. As an illustration, let us consider a hypothetical situation which is simpler than the situations which parasitoids like *Nasonia* and *Caraphractus* actually face.

Suppose parasitism can be double but no higher than double; that double parasitism brings no extra mortality; and that the parasitoid must lay all its eggs in the one host. All egg batches should then be of the same size, and the actual size, if it is not too small, can be left out of the argument. Suppose the parasitoid is ideally gifted and can detect not only previous parasitism but also the sex ratio (x_0) of the eggs previously laid. If it lays its own eggs to give a sex ratio x, we find, from relation (4), that its fitness is proportional to

$$\frac{x}{x+x_0}[(1-x)+(1-x_0)]+(1-x)$$

or

$$1-2x+\frac{2x}{x+x_0}. \tag{6}$$

By differentiating, the value of x which gives the highest possible value for expression 6 is found to be given by

$$x^* = (x_0)^{\frac{1}{2}} - x_0. \tag{7}$$

Thus, the highest value x^* itself can take is $\frac{1}{4}$, corresponding to $x_0 = \frac{1}{4}$. For $x_0 > \frac{1}{4}$, $x^* < \frac{1}{4}$; for $x_0 < \frac{1}{4}$, $x_0 < x^* < \frac{1}{4}$. Hence it is evident that, through trial and error, two naive players would quickly learn that constant playing of $\frac{1}{4}$ was the optimum-yielding strategy.

In the 'game' under discussion, however, the play of $\frac{1}{4}$, although ultimately optimum-yielding, should be described as 'unexploitable' rather than 'unbeatable'. This distinction arises because the game is not 'zero-sum'. Biologically each parasitoid succeeds in so far as it contributes to the gene pool at large. Evolved instincts will cause it to seek the highest payoff in the sense of expression (4); except, perhaps, when the population is very small, it could have no interest simply in outscoring its co-parasitoid. But if, on the contrary, the players of such a game are opponents motivated to outscore, they would find that $\frac{1}{4}$ is beaten by a higher ratio; $x^\dagger$, the value of x which gives its player the greatest possible advantage over the player playing x_0, is found to be given by the relationship

$$x^\dagger = (2x_0)^{\frac{1}{2}} - x_0, \tag{8}$$

and this shows $\frac{1}{2}$ to be the unbeatable play.

Relations (7) and (8) are shown graphically on Fig. 4.4. This figure also shows fitnesses (derived from expression (6)) as a superimposed table, and it can be seen that the curve of x^* traces the maxima with respect to x on the implied fitness surface. If the diagram is rotated clockwise through one right-angle the tabulation takes the conventional form of a payoff matrix, say $\mathbf{W}$, showing 'payoffs' to the 'player' who has choice of row. The transpose, say $\mathbf{W}_0$, of such a matrix would show payoffs to the player having choice of column, and in Fig. 4.4 the curve of $x^{\dagger}$ traces maxima with respect to x on the surface corresponding to the matrix of differences $\mathbf{W} - \mathbf{W}_0$.

Apart from the considerations just given, our earlier analysis (for $n = 2$) showed that, unless extra refinements of behaviour are added, inclinations to use sex ratios aggressively as directed by equation (8) have no prospects under natural selection. Nevertheless, an arithmetical illustration of the earlier result seems useful and leads on to an interesting point. Suppose sex ratios are fixed by genotype and suppose that only types giving $\frac{1}{2}$ and $\frac{1}{4}$ are present in the population. Then the average fitness of type $\frac{1}{4}$ will be some weighted average between $1\frac{1}{2}$ and $1\frac{1}{6}$. Similarly the average fitness of type $\frac{1}{2}$ will be some weighted average between $1\frac{1}{3}$ and 1. The weights in the two cases depend on the frequencies of the different types of pair. Thus, if pairing is random, the weights are the same for both averages, being simply the frequencies of the types, and it is clear that, for all frequencies, type $\frac{1}{4}$ will have the higher average.[42]

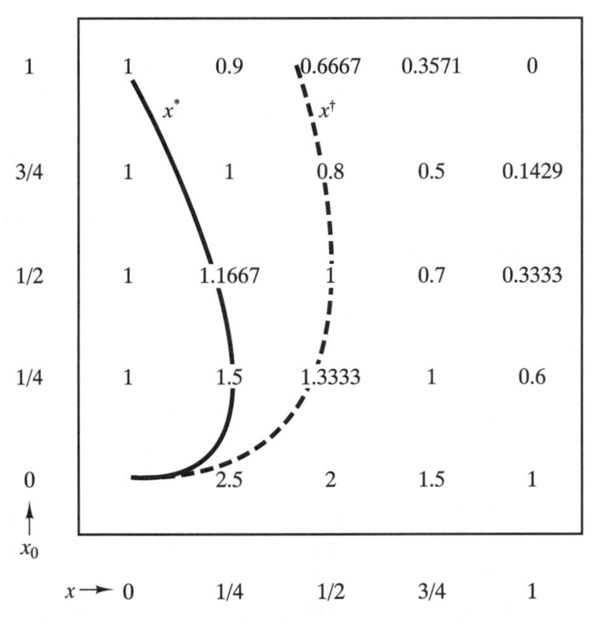

Figure 4.4 Sex-ratio-dependent fitness in a pair. Numbers in the box are values of expression (6), showing the fitness of a female using sex ratio x when her partner uses x_0. The fitness value at 0,0 is indeterminate; if sex ratio zero means strictly 'no males produced', the value actually must be between 0 and 1; but if it means merely 'extremely few males produced', the value depends on the ratio $x : x_0$ and on the male's ability to fertilize a large number of females—the theoretical upper limit of the fitness value is 3. The solid and dashed lines correspond to equations (7) and (8), respectively; their significance is explained in the text.

If pairing is non-random, however, unequal weightings result. Among non-random cases an interesting one is that in which $\frac{1}{2}$ contrives to pair off as often as possible with $\frac{1}{4}$: $\frac{1}{2}$ then spreads when $\frac{1}{2}$ and $\frac{1}{4}$ are equally common, although it is still unable to do so when it is very rare or very common. But if a genetic type could dissort in this way it should easily evolve the further ability to adjust its sex ratio according to whether it was paired with its own type or with the other type, and then if it played 0 in the former case and $\frac{1}{2}$ (or, even better, $(2x_0)^{\frac{1}{2}} - x_0$) in the latter, it would spread easily at all frequencies of occurrence against any non-discriminating type.

No such refined discrimination can be expected in parasitoids. Ability to detect whether a host is already parasitized is well known, but no clear case where a female is able to assess even the quantity of previous parasitism has yet been reported;[43] therefore, ability to assess sex ratio is extremely improbable. However, although this limits the precision with which a parasitoid can take advantage of a predecessor (in theory, by use of equation (7)), it does not lessen the game-like character of the situation, or even change the unbeatable ratio for the case of groups always of a fixed size.

With the case of occasional double parasitism likewise it may be seen that game-theory considerations apply and prescribe definite policies. Evolved tendencies will be such that first-comers to the host will use a sex ratio less female-biased than the workable extreme in order to allow for the chance that double parasitism will occur. The first-comers' best sex ratio will depend both on this chance (p, say), and on the sex ratio that second-comers can be expected to use. The second-comers' best sex ratio, assuming they can detect previous parasitism, will be based only on the expected sex ratio of the first-comers. Hence it can be shown either by game-theory reasoning (each individual supposedly considering the intelligent policy of the other) or, more realistically, by supposed behaviour gradually 'learned' by the species through natural selection, that, in the idealized conditions we are considering, the following responses will finally become established: first-comers will use sex ratio $p^2/(1+p)^2$ and second-comers will use $p/(1+p)^2$. Thus, a first-comer, because of the handicap of its uncertainty, has lower fitness than a second-comer.

It might be hoped that these two formulas would give a basis for checking the theory against experimental results. But, in its failure to take into account extra mortality due to superparasitism, the fact that females lay varying numbers of eggs and may move on from one host to another, and so on, the basis is certainly much too simple. Obviously the formulas do not check with Wylie's results,[40] since they cannot account for sex ratios above $\frac{1}{4}$. In fact the ratio of 0.59 Wylie obtained in one experiment is too high even to be the best ratio for the out-scoring game. But the variability of Wylie's ratios from ratios for controls and the fact that his experimental stock was not fresh from the wild (where triple parasitism should be very uncommon) also discourage hope of any detailed check. It can only be said that, in its direction, the effect that Wylie[40] and Jackson[41] have independently reported accords with the theory.

Acknowledgements

I thank Dr D. Bevan for information about bark beetles and Dr T. Lewis for information about thrips.

References and notes

1. R. A. Fisher, *The Genetical Theory of Natural Selection*, 2nd edn, p. 158 (Dover, New York, 1958; first published 1930).
2. W. F. Bodmer and A. W. F. Edwards, *Annals of Human Genetics* **24**, 239 (1960); W. A. Kolman, *American Naturalist* **94**, 373 (1960).
3. J. Verner, *American Naturalist* **99**, 419 (1965). Verner claims to show that, given factors causing fluctuations of the population's primary sex ratio, a 1:1 sex-ratio production proves the best overall genotypic strategy.
4. E. Novitski, *Genetics* **32**, 526 (1947).
5. L. C. Dunn, *Science* **144**, 260 (1964); L. Sandler and Y. Hiraizumi, *Canadian Journal of Genetics and Cytology* **3**, 34 (1961).
6. W. A. Hickey and G. B. Craig, Jr, *Genetics* **53**, 1177 (1966); *Canadian Journal of Genetics and Cytology* **8**, 260 (1966).
7. D. F. Owen, *Heredity* **21**, 443 (1966).
8. B. Wallace (in *Evolution* **2**, 189 (1948)) documents an actual case of X-linked sex-ratio drive. J. B. S. Haldane and S. D. Jayakar (in *Journal of Genetics* **59**, 29 (1964)) give conditions for equilibrium: a driving sex chromosome X' is not arrested unless XX' is more fit than $X'X'$ by at least $\frac{1}{3}$.
9. H. D. Stalker, *Genetics* **46**, 177 (1961).
10. S. W. Brown, *Science* **151**, 417 (1966).
11. G. D. Hanks, *Genetics* **50**, 123 (1964).
12. J. D. McCloskey, *American Naturalist* **100**, 211 (1966).
13. A. S. Parkes, in J. B. Cragg and N. W. Pirie (eds), *The Numbers of Man and of Animals* (Oliver and Boyd, Edinburgh, 1955); G. Sundell, *Journal of Embryology and Experimental Morphology* **10**, 58 (1962). The work of J. A. Weir (*Genetics* **45**, 1539 (1960) emphasizes the importance of the father in determining sex ratios in the mouse.
14. E. Schilling, *Journal of Reproduction and Fertility* **11**, 469 (1966).
15. D. Lewis, *Biological Reviews* **17**, 46 (1942).
16. A. Wilkes, *Science* **144**, 305 (1964); *Canadian Entomologist* **97**, 647 (1965).
17. If each haploid male can mate with k females and if mated females produce sex ratio e, some females fail to be mated if $k < (1 - e)/e$. Such cases finally have a regular alternation of sex ratios between e and $1 - ke$. Between these values there is an unstable sex-ratio equilibrium at $\frac{1}{2}\{2 + kf - [kf\,(4 + kf\,)^{\frac{1}{2}}]\}$, where $f = 1 - e$. In 1917, C. B. Williams gave arithmetical models showing alternation (*Journal of Genetics* **6**, 189 (1917) and A. F. Shull published observations suggesting that alternations may occur in thrips in the wild (*Genetics* **2**, 480 (1917)).

18. Bodmer and Edwards 1960 (ref. 2). They imply (I think incorrectly) that there is a possible situation for which their analysis is exact.
19. S. W. Brown, *Genetics* **49**, 797 (1964).
20. M. J. D. White, *Animal Cytology and Evolution*, pp. 239–46 (Cambridge University Press, Cambridge, 1954).
21. F. W. Edwards, *Entomologist's Monthly Magazine* **70**, 140 (1934).
22. C. W. Metz, *American Naturalist* **72**, 485 (1938).
23. F. W. Edwards, *Entomologist's Monthly Magazine* **49**, 209 (1913).
24. H. F. Barnes, *Gall Midges of Economic Importance*, Vols 1–6 (Crosby Lockwood, London, 1946–56).
25. S. Wright, *Ecology* **26**, 415 (1945).
26. C. Chararas, *Etude Biologique des Scolytides des Conifères, Encyclopédie Entomologique* (Lechevalier, Paris, 1962).
27. F. G. Browne, *Malayan Forest Record* **22**, 1 (1961).
28. P. F. Entwistle, *Proceedings of the Royal Entomological Society of London*, A **39**, 83 (1964); 14 per cent of broods had no male.
29. M. W. Blackman, *N. Y. State Colony Forestry Technical Publication* **16** (1), 11 (1915); D. L. Wood, *Pan-Pacific Entomologist* **38**, 141 (1962). Comparable evidence for polygynous birds is cited by J. Verner and M. F. Willson (*Ecology* **47**, 143 (1966)).
30. G. R. Hopping (in *Canadian Entomologist* **94**, 506 (1962)), reports the existence of a thelygenous strain in *Ips tridens*. Such strains are unlikely to be widespread in the genus (compare the 'sex-ratio' traits of *Drosophila pseudoobscura* and its allies).
31. E. A. Osgood, Jr and E. W. Clark, *Canadian Entomologist* **95**, 1106 (1963).
32. J. P. Vité and J. A. Rudinsky, *Forest Science* **3**, 156 (1957).
33. R. W. Reid, *Canadian Entomologist* **90**, 505 (1958).
34. R. Newstead and H. M. Duvall, *Royal Society Report No. 2 of Grain Pests (War) Committee* (London, 1918). A report by E. A. Elbadry and M. F. S. Tawfik (in *Annals of the Entomological Society of America* **59**, 458 (1966)) shows that *Adactylidium* has biology very similar to *Acarophenax*.
35. G. S. Graham-Smith, *Parasitology* **11**, 347 (1919).
36. F. Wilson, *Australian Journal of Zoology* **9**, 737 (1961).
37. F. Van Emden, *Zeitschrift für Morphologie und Ökologie der Tiere* **23**, 425 (1931). Even in the absence of occasional outbreeding, all female batches are not necessarily doomed. Unmated females of *Cephalonomia quadridentata* produce daughters after mating with their sons. This ability is widespread among arrhenotokous organisms, and is developed into a remarkable adaptation in *Melittobia acasta* (see F. B. Browne *Parasitology* **14**, 349 (1922)). *Xyleborus compactus* has the ability, and data suggest that it must be frequently used in nature (see note 28). In *Podapolipus diander* the female is regularly mated by a precocious son arising from her first egg (M. Volkonsky, *Archives de l'Institut Pasteur, Algerie* **18**, 321 (1940)).
38. C. P. Clausen, *Journal of the N.Y. Entomological Society* **47**, 1 (1939).

39. G. Salt, *Journal of Experimental Biology* **13**, 363 (1936). This effect accords with Fisher's principle. Superparasitism brings unlike genotypes together and causes at least partial outbreeding: therefore, since males are scarcer, they are under more intense selection for competition ability.

40. H. G. Wylie, *Canadian Entomologist* **98**, 645 (1966).

41. D. J. Jackson, *Transactions of the Royal Entomological Society of London* **118**, 23 (1966).

42. A similar argument shows that the play 0 also loses to the play $\frac{1}{4}$. The combination 0.0 gives the highest possible payoff to the group (namely, 4), so this 'solution' should be favoured by a 'group-selectionist'. From what has been said, the 'solution' $\frac{1}{2}, \frac{1}{2}$, with group payoff 2, should be favoured by the extreme believer in biological *bellum omnium contra omnes*. It is pleasing, therefore, to find that what turns out to be the true solution in this case, $\frac{1}{4}, \frac{1}{4}$, lies exactly midway between the others, both in position and in payoff.

43. G. Salt, *Symposia of the Society for General Microbiology* **15**, 96 (1961).

44. G. Grandi, *Bollettino del Laboratorio di Entomologia Instituto Superiore Agraria, Bologna* **2**, 1 (1929).

45. P. Rau, *Annals of the Entomological Society of America* **40**, 221 (1947); J. H. Fabre, *Souvenirs Entomologiques* **3**, 179 (1886).

46. A. A. Moursi (in *Bulletin de la Société Fouad de l'Institut Entomologique* **30**, 21 (1946); ibid., p. 39) reported very rare thelytoky.

47. M. F. Schuster, *Annals of the Entomological Society of America* **58**, 272 (1965).

48. C. P. Clausen, *University of California (Berkeley) Publications in Entomology* **3**, 253 (1924).

49. F. B. Browne, *Parasitology* **14**, 349 (1922). These are feeble, reluctant fliers.

50. E. R. Buckell, *Pan-Pacific Entomologist* **5**, 14 (1938); R. G. Scmieder, *Biological Bulletin* **74**, 256 (1938). The sex ratio is of Scmieder's second form.

51. T. H. C. Taylor, *The Biological Control of an Insect in Fiji: An Account of the Coconut Leaf-mining Beetle and its Parasite Complex* (Imperial Institute of Entomology, London, 1937).

52. G. Salt, *Parasitology* **29**, 539 (1937) and **30**, 511 (1938). The males are winged when bred on hosts other than *Sialis*.

53. F. Enoch, *Entomologist's Monthly Magazine* **34**, 152 (1898). M. Rimsky-Korsakov (*Russkoe Entomologicheskoe Obozrenie* **16**, 211 (1917)) reports that females are dimorphic; some are apterous. K. L. Henriksen (*Annales de Biologie Lacustre* **11**, 19 (1922) suggests the existence of a thelytokous race; like R. Heymons (*Deutsche Entomologische Zeitschrift* **1908**, 137 (1908)) he observed fewer males, and no mating in the host egg.

54. The sex ratio (from columns 3 and 4) is typical for third-instar hosts.

55. A. F. Satterthwait, *Journal of the N.Y. Entomological Society* **39**, 171 (1931).

56. M. C. Mossop, *Union of South Africa Department of Agriculture Science Bulletin* No. 81 (1929).

57. A. de Costa Lima, *Instituto Oswaldo Cruz Memórias* **21**, 201 (1928); A. Dreyfus and M. E. Breuer, *Genetics* **29**, 75 (1944). Sex ratio for *Triatoma megista* hosts.

58. Sex ratio for *Asolcus basalis*. A definite tendency towards the inclusion of at least one male among eggs laid in a host by *A. semistriatus* has been observed by M. Javahery (personal communication).

59. Van Emden 1931 (ref. 37). The males are dimorphic; some are apterous.

60. J. C. Bridwell, *Proceedings of the Hawaii Entomological Society* **4**, 291 (1920). Aptery occurs, but rarely. C. E. Keeler (*Psyche* **36**, 41 (1928) and ibid. p. 121) found a thelytokous race.

61. H. F. Willard, *U.S. Department of Agriculture Technical Bulletin* **19**, 1 (1927); A. Busck, *Insecutor Inscitiae menstruus* **5**, 3 (1917).

62. E. Pussard-Radulesco, *Annals de l'Ephiphyties* **16**, 103 (1930).

63. A. E. Treat, *Proceedings of the 10th International Congres of Entomology* **2**, 475 (1956).

64. K. W. Cooper, *Proceedings of the National Academy of Sciences, U.S.A.* **23**, 41 (1937).

65. E. E. Lindquist and W. D. Bedard, *Canadian Entomologist* **93**, 982 (1961).

66. The results of Bodmer and Edwards 1960 (ref. 18) on the rate of approach to equilibrium do not apply in detail to such extreme genetical cases as those of Fig. 4.1. In general it is true, as these workers concluded, that the rate of approach depends positively on the variance (in the cases shown, $\Delta X/(\frac{1}{2} - X)$ is ultimately 2.086V for normal reproduction and 2.88V for male haploidy). But if, for example, *Gg* gives all females, as in *Sciara coprophila* and *Chrysomyia rufifacies* (Metz 1938 (ref. 22); F. H. Ullerich, *Chromosoma* **14**, 45 (1963)), equilibrium is reached immediately.

The Thirteenth Fairy

Permanent narcoleptic seizure caused by the neglected fairy to Sleeping Beauty and all inmates of her palace was an extreme act normally thought of as spite. But by the stricter definitions of an evolutionist it probably wasn't. If the fairy lived on among others to whom her vengeful act became known, she probably was not passed over for the next party. Such benefit to herself dubs her act 'selfish', not 'spiteful'. The story may thus parallel the argument and examples of the coming paper which claims that in the real world adaptive true spite can hardly exist.

CHAPTER 5

SPITE AND PRICE

Selfish and Spiteful Behaviour in an Evolutionary Model

Go, go, go said the bird; human kind
Cannot bear very much reality.

T. S. ELIOT[1]

BY the time the paper of Chapter 4 came out in 1967 I had been living for a year in a flat in a converted mansion close to Silwood House itself. Silwood is in a belt outside London that is better described as countrified than as country, more as paddocky than ploughed, and is a belt where rambling Victorian and Edwardian houses are plentiful. Many are still homes, sometimes to well-known names. For example, John Lennon, the Beatle musician, owned the house just across the road from our Field Station, while Diana Dors, the film star, owned another just down the road. Ordinary mortals on salaries may be able to work in the area but have difficulty living there and most of the staff at Silwood either had houses many miles away or, like me, rented. My landlord at this time was Blue Star Garages Ltd, and the windows of my high-ceilinged flat looked out on their emblem reared over a filling station, one specimen of their chain. Other renting groups of graduate students were scattered through the same building and my flat was shared with Dr Yura Ulehla, a Czech plant physiologist who was then visiting the Field Station. Yura became and still is a close friend. By the time I began writing this paper I had also acquired another close friend, Christine Friess, with whom I was in love. When released at weekends from her studies in London, she usually joined us.

Most important for this chapter, however, is yet another character that I must introduce and for this must look back a little from the Sunninghill Blue Star Garage. First, my present senescent phase in Oxford needs to be distinguished carefully from what I will call my senescence phase (pre-Blue Star, see Chapter 3) at Silwood in the mid-1960s. It was actually during that senescence phase that an unusual man in his fifties came to London from New York. Oppositely to my efforts to enter marriage he had just been divorced. Also oppositely he was even a grandfather. Gradually by letters we became acquainted and by the time I wrote this paper a new thread that he brought into my scientific life, colourful and dreamlike but quite brief, had been half spun.

As with a dream I have difficulty with all the details. For a start I cannot quite remember when and how George Price first contacted me. I am almost certain he was one among the many writing to me for a copy of my 1964 paper and also am almost sure he accompanied his request with some information that he was busy with a related idea about which he would soon send me more information. A manuscript did eventually come from him but what I found set out was not any sort of a new derivation or correction of my 'kin selection' but rather a strange new formalism that was applicable to every kind of natural selection. Central to Price's approach was a covariance formula the like of which I had never seen. It presented a covariance between the expressed reproduction of an individual and the gene content of that individual with respect to a particular allele that affected fitness: divided by the mean fitness this covariance produced, for me at the time rather like a rabbit from a conjurer's hat, the change in frequency of the allele in the population. The formula was easily checked to be correct; yet the approach by which it came obviously owed nothing whatever to any previous account of selection theory I knew of. Price had not like the rest of us looked up the work of the pioneers when he first became interested in selection; instead he had worked out everything for himself. In doing so he had found himself on a new road and amid startling landscapes.[2]

Later I learned that he had been a physicist and chemist but had long ago turned from that to science journalism. His voice was squeaky and condescending, rather guarded, on the phone. I gathered he was living in central London and was already in contact with my former supervisor, Cedric Smith at the Galton Laboratory, about his new formula. Cedric was being very kind and had given him space to work there. He spoke of his formula as 'surprising for me too—quite a miracle'. I thought this reference to a 'miracle' was a mixture of metaphor and modesty at the time but I later realized he was more serious. When he started thinking about natural selection, he told me, he had known neither about population genetics nor statistics nor even what a covariance was. He had read my paper, and, shocked by it, as he admitted with strange emphasis, had determined at once that he would try to re-derive my conclusions and decide if they were right. He informed me solemnly that the check had been positive apart from small discrepancies that he would like sometime to take up with me—especially, he said, as there was something about spite. 'About what?', I said. 'Spitefulness—spiteful behaviour.' 'Oh—but you don't even cover selfishness in the paper you sent me,' I said, 'in fact you don't actually discuss kin selection at all.' 'That's true. That will all come later. Cedric has gotten me space to work . . . Have you seen how my formula works for group selection?' I told him, of course, no, and may have added something like: 'So you actually believe in that do you?' Up to this contact with Price, and indeed for some time after, I had regarded group selection as so ill defined, so woolly in the uses made by its proponents, and so generally powerless against selection at the individual and genic levels, that the idea might as well be omitted from the toolkit of a working evolutionist. If I did ask the question as to whether he believed in it I easily imagine George saying 'Oh, yes!', at once enthusiastic and imperious, followed probably by a long silence that would have been broken at last by some different topic.

He never did write down his views on group selection at all clearly but after I had finished following up the challenge of those similarly cryptic statements about the evolution of spite, with the outcome that makes up the paper in this chapter, I took up his claim about group selection with the result that is now Chapter 9. I am pleased to say that, amidst all else that I ought to have done and did not do, some months before he died I was on the phone telling him enthusiastically that through a 'group-level' extension of his formula I now had a far better understanding of group selection and was possessed of a far better tool for all forms of selection acting at one level or at many than I had ever had before. 'I thought you would see that', the squeaky laconic voice said, almost purring with approval for once. 'Then why aren't you working on it yourself, George? Why don't you publish it?', I asked. 'Oh, yes. Cedric wants me to also . . . But I have so many other things to do . . . population genetics is not my main work, as you know. But perhaps I should pray, see if I am mistaken.' In the end, as I say, the main explanation of the covariance approach as it applies to group selection was left to me.

Not long after the above conversation and just after Christmas 1974, in an abandoned building in Tolmers Square near Euston Road where he had been living as a squatter for several months, George jabbed nail scissors into his neck and snipped his carotid artery. He told me long ago of another failed attempt at suicide and the scar of that still showed. This second time he made sure. As a local and recent addressee on certain carbon copies of letters in the room, I found myself invited by the police to collect his effects. The room where he had been living was large and empty, up one flight of stairs and overlooking the street, a street which, however, could hardly be seen because of many strips of brown paper covering starred stone-holes in the windows and holding together the remaining slabs of glass. One bare bulb lit the room from mid-ceiling. Although the house was awaiting demolition the electricity was still on: it mightn't have been too freezing for George when he was there all alone over Christmas. I was told there had been an electric heater but after he died someone had taken it. A mattress on the floor, one chair, a table, and several ammunition boxes made the only furniture. Of all the books and furnishings that I remembered from our first meeting in his fairly luxurious flat near Oxford Circus there remained some clothes, a two-volume copy of Proust, and his typewriter. A cheap suitcase and some cardboard boxes contained most of his papers, others were scattered about on the ammunition chests. As I tidied what was worth taking into the suitcase, his dried blood crackled on the linoleum under my shoes: a basically tidy man, he had chosen to die on the open floor, not on his bed.

That is how his life became dreamlike for me and also how his colourful thread in my science and my life ran out. I tell the end of his story inappropriately here just to give substance to the earlier metaphors: I will say more about George Price, a man as deeply shocked by kin selection as the Victorian lady and her friends had been by evolution itself and yet so different in his reaction, in the introduction to Chapter 9, which will be more concerned with group selection. Meanwhile I return to the paper on spiteful behaviour that I wrote both for my own sake and to repay George for his hints. We were both in good spirits then, planning a Trojan horse to break into a dark castle. The castle was none other than the journal *Nature*, that whose portcullis clangs shut against so many hopes—even against hopes nurtured in some of the cleverest, hardest, and most insinuating minds of science, leave alone shutting against such spongy biological theorists and evolutionists as we were.

After what seemed a very long gestation (which puzzled me less after I had learned more detail of what the 'other work' he had referred to in his telephone conversations actually was) George had submitted a paper on his covariance selection formula to *Nature*. It was rejected but the referees seemed more confused about it than adamant and the editor was allowing a possibility of re-submission. George told me he didn't want to waste more time: he was for giving up *Nature* and trying somewhere easier—the *Journal of Theoretical Biology*, for example, would do. Not long before, however, his much earlier, rather patronizing remarks about kin selection and spite had induced me to work out the account that is in this paper. I had been delighted with the results not so much because of spite, which even now I regard as practically a non-starter for important evolutionary effects (but see Notes 3 and 4) but because I could reformulate all the results that I had had before in my 1964 paper in a vastly more economical and appropriate way.

I, too, of course, thought it would be nice to have my derivations published in *Nature* and I knew that the editor had changed from the one who had told me in 1962 to take my '$k > 1/r$' away to a 'socialogical journal'; I felt generously disposed to give the new editor a chance to publish both of our results together. I suggested a plan to George; he agreed, and very shortly we swung into action. His revised paper went to *Nature*; mine followed quickly; his came back not even reviewed; weeks later back came mine: mine was accepted! I now wrote to *Nature* regretting that I must withdraw it because the powerful new method I had used and cited had been recently refused by *Nature* and I could not proceed until that method was published somewhere. There followed a telephone call from the editor saying he was ringing me because he did not know how to contact Dr Price. In view of the support Price had had from my accepted paper he had decided it would be worthwhile to send out Price's manuscript for review after all; could I suggest how to find him or would I like to suggest to Price directly that he should submit yet again?

Nature has an editor who phones you, a human being! And an editor who can be swayed by a collusion! The happy ending is that *Nature* took both papers. George's came out in August, mine in December. A distinctive feature of George's is that, as correctly reflects his completely independent approach and the generality he had attained, there was not a single citation. I doubt *Nature* has ever regretted the two papers. Yet if in 1970 George and I had combined instead of splitting, if we had sent a single paper deriving the covariance formula, re-establishing kin selection with it, and finally exemplifying spite, all in one glorious synthesis, I am fairly sure that the joint effort would have been rejected.

In any case we certainly began no revolution in *Nature*'s policy: to this day it accepts extremely few evolutionary papers and those it does still mostly flow via favoured cliques and topics. For sure, we were neither *Nature*'s first nor last of colluding evolutionists.

References and notes

1. T. S. Eliot, *Collected Poems 1909–1962* (Faber and Faber, London, 1974); 'Burnt Norton', 1935.
2. S. A. Frank, George Price's contributions to evolutionary genetics, *Journal of Theoretical Biology* **175**, 373–88 (1995). Steven Frank has reminded me that the covariance part of Price's formula was published twice at about the same time that Price was deriving it. Once was by Alan Robertson (*Animal Production* **8**, 95–108 (1966)). After Price's own first publication someone drew his attention to this: in a letter to me he mentions the matter and notes his intention to acknowledge Robertson's priority in his next publication. He did not live to do this. C. C. Li in 1967 had also published the covariance part of Price's equation; again, Price discovered this after his own work was done. These prior formulations, however, deal only with the first covariance term of Price's equation and therefore completely fail to suggest the nested analysis of levels of selection that is a major beauty of the approach (see Chapter 9).
3. R. Gadagkar, Can animals be spiteful?, *Trends in Ecology and Evolution* **8**, 232–4 (1994).
4. L. Keller, *et al.*, Spiteful animals still to be discovered, *Trends in Ecology and Evolution* **9**, 103 (1994).

SELFISH AND SPITEFUL BEHAVIOUR IN AN EVOLUTIONARY MODEL†

W. D. HAMILTON

Incidents in which an animal attacks another of the same species, drives it from a territory, or even kills and devours it are commonplace. They may be described as examples of biological selfishness. The effect consists of two obvious parts: the gains (in fitness) of the victor and the losses of the victim. Attempts to secure the gains are easily understood to be adaptive: this is the fundamental response to what Darwin called the 'struggle for existence'. But, considering the more controversial catch-phrase of evolutionary theory—'the survival of the fittest'—it seems to be a neglected question whether the harm delivered to an adversary is always merely an unfortunate consequence of adaptations for survival. Could such harm ever be adaptive in itself? Or nearer, to the possibility of a test, would we ever expect an animal to be ready to harm itself in order to harm another more? Such behaviour could be called spite. Is it ever observed?

Previously (see Chapters 1 and 2) I showed that the average genetical relatedness of interacting individuals is an important factor in the evolution of social adaptations. In the model, selfishness within certain limits was readily accounted for; spite did not seem possible. But another line of reasoning (see Chapter 6) shows that spite can be selected. Independently, using his new formulation of natural selection[1] in a more general analysis, Dr G. R. Price reached the same conclusion. The original argument was based on a supposedly infinite population in which the total average relatedness of an individual to other members of the population was zero. With realistic finite populations this is not quite the case; the criteria for ideally adaptive behaviour are then more complex and involve consideration of the general average of relatedness. I report here a synthesis of unpublished work by Price and my results—including an extension to cover inbreeding—and demonstrate that although it can

†*Nature* **228**, 1218–20 (1970).

have a selective advantage, spite is unlikely to become permanent and elaborated into a complex adaptation. It should be looked for in dwindling panmictic species.

Let the n individuals of a population be numbered and let s_{ij} be an additive effect caused by individual i to the fitness of individual j. Considering the total of such effects, including the effect s_{ij}, which is equivalent to the 'selective advantage' of the classical theory, let individual j actually express the fitness

$$w_j = 1 + \sum_i s_{ij}.$$

The mean fitness in the population is

$$w = \frac{1}{n} \sum_j w_j.$$

Similarly, if q_j represents the gene frequency of an allele A in $j(q_j = 0, \frac{1}{2}$ or $1)$, the population gene frequency (in the continuous range $0 \rightarrow 1$) is

$$q = \frac{1}{n} \sum_j q_j.$$

Using Price's basic covariance formula, one generation of natural selection creates an expected new gene frequency q' by a step of magnitude

$$q' - q = \Delta q = \frac{1}{w} \operatorname{Cov}(q_j, \sum_i s_{ij}),$$

which can be put in the form

$$\Delta q = \frac{1}{w} \sum_i \left\{ \frac{1}{n} \sum_j (q_j - q) s_{ij} \right\}. \tag{1}$$

This form associates the social effects with the individuals affected. Considerations of genetical kinship can give a statistical re-association of the effects with the individuals that cause them.

The genotype of an individual j can be considered as consisting of two parts (in the sense of statistical expectations) with respect to the genotype of a relative i: one part consists of genes which are copies by direct replication of genes in i; the other part consists of non-replica genes. The replica part has gene frequency q_i. The non-replica part has a gene frequency which will generally be close to the population gene frequency q. Exactly, b_{ij}, representing the replica fraction, is to satisfy

$$E(q_j) = \frac{1}{1 - b_i} \{ (b_{ij} - b_i) q_i + (1 - b_{ij}) q \}, \tag{2}$$

where

$$b_i = \frac{1}{n}\sum_j b_{ij}.$$

With the further abbreviation $b_{ijS} = (b_{ij} - b_i)/(1 - b_i)$ (2) corresponds to

$$E(q_j) - q = b_{ijS}(q_i - q),$$

which, because Δq is itself an expected change, gives a legitimate substitution for $(q_j - q)$ in (1). The result is

$$\Delta q = \frac{1}{w}\ \mathrm{Cov}(q_i, x_i), \tag{3}$$

where

$$x_i = \sum_j b_{ijS}\ s_{ij}. \tag{4}$$

As $b_i \to 0$, this definition of x_i becomes the same as that given for the quantity called 'the inclusive fitness effect' previously (Chapter 2), but even in the limit it is incorrect to identify b_{ijS} with Wright's coefficient of relationship. Wright's coefficient is a correlation coefficient[2] whereas the coefficient required here is the corresponding regression coefficient. The following formula can be used to approximate it from a pedigree diagram (the approximation being exact only in a population not undergoing selection):

$$b_{ij} = \frac{2r_{ij}}{1 + F_i},$$

where F_i and r_{ij} are probabilities of alikeness by direct replication for pairs of allelic genes which are, respectively, the pair which constitutes the genotype of i, and two randomly selected genes drawn one from i and one from j. If, following Wright, r and F are defined as correlations instead of probabilities, the same formula gives b_{ijS} directly. The subscript S is chosen to accord with Wright's later notation.[3] His analysis suggests extensions of (3) when there is more than just one population, but here I continue to consider just one.

Now suppose behaviour is determined by genotype. Consider first a case with two alleles A and $+$ and let $\overline{AA}$ and $\overline{++}$ denote inbred homozygotes which, having a different set of relationships with other members of the population, may have different average inclusive effects from the non-inbred homozygotes denoted AA and $++$. Grouping the contributions to the covariance in (3) according to the five 'genotypes' so specified, we find average product contributions of $(1 - q)x_{\overline{AA}}, (1 - q)x_{AA}, (\frac{1}{2} - q)x_{A+}, (0 - q)x_{++}$ and $(0 - q)x_{\overline{++}}$ made with expected frequencies $Fq, (1 - F)q^2$, $(1 - F)2pq, (1 - F)p^2$, and Fp, respectively,[4] where $p = 1 - q$. Forming the covariance, substituting in (3) and rearranging:

$$\Delta q = \frac{pq}{w}[(1 - F)\{q(x_{AA} - x_{A+}) + p(x_{A+} - x_{++})\} + F(x_{\overline{AA}} - x_{\overline{++}})] \tag{5}$$

In the sense of expected change this would be exact if the b_{ij} were really known, and if x_{AA} and the others are averages calculated for the current generation. The x cannot represent a perfectly constant inclusive fitnesses associated with a genotype as is easily seen in the case of a population in which the level of inbreeding tends to no limit short of complete fixation. Passing over this unsolved difficulty—which is probably not very serious in practice—if '+' refers to a set of alleles having differing connections with inclusive fitness, and therefore to a set having changing frequencies among themselves, (5) remains valid provided the averages involved in x_{A+}, x_{++} and $x_{\overline{++}}$ are broadened accordingly. The following compact form also holds for multiple alleles:

$$\Delta q = \frac{pq}{2w}\frac{\mathrm{d}(x + Fx_F)}{\mathrm{d}q},$$

where x and x_F are the mean inclusive fitness effects in the whole population and in the inbred part of it, respectively. The case of this applying in the classical theory has long been known.[5] An equivalent form, less the involvement of inbreeding, has been given in a previous article and elsewhere, and it was suggested that some kind of a mean inclusive fitness tends to maximize under natural selection.

Equations (3) and (4) show more directly that genes whose concentration in individuals tends to be correlated with high inclusive fitness are favoured in natural selection. An omniscient animal i choosing behavioural strategies to maximize his inclusive fitness would seem to value the fitness of all other individuals against his own according to the value of b_{ij}. Considering actions which cost him nothing, for example, he would treat all individuals more related to him than the average ($b_{ij} > b_i$) as 'relatives and friends' and would benefit them when he could; he would be indifferent to any individuals having the average relationship itself; and he would treat as 'enemies', harming them when he could, all individuals having less than average relationship. If social effects are difficult to achieve, and do cost something, the animal should be less indiscriminate: his most striking responses would be to the closest and least related individuals with risks being taken to benefit the former and to harm the latter.

Behaviour which harms others without benefit to the self may well be called 'spiteful'. The word has already been used in this sense in connection with behaviour in more elementary evolutionary models (Chapter 6). Seemingly spiteful actions are sometimes observed[3]. Most of these concern pairs of different species or subspecies. Relationship is zero but the spite is probably best considered from the point of view of interspecies competition (for example, egg-chipping by wrens[6]). Of intraspecific cases it is doubtful if any can be considered confirmation of the present model. 'Merely selfish' interpretations are too readily available, as the following three cases show.

1. The apparently spiteful wrecking practised by male bowerbirds in bowers of their neighbours[7] probably increases the number of matings that a male achieves; it is known that bowerbirds are polygamous and promiscuous, and the male who wrecks neighbouring bowers increases the relative attractiveness of his own.
2. Larvae of two species of mosquitoes (*Toxorhynchites*) are predatory and cannibalistic in all stages. Just before pupating a larva tries to kill, but does not eat, all other larvae in its tree hole or bamboo hollow. This apparently wanton killing is probably self-protective; the pupa is relatively helpless and would be likely to fall a prey to any larva that had been spared.[8]
3. The first caterpillar of the corn ear worm (*Heliothis zea*) to establish itself inside an ear of maize 'usually eats all subsequent arrivals. It is very uncommon for more than one ear worm to survive in each cob although the food would be sufficient for 2 or more'.[9] Such behaviour seems truly spiteful; nevertheless it may well have evolved in smaller flower heads or ancestral maize cobs that were not ample food for two larvae.

Irrespective of spite, the present theory predicts that in cases (2) and (3) the female normally lays only one egg per cob or waterhole; if this were not the case even extreme selfishness could not be favoured by selection. *Heliothis* seems to have this habit in nature[10] in spite of a contrary impression given by Kirkpatrick.[9] But why, if the model is correct, are more convincing examples of spite hard to find? First, I suggest, because all actions do cost something. Second, because an animal will not normally have any way of recognizing which other members of its species have less than average relationships. Memory, local odour, and nearness to place of birth may permit recognition of close relatives, but this leaves a remainder whose average relationship will be only very slightly less (in large populations) than the total average. Moreover, with common patterns of dispersal the unrecognized individuals most likely to be encountered are the ones likely to be most related. Third, because single populations that are so small in 'effective size' as to have b_i generally much different from zero must be in a precarious position already, and the selection of a gene causing spite can only hasten their extinction. If, on the other hand, there is a composite population made up of many such small precarious elements, either we must recognize a process of group selection in which those elements that have not yet been overwhelmed by selfish and spiteful mutations replace those that have, or we must consider relationship against the background of the total population, as seems the natural point of view if the subpopulations are emitting and receiving migrants.

Taken together these points suggest that my previous account is not seriously inaccurate for social behaviour in realistic circumstances. The point that has now been brought into focus, the adaptiveness of selfishness and spite in panmictic populations, suggests one reason why panmixia hardly occurs. When it does occur, the selection of selfish traits reduces mean fitness and the population is forced to retreat from its less favourable habitats. If it remains panmictic the trend may continue and lead to extinction. If, on the other hand, the population breaks into isolates as population declines, a breeding structure may be achieved in which further spread of selfishness, and hence further decline, can be resisted.

References

1. G. R. Price, *Nature* **227**, 520 (1970).
2. S. Wright, *American Naturalist* **56**, 330 (1922).
3. S. Wright, *Genetics* **19**, 395 (1965).
4. S. Wright, *Annals of Eugenics* **15**, 323 (1951).
5. S. Wright, *Bulletin of the American Mathematical Society* **48**, 223 (1942).
6. G. H. Orians, and M. F. Willson, *Ecology* **45**, 736 (1964).
7. A. J. Marshall, *Bowerbirds* (Oxford University Press, 1954).
8. P. S. Corbet, and A. Griffiths, *Proceedings of the Royal Entomological Society* A, **38**, 125 (1963).
9. T. W. Kirkpatrick, *Insect Life in the Tropics* (Longmans, London, 1957).
10. M. G. Walker, *Canadian Journal of Research* **20**, 235 (1942).

CHAPTER 6

A new shield for Achilles

As described by Homer in the *Iliad*, the blank surface that the smith god Hephaestus here polishes is to receive an intricate design which will include, amidst images of both cities and countryside, vivid depictions of Hellenes at War and Hellenes at Peace. No graphic artist ancient or modern has attempted representation of all Homer's detail but the draughtsman of this red-figure amphora hints at what he can. A simply dressed mother (Thetis) commissions the work that will protect her son in battle; on the wall behind a weakling smith (Hephaestus) hang tokens for the scenes he is about to engrave: the armour of a soldier, and the tools of the artisan that the soldier fights to protect or enslave.

CHAPTER 6

AMERICA

Selection of Selfish and Altruistic Behaviour in Some Extreme Models

Then fly and do not weaken. They will hound you yet
Through seas and island cities, over the vast continent,
Wherever the earth's face is hard with wanderers' feet.
AESCHYLUS[1]

It was the curse of mankind that these incongruous fagots were
thus bound together—that in the agonised womb of consciousness,
these polar twins should continually be struggling.
ROBERT LOUIS STEVENSON[2]

THE paper of this chapter was my first intellectual wild oat, the first for publication in a symposium volume. Others were to follow and my promiscuity of the kind has increased steadily. The oats are incited by such temptations as reputation, money (honoraria), and travel to exciting places. Defects of the present example will prepare the reader for the regular shortcomings which are haste, lack of focus, and lack of anything new to say.

The symposium itself, also the first I ever attended, was a rather grand and international affair held in 1969 in Washington DC, and titled 'Man and Beast: Comparative Social Behavior'. It was organized by the Smithsonian Institution seemingly with the object of impressing politicians and the public with current wonders of behavioural, evolutionary, and human biology, supposedly convincing them that even these sometimes derided and ragged fringes of science were making progress and deserved continuing support. I was, of course, delighted with the honour of being one of the few principal speakers.

With the arrival of the unexpected letter and invitation, the crank or trivialist I had once feared myself to be shrank back farther still into the shadows and from this meeting on almost ceased to trouble me. Clearly, my papers on kinship theory were gaining recognition, at least in the US.

When the symposium took place I was only recently back from my second, almost year-long visit to Brazil. Most of my time there I had been travelling with my wife in the remote interior collecting social wasps. Even in the civilized south of Brazil, in imperial Rio and industrial São Paulo, such words as 'interior' and 'sertão' still evoked distance and danger, a life that was dusty and basic, ruled by heat and bugs. Freed only a few weeks from that life I found myself not just in a hotel of São Paulo but of Washington, DC. Still vivid in my mind were the hotels of the BR14 (the Belém–Brasilia), along which my wife and I had travelled again. I thought of their earth floors, the bedrooms empty but for the hooks on the walls whence you might hang your hammock. I thought of how out in the cerrado, in our more usual campsites, wiry grass wasn't deep or bare sand soft like the carpet in this Washington room, and how when tussocks tripped you, they didn't cannon you to a humming refrigerator tinkling with glass and (as would appear at check-out time) unbelievably expensive fluids. I thought how in this hotel it wasn't mosquitoes whining through paneless windows, nor dreaded barbeiro bugs, bearers of Chagas' trypanosomes and sneaking from cracks in clay walls, that came to bite your fingers; instead it was simple and physical static hitting you from door handles and elevator buttons . . . Never had I imagined newness, tidiness, and luxury such as I found around me; even a swearing and sharing Australian was hardly able to spoil the novelty. Yet musing on all this as I sat on my bed with the Australian blessedly absent for a half hour, I was remembering a hardness, the chill of plateau dawns, dew on cerrado flowers, first sunlit and 'backward-flying' toucans crossing from copse to copse through an infinite arch and the small hills hiding also, a little out of sight, a seriema pealing its dawn call like cracked bells. My socks scattered on the carpet set me thinking of others I had left carelessly under my hammock in Brazil and of the swarms of termites dimly perceived in the dawn half light consuming their last remains—and then of the same termites starting on my bare feet (actually probably biting more in self-defence) when I stepped out to investigate. Here I could be lazy, untidy, unbitten, at ease—and yet, comparing the luxury of the A—— Hotel to a dawn in a cerrado camp site, which did I prefer?

Of the meeting itself my memory brings talks in huge dark halls, speaking myself into the darkness to dim ranks of faces, food, and flowers banked high in zoos, museums, and even in some other massive building where I was told President Lincoln once held a ball (I believe that was where I spoke). I recall talking to senators who seemed proportioned in size to the halls where they held court. More vividly still I recall a more European-sized wife of one senator pinning me at once with her chin and her fierce eyes as she asked me how my theory could help to reduce violence and crime in America. I recall names dropping round me like warm snow flakes (a special Washington style I was to decide after a second visit), and just a few of those names being known to me and the rest potentially officers of Xerxes' Persian army as far as they mattered to or impressed me. I recall scientists of every type but most of them larger both physically and, seemingly, in spirit than those I knew in England. Here were the laureates, the pompous, the hirsute, the fantastical, the unbelievably industrious, the funny, pugilistic, queer . . . ; Tolstoyans like Alexander, Rabelaisians like Chagnon. Perhaps I would have been less amazed and less entertained if I had been to a scientific meeting in England, but this was my first. Now that I think of it, my sensation of my hotel being dull certainly did not apply to the meetings even if their liveliness was caused only by personalities and the style of talk.

I think that some of my naivety and confusion at the meeting, reflecting a deep-rooted uncertainty as to why so much money was being spent and what I was there for, appears in the paper that follows. It was drafted quickly before the symposium and revised after it, and it turned into a rambling miscellany of what I recently had been thinking about combined with an artificial effort to draw out anything that appeared relevant to an understanding of Man. There is little wrong that I know of but nothing particularly exciting or right either. The most definite mistake I have noticed—two typos in the first equation—is corrected in the version reprinted here. An emphasis that I now think may be quite mistaken affects my treatment of animal fights in the opening paragraphs. Matching his presentation with the idea he aimed to impart, including some vivid sketches of the actions of two boxers, Richard Alexander explained to me that there was less that needs to be interpreted as restraint in a typical fight than I had assumed; more, what I was seeing was just prudence regarding the damage a losing and desperate adversary may still inflict. Most of this I accepted (one almost had to accept with Alexander) but I stayed convinced that some combatants, especially among the social hunters such as wolves, hyenas, and mongooses, sometimes pull their punches in their interactions of dominance in the interest not of themselves directly but of the strength of their group. This was how the behaviour had looked to me in, say, fighting Brazilian street dogs and it was also how I had seen it in *Polistes* wasps. Restraint was also sound theoretically by my arguments of inclusive fitness. It would be foolish for the group, and hence ultimately for every individual and gene contained in that group, if the behaviour should be otherwise.

From the positive side of my miscellany, what seems worth noting? Perhaps the most novel and useful formula is that in the first appendix, which sketches the handling of relatedness in cases where there is inbreeding (a topic soon to be expanded in the paper of Chapter 8). Care is needed with the formula, however: the inbreeding coefficients must be based at an appropriate population scale.[3] One of the most surprising conceptual points in the paper is also concerned with altruism under inbreeding and may seem even contradictory to altruism being correlated with relatedness. This is in my verbal caution that low dispersal by itself (population 'viscosity') as a way of reaching high relatedness has snags. The point is that, to be effective, altruism must put offspring into competition with non-altruists, not bunch them in a wasteful competition with their own kind. Recently this point has been treated explicitly and quantitatively with some quite surprising conclusions.[3–5]

If relatives cannot avoid the bunching then the group itself must expand at the expense of others. For this to work most effectively expansive groups need to replace other groups; in essence, members need and are expected to evolve a degree of xenophobia.[6] This line leads on to thoughts that for the re-slanted spiritual descendants of the prim Victorians of my second introduction remain quite paralysing and

I have to admit that at the time the thoughts were painful enough to me. It is hard even to feel and harder still to write in a way that runs counter to a current world view, especially a moral one, and it is all the harder when the way is re-shaping a plane of perfection to which all civilized cultures are thought to be striving. A scientist or philosopher with a programme of such heresy has to be tough if he or she is to communicate it and, while doing so and for long after, must endure the tortures of Orestes.[1,7] For me it was the discussion of the darker of the 'innate aptitudes' that I believed must exist in all human populations and most individual humans, and of the selection our forbears must have undergone through competition between populations, creating their warlike inclinations (including as a sideline, their relish in cruelty) that caused me most pain to write. The feelings remained acute when I had to write similarly for my next symposium contribution (Chapter 9): yet I am glad I gave the discussions I did. Theory concerning the nature of the 'beast within' was why I had been invited and I continue to believe that only from a basis of honest description can there be hope of taming what we have and may not like.

At the time I had only shreds of evidence for proto-warlike activities in vertebrate animals, one of them being from writings of my co-speaker at the conference, Hans Kummer, who had occasionally observed mass hostilities between two baboon troops. Since then group antagonisms discovered in our much nearer relative, the chimpanzee, fit in surprising detail the prototype for hominid hunter-gatherer warfare that I sketched in this paper.[7,8] Like propensities have also been discovered in other group-hunting carnivores and a murderous hostility to outsiders is known in some social omnivores as well. Since the symposium it has surprised me to see the growing strength of this evidence, especially the dawn of a possibility that an unbroken thread of such violence may trace back to our common ancestor with chimpanzees. It now seems generally accepted that humans probably had warlike propensities as hunter-gatherers throughout the long reaches of the palaeolithic—that is, during all that time when most of our psychic biases were in formation or being strengthened.[9] Previous to all the recent evidence I had been content with the argument that chimpanzees, which surely on a basis of brain size alone are bound to be less able 'machiavellians' than *Homo* (whether the latter is at the *erectus* or the *sapiens* level), would not find themselves gaining much by group violence. Therefore I believed that the violent and sadistic ideas, which seemed to arise so easily in my own psyche in certain moods, must be vestiges from a period that occurred subsequent to the separation of the chimpanzee line. I have rather expected evidence to come from fossil finds of *Australopithecus* or early *Homo*. Although it is hard to imagine what this evidence could be, I still suspect that indications of particularly dark periods may appear. Again novelists, tapping biases of human psychology in their search for themes to engage their readers, may be in the van in presenting some of the grimmest truths about our ancestors: thus it would not surprise me to learn of palaeoarchaeological evidence of torture or of cruel performances by man-apes similar to that which Arthur Conan Doyle put long ago into his novel, *The Lost World*.

The very isolation of *Homo sapiens* from closely allied forms hints at such competitive intolerance. It has also occurred to me that important parts of the human record may lie concealed not just by special local ill-chances of palaeontology but more systematically. For example, bones and artefacts of an almost unknown rich seashore phase of human evolution could have been submerged wholesale by the deepening of the oceans during our present interglacial. With the recent evidence from chimpanzees, however, such speculation is hardly needed. The way the gangs of our childhood connect with forays away from home, with tree houses and with dens in woods, ought to have prepared me to think of chimpanzees and of much more ancient origins than I did. Now it seems to me that, kinship theory notwithstanding, metaphorically we climb down out of our forests of xenophobia and thence come out much more directly than I had imagined into open plains of commerce and relative toleration. Unfortunately all is still not bright out in the wild grasslands, and moving beyond them to the even more fertile plains of the neolithic and to agriculture, we arrive, as is increasingly endorsed by discovery and translation of early written records, at the use of a new style of warfare, one that is both wider in scale and more mercenary.

It can be seen that the ideas I put into the paper are not just from population and natural selection theory. For better or worse some originate out of what I hope to have been an honest analysis of my own psyche. Possibly I am just unlucky in my variant of this organ and introspection for others can be sunnier; however, I seem to see very numerous signs that I am not alone. If so (and here I summarize), we may have good reason to consider some aspects of human character inappropriate in the modern world and to wish certain traits would disappear. If we decide we need policy for assisting the disappearance, it will be well to understand how human nature was constructed. To gain this insight we must use the best factual and theoretical evidence available. I first read the passage of Aeschylus that I have set at the head of this chapter in a secondhand book shop in central São Paulo in 1964. In the dark alcove shut under the 20 teeming downtown storeys, applying the passage to myself, I even cried: this was my relief, realizing how others had known my troubles so long before. But later, following the advice given, I became hardened again and since then have realized slowly how Orestes is actually all of humankind and Athens, city of reason, the goal we are all seeking where our past can be forgiven.

A point also near the end of the paper about how a Prisoner's Dilemma (PD) can be resolved via relatedness seems trivial once the idea of inclusive fitness has been grasped but appeared worth making in a context where few before the meeting would have heard either of the dilemma or of kin selection. Possibly it isn't so obvious either[10] and, extending this point, I am glad that I emphasized that such resolution would be partial: given relatedness within a group the dilemma simply reappears at the intergroup level where, alas, as I already emphasized, it has only increased its potential to be destructive. The sequence runs roughly thus: the first individual cheating and the response it draws, then the feud, then the tribal fight and the small war, . . . lastly, in the final stages, the imperial or the superpower war. This protean, multilevel aspect of the dilemma was, I think, familiar to games theorists well before I came to it.[11]

To summarize, I am as certain now as I was in 1969 that reliance on human nature as currently evolved, the instincts of a supposed 'noble savage', is no answer to the Prisoner's Dilemma or to Malthus. Group selection[12] (an idea needing much caution in the first place because it is apt to be a weak force) doesn't exorcise harsher facets of natural selection. 'Liberal' thinkers should realize that fervent 'belief' in evolution at the group level, and especially any idea that group selection obviates supposedly unnecessary or non-existent harsh aspects of natural selection, actually starts them at once on a course that heads straight towards Fascist ideology. This is not difficult to see from Fascist propaganda and, reading a little more between the lines, a route that is similar and was perhaps initially even identical, has always also been signposted from Marxist propaganda[13]—and that signpost often followed. But, if not in human nature, where else can we place trust, our hopes for a peaceful and creative future? Is a steadily evolving economic motivation (reciprocation) already in the long haul of rescuing us from essentially tribal 'final solutions' of the past? Or is this new feature really going to do no more than offer us new kinds of cruelty and wastefulness to take the place of the old? For grim realism on this question from far back and for other generalities which are hard to dismiss even today, the reader may be referred to a little known poem by Robert Louis Stevenson called 'The Woodman'.[14]

This is no place to go into such questions in any detail, but the sort of new dilemma that is being created through the 'economizing' of humanity may be illustrated in the exchange I had with an editor or subeditor over my use in the paper of the adjectival combination in 'the chafing litigious fairness of would-be thieves' (page 203) in this paper. The request for change of this was, I think, the only one of his suggestions I declined to follow and I am glad I did decline. The phrase isn't ideal but none better has ever came to mind. In retrospect, the word 'litigious' seems to me actually fairly appropriate to a public meeting in a great country that is proceeding as if determined to submerge its greatness beneath a crippling combination of parasitical law suits and insurance claims. The pressures that cause this trend are, of course, democratic and understandable and yet it is obviously a path via greed to poverty, not as might have been initially hoped to justice and wealth. Again, of course, it is just a case of the bad choice in a PD. It is only necessary to contrast the frequency of professional lawyers in the US and in Japan to see a hint of the international aspect of the dilemma. I don't know about relative sizes of insurance industries in these two countries but suspect that this comparison points the same way.

Further thoughts on puzzles of these kinds will appear in later chapters but I confess I still doubt that I have even now much of a guiding theory, leave alone a policy, that answers the practical request from the senator's wife in Washington. The only sure recipe I myself believe in requires as a major ingredient my old 'number first thought of': eugenics[15] (see also the introduction to Chapter 2). For me, either it is that or else, not many generations hence, tranquillizers and other mood drugs and unlimited medical patches for every one of us all of the time, combined with general submission to the idea that being no longer capable of free life we are destined, as individuals, to submerge indefinitely, to take status as mere executive cells within superorganisms that are forming around us and which we serve. Our controllers if this happens will be simply the civilized systems already established, although I foresee hospitals becoming much larger and more prominent components. In effect these will become major centres of the immune system of the superorganisms and they will also make up for the growing medical incompetence of individual 'cells'—that is, us.[16] Mass media will simulate the endocrines, networked computers and telephones merge to become our controlling nervous system, and computers will double again through their attachment to huge data bases of knowledge that increasingly only they 'understand', to link together the 'superbrains' controlling all that we do.[17] In this general direction there certainly can be and has begun a viable programme. However, it must not be imagined that the harshness of natural selection can ultimately be evaded by this route. In addition to what a superbrain might decide to be proper economy within its own 'body', there will eventually begin to be nasty die offs of whole systems as superorganisms themselves have to compete. Still, in that coming world individual deaths will not necessarily be so much to be feared as they are now: illusions of blissful heroism (or simply illusions—not fearsome, illusions of anything but death) created in us by the cocktails of drugs that all 'cells' are receiving, should take care of that. Thus under this parallel, I do not believe my individual skin cells 'fear' their inevitable coming deaths as they exfoliate my skin or settle to form a scab even though there is a sense in which I do believe my whole body may fear, and my skin with it, when an increasingly uncontrolled dying of cells is happening—as may be the case (say) if I begin to show the symptoms of AIDS. Even then I doubt if the individual skin cells are unhappy. This is not because I think they are being kept high on drugs but rather because they are evolved to be wholly altruistic towards my body anyway. This will not be case with us as cells, so we will need more 'persuasion'. But still no great hardship; all will be under control. The appropriate comparison here would be slave ants working for alien colonies; by now it should be clear, however, that my background model images an event much older than slave ants. The event happened in the Precambrian as 'super' metazoan organisms began to take the lead in a world previously ruled by independent unicells.

I see the 'superorganism' alternative to eugenics as really the more feasible and the more likely course for humanity to take (see later chapters, especially in Volume 2). It is a relatively easy and seemingly uncontroversial road at the start and indeed the move on to this road is already in full swing. Yet I still rebel, still dream foolishly of my harsh cerrado dawns even when I am plainly not choosing to live in them. I forbear, therefore, in my imagination, to tell the senator's wife to think in terms of tranquillizers and happiness drugs that are first to be tested in the water supplies to the psychiatric hospital (the precedent here, of course, is the proverbial bromide of British soldiers' tea!) and later provided to the public more generally. Instead I still give her only my muddled thoughts concerning the other course, the one that keeps us all still potentially free-living. In this line (again in my imagination only) I tell her that she should assume that human heredity in general is as strong as in her worst nightmare, as weird as in the strangest 'coincidences' of Bouchard's twin pairs,[18] and, because of a good possibility that the nightmare suggested in those results will prove itself reality, she should try to see that her husband the senator attends not just to immediate issues but to foundations and distant problems of national well-being as well. She should make sure the laws he helps to enact are always such as to encourage not merely the direct influence but also the breeding of citizens democratically considered above average in ability, kindliness, and health, while discouraging influence and breeding from opposites. I think it is hard, perhaps impossible, to define what is best, but not so hard to say what is simply moderately good and that is all we need; on the other side it is less difficult than most seem to think to define what are bad and worst, although even here we need caution.[19] Nor is it so serious as most people think if there are occasional mistakes; financial as well as public apologies to the injured person and his or her relatives, guided by inclusive fitness, could be considered, for example, where mistakes are proven. Finally, I would tell her to be very patient and wait until long after she is dead.[20]

Possibly what I did tell her in the hall where we met in 1969 was not so different, but I doubt that I had the courage.

References and notes

1. P. Vellacott (ed.), *Aeschylus: The Orestian Trilogy* (Penguin, London, 1956).
2. R. L. Stevenson, *The Strange Case of Dr Jekyll and Mr Hyde* (Longmans, London, 1886).
3. D. C. Queller, Genetic relatedness in viscous populations, *Evolutionary Ecology* **8**, 70–3 (1994).
4. D. S. Wilson, G. B. Pollock, and L. A. Dugatkin, Can altruism evolve in purely viscous populations? *Evolutionary Ecology* **6**, 331–41 (1992).
5. P. D. Taylor, Altruism in viscous populations—an inclusive fitness model, *Evolutionary Ecology* **6**, 352–6 (1992).
6. D. R. Vining Jr, Group selection via genocide, *Mankind Quarterly* **21**, 27–41 (1982).
7. W. Boyd, *Brazzaville Beach* (Hamish Hamilton, London, 1990).
8. J. Goodall, *Through a Window* (Houghton Mifflin, Boston, 1990).
9. L. Dayton, Humans like warfare from the start, *New Scientist* **144**, 7 (only) (1994).
10. M. J. Wade and F. Breden, The evolution of cheating and selfish behavior, *Behavioral Ecology and Sociobiology* **7**, 167–72 (1980).
11. W. Poundstone, *Prisoner's Dilemma* (Oxford University Press, Oxford, 1993).
12. D. S. Wilson and E. Sober, Re-introducing group selection to the human and behavioral sciences, *Behavioral and Brain Sciences* **17**, 585–654 (1994).
13. G. Watson, *The Idea of Liberalism: Studies for a New Map of Politics* (Macmillan, London, 1985).
14. R. L. Stevenson, *Poems* (Chatto and Windus, London, 1910).
15. W. D. Hamilton, review of *Human Diversity* by K. Mather, *Population Studies* **19**, 203–5 (1965).
16. R. G. Evans, Health care as a threat to health: defense, opulence, and the social environment, *Daedalus* **123**, 21–42 (1994).
17. Use of the word 'superbrain' conjures the idea of computers or networks designed for a controlling function, but this is not what is imagined. The idea is rather that control will arise spontaneously out of, at first, organizations of people. People will still feel themselves to be independent but actually will be entering increasing dependence on each other, on interactions with their machines (especially those for communication, computation, and information retrieval), and on support from 'outside' (e.g. from still relatively healthy manual workers and traditional hardware). Ever more decisions will be being taken, in effect, by computers, and supposedly controlling human elites will serve more and more

like glial cells to the computers' neurones, although this will probably not be a typical perception by the humans. At all times decision and control functions will be dispersed, more nearly analogous to the nervous system of a multigangliate sedentary invertebrate than to a highly cephalized vertebrate. (Traditional evolutionary cephalization and its placement is connected with movement; yet even in large moving spaceships it is hard to imagine any advantage to having control at a 'front end'.)

18. T. J. Bouchard Jr, D. T. Lykken, M. McGue, N. L. Segal, and A. Tellegen, Sources of human psychological differences: the Minnesota study of twins reared apart, *Science* **250**, 223–8 (1990).
19. W. D. Hamilton, Population control, *New Scientist* **44** (20 October), 260–1 (1969). The text of this letter was as follows:

> Sir,—Mr Benham's argument ('Letters', 16 October, p. 156) is that human population control through birth control cannot be both biologically feasible and ethically acceptable at once. His points are serious but not impassable. It is true, as he says, that an animal producing exactly two offspring from every mated pair is biologically impossible: it would cease to evolve and would be slowly destroyed by mutation. This is a valid criticism of John Tinker's implication ('man as Epidemic', 2 October, p. 18) that, for a stable population *all* parents having more than two children should be censured. The idea that some are to be censured and not others evidently provokes Mr Benham's fear that in attempting to maintain our health and adaptive versatility by variation around an average family size of two we will have to submit to oppressive legislation, compulsory sterilization, decisions by eugenic 'experts', and the like.
>
> Regarding the first of these phantoms, our ancestors might well have dreaded the authoritarian system under which we already live. The net of petty necessary laws draws ever tighter as population increases. Except through population control we can withdraw from this situation only through a return to complete social laissez faire—abandonment of the welfare state and perhaps even of the very idea that we ought to care about the well being of our fellows. This seems to be the line Mr Benham hopes for. It leads to the various kinds of oppression that are historically familiar.
>
> In the other direction at worst there is no reason why population control should be less democratic than any other activity of government. If this seems not enough politics can be kept almost out of the vital field through selection of a more constitutional system. The following, for example, would also deal with Mr Benham's other fears. Every individual is given rights to participate in the parenthood of two children. A person acquires his rights at the minimum legal age for marriage, and anyone can give away, at any time, to anyone he chooses, a right which he has not yet invested in a child. Under such rules probably most gifts would go between relatives and would simply accentuate the already existing pattern of family aid that flows through bachelor uncles and maiden aunts. No one need be forcibly sterilized and no social disapprobation would cause anyone to lose his rights. The aim would be simply to keep mankind as he most desires himself to be. A criminal in jail, for example, might give his rights to an ex fellow prisoner. Good: following Mr Benham's point, we do not know for certain that a criminal's genes or traditions may not some time help us to break out of enslavement by the Martians.

Obviously there are many grave problems for such a system and no one would wish to see it started overnight. Some of the issues it raises are perhaps unpleasant to contemplate, but, from a humane point of view, so is natural selection. The advocates of natural selection must recognize that until we evolve the mental outlook of barnacles life is going to be less pleasant the more we become crowded. This must apply even to the survivors; simultaneously large numbers of people will be dying, and not dying of old age.

W. D. Hamilton

20. I believe it will be a long time before democratic governments are able to take other than minor steps in eugenics, because popular forces of tradition and emotion towards saving all neonates and even all foetuses will be too strong to allow it. It seems likely, however, that, as the general decline of health concurrent with the 'superorganism route' becomes more perceptible (see A. S. Kondrashov, Deleterious mutations and the evolution of sexual reproduction, *Nature* **336**, 435–40 (1990)), new kinds of eugenics societies may arise practising within themselves the disciplines of breeding, non-intervention, etc., that can preserve the physical integrity of individuals, while also, perhaps, diminishing or advancing some other more 'normal' traits. After several generations the health and physical independence among descendants in such societies may encourage the general public to desire similar rules to be more widely applied; or alternatively more and more people may seek to join the societies. Some religious sects already refuse medical intervention, but this seems to be not at all with a eugenic motivation; indeed such sects typically reject evolution by natural selection completely. Nevertheless, I take their existence to show that, for a committed group, making a difficult decision is not impossible, and I record here my admiration for their courage in resisting a current emotional vogue.

SELECTION OF SELFISH AND ALTRUISTIC BEHAVIOUR IN SOME EXTREME MODELS†

W. D. HAMILTON

> Anti-Darwin. *As for the famous 'struggle for existence', so far it seems to me to be asserted rather than proved. It occurs, but as an exception; the total appearance of life is not the extremity, not starvation, but rather riches, profusion, even absurd squandering—and where there is struggle, it is a struggle for* power. *One should not mistake Malthus for nature.*
>
> NIETZSCHE (1889)[1]

It seems certain that both selfish and altruistic adaptations exist. That is, there are situations in which animals increase personal fitness by harming their fellows, and there are others in which they regularly take risks in order to benefit their fellows. It is always very difficult to know what particular traits, especially behavioural traits, do for biological fitness, and it may be, therefore, difficult to convince a sceptic of this by means of one or two examples. Nevertheless I think that careful consideration of a typical animal fight will reveal evidence of both selfish and altruistic adaptation.

Apart from loss of status or territory by the loser, fights do sometimes cause injury or death. The armour on the shoulders of boars would not have evolved unless injury were fairly frequent in this species.[2–4] Usually, however, as soon as it becomes obvious which combatant is likely to win, and thus just when serious wounding might be expected to occur, the fight stops and after a brief pause in which the loser shows submission in some way, he is allowed to retire. Although it is not quite what we mean by altruism in a human sense, the restraint of the victor in this context is altruism: because the loser has not been killed he may have to be fought again in the distant future. Although the loser may have been ousted from the immediate neighbourhood, the winner

---

†*In* J. F. Eisenberg and W. S. Dillon (eds), *Man and Beast: Comparative Social Behavior*, pp. 57–91 (Smithsonian Press, Washington DC, 1971).

may shortly suffer a slightly increased territorial pressure because the loser reinserts himself elsewhere in the colony, and so on. Yet, according to Konrad Lorenz, the inhibition of the winner when the loser submits is instinctive and therefore of genetic rather than cultural origin. In particular, the inhibition against inflicting a fatal wound when this has become easily possible, seems to be so strong and reliable that the loser's submissive posture can actually exploit it so as to make any further hurt less likely.[5–7] Such 'ritualized' ending of fights plus the existence of armour evolved for intraspecific fighting is a combination of facts which will be very hard to explain in any single version of a 'group selection' theory, and also equally hard to explain with the assumption that all adaptation is concerned with individual fitness.

Selfishness creates less difficulty than altruism for the classical models of natural selection; therefore, I propose to consider first the possible extremes of selfishness. It must be remembered, however, that if one trait is treated as 'selfish', the alternative traits must be relatively altruistic, so that circumstances that are found to be unfavourable to selfishness must be at least relatively favourable to altruism. The classical models were based on the concept of individual reproduction and so it is not surprising to find that selfishness has an easy field. The simplest model, that of random mating in a single very large population, allows free play to a rather unrealistic ferocity of selfishness. The individual will do all the injuries of which he is capable if he gains thereby even the slightest increase to his personal fitness. The only limit to selection of this kind is set by the extinction of the population. This could occur quickly if the selfishness is severe. For example, consider a recessive type which has a slight advantage over other types in the number (n) of offspring which it rears, but gets this advantage through the killing of more than n other adults. Clearly at some point during the spread of such a genotype the number killed will reach the total of the current generation. Yet such a type certainly has a selective advantage, as can be seen directly when we consider that the killings are subtractions made from the gene pool at random. They involve a set of genes having the characteristic ratio of the gene pool, and their removal, therefore, does not alter the frequencies in that pool, whereas the selfish gains are additions to the gene pool coming from the particular selfish genotype: the additions change gene frequency and this is natural selection.

Even more destructive behaviour can be selected if the random mating population is not 'very large'. Let us call an action which harms others without benefiting the self 'spiteful', and if a spiteful action involves harm also to the self, let us call it 'strongly spiteful'. The condition for the selection of a strongly spiteful trait is that the total of harm (in units of fitness) done to others shall exceed the harm done to the self by a factor almost as large as the size of the population (exactly, by a factor greater than $N - 1$). A population which is small enough, and sufficiently bunched together, to make possible the distribution of such extensive harm must be in danger already, and the spreading of any strongly spiteful mutation is very likely to cause its extinction. Such trends of selection in small populations, if they occur at all, must act like a final infection that kills failing twigs of the evolutionary tree. The word 'adaptation' seems to imply at least a little complexity in the achievement of some end, and complexity is unlikely to be based upon a single mutation. Accepting this, it seems unlikely that a multigenic spiteful adaptation could evolve.

No selective trend of any of the kinds yet discussed has ever been observed. Probably this is not only because of the practical difficulty of expressing murderous selfishness and pervasive spite; it is probably partly owing to the unrealistic 'panmixia' of the classical model. Panmixia is usually used to imply random mating; I intend it here to imply that the distribution of social effects to other members of the population is also random. If mating is not random, then presumably it is also likely that harms and benefits will not be distributed randomly. If they are not, or if harms are somehow moderated when directed at like genotypes, then the evolution of spitefulness is a more realistic possibility, in that selection will be less likely to bring demographic disaster.

I propose to leave aside for the present the question of whether mating is random. Indeed I propose to start by avoiding it altogether, by letting a model of competing asexual strains do duty for a model of Mendelian population. This may seem to confine me to the evolution of protozoa and bacteria for which the social possibilities seem rather limited; but the analysis is much simpler, and, so far as I can see, using reasonable assumptions, the main features of the sexual model can be predicted from it. Moreover, the two models do really approximate very closely for the sexual case where the traits of heterozygotes are arithmetic means between the traits of their homozygotes.

INTERACTIONS IN PAIRS

As a further simplification it will be supposed that the social trait affects only one other individual. Again, in many contexts this may seem unrealistic, but once more, we may hope that the simplest model will illustrate the sorts of things that can happen in more complex models.

Inherited traits are supposed to be expressed when organisms meet in pairs. The pairing can either be of a type lasting a considerable time or a mere passing encounter. Suppose there are two types in the population, a 'normal' N, and a 'mutant' M, and their frequencies are p and q. The outcome of the three possible types of encounter can be set out in a matrix:

	N	M
N	a	b
M	c	d

By convention the values in the matrix are those received by the type at the left. Thus in an encounter of M with N, M gets c and N gets b.

If pairing is at random we have the average value obtained from a pair, which is $(pa + qb)$ for N and $(pc + qd)$ for M. If the values in the matrix are taken to be actual fitness and not merely parts of fitness, then selection proceeds in favour of the type having the greater average. The equation for change in strain frequency expresses this fact:

$$\Delta q = pq\{(pc + qd) - (pa + qb)\}/W,$$

where W is the mean fitness.

If the mutant is of the strongly spiteful type, and if when paired together two mutants harm each other as much as a mutant harms a normal, then $c < a$ and $d < b$ (and by equal amounts), so that the mutant can never progress.

If such a mutant could discriminate and moderate the harm which it causes to its own type to such an extent that $b < d$, then its prospects are slightly better in that it would come under positive selection if its frequency exceeds the value

$$\frac{(a - c)}{(d - b) + (a - c)}.$$

This could happen through a chance concurrence of mutants in a small group founding a new population. Even so the case can be dismissed as very unrealistic; it is mentioned as showing the most extreme case for dysgenic natural selection in the case of pairs.

Returning to merely selfish traits, it is fair to assume that in all cases $c > a > b$; that is, when paired with a normal the selfish individual increases its own fitness at its partner's expense. These relations are in themselves sufficient to ensure that the mutant starts to progress in frequency. As selection proceeds mutant–mutant pairs become commoner (as q^2), and the finish depends on what is achieved in these pairs.

If $d > b$ the spread of the selfish strain continues to completion. In the reasonable case where the paired selfish mutants do not do so well as the paired normals we have the overall arrangement:

$$\begin{array}{ccc} a & & b \\ \wedge & \searrow & \wedge \\ c & & d \end{array}$$

During selection of the selfish mutant the mean fitness of the population falls from a to d. In game theory a 'payoff' matrix having this form is associated with a controversial problem known as 'Prisoner's Dilemma'.[8,9] On the basis of this parallel I propose to refer to a biological situation having the above pattern of fitnesses as a 'PD'. The question of whether the outcome in natural selection has any bearing on the controversy about the game will be briefly touched on later. Situations of the PD type are probably quite common in nature. I will cite two possible cases.

One is involved in the conditions of sex-ratio selection that sometimes occur in sperm-storing and readily incestuous arthropods. I have discussed these conditions at some length in another paper (see Chapter 4). Although there is not much evidence directly relevant to the special case which involves the PD, the overall correspondence between fact and theory seems good. Since the competing strain model is not directly applicable to sexual organisms, a parallel study of a diploid model was carried out by computer simulation, and this gave an optimum sex ratio which differed only slightly from that predicted by the strain model. Both models confirm that the more selfish type wins in natural selection in a PD situation, so that mean fitness may fall; and, as stated, the data broadly confirm the models.

More tentatively, I suggest that a PD situation is involved during evolutionary trends to polygamy. Such trends seem to have been quite numerous, for example, in birds.[10] Successful polygamous males must have higher reproductive expectations than they would have if monogamous, and the increase in their fitness must come at the expense of males that are barred from breeding. The number of offspring reared per female is probably reduced relative to monogamy. Consider a generally monogamous species in which males become bigamous if they can capture the territory of a neighbouring male. A mutant type whose increased pugnacity and skill in territorial encounters often enables it to oust one of its male neighbours would spread through the population in much the way predicted in our model for pairs. When mutants tried to oust each other it is easy to imagine that through real fighting and the probably reduced per-female fitness of the winner, they would end up with lower average fitness than a contentedly monogamous male. Yet their average fitness would certainly be greater than that of the ousted non-mutant males ($d > b$) and this gives the condition for monogamists to be supplanted completely.

In other situations moderately relevant to the model for pairs it is clear that a genotypically controlled selfish tendency must reach an equilibrium because $b > d$. Consider, for example, tendencies to usurp nest burrows that are very widespread among solitary bees and wasps (see Chapter 2), and tendencies to dump eggs in other nests of the same species in birds.[11,12] These tendencies are selfish in a way that can be called *parasitical*, and it is obvious that the population cannot come to consist entirely of usurpers or dumpers who are unable to make nests for themselves. This kind of situation will be referred to as an EP (equilibrium of parasitism).

In considering the equilibrium frequencies in intraspecific situations of this kind caution is again necessary in generalizing to sexual organisms from the case of competing asexual strains. Nevertheless, if the heterozygous class behaves exactly like one of the homozygous classes (i.e. there is complete dominance) the equilibria at least remain unchanged.

Actually it is not known whether the not uncommon parasitical variants of nesting behaviour are based on genetical polymorphism or whether these are merely built-in potentialities of which the realization depends on environmental factors. Females of the sand wasp *Mellinus arvensis* may be observed repeatedly trying to force entry into burrows occupied by other females, and going from one burrow to another in their attempts to do this. Fierce fights sometimes occur and these may lead to fatalities.[13] Simultaneously other females in the same populations can be observed actively digging their burrows. There are various possibilities as to the nature of the parasitical variants. They may be genetically different from the others. If so, their parasitic inclination may be obligate, or it may be facultative in the sense that they would dig their own burrows after some initial trials have shown that opposition to parasitism is stiff, or that fellow parasites are numerous. On the other hand the females may be constitutionally no different from the others and may be merely behaving the way they are because some accident has destroyed their own burrows and thus caught them in the wrong physiological state or too late in the season to begin burrowing again. Some observations of Fabre[14] show that this can happen in a mason bee, *Chalicodoma muraria*. In general, however, the truth is unknown, but there is a strong likelihood that something like the described EP either is now involved or has been involved in the evolutionary past.

When any such equilibrium occurs it is likely that selection of modifiers that cause a changed reaction when like meets like will eventually resolve it; that is, will allow the selfish gene to complete its spread. The various possibilities could not be discussed in detail for the case of *Mellinus* without reference to a far more complex model. But probably the evolutionary avenues that we see open in the elementary pair model correspond fairly well to avenues that are open in the more complex real cases.

DISCRIMINATION AND SELECTIVE PAIRING

If while held in equilibrium the selfish type evolves the ability in some way to moderate its selfishness or aggressiveness when paired with its own type, then d rises and we tend to return to the PD situation. As soon as d exceeds b, the selfish type is able to supplant the other completely. Fairness returns, but each time this happens we expect slightly more the chafing litigious fairness of would-be thieves. Resolution of equilibria in this way is likely to be common with intelligent teachable species. With humans, for example, I think it is clear that most selfishness and much criminality arises from evolved selfish potentialities which are latent in all of us and which come to be expressed either owing to lack of opposition, as through a 'spoiled' upbringing, or under stress, as in poverty.

It is similarly adaptive for the more altruistic type to react selfishly where selfishness is expected in the partner. In theory such adaptation by both types tends to bring on a situation with an unstable equilibrium, but in practice the type which is slower in evolving discrimination is simply eliminated from the population.

Another course open to both types when held up in an EP involves non-random pairing. Rather than continue in a jangling partnership, the disillusioned co-operator can part quietly from a selfish companion at the first clear sign of unfairness and try his luck in another union. The result would be some degree of assortative pairing.

In accordance with the common aims of the 'altruists', their assortative pairing would leave selfish individuals with no alternative but to injure each other. The selfish strain, however, would be evolving the opposite way, with its members tending to seek unlike pairings. For example, in the EP situation, instead of moderating their selfish action when they find themselves paired together—the road we have already seen to lead to a kind of universal 'equity of thieves'—the members of an early formed selfish pair could part and seek contact with other unpaired individuals who might prove to be unselfish. Insofar as they are successful, dissortative pairing would result. Of course, attempts by 'parasitical' individuals to attach themselves *always* to suitable hosts have to fail when, through initial success, they have become too numerous. So in a sense this is a weaker course than that which leads to the 'equity of thieves'. Nevertheless if a dissorting selfish mutant is introduced into a random-pairing EP situation it is clear that it would replace the non-dissorting selfish type and end up equilibrated at a higher frequency (see Appendix A, p. 219).

The biological literature has many examples of pairs of fairly closely related species of which one has become parasitical on the constructive labours of the other. Any species which makes or stores things for future use seems open to the possibility of such parasitism. Presumably in these cases sporadic traits like those evident in *Osmia*,[15] *Halictus*,[16] *Mellinus*,[13] *Psammochares*,[17] *Polistes*[18]—to mention insect examples only—have evolved via dissortation into a much more permanent condition.[19,20] In such sexual species, however, assortative mating must at some stage have accompanied the dissortative pairing, and on this account it is difficult to imagine the sympatric speciation of a host–parasite sibling pair from a single species. Rejoining of allopatric subspecies or sibling species could provide suitable conditions in both respects. It was on this basis that Richards[21] outlined a plausible course for the divergence of the non-social cuckoo bees, *Psithyrus*, from a social industrious group ancestral to their present bumblebee hosts, *Bombus*, using as illustration the sporadic parasitism practised by the more southerly *Bombus terrestis* on its more northerly sibling *B. lucorum* in Britain. A striking parallel to this illustration has since been documented in a different subgenus of *Bombus*.[22] The fact that the less-specialized nest parasites are usually found parasitizing closely related species suggests that this was the usual beginning even for the highly specialized examples which are now found imposing on distant genera.

As is shown in detail in Appendix A, whereas dissortation is favourable in only a limited way to selfish traits, assortation is very definitely unfavourable to them, which means, of course, that it is favourable to altruistic traits. Full assortation eliminates all equilibria; it eliminates unlike pairs completely and selection proceeds in favour of the type with the fitter 'homopair'. Thus in the case of a PD situation, in going from no assortation to full assortation the direction of selection must completely reverse.

Assortation makes altruists enjoy the benefits of their altruism, and this obviously applies to assortation of any kind, not just to that which can occur among pairs. From here on I will be concerned mainly with the various ways in which assortation can occur and the degrees of altruism that can be supported. With my emphasis now shifting from the negative to the positive aspects of sociability, however, it seems an appropriate point for a brief digression on a conspicuous kind of animal sociability which I think ought to be left behind here as mildly selfish, although many writers seem to regard it as positively social.

GREGARIOUS BEHAVIOUR

G. C. Williams[23,24] has suggested that the schooling of fish is to be regarded as the outcome of cover-seeking instincts of the individual fish: each fish behaves as if trying to put other fish between itself and the potential predator. If a fish succeeds in this it is behaving selfishly: the school is more conspicuous and less manoeuverable than a lone fish so presumably the predator gets his meal more easily. In other words mean fitness in the school is probably lower than it would be if the fishes did not school. The school may be the 'many body' analogue of the harm-exchanging pairs that we have seen to be selected under PD conditions.

I think that Williams's experiments demonstrate that seeking cover is certainly an important element in schooling behaviour and I am less cautious than he about generalizing the idea to cover most occurrences of massive animal aggregations, including relatively static ones like nest aggregations of birds, and even, by further extrapolation, aggregations in timing such as those shown in the sudden mass emergence of bats from caves.[25,26] There is evidence in most of such cases that predators do tend to take peripheral individuals and that in the actual presence of these predators, the individuals of an aggregation react in the way the hypothesis leads us to expect.[27–35]

A hint of this idea with regard to gregarious behaviour is recorded in an essay by Francis Galton[36] on the behaviour of half-wild cattle in South Africa. Reading this passage stimulated me to consider some geometrical models which should help to show how readily gregarious behaviour could originate through predation if it was initially completely absent. I considered an edgeless distribution of cattle feeding in long grass which could conceal at any point a sleeping lion. If the cattle became aware of the presence of the lion, although completely unable to guess its position, and if each cow then moved in a way tending to reduce its chance of being the cow closest to the lion, it appeared that any initially random or spaced-out arrangement of the cattle would undergo progressive condensation into close-packed groups of ever-increasing size (Fig. 6.1). Of course the perfect concealment of the lion combined with a

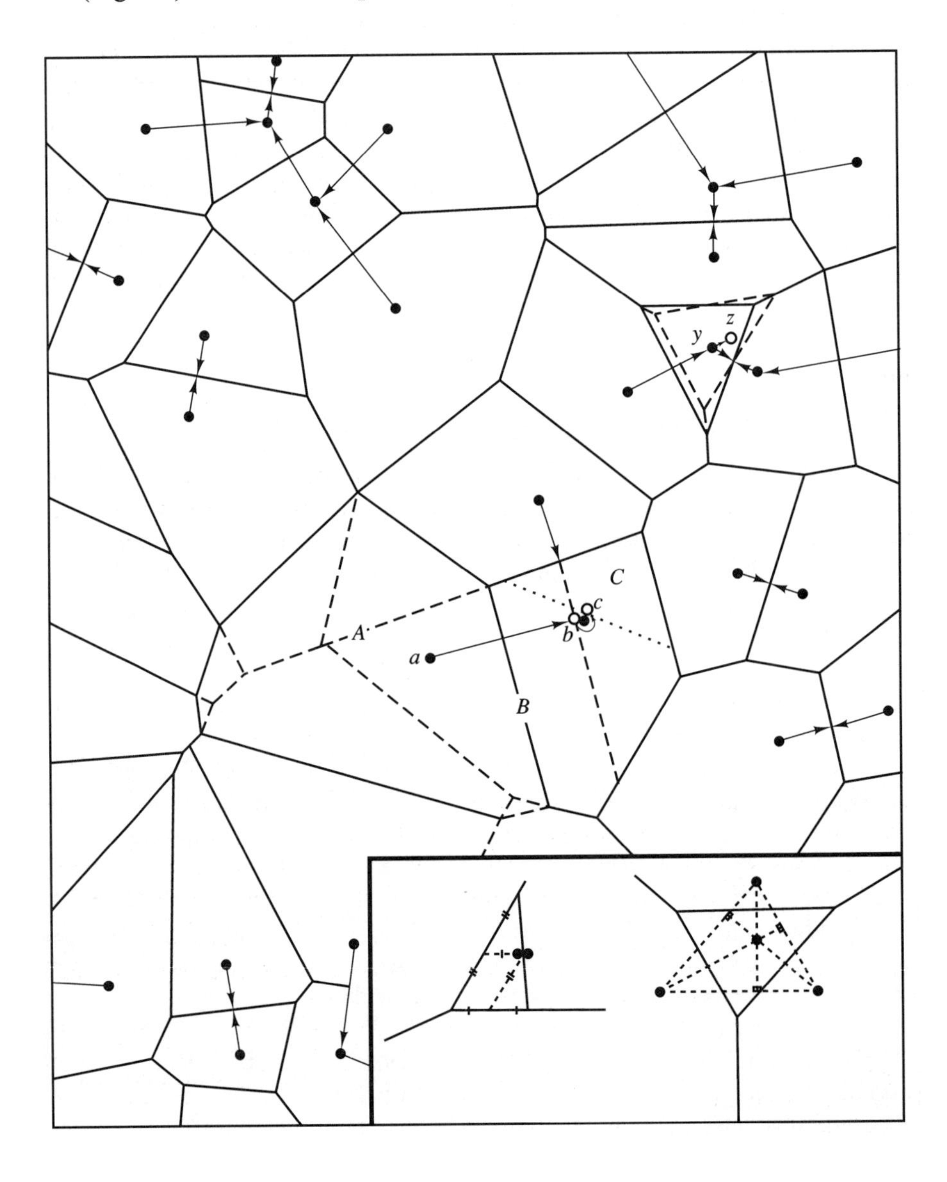

sudden universal awareness of his existence is very unrealistic in this case. A slightly better example in which the enemy would be concealed almost anywhere up to the moment when it begins to move towards its prey would be that of oestrid flies attacking reindeer.[37]

The suggestion that selfish cover-seeking has been an important factor in the evolution of gregarious instincts does not oppose the possibility that other forms of selection may also favour gregarious behaviour either working alone or in conjunction with selection for seeking cover nor does it deny that social structure and co-operative relations may also exist in the aggregations. It does predict, however, that social structure should be less evident during phases of panic than at other times. Galton[36] claimed that the cattle herds had leaders and attributed some altruism to them—this is implied, indeed, by their very willingness to move outside the herd. (Galton thought that leaders both of cattle and of men were hereditarily endowed, and he outlined, very vaguely, factors of selection that might stabilize a leader-like disposition at low frequency. To judge from sheep,[38] he was probably wrong about leaders of ungulate herds. Regarding the usually male leaders of macaques, baboons, and humans he may have been more nearly right. Various data are reviewed by Russell and Russell[39] and by Crook.[40]) Fairly elaborate intrapopulation structure based on components of hierarchy, territory, and family groups is also often detectable in mammalian aggregations, especially when unmolested. Altruism, however, only becomes conspicuous in certain types of small groups. We have now to ask, what is it that makes musk oxen, for example, form outwardly facing mutually defensive stands against a predator when their near-relatives sheep, under similar threat, not only tend to face inwards but

---

Figure 6.1 The main diagram shows part of a random dispersion of cattle, represented by black dots. A lion is imagined to be concealed somewhere among the cattle. It will suddenly attack the nearest cow. This situation specifies a danger domain for each cow as a complex polygon bounded by right bisectors between a cow and certain neighbours. It contains all points nearer to that cow than to any other.

If a cow becomes aware of the presence of the lion without being able to guess its position, it usually decreases its danger domain by approaching its nearest neighbour. For example, movement from a to b changes the domain from A to B. Having reached the nearest neighbour in general a further decrease of domain can be achieved by moving to a particular side of the neighbour (position c and curved arrowed line). Thinner arrowed lines indicate nearest-neighbour relations for all the cattle and link primary groups that should form if all cattle are alerted simultaneously and move towards nearest neighbours.

In the rare case of a domain completely determined by three neighbours in an acute-angled triangle (frequency 0.009), as at y, movement to the nearest neighbour, or to any side of it, may not decrease the domain; the minimum domain is equal to the triangle of neighbours and is obtained from its orthocentre, as at z.

The inset illustrates the two geometrical principles of minimum domains.

may attempt to butt or jump themselves into the flock—inwards from the perilous outside ranks? Here we return to the problem of conditions for altruism.

RECOGNITION OF RELATIVES

Galton[36] stated that cows that have dropped out of a moving herd in order to calve will attempt to fight off predators, although a solitary condition is one which usually causes great terror. This phenomenon, of course, is well known with many normally timid species and as an aspect of parental care it is easily understood even in the classical models of natural selection. But the case can also be viewed as conditional altruism encouraged by a particular kind of assortation of like genotypes; the mother and offspring have half their genes in common by direct replication.

Other cases of relationship may be viewed in the same way. The relatives of an individual can be considered to carry his genes in a statistically diluted state with the dilution depending in a definite way on the structure of the relationship (see Part I of the paper in Chapter 2). Distant relationships are obviously more dilute. Indeed, the required precise measure of relationship corresponds more or less to the vague popular notion of 'blood' similarity.

These ideas are illustrated in Fig. 6.2. In general, if we know all the paths that connect two relatives through their common ancestors, we can say what fraction of the genes carried by one can be expected to occur as replicas in the other. When we are concerned with selection at a particular locus it may be useful to think of this fraction, which is probabilistic for any particular pair of relatives, as being the expectation applying to a large set of relatives having like pedigree connections. The fraction can be called the *coefficient of relatedness*. For relationships which do not involve inbreeding it is the same as Wright's coefficient of relationship, which is the correlation between relatives on the basis of wholly additive gene effects.

Consider a relationship with coefficient b between two individuals A and B. If A knew the relationship and could intelligently consider the consequences of a selfish or altruistic act toward B, he would see that a fraction b of the effect he caused to B's fitness consisted, in a statistical sense, of pure loss or gain to genes which he himself carries. The remaining fraction of the effect he would see as a loss or gain to the undifferentiated fragment of the gene pool that makes up the rest of B's 'expected genotype'. If A was interested in the natural selection of the genes which he carries he would regard the effect which might be caused to this second part with indifference: it is merely an increment or decrement to the gene pool as it already stands. The effect which he might cause to the 'related' (effectively identical) part of B's genotype he has to weigh against the magnitude of the effect his action has on his own fitness. In the case

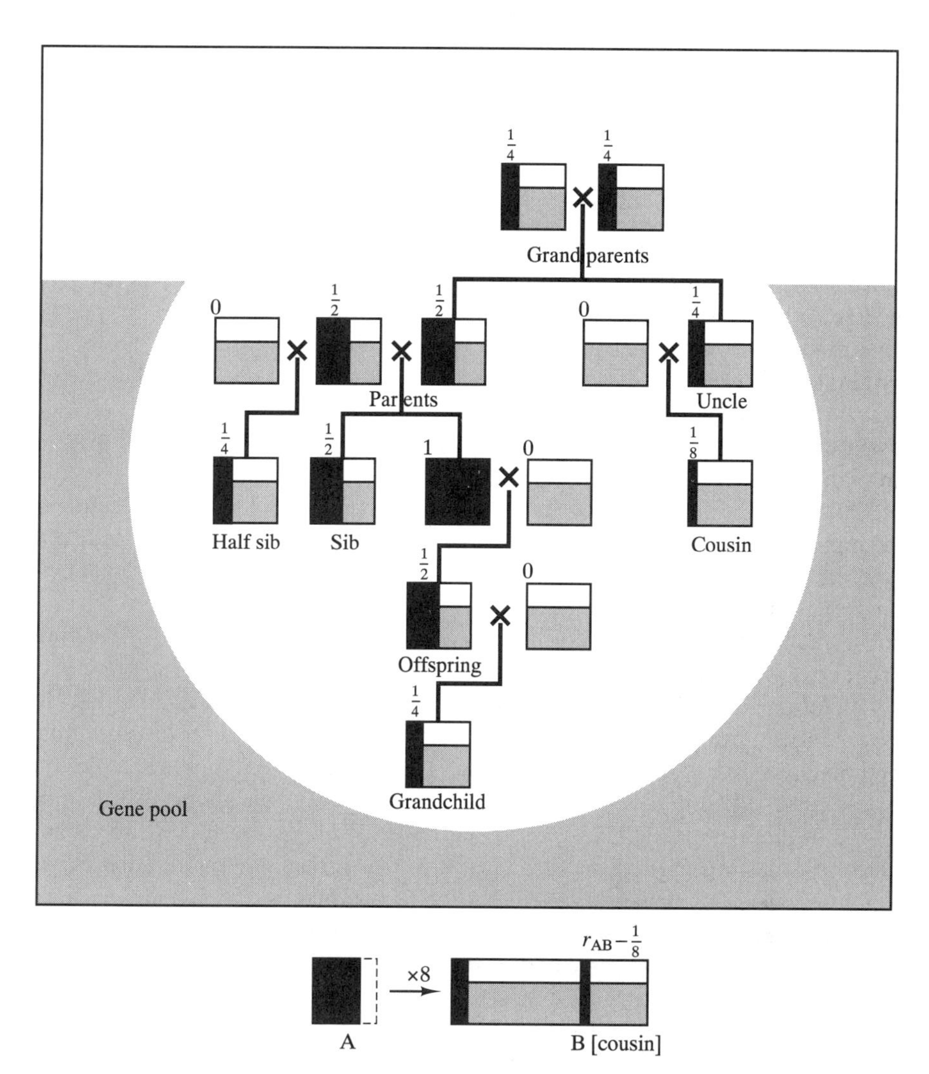

Figure 6.2 Related individuals have genes in common due to common ancestry. The pedigree diagram shows certain near relations of individual A and the blackened fraction shows their probable content of genes, which are replicas of those carried by A. Non-replica genes can be supposed to derive from individuals unrelated to A and therefore to have a probable genetic constitution the same as that of the gene pool, as indicated by the level of horizontal shading in the unblackened fractions.

The diagram below represents an altruistic action by A that has null selection. The cousin B receives eight times as much reproductive potential as is sacrificed by A through his altruistic action. The expected additional replicas of A's genes through the additional potential of B exactly make up for A's loss. There is also a non-selective addition to the gene pool.

of a selfish act, for example, he has to consider whether the gain to personal fitness is going to be greater than a fraction b of the amount of fitness which he destroys in his relative. If it is greater, then the concentration of his own genes in the gene pool will be increased. The result will be either a contribution to selective advance or perhaps, if A is a heterozygote, a contribution to the maintenance of equilibrium.

Of course there is no need to suppose that individuals think intelligently about natural selection or about the genetic constitution of their relatives: this notion is brought in merely to aid explanation. The model we are concerned with here supposes that A's genotype causes him to act in a particular way toward B; if δ_A and δ_B are additive effects to fitness of A and B due to A's action, then A's genotype is positively selected if $(\delta_A + b\delta_B) > 0$. On this basis I have suggested that it may be useful in contexts of social evolution to replace the classical concept of individual fitness with a concept of 'inclusive' fitness, consisting of A's own fitness plus the sum of all the effects he causes to the related parts of the fitnesses of his relatives. (Note that A's own fitness in this aggregate has to be taken not as the fitness A expects to express but as what he would expect to express were he not subject to the social effects that come from his relatives.)

The evolutionary outcome of such selection might well be that A appeared in his social behaviour to value his relatives' fitness against his own according to weightings given by b_{AB}. He would always value a unit of fitness in a relative less than a unit of his own fitness except in the special case in which the relative is clonal (as in the case of an identical twin): then $b_{AB} = 1$. In humans, twinning is too rare for any special social adaptations to have arisen upon this relationship, but the idea that clonal groups should show almost perfect co-operation helps to explain other contrasts in nature—for example, the contrasting behaviour in budded clones of sessile marine organisms and in sessile colonies derived from settling larvae; and the similar but weaker contrast between the perfect co-operation of cells within a multicellular organism and the imperfect co-operation among the sexually produced individuals of the colonies of social insects.

People have a great interest in their blood kin. Popular terminology reflects that interest. Human knowledge of human pedigrees is sometimes amazingly extensive, and it is so especially, considering the dependence on oral tradition, in primitive societies. There is no doubt that an appeal to kinship in general does tend to moderate selfishness and encourage generosity in human social interactions; but, at the same time, human sociability is certainly far from being regulated wholly in accordance with the above principle. Human generosity seems also to depend partly on idealism and partly on correspondences of personality that encourage friendship. Moreover, it is well known that enmities between relatives can be exceptionally bitter and destructive. Progressive human cultures seem to have been rather inclined to reject nepotism. The growing importance in human social life of what might be described as a symbiosis of aptitudes, based in part on training but also in part on genetical endowment; in other words, the growing importance of the civilized socioeconomic system, may play some part in this reaction. On the whole, if not in all details, the trend seems to be as one should wish.

It is certain that no other animal has any abstract conception of relationship. But many higher animals can recognize individuals by personal attributes, and this permits some discrimination towards very close relatives, possibly to the distance of nephews and grandchildren. In the Japanese macaque a male's position in the rather elaborate social hierarchy seems to depend partly on nepotism and partly on ability as a leader, much as with humans. This and other cases are considered in Crook.[40] Less-intelligent animals, which would be unable to remember their siblings or offspring individually, may still be able to differentiate between relatives on the basis of physical similarity and nearness to the place of birth.

There is some evidence, but no proof, that some social insects are hostile to aliens on the basis of physically inherited traits.[41] Such a system would have the advantage of making enslavement and usurpation by parasite queens very difficult, but presumably would have disadvantages through the occasional occurrence of segregating colonies which would fight within themselves. An acquired colony odour homogenized by food exchanges seems to be a more common method by which social insects distinguish the closely related members of their own colony from aliens. Odour is evidently important as a means of group identification in many species of mammals, especially in rodents.[5,42,43] The odours discriminated appear to be acquired rather than inherited.

The phenomenon of visual imprinting suggests how an individual could discriminate on a basis which would be largely genetic. Auditory imprinting could work the same way, but aptitude for vocal mimicry, such as exists in birds, would lessen the reliability of this method.

The simplest case where nearness to place of birth implies relationship is that of nestlings and litter mates. I think the evidence in general shows that agonistic behaviour in groups of siblings is mild compared to what takes place after they leave the nest. There are various seemingly contrary cases where extensive cannibalism takes place among sibling groups—for example, in the embryonic stages of some molluscs. I have suggested reasons for such exceptions elsewhere (see Chapter 2). A point worth noting here is that structural adaptation for attack and defence in these situations has not evolved, whereas it has done so, for attack at least, in cases where competing young are unrelated.[44]

POPULATION MODELS

If at maturity panmictic dispersal takes place, there is no further distinction to be made; the individuals can be reckoned never to encounter their more distant relatives. If, however, dispersal distances tend to be limited, then neighbouring individuals will tend also to be relatives of varying degree. So also will mating pairs; some degree of inbreeding follows almost inevitably from limited dispersion. (Panmictic dispersal by only one sex (or type of gamete) is sufficient to ensure random mating. Thus in theory it is possible to have a population structure in which neighbouring individuals have non-inbred relationships through single common ancestors—half sibs, half cousins, and so on. Wright refers to models of this type as involving 'isolation by distance'; more briefly, they can be referred to as 'viscous'. The population is a continuum but in terms of the gene flow of a few generations, distant parts are isolated from one another. Most generally realistic are the cases in which dispersal has something like a normal form with the mode at the place of birth.

Another type of model which allows only limited dispersal is the 'stepping-stone' model of Kimura and Weiss.[45] Here internally random-mating populations of finite size are arranged on a line or in a lattice and certain fractions of individuals migrate reciprocally between populations in each generation. This model suggests a very general and realistic one which would include both Wright's 'isolation by distance' and his 'island' models as special cases (see Fig. 6.3). The effect of having a distribution of migration (instead of migration to the nearest population only) has not been analysed in detail, although it is not beyond the scope of the methods used. The generalization required in abandoning the lattice arrangement has not been considered.

In all models which lack panmictic dispersal it appears that genetic drift reaches no equilibrium except for that holding in local patches of gene fixation, with the mean size of patches ever increasing. This is so, at least, so long as the factor of mutation is not considered. In the absence of mutation, ultimately each individual should become related with almost unit coefficient to all its neighbours. But during progress to the final state, after a population has been sown from a random mixture, and also subject to the occurrence of mutation (which, of course, an evolutionary model must allow), relationship can be expected to fall off with increasing distance, gradually in the case of the viscous models and stepwise in the case of the stepping-stone models.

No detailed study of social selection in such models has yet been made. There are various difficulties: the analytical complexity of the models, the complexity of the coefficient of inclusive relationship under inbreeding, and even doubt as to the validity of coefficients based purely on pedigrees when there is selection in the system. It seems worthwhile, however, to make some tentative remarks on the basis of a coefficient which is easily obtained from

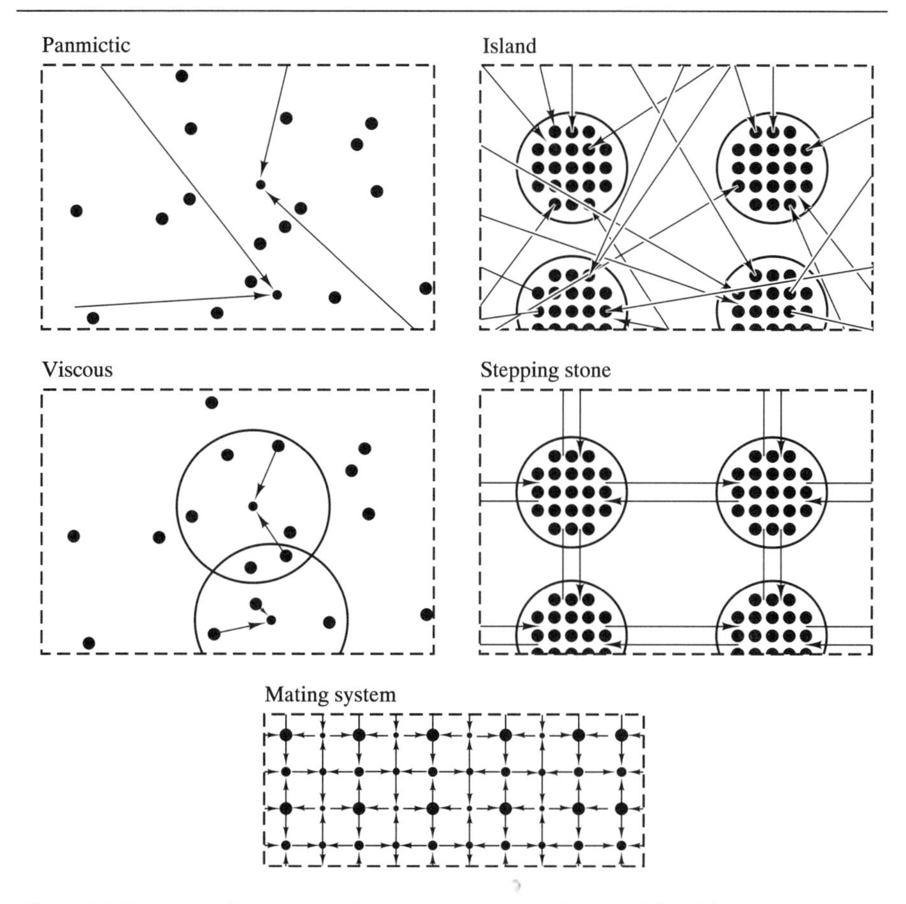

Figure 6.3 Examples of simple breeding structures. In the 'panmictic' and 'viscous' structures and the example of a 'mating system', smaller dots indicate newly formed, or younger, zygotes; in the 'island' and 'stepping-stone' structures newly formed zygotes are not shown. Arrows indicate particular parenthoods, particular migrations between subpopulations ('island'), or regular paths of migration ('stepping stone'). In the island and stepping-stone structures most reproduction is supposed to occur through random unions within subpopulations. The mating system diagram illustrates a cycle of four generations. Most analyses of mating systems have concerned linear systems: to my knowledge the one shown has not been analysed

coefficients discussed in the pioneer accounts of the models, and which should certainly serve to point out the main social implications. This coefficient is

$$b_{AB} = \frac{2r_{AB}}{1 + F_A}, \tag{1}$$

where r_{AB} is the probability of 'identity by descent', or correlation, of random gametes drawn from the two individuals A and B, and F_A is the correlation of the homologous genes of A, whose social trait is being considered. To see a rough justification for this formula, it may help to imagine oneself standing in the place of one of the genes of A in the role of matchmaker and considering arranging a successful mating for either A or B; the chances that genes passed on through such matings replicas of the 'matchmaker' gene are $\frac{1}{2}(1 + F_A)$ and r_{AB}, respectively. Alternatively, b_{AB} can be considered as the regression coefficient of the additive genotypic value of B on that of A.

Unless A is specially inbred, F_A will be the mean coefficient of inbreeding (F). Wright's work has given much attention to coefficients like F and r. His papers of 1951 and 1965[46,47] review a form of analysis of population structure in terms of 'F-statistics' which is particularly useful from the present point of view. This touches, however, a rather technical field and for further information the reader is referred to Appendix B. The most useful predictions to which the analyses give rise can be briefly summarized as follows (F symbols used in this section are explained in Appendix B):

1. Highly dispersive species (F_{IT} low) will show little positive sociability, although they may be gregarious. They are more likely than indispersive species to be polygamous.
2. Within species, fights will be most damaging when combatants: (a) differ most in heritable characteristics perceptible to the opponent, and (b) derive from distant regions or from different subpopulations.
3. Among species, fighting will be least damaging in species with the most viscous or the most subdivided populations.
4. Co-operative relations and weak social hierarchy within groups will correlate with the following interrelated factors: (a) small group size, (b) hostility to strangers, (c) endogamy, and (d) high F_{ST}.
5. Conservation of local resources by populations will correlate with the factors cited in prediction 4, subject to the requirement that a population remains in and defends a group territory for many generations.
6. The general level of aggressiveness in a population will rise during periods of increased dispersal and will remain raised until a higher average relationship between neighbours has been re-established. The persistence of aggressiveness after a population crash may be relevant here, but study of F_{IS} and F_{ST} over the whole period of a population cycle would be necessary to confirm this.

INTERGROUP HOSTILITY, WARS, AND CRUELTY

In a recent study of biochemical polymorphism in farm populations of the house mouse, Petras[48] found no evidence of other than random mating within local groups, but on the basis of gene-frequency differences between groups he estimated $F_{ST} = 0.18$. Substituting in the formula explained in Appendix B (page 221) this gives $b = 0.305$, which shows that within groups mice should treat the average individual encountered as a realtive closer than a grandchild (or half sib) but more distant than an offspring (or full sib), referring to an outbred population.

It is well known that mice and rats in the wild have unusually united social groups in which dominance behaviour is very restrained and polygamy moderate. This mutualism is in strong contrast to the treatment of strangers which is rather vicious; in rats at least, aliens are often killed if they cannot escape.[5] The variability of aggressiveness in captive colonies of mice[49] perhaps depends partly on the extent to which the founding animals came from different wild groups.

Some birds seem to have gone further than rodents in evolving small co-operative groups.[35,50] In some cases several females may contribute eggs to a common nest and take turns in brooding them, or offspring may help their parents to rear subsequent broods, so approaching the situation of some primitively social bees and wasps.[51,52] But as with these bees and wasps the stage has not been reached without the appearance of signs of parasitic tendencies (see Part II of the paper in Chapter 2).

In all these cases group territories are defended. From what has been said it would not be surprising to find the demographically stronger groups pushing into the territories of their weaker neighbours. Fission could either precede or follow the taking over of territory. No doubt strong groups would also have higher rates of emission of long-range migrants looking for fresh land to colonize or seeking assimilation into other groups, but for the reasons already given I suspect that production of new colonies by budding is relatively more important in the most mutualistic species.

Similar rather closed and highly co-operative societies appear in primates. As would be expected, with increased intelligence, they are much more complex, but intergroup hostility in varying degrees still occurs.[53]

With still further increase in intelligence, with increase in ability to communicate (and hence also to organize), with invention of new weapons (primarily for hunting) and ability to transmit culturally the techniques acquired, and with increase in possessions that could be carried off or usurped and used *in situ*, I find no difficulty in imagining that it could become advantageous for groups to make organized forcible incursions into the territory of their weaker neighbours. In other words, I suggest that warfare was a natural development from the evolutionary trends taking place in the hominid stock. (The origin of the sting in Hymenoptera provides an evolutionary precedent for Ardrey's suggestion[54] that in hominids the use of tools for hunting was quickly followed by their use as weapons of combat. The sting was originally the ovipositor of a vegetarian sawfly.[55] It then became a paralyser of insect prey. Finally, in many species it has become a lethal weapon against conspecifics, used both in aggression and defence. The structure and use of the sting depends almost wholly on genotype. The human use of tools depends largely on intelligence; thus the parallel conversions of hand-plus-tool in hominids could be expected to occur much more rapidly. There are other examples of appendages converted as murderous weapons in arthropods.[56] Among arthropods, however, mortal combats are more usual than among vertebrates.)

As mentioned by Wilson,[57] something very like warfare occurs within many ant species. The human phenomenon of slavery also has a parallel in ants; and robbery, in various forms, occurs in bees, wasps, and ants, with examples both within and between species. Social, insect colonies with multiple queens also parallel humans in having a 'tribalistic' breeding structure, and whether or not they have multiple queens they are certainly highly intrarelated and show the expected marked contrast between inward co-operation and outward hostility.

The social insects also parallel humans in two out of four of the extrinsic trends mentioned above as occurring in hominids (communication and possessions). If we accept that the elaborate instinctive patterns involved in the 'war', 'slavery', and 'robbery' of the social insects are evolved by natural selection, can we consider it unlikely that in humans also the corresponding phenomena have a natural basis? In humans certainly we are concerned with amorphous and variable inclinations rather than instincts; but, of course, considering what a newcomer our species is to our present ecological situation, compared with any of the highly social ants, bees, or wasps, and how differently the ontogeny of human behaviour is planned, this is what we expect.

I hope that by now the relevance to my theme of the quotation from Nietzsche will have become apparent. Populations are usually viscous and subdivided. On considerations of inclusive fitness we do not expect to find everywhere a Hobbesian war of all against all for the necessities of life—the 'Malthus' of Nietzsche's phrase. There should be restraint in the struggle within groups and within local areas in the interest of maintaining strength for the intergroup struggle or for the united repression of outsiders. We now see at least one reason why the 'struggle for existence' should often become restrained and conventional. I suspect that what Nietzsche meant by a 'struggle for power' is what we would now refer to as striving for social dominance, for territory, or for a place in an 'establishment'. From his all-doubting eminence, relying on common observation, Nietzsche seems to have noticed an important discrepancy in Darwin's theory. At the same time, as appears from the rather muddled paragraph following the one quoted, he was probably far from being able himself to explain what he saw, and he offered no amendment to Darwin's argument. Neither for a biological nor for a human context did he indicate how Malthus is to be answered.

The evolutionary process certainly has no regard for humanitarian principles. Pain is itself evolved to teach the animal to avoid harmful stimuli. In the immediate situation pain warns of a threat to fitness; when a mortally threatened animal cannot extricate itself, it may be expected to experience pain until, approximately, its situation is such that further efforts could not possible avail. Thus whatever the major agencies of natural selection happen to be—whatever kind of 'struggle' occurs—pain must be a usual prelude to elimination, and, as determined by the birthrate, elimination has to go on. Darwin, of course, was aware of this and recently the point has been re-emphasized by Williams[24] and Lack.[58]

Unfortunately with human beings there may be additions to the basic toll of suffering which the nature of life exacts. We are certainly able to learn to exploit the pain of others selfishly and seem even to have instincts acting in this direction. In the light of the evidence for a cruel streak in human nature, 'natural' systems of ethics, including Nietzsche's own, look less attractive. If as I believe, and as others have thought likely,[59–61] vicious and warlike tendencies are natural in humans and were formerly (at least) adaptive, Nietzsche's exhortation to follow one's instincts makes it not surprising that some of his admirers developed a cruel warlike ethos. Along with a seeming gloss over the Malthusian issue, Nietzsche's attraction to early Hellenic culture perhaps also helped him to overlook the worst aspects of human 'nature'. The Hellenes in Homer were readily warlike but not wantonly cruel. In the *Iliad*, for example, the pictures of two cities which Hephaestus the smith god wrought on Achilles's shield well represent the two aspects of tribalistic society that I have described: the inward harmony and the outward snarl. Yet as described by Homer, the warfare around the besieged city is clean: no captives are being tortured for information, no peace emissaries are being sent back blinded and mutilated into the town. In the siege of Troy likewise, revenge is ruthless but the killing of captives is mentioned with censure and there is no torture. There are none of the atrocities of conquest which a little later the Assyrians proudly perpetuated in their bas-reliefs. Hopefully, these traces of relatively humane conduct in the bloody dark-age tribes who fought against Troy may be connected with the marvellous contributions shortly to be made by their descendants to human culture; but less hopefully, the restraint may have been due to the fact that they were fighting against their not-very-distant kinsmen. In general, as Freeman has emphasized,[59] the atrocities recorded in human history appear to bear witness to the existence of a spontaneous drive to imagine and carry out cruelty which, at least in some people, in some circumstances, has been given free play and public approval.

Bringing together evidence from social anthropology, from early historical literature, from the fossil record of hominid violence and cannibalism,[54] from studies of primate behaviour, and from theoretical considerations of social evolution, I find it only too easy to imagine that the genes that reared cruelty out of the primate's aggressive drive have been favoured by natural selection in the hominid line. For the selection of cruelty, indeed, it is unnecessary even to consider inclusive fitness, except insofar as this may have been involved in the speeding-up of progress in mental and linguistic ability. With animals able to communicate and intelligent enough to see distant objectives, there would be nothing surprising in one refusing to communicate some information useful to another; nor would it be unnatural for the other to try to obtain the information by force. With pleasure in cruelty as the motivation to punishment,[62] and with cruelty to create terror, an individualistic argument sounds less plausible; but considering inclusive fitness, or group selection (which may be a really appropriate term for many human situations), there is no theoretical difficulty. For example, *a priori*, terror seems as good a weapon as false promises in bringing about the early submission or removal of a threatened tribe.

Against the last-mentioned mode of selection, terror tends to unite otherwise dissident enemies. Also, against all modes of selection, the intrinsic resistance to cruelty must continually increase. As was pointed out in connection with simpler cases at the beginning of this presentation, evolution produces shields to its own weapons. Perhaps in a slow natural course of events we may become as unconcerned about cruelty and terror as we are now about lies, given and experienced—but this will be a long time to wait and in any case 'natural' *Homo sapiens* will have invented fresh horrors by that time.

PRISONER'S DILEMMA

The increasing subtlety of the process of natural selection in the hominid line, which has been far from wholly eugenic if the above suggestions are right, and the corresponding subtlety of its product, bring me back to a topic touched on earlier, the theory of games. This theory lights up fascinating and disturbing problems in human biology. But in non-human biology, at least as regards preconsidered strategies, its relevance must be very limited. The theory presupposes beings able to think, and, potentially, to communicate. Regarding the first attribute, already the theory has provided problems insoluble to *Homo sapiens*. Regarding the second, I suspect that animals fall short not so much in ability to communicate as in ability to deceive; by our lofty standards animals are poor liars.

The implications of game-like situations in ecology are not so difficult to see as the implications of the corresponding 'game' in game theory. For example, if Prisoner's Dilemma is played between individuals meeting at random and if the payoffs are fitnesses, we have seen that it 'pays' in natural selection to take the selfish course consistently. This is because the type which does so gets greater-than-average fitness when associated with any other type, in no matter what ratio. If assortation occurs, however, this outcome is not certain; increasing correlation of partners must eventually reach a point where fitness in the 'homopairs' dominates the mean fitnesses of the types. The concept of inclusive fitness provides a simple test for the resolution of games in this way. The test consists in adding to the expressed fitnesses a fraction b of the fitness of the partner where b is the coefficient of relatedness of the partner. The differences between the totals so formed are differences in inclusive fitness. For example, if the partners are sibs

$$\begin{matrix} 2 & 4 \\ 1 & 3 \end{matrix} \quad \text{gives rise to} \quad \begin{matrix} 3 & 4\frac{1}{2} \\ 3 & 4\frac{1}{2} \end{matrix}$$

showing that with this degree of relationship the incentive to 'let the partner down' has become zero. If the partners were full sisters in a male-haploid species ($b = \frac{3}{4}$), the matrix would become

$$\begin{matrix} 3\frac{1}{2} & 4\frac{3}{4} \\ 4 & 5\frac{1}{4} \end{matrix}$$

and the altruistic course is then definitely favoured.

The possibility that the real payoffs might be other than as they are represented in the matrix of a game is carefully precluded in the formal development of game theory: the payoffs must show what the players really prefer. Nevertheless, the point from natural selection serves to emphasize that it is unrealistic to suppose that there exist real situations which present the same game to all pairs or groups of people. It is quite unrealistic, for example, to represent the payoffs of Prisoner's Dilemma in terms of years in prison, or as sums of money expected from a partnership, without making allowance for the degree of fellow feeling that exists between the 'players'.

I am doubtful whether the findings from natural selection throw any light on the problem of how it is rational to act when the desirabilities of outcomes are really in the pattern of Prisoner's Dilemma. But natural selection, the process which has made us almost all that we are, seems to give one clear warning about situations of this general kind. When payoffs are connected with fitness, the animal part of our nature is expected to be more concerned with getting 'more than the average' than with getting 'the maximum possible'. Little encouragement, I think, can be drawn from the fact that this may, in some cases, imply less than maximum population densities; it implies concurrently a complete disregard for any values, either of individuals or of groups, which do not serve competitive breeding. This being so, the animal in our nature cannot be regarded as a fit custodian for the values of humanity.

APPENDIX A

Interactions in pairs when pairing is non-random

There is a close analogy between the model for paired asexual organisms and a model for selection in diploids: the asymmetry of the fitness matrix (see page

200) can correspond to meiotic drive. Random pairing corresponds to random assortment of gametes. For non-random pairing the frequencies of normal–normal, normal–mutant, and mutant–mutant pairs may be represented

$$P^2 + pqA, \qquad 2pq(1 - A), \quad \text{and} \quad q^2 + pqA,$$

where p and q are, as before, the frequencies of the normal and mutant strains, and A is a parameter exactly analogous to the inbreeding coefficient, F, used to measure departure from Hardy–Weinberg genotype distribution.

Association of like with like (assortation) is indicated by positive A. When $A = 1$, unlike pairs do not occur; this can happen at any frequency. Association of unlikes (dissortation) is indicated by negative A, but in this case the minimum value of A depends on the frequencies of the strains and can reach -1 only when they are equally frequent. In general the minimum is q/p or p/q, whichever is the greater. Thus selection under 'maximum dissortation' will imply selection under progressively changing negative values of A.

With fitness matrix as before, the change in strain frequency in one generation is

$$\Delta q = pq[(pc + qd) - (pa + qb) + A\{p(d - c) + q(b - a)\}]/W.$$

Table 6A.1 shows the effect of maximum dissortation, and of half and of full assortation, on examples of model situations. The four examples that have been chosen are ones discussed in the text; out of 12 possible significantly different situations these four seem to have the most biological meaning.

Table 6A.1

Matrix	**Type of behaviour**	**Maximum dissortation**	**A = 0**	**A = $\frac{1}{2}$**	**A = 1**
4 2 3 1	Spiteful, undiscriminating	$\check{q} = \frac{1}{3}$, $\hat{q} = \frac{2}{3}$	↖	↖	↖
4 1 3 2	Altruistic, discriminating Spiteful, discriminating	$\check{q} = \frac{1}{4}$	$\check{q} = \frac{1}{2}$	↖	↖
3 2 4 1	Selfish, undiscriminating	$\hat{q} = \frac{3}{4}$	$\hat{q} = \frac{1}{2}$	↖	↖
4 1 4 2	Altruistic, undiscriminating Selfish discriminating	↘	↘	$\Delta q = 0$	↖

Arrows show the direction of selection; ˆ and ˇ symbolize stable and unstable equilibria, respectively.

APPENDIX B

Viscous and subdivided population models

An outline of this analysis can be introduced by quoting an identity, which Wright proves:

$$(1 - F_{\mathrm{IT}}) = (1 - F_{\mathrm{ST}})(1 - F_{\mathrm{IS}}).$$

In this identity, F_{IT} is the correlation of uniting gametes relative to the array of gametes of the whole population, F_{IS} is the correlation of uniting gametes relative to the array of gametes of their own subdivision, and F_{ST} is the correlation of gametes drawn randomly from a subdivision relative to the array of gametes from the whole population.

For evaluation of b from equation (1) (page 213), F_{IT} must be used in the calculation of both r_{AB} and F_{A}. If instead F_{IS} is used, an inclusive coefficient is obtained relevant only to selection within the subdivision. It could easily happen that an altruistic trait that was 'too generous' to increase in frequency within its subdivision (except by drift), increases, nevertheless, in the population as a whole due to the more rapid expansion of those subdivisions which contain altruists in higher frequencies.

Consider, for example, the straightforward and often realistic case where the altruistic trait is expressed not towards particular relatives but in all social encounters. If there is free mixing within subdivisions an encounter concerns a randomly selected pair from the subdivision. The correlation of gametes from such a pair is zero with respect to their subdivision. Thus an altruistic trait expressed in random encounters is certainly counterselected within the subdivision. The correlation of gametes with respect to the whole population is F_{ST}, which is always greater than zero, depending on the degree to which the gene frequencies of the isolates have differentiated. Thus if there is a gain to inclusive fitness on the basis of the coefficient

$$\frac{2F_{\mathrm{ST}}}{1 + F_{\mathrm{IT}}}$$

the genes for the trait are positively selected in the population as a whole. If mating is random within subdivisions $F_{\mathrm{IS}} = 0$, so that $F_{\mathrm{ST}} = F_{\mathrm{IT}} = F$ (say), and the coefficient is simply $2F/(1 + F)$.

The formulas, graphs, and tables of Wright's papers[46,47,63] will permit the estimation of b on the above lines for a wide range of cases in his 'island' and his 'isolation by distance' models and in various 'mating systems'. How far the inclusive coefficients so estimated will give useful criteria for social selection is unfortunately not very clear. Wright's analysis is worked out for the case where there is no selection. He has used it as a basis for qualitative discussion of the implications of breeding structure for ordinary selection but we cannot follow with quite equal confidence using concepts of social selection and inclusive fitness. There are added doubts: first, about the validity and sufficiency of the single coefficient under selection and inbreeding, and, second, because social selection will tend much more than classical selection to change the breeding structure as it progresses. Nevertheless, I feel confident that by linking the concept of inclusive fitness to Wright's type of analysis we can come nearer to understanding genetical social evolution than we could do with no quantitative ideas at all.

The favouring of altruism in subdivided populations can be seen as due to the fact that drift tends to bring about assortation. Chance concentrations of altruists, due to drift, will tend to grow rapidly. Within subdivisions, apart from the influence of immigration, the frequency of altruists may be everywhere tending to decrease, but if the vigour of the altruistic groups can somehow counteract the backward drift and selection in other groups, positive selection may hold overall. How can such general enrichment take place? And how can it avoid altering too much the general structural situation?

In models with discrete subdivisions ('island' and 'stepping-stone'), there are two reasonable ways in which the vigour of altruistic groups could manifest without causing too much structural alteration.

1. The size of groups could be kept constant by different rates of emission of migrants; with the condition for an advantage in inclusive fitness satisfied, the population of migrants would be richer in altruists than the population of non-migrants, and the random acceptance of migrants into groups would lead to general enrichment. Wright[63] drew attention to the possibility that altruism could be established by a process of this kind.
2. Vigorous groups could grow in numbers and then divide when some critical size was reached. To keep the total population and the number of groups roughly constant, the smallest groups could be removed, or, in the 'stepping-stone' model, the smallest neighbouring group could be 'overrun' and replaced by an emigrant portion of the vigorous neighbour. Groups which increase rapidly, due to fortunes of initial constitution, immigration, and drift also divide often and so spread the gene in spite of its average tendency to decrease within groups. This is the one type of selection process that can reasonably be called 'group selection'. Haldane[64] discussed the selection of altruism with this sort of process in mind, but accompanied his verbal discussion by a strangely irrelevant mathematical analysis in which he considered numbers of genes rather than gene frequencies (see Chapter 1).

For the 'isolation-by-distance' model, parallel alternatives can be seen. Corresponding to (1), local regulation of density by more long-range migration out of areas of increasing density will have a powerful effect in limiting the relationship of neighbours and is correspondingly against altruism, especially its initial establishment. Corresponding to (2), regulation by local adjustments of the spatial distribution (for example, some sort of elastic slumping from high-density areas), together with random elimination to keep overall density constant, will be more favourable to altruism.

As in all cases the altruistic mutation still faces the initial difficulty that at its first occurrence it tends to waste its altruism on individuals unrelated at the locus in question. This difficulty is particularly severe, of course, for a dominant mutation. But even for a dominant, provided the altruism is not suicidal, the recurrence of a mutation plus the possibility of local increase by drift can overcome this barrier. Eventually there will occur a clump of altruists of such size and purity as to reverse the local counterselection, and with further growth increasingly determinate progress can take hold.

Social selection cannot occur in completely regular 'mating systems'. It may be noted here, however, that the most 'system-like' version of an 'isolation-by-distance' model, which is supposed to preclude long-range migration and elastic expansion from vigorous areas, is rather hostile to altruism. Inside the patches which form, the average reproductive performance is the same whether the patch is of altruists or of egotists; on the margins altruism tends to benefit both types equally; and when, more rarely, altruists occur in isolation, they benefit mainly egotists.

The occurrence of negative F_{IS} in some mating systems, demonstrated by Wright,[47] raises the interesting point that with avoidance of consanguine mating within subdivisions, negative b for mates relative to their subdivision could arise in more natural systems. This implies that spite between mates could be positively selected within subdivisions. It could take the form of preventing a mate from having offspring by unions with other individuals. (The only possible case of such spite known to me is the insemination reaction caused by some male *Drosophila* in their mates. This is but one of many examples of males attempting to prevent re-insemination,[65] but generally such action does not lower fertility of the mate and, considering the possibility that the last-given sperm will have precedence in fertilization, it has obvious adaptive value in terms of individual fitness.) For the population as a whole, however, such episodes would be transitory; as we found earlier when considering supposedly pervasive spite not connected with dissortation, the selection of spite will tend to cause the collapse and extinction of the groups in which it occurs.

References and notes

1. F. Nietzsche, *Twilight of the Idols: The Portable Nietzsche*, translated by W. Kaufmann (Viking Press, New York, 1954; first published 1889).
2. E. Darwin, *Zoonomia, or the Laws of Organic Life* (Johnson, London, 1796).
3. C. Darwin, *The Descent of Man* (Murray, London, 1871).

4. V. Geist, The evolution of horn-like organs, *Behaviour* **27**, 175–214 (1966).
5. K. Lorenz, *On Aggression* (Methuen, London, 1966).
6. I. Eibl-Eibesfeldt, The fighting behaviour of animals, *Scientific American* **205** (6), 112–22 (1961).
7. H. Kummer, *Social Organization of Hamadryas Baboons* (University of Chicago Press, 1968).
8. R. D. Luce and H. Raiffa, *Games and Decisions* (Wiley, New York, 1957).
9. A. Rapoport, *Two-person Game Theory* (University of Michigan Press, Ann Arbor, 1966).
10. J. Verner and M. F. Willson, The influence of habitats on mating systems of North American passerine birds, *Ecology* **47**, 143–7 (1966).
11. H. Friedman, The parasitic habit in the ducks: theoretical considerations, *Proceedings of the US National Museum* **80** (18), 1–7 (1932).
12. W. J. Hamilton, III and G. H. Orians, Evolution of brood parasitism in altricial birds, *Condor* **67**, 361–82 (1965).
13. B. M. Hobby, Evidence of conflict between females of the fossorial wasp *Mellinus arvensis* L. at the burrow, *Proceedings of the Royal Entomological Society of London* **5**, 110 (1930).
14. J. H. Fabre, *The Mason Bees*, translated by A. T. de Mattos (Hodder & Stoughton, London, 1914; first published 1879).
15. A. Desey, Recherches sur la sexualité et l'instinct chez les hymenoptères, *Bulletin Biologique de la France et de la Belgique* **58**, 1–37 (1924).
16. C. Plateaux-Quénu, Utilisation d'un nid de *Halictus marginatus* par une fondatrice de *Halictus malachurus, Insectes Sociaux* **7**, 349–52 (1960).
17. A. Crèvecoeur, Le maraudage occasionel: tendence au cleptoparasitisme chez divers Psammocharidae, *Mémoires de la Société Royal Entomologique de Belgique* **23**, 183–7 (1931).
18. K. Yoshikawa, A polistine colony usurped by a foreign queen, *Insectes Sociaux* **2**, 255–60 (1955).
19. W. M. Wheeler, The parasitic Aculeata: a study in evolution. *Proceedings of the American Philosophical Society* **58**, 1–40 (1919).
20. J. Scheven, Beitrag zur Biologie der Schmarotzerfeldwespen *Sulcopolistes atrimandibularis Zimm., S. semenowi* F. Morawitz und *S. sulcifer* Zimm., *Insectes Sociaux* **4**, 409–37 (1957).
21. O. W. Richards, The specific characters of the British bumble-bees, *Transactions of the Entomological Society of London* **75**, 233–68 (1927).
22. H. E. Milliron and D. R. Oliver, Bumblebees from North Ellesmere Island with observations on usurpation by *Megabombus hyperboreus* (Schonh.) (Hymenoptera: Apidae), *Canadian Entomologist* **98**, 207–13 (1966).
23. G. C. Williams, Measurement of consociation among fishes and comments on the evolution of schooling. *Michigan State University Museum Publications, Biological Series* **2**, 351–83 (1964).

24. G. C. Williams, *Adaptation and Natural Selection: A Critique of Some Current Evolutionary Thought* (Princeton University Press, Princeton, NJ, 1966).

25. W. G. Moore, Bat caves and bat bombs, *Turtox News, Chicago* **26**, 262–5 (1948).

26. H. Pryer, An account of a visit to the bird's-nest caves of British North Borneo, *Proceedings of the Zoological Society of London*, 532–8 (1884).

27. F. F. Darling, *A Herd of Red Deer* (Oxford University Press, Oxford, 1937).

28. G. P. Baerends and J. M. Baerends-van Roon, An introduction to the ethology of cichlid fishes, *Behaviour*, Suppl. **1**, 1–242 (1950).

29. E. Horstmann, Form und Struktur von Starenschwarmen, *Verhandlungen der Deutschen Zoologischen Gesellschaft, Freiburg*, 153–59 (1952).

30. N. Tinbergen, *Social Behaviour in Animals* (Methuen, London, 1953).

31. J. A. Hudleston, Some notes on the effects of bird predators on hopper bands of the desert locust (*Schistocerca gregaria* Forskal), *Entomologist's Monthly Magazine* **94**, 210–14 (1958).

32. J. H. Crook, The adaptive significance of avian social organisation. *Symposia of the Zoological Society of London* **14**, 181–218 (1965).

33. H. Kruuk, Predators and antipredator behaviour in the black-headed gull (*Larus ridibundus* L.), *Behaviour*, Suppl. **11**, 1–129 (1964).

34. J. C. Coulson, Differences in the quality of birds nesting in the centre and on the edges of a colony, *Nature* **217**, 478–9 (1968).

35. D. Lack, *Ecological Adaptations for Breeding in Birds* (London, Methuen, 1968).

36. F. Galton, Gregariousness in cattle and men, *Macmillan's Magazine* **23**, 353–7 (1871).

37. Y. Espmark, Defence reactions to oestrid flies by forest reindeer, *Zoologische Beitrage* **14**, 155–67 (1968).

38. J. P. Scott, Social behaviour, organisation and leadership in a small flock of domestic sheep, *Comparative Psychology Monographs* **18** (96), 1–29 (1945).

39. C. Russell and W. M. S. Russell, *Violence, Monkeys and Man* (Macmillan, London, 1968).

40. J. H. Crook, Sources of cooperation in animals and man, in J. F. Eisenberg and W. S. Dillon (eds), *Man and Beast: Comparative Social Behavior*, pp. 237–72 (Smithsonian Press, Washington DC, 1971).

41. C. R. Ribbands, The role of recognition of comrades in the defence of social insect communities, *Symposia of the Zoological Society of London* **14**, 159–68 (1965).

42. S. A. Barnett, Rats, *Scientific American* **216** (1), 79–85 (1967).

43. J. M. Bowers and B. K. Alexander, Mice: individual recognition by olfactory cues, *Science* **158**, 1208–10 (1967).

44. See E. O. Wilson, Competitive and aggressive behavior, in J. F. Eisenberg and W. S. Wilson (eds), *Man and Beast: Comparative Social Behavior*, pp. 183–217 (Smithsonian Press, Washington DC, 1971) for cases in parasitoid Hymenoptera; and Lack 1968 (ref. 35) for cuckoo parasitism in birds.

45. M. Kimura and G. H. Weiss, The stepping stone model of population structure and the decrease of the genetic correlation with distance, *Genetics* **49**, 561–76 (1964).
46. S. Wright, The genetical structure of population, *Annals of Eugenics* **15**, 323–54 (1951).
47. S. Wright, The interpretation of population structure by F-statistics with special regard to systems of mating, *Evolution* **19**, 395–420 (1965).
48. N. L. Petras, Studies of natural populations of *Mus*, I: biochemical polymorphisms and their bearing on breeding structure, *Evolution* **21**, 259–74 (1967).
49. C. H. Southwick, The population dynamics of confined house mice supplied with unlimited food, *Ecology* **36**, 212–25 (1955).
50. J. H. Crook, The adapative significance of avian social organisation, *Symposia of the Zoological Society of London* **14**, 181–218 (1965).
51. C. D. Michener, The evolution of social behaviour in bees, *Proceedings of the 10th International Congress of Entomology* **2**, 441–7 (1958).
52. S. R. Sakagami, Ethological peculiarities of the primitive social bees *Allodape lepeletier*, and allied genera, *Insectes Sociaux* **7**, 231–49 (1960).
53. See, for example, Kummer 1968 (ref. 7); C. H. Southwick, M. A. Beg, and M. R. Siddiqi, Rhesus monkeys in North India, in I. DeVore (ed.), *Primate Behavior*, pp. 111–59 (Holt, Rinehart & Winston, New York, 1965).
54. R. Ardrey, *African Genesis* (Collins, London, 1961).
55. S. I. Malyshev, *Genesis of the Hymenoptera and the Phases of their Evolution* (Methuen, London, 1969; transl. from the Russian).
56. See, for example, P. S. Corbet and A. Griffiths, Observations on aquatic stages of two species of Toxorhynchites (Diptera: Culicidae) in Uganda, *Proceedings of the Royal Entomological Society of London* A, **38**, 125–35 (1963); J. P. Woodring, Observation on the biology of six species of acarid mites, *Annals of the Entomological Society of America* **62**, 102–8 (1969).
57. Wilson 1971 (see ref. 44).
58. D. Lack, Of birds and men, *New Scientist* **41**, 121–2 (1969).
59. D. Freeman, Human aggression in anthropological perspective, in J. D. Carthy and F. J. Ebling (eds), *The Natural History of Aggression*, pp. 109–10 (Academic Press, New York, 1964).
60. J. M. Emlen, Natural selection and human behaviour, *Journal of Theoretical Biology* **12**, 410–18 (1966).
61. R. D. Alexander and D. W. Tinkle, A comparative review, *Bioscience* **18**, 245–8 (1968).
62. W. H. Cazaly, A theory of cruelty, in I. J. Good (ed.), *The Scientist Speculates: An Anthology of Partly Baked Ideas* (Heinemann, London, 1962).
63. S. Wright, Tempo and mode in evolution: a critical review, *Ecology* **26**, 415–19 (1945).
64. J. B. S. Haldane, *The Causes of Evolution* (Cornell University Press, New York, 1932).

65. See, for example, P. A. Labine, Population biology of the butterfly *Euphydryas editha*, 1; barriers to multiple insemination, *Evolution* **18**, 335–6 (1964); H. K. Buechner, J. A. Morrison, and W. Leuthold, Reproduction in Uganda kob with special reference to behavior, in I. W. Rowlands (ed.), *Comparative Biology of Reproduction in Mammals, Symposia of the Zoological Society of London* **15**, 69–88 (1966); W. A. Foster, Cooperation by male protection of ovipositing female in the Diptera, *Nature* **214**, 1035–6 (1967).

Merino sheep in New Zealand bunch to avoid a dog

This sheep farmer sometimes deploys (not shown) a second dog as a 'barker', whose specific effect is to group the sheep tighter still. An instinctive reaction? Or have they been conditioned by the barking of tourists' dogs who may really harm them?

CHAPTER 7

PANIC STATIONS

Geometry for the Selfish Herd

It is particularly striking how often animals congregate in herds on grasslands, much more so than in the forest.

R. HESSE, W. C. ALLEE, and K. P. SCHMIDT[1]

THIS was a theme that I had worked out in note form long before I tried to publish it. Once again the idea began in an irritation while I was still an undergraduate reading and hearing about schooling, flocking, and herding habits of animals. My doubts that these phenomena were being correctly interpreted became more focused when V. C. Wynne-Edwards of the University of Aberdeen published in 1962 his book *Animal Dispersion in Relation to Social Behaviour*, which suggested that a population display and self-census function was involved in many instances of aggregation. He unhesitatingly appealed to selection at the population level to explain how such phenomena had arrived. I could not see that this view was either necessary or likely.

Ever since I was a boy I have loved secondhand bookshops. A lot of my early reading in fiction and science came out of such shops because almost there alone the books were within my means. On the emotional side I believe this made a last dinosaur of me, a ponderous romantic doomed to live through decades when romanticism was in steep decline. On the scientific side it simply made me out of date in my interests. Nevertheless, because topics in science often fade for no good reasons—certainly not because of error—it may sometimes help inspiration to be reminded of forgotten facts or to see new ones from an old point of view. During a pre-Chiswick period of my stay in London there was in Marchmont Street, near also to King's Cross and the restaurants mentioned in the introduction to Chapter 2, a secondhand bookshop that was not far from my zig-zagging course down to LSE from my digs (these were then in Cruikshank Street, just off Claremont Square in Clerkenwell). In the recesses of this shop I obtained a copy of a battered Everyman's Library edition of Galton's *Inquiry into Human Faculty*. Re-reading it after my interest (or irritation) over aggregation had been re-stimulated by other more holistic interpretations, I noticed some remarks about cattle in South Africa. These led me to an obscure paper by Galton, which I think few can have read. His argument was on the line I was already inclined to. Galton, I was surprised to be reminded in the book, had personally explored Damaraland in southern Africa and had observed for himself half-wild cattle in a terrain where lions and other large predators were still common. To see is to sympathize, especially when the observed animal is a mammal; moreover, Galton probably personally experienced the sensation of safety from surprise attack when walking in the savannah among cows as compared to when walking alone. The idea itself was little more than common sense but I had realized from experience that university people sometimes don't react well to common sense and in any case most of them listen to it harder if you first intimidate them with equations. Their extra attention still holds when the equations are quite unimportant for the conclusion. I started to devise geometrical models and quite soon these took on a life of their own. I wanted to get some generalities about aggregation that arises through cowardly cover-seeking. Some of the geometrical puzzles I encountered were not only engrossing but funny; hence eventually the title that I chose.

Shortly after starting to think about all this and reading Galton, I found that George Williams as usual had anticipated me in studying this egoistic interpretation of aggregation. By experimental work as well as by argument he had shown quite convincingly that much of the schooling of fish could be parsimoniously explained by a selfish and cover-seeking motivation.[2] So far, so good; the battle for a proper assessment of aggregation ought, then, to be already in progress even if it seemed not to be. Given these forerunners, there seemed to be little left for me to do except perhaps to draw attention to the ideas and stir lovers of geometrical puzzles to join the fray. I saw such work as part of a general crusade to make biologists attend to the important book Williams had already written,[3] and also, of course, to induce them to read my own earlier writings. People must see that animals are neither mindless bouncing molecules on the one hand nor benevolent angels all determined on mutual aid on the other. Rather they are creatures of mixed motives much like ourselves. When a crowd rushes for the exit of a cinema in a fire, trampling each other underfoot, are they acting efficiently, with group survival in mind? Did we need to postulate any 'emergent organization'? I was sure that generally we didn't: why was it any different, then, with sheep or starlings? Thus far the matter was really just common sense, but other aspects of the problem continued to tease me, and some still do. Fish in the presence of a predator don't always 'school'; sometimes instead they explode—flee centripetally in every direction. Likewise the starlings don't have to have a hawk anywhere near when they begin their massive manoeuvres. On the more purely geometrical side, in the rush for spatial safety as the theory specifies it, when, if ever, is it best to stay where you are? I found two cases and showed them in the paper. Could I find others? So it went on.

It seemed to me that some of the maths that I got into might have applications in other spheres of biology. This proved to be the case but here I found myself long forestalled. The Voronoi polygons and the groups that are linked and separated from others by nearest-neighbour relations I found for myself and only later discovered the same problems to have generated page after daunting page of deep analytical development in the mathematical journals. The backgrounds there had been crystal structure, the interpretations to be put upon noisy signals, and others. Clearly mathematicians had already been interested in my problem but for quite different reasons. This line of study still continues.[4] In the sphere of biology most of what I was doing I thought of as being at best a parable for the problem of gregarious instincts of real animals; otherwise I thought of it as more of a mathematical game. Some of the ideas that arose were certainly picturesque and by about 1966 I had a miscellany of conclusions all favourable to my original suspicion that moving closer to a nearby other potential prey would, in general, reduce risk in the presence of a predator.

My circular lily pond with its cowardly frogs became a popular if simplistic demonstration of such selfish risk reduction in one dimension.[5] My own favourite among my models, however, was that at the opposite extreme, of the potential herding behaviour of cattle in what I might call *N*-Damaraland (i.e. behaviour in an *N*-dimensional, open, grassy plain). I show that the primary clumps of prey that form when the *N*-lion roars have, even when *N* is approaching infinity, an average size of no greater than four. Deep in such infinitely hyperspatial grass just the tip of the lion's tail twitches, not one sees it, he is ready, then . . . But what if today he is hunting groups, not individuals? Four prey at a spring I suppose would be an easy feat for such a lion; but to what quartile of the new distribution of the clumps that are already gathered, can he be expected to go? Somehow such childish images made alternative universes, the hyperspaces, of the mathematician and the physicist seem less awesome to me. Eventually I thought I would try some of my more amusing conclusions on Martin Gardner, editor of the 'Mathematical Games' section of *Scientific American*.

I sent Gardner a lighter version than the paper here, containing no maths. I received in return a pleasant personal letter saying that he found my theme fascinating but regretted *Scientific American* couldn't publish it for two reasons. One was that I had said that none of it had been published elsewhere and it was a rule that *Scientific American* didn't publish original work. Second, the attention span of his readers was too short for them to take in the ideas—some rather scathing number of seconds was mentioned for this span if I remember. He remarked that requiring readers to reach for pencil and paper was the limit he could expect. I decided later that all this simply meant I hadn't written my piece well enough. Gardner didn't say so but it was probably also the case that my account did not fall well into any established category of the journal, being rather too short and too frivolous for a main article but too long, serious, and new for him to abstract a commentary that could be put in his column. I was disappointed by the outcome but not much, and I shelved my manuscript. When I started thinking of possible themes for the Smithsonian symposium (Chapter 6) I foresaw an opportunity to mention some of the ideas there, but, given the nature of the paper requested of me, this mention had to be very brief and popular even compared to what I had intended for the *Scientific American* article.

About the same time a colleague at Silwood to whom I had talked about my theme mentioned my findings to Ian Vine whom he knew to have become interested in the evolution of aggregation and likewise to have had critical thoughts concerning existing views. I sent Vine my rejected *Scientific American* article. His comment was quite enthusiastic; he suggested that I should seek to publish in a more normal journal and stated that he would like in any case to cite it in a paper he was writing. I decided to try whether my old outlet, *Journal of Theoretical Biology*, might serve me yet again.

As can be seen from the style and title of the paper I did not completely eliminate the mathematical-game levity I had drafted for *Scientific American*, and I like to think that this makes the present version more readable than most of my papers and also more than others that are at the same level of content. Certainly it seems to have been read a lot because some time ago I learned that it reached the status of a 'Citation Classic' for the journal *Current Contents*. I regret that I have not, previous to now, had time to write up and submit any sketch to that journal concerning its history. A modern review of its ideas and evidence can be found in a paper by Mooring and Hart.[6] I think it can be said that since it was written the emphasis that I gave in the paper to attack by parasites has held up very strongly,[5] although there are now good examples for predators also.[7,8] In other papers worth seeing the motive of pure selfish herding[9] may come to be mixed a little with kin-based altruism—for example, this is implied in the theme of teaching predators to avoid distasteful relatives[10] (see Chapter 2).

I have forgotten where I was living when writing this paper. My notes on the problem were begun in my Blue-Star-Garage days but I had moved on from there before writing even the first version. The two subsequent possibilities are a top-floor flat of an old terrace in Elsham Road, West Kensington, or the house that my wife and I bought in the early 1970s in Shurlock Row, near Twyford in Berkshire. The Elsham Road flat became our base after return from our first joint trip to Brazil in 1968–69 and we lived there while looking for a house near to Silwood. We ended being 12 miles away, I with a daily commute by car. However, I had a very pleasant road entirely through countryside and it passed, as I came into the Windsor–Ascot area, royal estates including, right at my roadside, stretches of the very old woodland of Windsor Forest—quite a treasure of wildlife. More of Windsor Forest will come in a later introduction and it is not much connected with themes of this paper. The long, straight, back-country road I drove was called locally 'The Drift Road'. 'Drift' implies that 100 years ago it would have been a 'green' road for droving—that is, for driving farmers' stock to distant markets. Had I been riding or walking the road then I doubtless would have seen almost daily flocks of sheep bunching as each tried to ensure that it was not in the back rank and having its ankles nipped by the dog. I did occasionally still encounter flocks of sheep or cows on the Drift Road but these animals were being moved only from one field to another; it was much longer ago that I had first seen sheep being formed by shepherd and dog into those tight and seething masses that told me, from the often frenzied attempts to avoid outside positions, that the underlying motivation had to be selfish. Our house in Shurlock Row also overlooked pastures but those were most often tenanted by rather placid and ungregarious cows from the farm that was the next-but-one-house to ours on one side.

Was the paper, or at least its '*Sci. Amer.*' predecessor, written while we were still in Elsham Road? There, if I had seen actual selfish herds at all they would certainly have to have been mainly starlings but perhaps with smaller groups of town pigeons contributing. This would be partly because we were in London, where, reversing the usual human commuters trek, starlings like to come in vast numbers from the distant fields to roost (while pigeons are always present) and partly because our attic windows were so high in the old Victorian building that they normally revealed little of wildlife outside except starlings and pigeons, and more occasionally gulls, and swifts—and all these mostly seen against the sky. It was the swifts there that I liked the best. One day standing at the window that opened over a chasm of railways directly beneath me and towards Olympia they became especially interesting. I noticed how, from the background of the railway sidings, these dull sickle-winged crosses came up hurtling and screaming towards my window and then at the last moment, just as they appeared about to hit the pane by my face, they mysteriously vanished.

Watching more carefully I realized their magic to be somehow connected to the lead of the outside window sill and soon further that they must be attending nests in cavities under that sill and only a few inches from my chest and just below the high window.

This discovery accounted not only for some of the 'ghost' noises the home had been giving us by both day and night, which we had hitherto guessed to be mice, but also, via nest occupants and debris, it explained some of the very interesting and unexpected small fauna I had been finding in the flat. There were the pseudoscorpions on a high shelf, *Dermestes haemorrhoidalis* (which I suppose translates as the 'as-if-pile-afflicted larder beetle'—it has a red back end) as well as other rarer beetles on the floor and under the carpet, many mites and spiders, and all round a general air of minute shuffling activity, which fortunately for us (although I state this almost sadly) mostly didn't quite reach the scale of herding and flocking.[11] Such living overspill put me in mind how birds nests, especially those of hollows, make an important road by which arthropods evolve to be 'anthropophilic' (i.e. to be inhabitants of our buildings and homes), with a few cases going further to a more sinister alimentary interest in our bodies.[12] Via this parallel of the sub-sill nests with tree holes, my life in the Elsham Road flat was leading away from herding and towards my later interest in rotting wood (see Chapter 12). However, on the theoretical side I was certainly working on the ideas of herding, on the Voronoi polygons, etc., at the time I was living there.

The Elsham Road terrace has been demolished. Trains now rumble past the new posh flats that I could never rent; but the screaming swifts, those atoms of the sky that I felt so proud would choose my window for their fleeting summer contact with the land, like me, those have gone—they flash to no sills of the new building I'm sure. Marvellous birds, permanent to skies over land as albatrosses and frigate birds are to the skies of ocean, and as light and as truly swift as those mariners are ponderous, I hope that mine have found other hospitable lead-work somewhere in that dear older London where I used to live; and I hope also that the satellite life that they once passed on to me through the walls has found some way to fly or shuffle along there to rejoin them, although I don't suppose that the swifts themselves care too much about that.

References and notes

1. R. Hesse, W. C. Allee, and K. P. Schmidt, *Ecological Animal Geography* (Wiley/Chapman and Hall, New York/London, 1937); see pp. 450–1.
2. Overlooked by both Williams and me, the following paper may have been the first to deal with cover-seeking in fish schooling: V. E. Brock and R. H. Riffenburgh, Fish schooling: a possible factor in reducing predation, *Journal Conseil Permanent Internationale pour l'Exploration de la Mer* **25**, 307–17 (1960).
3. G. C. Williams, *Adaptation and Natural Selection: A Critique of Some Current Evolutionary Thought* (Princeton University Press, Princeton, NJ, 1966).
4. A. Okabe, B. Boots, and K. Sugihara, *Spatial Tesselations: Concepts and Applications of Voronoi Diagrams* (Wiley, New York, 1992).
5. The frog model may have been less pure parable than I thought, as suggested in the following paper where a real snake does prey on frogs and toads and they do sometimes, in their terror, pile up in heaps: S. J. Arnold and R. J. Wassersug, Differential predation on metamorphic anurans by garter snakes (*Thamnophis*): social behavior as a possible defence, *Ecology* **59**, 1014–22 (1978).
6. M. S. Mooring and B. L. Hart, Animal grouping for protection from parasites: selfish herd and encounter dilution effects, *Behaviour* **123**, 173–93 (1992).
7. R. D. Estes, Social organisation of the African birds; in V. Geist and F. Walther (eds), *The Behaviour of Ungulates in its Relation to Management*, Vol. 1 (International Union for the Conservation of Nature and Natural Resources, 1974).
8. J. E. Treherne and W. A. Foster, Group size and anti-predator strategies in a marine insect, *Animal Behaviour* **32**, 536–42 (1982).
9. W. H. Calvert, L. E. Hedrick, and L. P. Brower, Mortality of the Monarch Butterfly (*Danaus plexippus* L.): avian predation at the overwintering sites in Mexico, *Science* **204**, 847–51 (1979).
10. B. Heinrich and F. D. Vogt, Aggregation and foraging behavior of whirligig beetles (Gyrinidae), *Behavioral Ecology and Sociobiology* **7**, 179–86 (1980).
11. H. Mourier, O. Winding, and E. Sunesen, *Collins Guide to Life in House and Home* (Collins, London, 1977).
12. M. Andrews, *The Life that Lives on Man* (Taplinger, New York, 1977).

GEOMETRY FOR THE SELFISH HERD†

W. D. HAMILTON

This paper presents an antithesis to the view that gregarious behaviour is evolved through benefits to the population or species. Following Galton[1] *and Williams*[2] *gregarious behaviour is considered as a form of cover-seeking in which each animal tries to reduce its chance of being caught by a predator.*

It is easy to see how pruning of marginal individuals can maintain centripetal instincts in already gregarious species; some evidence that marginal pruning actually occurs is summarized. Besides this, simply defined models are used to show that even in non-gregarious species selection is likely to favour individuals who stay close to others.

Although not universal or unipotent, cover-seeking is a widespread and important element in animal aggregation, as the literature shows. Neglect of the idea has probably followed from a general disbelief that evolution can be dysgenic for a species. Nevertheless, selection theory provides no support for such disbelief in the case of species with outbreeding or unsubdivided populations.

The model for two dimensions involves a complex problem in geometrical probability which has relevance also in metallurgy and communication science. Some empirical data on this, gathered from random number plots, is presented as of possible heuristic value.

1. A MODEL OF PREDATION IN ONE DIMENSION

Imagine a circular lily pond. Imagine that the pond shelters a colony of frogs and a water-snake. The snake preys on the frogs but only does so at a certain time of day—up to this time it sleeps on the bottom of the pond. Shortly before the snake is due to wake up all the frogs climb out onto the rim of the pond. This is because the snake prefers to catch frogs in the water. If it can't find any, however, it rears its head out of the water and surveys the disconsolate line sitting on the rim—it is supposed that fear of terrestial predators prevents the frogs from going back from the rim; the snake surveys this line and snatches *the nearest one*.

†*Journal of Theoretical Biology* **31**, 295–311 (1971).

Now suppose that the frogs are given opportunity to move about on the rim before the snake appears, and suppose that initially they are dispersed in some rather random way. Knowing that the snake is about to appear, will all the frogs be content with their initial positions? No; each will have a better chance of not being nearest to the snake if he is situated in a narrow gap between two others. One can imagine that a frog that happens to have climbed out into a wide open space will want to improve his position. The part of the pond's perimeter on which the snake could appear and find a certain frog to be nearest to him may be termed that frog's 'domain of danger': its length is half that of the gap between the neighbours on either side. Figure 7.1a shows the best move for one particular frog and how his domain of danger is diminished by it.

But usually neighbours will be moving as well and one can imagine a confused toing-and-froing in which the desirable narrow gaps are as elusive as the croquet hoops in Alice's game in Wonderland. From the positions in Fig. 7.1a, assuming the outside frogs to be in gaps larger than any others shown, the moves in Fig. 7.1b may be expected.

What will be the result of this communal exercise? Devious and unfair as usual, natural justice does not, in general, equalize the risks of these selfish frogs by spacing them out. On the contrary, with any reasonable assumptions about the exact jumping behaviour, they quickly collect in heaps. Except in the

Figure 7.1 Selfish prey movements through jumping given nearest-prey predation in a one-dimensional habitat. (a) A particular prey's *domain of danger* is shown by the solid bar; this includes all points nearer to the given prey than to any other. Independent of position within a gap between neighbour prey, the domain is equal to half of the gap. The arrowed movement illustrates a jump into a narrower gap beyond a neighbour, thereby diminshing a domain. (b) All prey are assumed moving on the principle of (a), with each jumper passing its neighbour's position by one-third of the gap length beyond. Increase of aggregation can be seen.

case of three frogs who start spaced out in an acute-angled triangle I know of no rule of jumping that can prevent them aggregating. Some occupy protected central positions from the start; some are protected only initially in groups destined to dissolve; some, on the margins of groups, commute wildly from one heap to another and yet continue to bear most of the risk. Figure 7.2 shows the result of a computer simulation experiment in which 100 frogs are initially spaced randomly round the pool. In each 'round' of jumping a frog stays put only if the 'gap' it occupies is smaller than both neighbouring gaps; otherwise it jumps into the smaller of these gaps, passing the neighbour's position by one-third of the gap-length. Note that at the termination of the experiment only the largest group is growing rapidly.

The idea of this round pond and its circular rim is to study cover-seeking behaviour in an edgeless universe. No apology, therefore, need be made even for the rather ridiculous behaviour that tends to arise in the later stages of the model process, in which frogs supposedly fly right round the circular rim to 'jump into' a gap on the other side of the aggregation. The model gives the hint which I wish to develop: that even when one starts with an edgeless group of animals, randomly or evenly spaced, the *selfish avoidance of a predator can lead to aggregation.*

10° segments of pool margin (degrees)

Position number	0									90									180									270								360
1	2	3	3	3		6	1	3	3		4	1	3	5			1	3	4	5	2	9	4		5	6	3	5	4		1	2	2		4	3
2		5	2	2		8		3	2		6	1	1	7				2	4	7		11	2		7	5	2	5	5			3	2		4	4
3		6	1	2		8		3	1		8			9					4	9		11			9	4	1	7	5			2	2		4	4
4		6	1			9		3			9			8					4	11		10			10	3		8	6			1	2		4	5
5		7				9		2			9			8					5	12		8			12	1		9	6				3		4	5
6		7				9					9			8					5	14		7			13			9	5				5		3	6
7		8				7					9			8					5	16		5			15			9	3				6		3	6
8		9				5					9			9					4	18		3			17			9	1				6		5	5
9		10				4					8			10					3	20		1			19			8					6		7	4
10		12				3					7			11					2	22					20			6					6		9	2
11		13				2					6			12					1	22					22			4					7		9	2
12		14									6			13						22					24			3					7		9	2
13		14									5			13						22					26			2					8		8	2
14		15									3			13						22					28			1					9		7	2
15		16									1			13						22					30								10		6	2
16		17												12						22					32								9		6	2
17		17												11						22					33								8		7	2
18		18												9						22					35								6		8	2
19		19												7						22					37								5		9	1

Figure 7.2 Gregarious behaviour of 100 frogs is shown in terms of the numbers found successively within 10° segments on the margin of the pool. The initial scatter (position 1) is random. Frogs jump simultaneously giving the series of positions shown. They pass neighbours' positions by one-third of the width of the gap. For further explanation, see text.

2. AGGREGATIONS AND PREDATORS

It may seem a far cry from such a phantasy to the realities of natural selection. Nevertheless, I think there can be little doubt that behaviour which is similar in biological intention to that of the hypothetical frogs is an important factor in the gregarious tendencies of a very wide variety of animals. Most of the herds and flocks with which one is familiar show a visible closing-in of the aggregation in the presence of their common predators. Starlings do this in the presence of a sparrowhawk;[3–6] sheep in the presence of a dog, or, indeed, any frightening stimulus.[7] Parallel observations are available for the vast flocks of the quelea[8] and for deer.[9] No doubt a thorough search of the literature would reveal many other examples. The phenomenon in fish must be familiar to anyone who has tried to catch minnows or sand eels with a net in British waters. Almost any sudden stimulus causes schooling fish to cluster more tightly,[10] and fish have been described as packing, in the presence of predators, into balls so tight that they cannot swim and such that some on top are thrust above the surface of the water.[11] A shark has been described as biting mouthfuls from a school of fish 'much in the manner of a person eating an apple'.[12]

G. C. Williams, originator of the theory of fish schooling that I am here supporting,[2,13] points out that schooling is particularly evident in the fish that inhabit open waters. This fits with the view that schooling is similar to cover-seeking in its motivation. His experiments showed that fish species whose normal environment afforded cover in the form of weeds and rocks had generally less-marked schooling tendencies. Among mammals, similarly, the most gregarious species are inhabitants of open grassy plains rather than of forest.[14] With fish schools observers have noted the apparent uneasiness of the outside fish and their eagerness for an opportunity to bury themselves in the throng[11] and a parallel to this is commonly seen in the behaviour of the hindmost sheep that a sheepdog has driven into an enclosure: such sheep try to butt or to jump their way into the close packed ranks in front. Behaviour of this kind certainly cannot be regarded as showing an unselfish concern for the welfare of the whole group.

With ungulate herds,[1,15] with bird flocks,[6,16] and with the dense and sudden-emerging columns of bats that have been described issuing at dusk from great bat caves[17,18] observations that predators do often take isolated and marginal individuals have frequently been recorded. Nor are such observations confined to vertebrate or to mobile aggregations. Similar observations have been recorded for locusts,[19] for gregarious caterpillars,[20] and, as various entomologists have told me, for aphids.

For the aphids some of the agents concerned are not predators in a strict sense but fatal parasites. Insect parasites of vertebrates are seldom directly fatal but, through transmitted diseases or the weakening caused by the activities of endoparasitic larvae, must often cause death indirectly nevertheless. Thus escape from insect attack is another possible reward for gregarious instincts. From observations in Russia, Sdobnikov[15] has stated that when reindeer are standing in dense herds only the outermost animals are much attacked by insects. Among the species which he observed attacking the reindeer, those which produced the most serious affliction were nose flies and warble flies, larvae of which are endoparasites of the nasal passages and the skin, respectively. Espmark[21] has verified and extended most of Sdobnikov's information. His work reinforces the view that such oestrid flies are important and ancient enemies of their various ungulate hosts, as is suggested by the fact that their presence induces a seemingly *instinctive* terror. Reindeer seem to be almost as terrified of them as they are of wolves, and cattle react as though they feared the certainly painless egg-laying of warble flies far more than they fear the painful bites of large blood-suckers.[22]

The occurrence of marginal predation has also been recorded for some of the aggregations formed by otherwise not very gregarious animals for the purpose of breeding. The best data known to me concern nesting black-headed gulls. The work of Kruuk[23] has been reinforced by further studies, summarized by Lack.[24] The latter seem to have shown that all marginal nests failed to rear young, mainly due to predation. Perhaps, nevertheless, the gulls that could not get places in the centre of the colony were right in nesting on the edge rather than in isolation where, for a conspicuous bird like a gull, the chances would have been even worse.

It is perhaps worth digressing here to mention the temporal aspect of marginal predation. The 'aggregation' in timing already alluded to for bats issuing from bat caves parallels the marked synchrony in breeding activities which is seen in most aggregations and which has been called the 'Fraser Darling effect'. In explanation of this, Lack[24] points out that late and early breeders in terns and the black-headed gull do worse in terms of young raised than those best-synchronized with the mass, and he implies that this is mainly due to predation. Individuals coming into breeding condition late or early may also have a problem in sexual selection—that of finding a mate. This point will be touched on later. There are similar influences of temporal selection for flowering plants.

The securing of a nest site in the middle of a colony area is certainly likely to be an achievement of protected position in a sense related to that explained in the story of the frogs, but in the relative immobility of such positions the case diverges somewhat from the initial theme. In all the foregoing examples, except perhaps that of the insect parasites of vertebrates, close analogy to that theme has also been lost through the assumption that the predator is likely to approach from outside the group. This is difficult to avoid.[25]

When a predator habitually approaches from outside it is comparatively obvious how marginal pruning will at least maintain the centripetal instincts of the prey species. Whether predation could also initiate gregariousness in an originally non-gregarious species is another matter. As mentioned before, this will begin to seem likely if it can be shown that when predators tend to appear within a non-aggregated field of prey geometrical principles of self-protection still orientate towards gregarious behaviour. So far I have shown this only for one highly artificial case: that of jumping organisms in a one-dimensional universe. For the case in two dimensions a more realistic story can be given. The most realistic would, perhaps, take reindeer and warble flies as its subject animals, but in order to follow an interesting historical precedent, which has now to be mentioned, the animals chosen will be cattle and lions.

In 1871, Francis Galton published in *Macmillan's Magazine* an article entitled 'Gregariousness in cattle and in men'.[1] In it he outlined a theory of the evolution of gregarious behaviour based on his own observations of the behaviour of the half-wild herds of cattle owned by the Damaras in South Africa. In spite of the characteristically forceful and persuasive style of his writing, Galton's argument is not entirely clear and consistent. Some specific criticisms will be mentioned shortly. Nevertheless, it does contain in embryo the idea of marginal predation as a force of natural selection leading to the evolution of gregarious behaviour. The main predators of the Damaraland cattle, according to Galton, were lions, and he states clearly that these did prefer to take the isolated and marginal beasts. The following passage shows sufficiently well his line of thought. After stating that the cattle are unamiable to one another and do not seem to have come together due to any 'ordinary social desires', he writes:

> Yet although the ox has so little affection for, or individual interest in, his fellows, he cannot endure even a momentary severance from his herd. If he be separated from it by strategem or force, he exhibits every sign of mental agony; he strives with all his might to get back again and when he succeeds, he plunges into its middle, to bathe his whole body with the comfort of closest companionship.

3. A MODEL OF PREDATION IN TWO DIMENSIONS

Although as Galton implies, lions, like most other predators, usually attack from outside the herd, it is possible to imagine that in some circumstances a lion may remain hidden until the cattle are feeding on all sides of it. Consider therefore a herd grazing on a plain and suppose that its deep grass may conceal—anywhere—a lion. The cattle are unaware of danger until suddenly the lion is heard to roar. By reason of some peculiar imaginary quality the sound gives no hint of the whereabouts of the lion but it informs the cattle of danger. At any moment, at any point in the terrain the cattle are traversing, the lion may suddenly appear and attack the nearest cow.

As in the case of the 'frog' model, the rule that the predator attacks the nearest prey specifies a 'domain of danger' for each individual. Each domain contains all points nearer to the owner of the domain than to any other individual. In the present case such domains are polygonal (Fig. 7.3). Each polygon is bounded by lines which bisect at right angles the lines which join the owner to certain neighbours; boundaries meet three at a point and an irregular tesselation of polygons covers the whole plane. On hearing the lion roar each beast will want to move in a way that will cause its polygonal domain to decrease. Not all domains can decrease at once of course: as in the case of the frogs, if some decrease others must grow larger. Nevertheless, if one cow moves while others remain stationary the one moving can very definitely improve its position. Hence it can be assumed that inclinations to attempt some adaptive change of position will be established by natural selection. The optimal strategy of movement for any situation is far from obvious, and before discussing even certain better-than-nothing principles that are easily seen it will be a cautionary digression to consider what is already known about a particular and important case of such a tesselation of polygons, that in which the 'centres' of polygons are scattered at random.

Patterns of a more concrete nature which are closely analogous to the tesselation defined certainly exist in nature. In two dimensions they may be seen in the patterns formed by encrusting lichens on rocks, and in the cross-sectional patterns of cracks in columnar basalt. The corresponding pattern in three dimensions, consisting of polyhedra, is closely imitated in the crystal grains of some metals and other materials formed by solidification of liquids. The problem of the statistical description of the pattern in the case where centres are distributed at random was first attacked with reference to the grain structure of metals.[26] More recently the N-dimensional analogue has been studied by Gilbert[27] on the incentive of a problem arising in communication science. Yet in spite of great expertise many simple facts about even the two-dimensional case remain unknown. The distribution of the angles of the polygons is known;[28] something is known of the lengths of their sides, and Gilbert has found the variance of their areas.[29] Nothing, apart from the fairly obvious value of six for the mean, is known of the numbers of sides. Since a biologist might be interested to know whether, in the ideal territory system of these polygons, six neighbours is the most likely number, or whether five is more likely than seven, some data gathered from plots of random numbers are given in Table 7.1.

After glancing at Gilbert's paper a non-mathematical biologist may well despair of finding a theory of ideal self-protective movement for the random cattle. Nevertheless, experimentation with ruler and compasses and one small evasion quickly reveal a plausible working principle. As a preliminary it may be pointed out that the problem differs radically from the linear problem in that domains of danger change continuously in size with movements of their owner. (In the linear case no change takes place until an individual actually passes his neighbour; this was the reason for choosing organisms that jump.) Now suppose that one cow alone has sensed the presence of the lion and is hastily

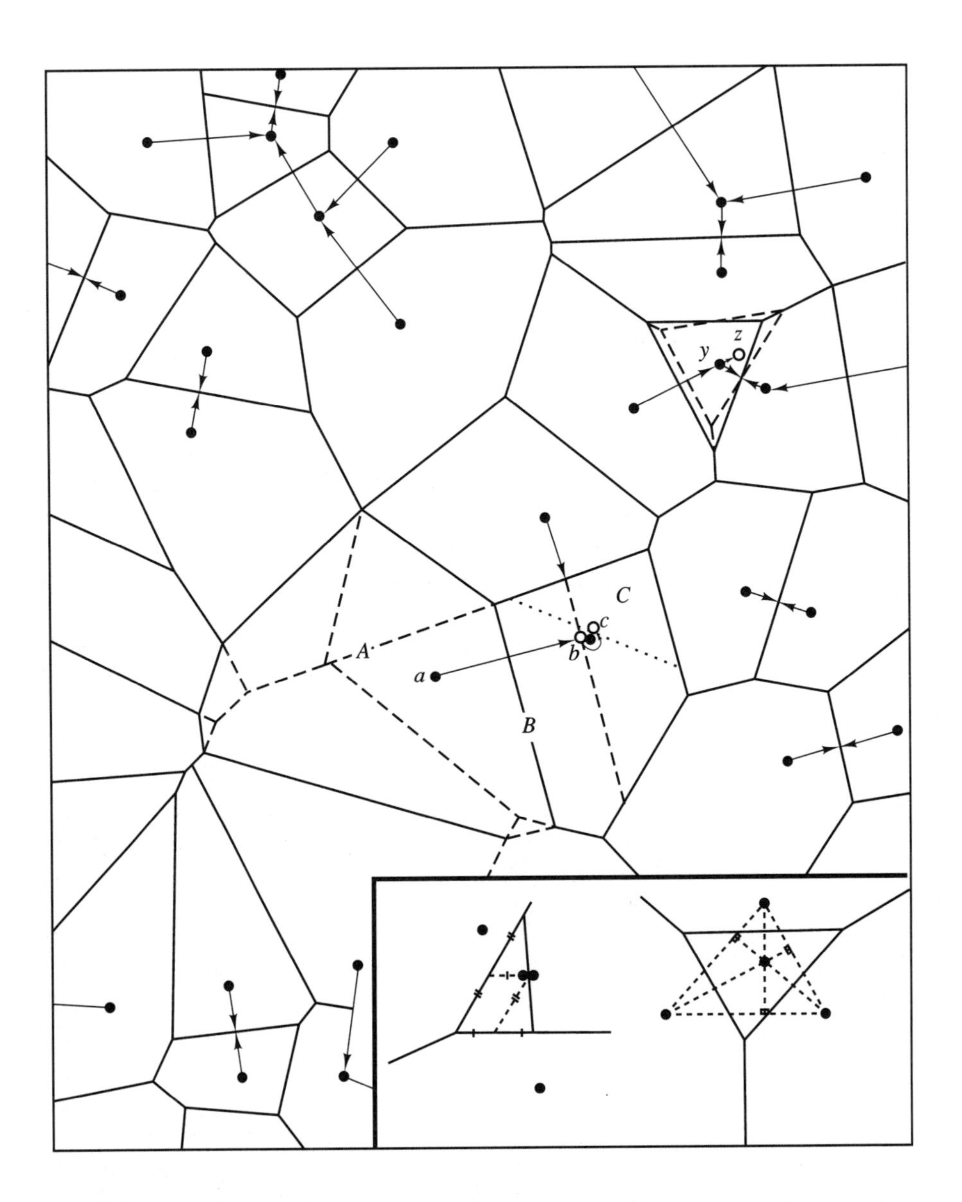

Figure 7.3 Domains of danger for a randomly dispersed prey population when a hidden predator will attack the nearest animal. Thinner arrowed lines indicate the nearest neighbour of each prey. Thicker arrowed lines show supposed movements of particular prey. The position of prey is given a lowercase letter and the domain obtained from each position is given the corresponding uppercase letter. Dashed lines are used for the boundaries of domains that come into existence after the first movement.

Approach to a nearest neighbour usually diminishes the domain of danger (as $a \rightarrow b$), but not always (as $y \rightarrow z$). On reaching a neighbour the domain may, in theory, be minimized by moving round to a particular side (as $b \rightarrow c$). The inset shows geometrical algorithms for minimum domains attainable by a (on reaching c), and by y. In the case of y the minimum domain is equal to the triangle of neighbours and is obtained from the orthocentre of that triangle.

Table 7.1 Distribution of the number of sides of 376 polygons constructed from random number plots

No. of sides		3	4	5	6	7	8	9	10	11
No. observed		3	38	97	123	70	29	11	4	1
Total no. observed	376									

Data for 100 of these polygons were received from E. N. Gilbert. The rarity of triangles indicates that the sample area chosen for Fig. 7.3 is not quite typical.

moving through an otherwise motionless herd. In most circumstances such a cow will diminish its domain of danger by approaching its nearest neighbour. A direct approach can seldom be the path it would choose if it requires that every step should bring the maximum decrease of danger, and when the nearest neighbour is a rather isolated individual it may even happen that a domain increases during approach to a nearest neighbour (Fig. 7.3, $y \rightarrow z$). Such increases will tend to be associated with polygons with very low side number. When an individual is enclosed in a ring of others and consequently owns a many-sided domain, decrease in area is almost inevitable (Fig. 7.3, $a \rightarrow b$). Since the average number of sides is six and triangles are rare (Table 7.1), it must be a generally useful rule for a cow to approach its nearest neighbour. This is a rule for which natural selection could easily build the necessary instincts. Behaviour in accord with it has been reported in sheep.[7] The evasion, then, is to imagine that these imaginary cattle are too slow-witted to do anything better. Most readers will agree, I hope, that there is no need to call them stupid on this account.

If the rest of the herd remains stationary and the alarmed cow reaches its nearest neighbour it will usually further decrease its domain of danger by moving round to a side that gives it a minimal 'corner' of the neighbour's now enlarged domain (Fig. 7.3, $b \rightarrow c$; left inset diagram shows the principle for finding such a minimal 'corner' domain). In this position the cow of the story may be supposed to reach equilibrium; it is some sort of equivalent, albeit not a close one, of the stable non-jump positions that sometimes occur with the frogs. Even more than with the stable positions of the frogs, however, the stable positions of this model must be almost impossible to attain. As in the case of the frogs this is seen as soon as we imagine that all the other cows are moving and have similar aims in view.

Nearest-neighbour relationships connect up points into groups (Fig. 7.3). Every group has somewhere in it a 'reflexive pair'; that is, a pair for which each is the nearest neighbour of the other. A formula for the frequency of membership of reflexive pairs when the points are randomly dispersed has been derived by Clark and Evans.[30] It is $6\pi/(8\pi + \sqrt{27})$. Halving and inverting this expression gives us the average size of groups as 3.218. No further facts are known mathematically about the relative frequencies of different sizes of groups, but a summary of the forms and sizes found in the random number plots is given in Table 7.2.

Clearly, if randomly dispersed cattle condense according to the statistical pattern indicated in Table 7.2, the groups so formed can hardly be described as herds. Nevertheless, there is undoubtedly a gregarious tendency of a kind, and once such primary groups have formed the cattle in each group will see a common advantage in moving through the field of groups using the same principles as before. Thus an indefinite series of such condensations will take place and eventually large herds will appear. Condensation is not prevented by having the points initially more evenly spaced than in the random dispersion. For example, when the points are initially in some lattice formation the inherent instability of the system is actually easier to see than it is for the random dispersion: every point, if moving alone, can definitely diminish its domain of danger by approaching any one of its equidistant neighbours. Thus even for the

Table 7.2

No. in group	2	3	4	5	6	7
No. of cases found	46	37	24	8	5	1
Configurations found	 46	37	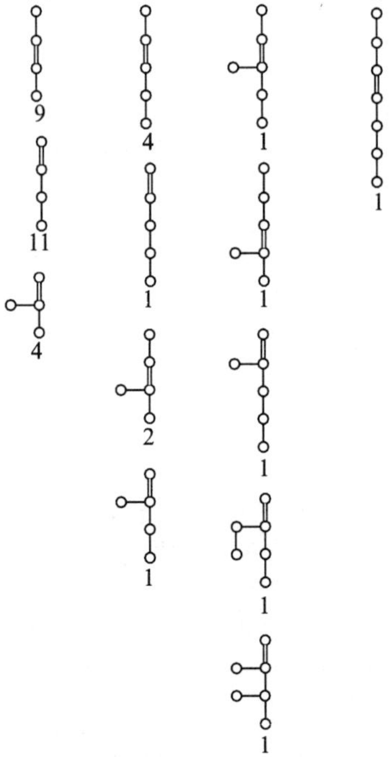 9	4	1	1
			11	1	1	
			4	2	1	
				1	1	
					1	

Mean N in group { observed : 3.107
expected : 3.218

The nearest-neighbour relation allocates points of a random dispersion to groups (see arrowed lines on Fig. 7.3). Here the configurations found in 121 groups formed by 376 points are recorded together with their frequencies of occurrence. Double connecting lines indicate the reflexive nearest-neighbour relationship.

most unpromising initial conditions it remains evident that predation should lead to the evolution of gregarious behaviour.

No story even as poorly realistic as the foregoing can be given for the case of predation in three dimensions. In the air and in water there is still less cover in which the predator can hide. If there were a hiding place, however, for what it is worth the argument can be repeated almost exactly and we arrive at $3\frac{3}{8}$ for the mean size of primary groups. For celestial cattle in a space of infinitely many dimensions the mean group size attains only to 4. The general formula, where n is the number of dimensions, is

$$\frac{4n}{n+1}+\frac{2}{(n-1)\pi^{\frac{1}{2}}}\left(\frac{3^{\frac{1}{2}}}{2}\right)^{n-1}\frac{\Gamma\{n/2\}}{\Gamma\{(n-1)/2\}}.$$

Whether or not marginal predation is, as I believe it to be, a common primary cause of the evolution of gregariousness, it is surprising that the idea of such a cause has received so little attention from biologists. The reason for the neglect of Galton's views cannot lie in the non-scientific nature of the journal in which it was originally published. Galton repeated them in his book *Inquiries into Human Faculty* (1883),[31] and this is one of his best-known works. Yet both the hints of the present hypothesis in writings on the schooling of fishes and the admirable study and discussion of this hypothesis by G. C. Williams[2] were completely independent of Galton's publications. Galton himself is certainly largely to blame for this. With one exception he did not relate his idea to species other than cattle and sheep, and the one exception was *Homo sapiens*. Something of the tenor of his views on the manifestations of human gregariousness may be gathered from the fact that the relevant passage in the book mentioned above is headed: 'On gregarious and slavish instincts'. Needless to say his analogy between human and bovine behaviour now seems somewhat naive. His views remain, as always, interesting and evocative, but their dogmatic and moralizing tone and their obvious connection with their author's widely distrusted line on eugenics can be imagined to have scared off many potentially interested zoologists. Another probable reason for the neglect of Galton's idea is that he himself presented it mixed up with another really quite separate idea which he treated as if it were simply another aspect of the same thing. This was that every cow, whether marginal or interior, benefited by being part of a herd, and that therefore herding was beneficial to the species. His supporting points are undoubtedly forcible; he mentions mutual warning and the idea that by forming at bay in outward-facing bands the cattle can present a really formidable defence against the lions. However, whether or not the cattle actually do, in the last resort, overcome their centripetal inclination and turn to face the predator (as smaller bands of 'musk ox' certainly do), these points raise a different issue, as do the mass attacks on predators by gregarious nesting birds. My models certainly give no indication that such mutualistic defence is a necessary part of gregarious behaviour. Moreover, mutual defence and warning can hardly be described as 'slavish' behaviour and so seem not to be covered by Galton's heading. There is no doubt, of course, that mutual warning and occasional unselfish defence of others are sometimes shown by gregarious mammals and birds, but where such actions occur they are probably connected on the one hand with the smallness of the risk taken and, on the other, with the closeness of the genetical relationship of the animals benefited (Chapter 1). With the musk ox, for example, the bands are small and clearly based on close family relationships. Sheep and cattle also take risks in defending their young (as Galton pointed out), but this is based on recognition of their own offspring. Apart from the forced circumstances, and the unnatural dispositions engendered by domestication, females usually do not associate with, still less defend, young which are not their own.[13] The ability to recognize particular individuals of the same species is highly developed in mammals and birds and there is no difficulty either practical or theoretical in supposing that the mutualistic behaviour of adult musk oxen, for instance, is an evolutionary development from the altruism involved in parental care. Positive social relations between members of family groups probably exist submerged in most manifestations of mass gregariousness in animals higher than fish. For example, the flock of sheep driven by a shepherd owes its compactness and apparent homogeneity to the presence of the sheep dogs: when it is left undisturbed on the mountain it arranges itself into a loosely clustered and loosely territorial system. The clusters are usually based on kinship and the sheep of a cluster are antagonistic to any strange sheep that attempts to feed on their ground.[32] It accords well with the present theory that the breeds of sheep that are most readily driven into large flocks are those which derive from the Merino breed of Spain,[9] which is an area where, until very recently, predation by wolves was common. Galton likewise gave the relaxation of predation as the reason for European cattle being very much less strongly gregarious than the cattle he observed in Africa. He noted that the centripetal inclinations evolved through predation would be opposed continually by the need to find ungrazed pasture, so that when predation is relaxed gregarious instincts would be selected against. This point has also been made by Williams.

Most writers on the subject of animal aggregation seem to have believed that the evolution of gregarious behaviour must be based on some advantage to the aggregation as a whole. Many well-known biologists have subscribed, outspokenly or by implication, to this view. At the same time some—for example, Hesse *et al.*,[14] Fisher,[33] and Lorenz[5]—have admitted that the nature of the group advantage remains obscure in many cases. On the hypothesis of gregarious behaviour presented here the apparent absence of a group benefit is no cause for surprise. On the contrary, the hypothesis suggests that the evolution of the gregarious tendency may go on even though the result is a considerable lowering of the overall mean fitness. At the end of our one-dimensional fairy tale, it will be remembered, only the snake lives happily, taking his meal at leisure from the scrambling heaps of frogs which the mere thought of his existence has brought into being. The cases of predators feeding on apparently helpless balls of fish seem parallel to this phantasy: here gregariousness seems much more to the advantage of the species of the predator than to that of the prey. Certain predators of fish appear to have evolved adaptations for exploiting the schooling tendencies of their prey. These cases are not only unfavourable to the hypothesis of a group advantage in schooling but also somewhat unfavourable to the present hypothesis, since it begins to seem that there must be an advantage to fish which do not join the school. The thresher shark is said to use its much elongated tail to round up fish into easily eaten schools.[2] Even worse, the swordfish is recorded as feeding on schools by first immobilizing large numbers with blows from its sharp-edged sword. This may be its usual method of feeding, and the sawfish may use its weapon similarly. Such predators presumbly do best by striking through the middle of a school; lone fish might fail to attract them. It is obvious, however, that such cases should not be assessed on the assumption that a particular predator is the only one.[12] At the time of attack by a swordfish there may be other important predators round about that are still concentrating on the isolated and peripheral fish. In the case of locusts the occurrence of such contrary influences from predation is recorded:[19] large bird predators (ravens and hornbills) attacking bands of hoppers of the desert locust did tend to disperse the bands, and to decimate their central members, by settling in their midst, but at the same time smaller birds (chats and warblers) captured only marginal hoppers and stragglers.

Adaptations of the predator to exploit the gregariousness of its prey certainly suggests the possibility of a changeover to a disaggregating phase of selection. In terms of the earlier story, if the snake evolves such a lazy preference for taking its prey from a heap that it comes to overlook a nearer lone frog, then it may come to be the mutant gregariphobe which survives best to propagate its kind. There are in fact many examples of species which are gregarious in only part of their range or their life cycle and possibly this may reflect the differing influences of different predators.

A surprising number of discussions of animal aggregation have mentioned the occurrence of marginal predation and yet shown apparently no appreciation of its possible evolutionary significance. Whether marginal predation or similarly orientated pressure of selection on individuals is sufficiently powerful to account for the gregariousness observed in any particular case cannot be decided by any already existing body of data. Nevertheless, it can be claimed that for none of the cases so far discussed do the data exclude the possibility. It can also be claimed that all the rival theories based on the idea of 'group selection' are theoretically insecure. Consider, for example, the theory of Wynne-Edwards[16] which, from this basis, attempts to bring many facts of animal aggregation under a common explanation. This theory suggests that aggregations serve to make individuals aware of the current level of population density and that this awareness reacts on reproductive performance in a way that holds off the possibility of a disastrous crash due to overexploitation of the food supply or other limited resources. Except in cases where groups are exclusive and mutually competitive for territory it is very difficult to see how there can be positive selection of any tendency by an individual to reproduce less effectively than it is able (see Chapter 6). Wynne-Edwards's cases of massive aggregations, even those having a kin-based substructure as in the gregarious ungulates, are unlikely candidates for the class of exceptions. The alternative theory supported here has no parallel difficulty as to the underlying processes of selection since it can rest firmly on the theory of genetical natural selection.

Certain other versions of the idea of a group benefit, unlike the Wynne-Edwards theory, do not require the concept of self-disciplined restraint by individuals. With such versions, statements that aggregation has evolved because it aids the survival of the species may be treated as merely errors of expression, in which a possibly genuine effect is given the status of a cause. The factor of communal alertness, for instance, may really make life more difficult for a predator and safer for a gregarious prey, especially if the predator is one which relies on stealth rather than speed. In Galton's perhaps over-persuasive words,

> To live gregariously is to become a fibre in a vast sentient web overspreading many acres; it is to become the possessor of faculties always awake, of eyes that see in all directions, of ears and nostrils that explore a broad belt of air; it is to become the occupier of every bit of vantage ground whence the approach of a lurking enemy might be overlooked.[1]

But there is, of course, nothing in the least altruistic in keeping alert for signs of nervousness in companions as well as for signs of the predator itself, and there is, correspondingly, no difficulty in explaining how gregariousness on this basis could be evolved.

Returning to the more interesting and controversial postulate of a population regulatory function for mass aggregation, consider, as a point of detailed criticism, the problem presented by aggregations in which several species are mixed. All the main classes of aggregation that have so far been mentioned—fish schools, bird flocks (both mobile and nesting), and grazing herds—provide numerous examples of species mixture. The species involved may be as widely related as ostriches and antelopes.[14] If it is difficult to see how the supposed group selection can work within a species it is certainly even more difficult when the groups are mixed. On the other hand, the theory that gregariousness is essentially due to the need for cover finds no difficulty over this point provided that the species which mix have at least one important predator in common. None of the vertebrate predators that have been mentioned as possible agents for the moulding of gregarious inclinations in their prey seems so specialized in their hunting as to make it improbable that individuals of one prey species could expect to gain some protection by immersing themselves in an aggregation of another species.

Perhaps most of the examples of mixed aggregation lie outside the class for which Wynne-Edwards and his supporters would claim a population regulatory function. The examples which have been cited in support of the theory are, in general, more directly concerned with reproduction. Apart from the case of nest aggregations, which has been discussed, there remain Wynne-Edwards's numerous citations of nuptial gatherings. With these the case for the effectiveness of marginal predation is admittedly weaker. Marginal predation seems somewhat less likely on general grounds and there is little evidence. However, there are other ways in which selection is likely to favour individuals which are in the nuptial gathering (or at least on its margin) over those that are isolated. Such selection may be sexual; it may work through differences in the chances of obtaining a mate. As one example, consider the swarms of midges that are so common in damp, still, vegetated places in summer. Many species of nematocerous flies have the habit of forming such swarms. Each swarm usually consists of males of a single species and tends to hover in a fixed spot, often near to some conspicuous object. Females come to the swarm and on arrival each is seized by a male. Passing over possible stages in the initiation of such habits, it is at least clear that as soon as proximity to the swarm itself becomes a key to the female's further co-operation in copulation there is likely to be little chance of mating for the male which does not join the swarm. In such a case the optimal position for a male is probably not, as it is under predation, at the centre of the throng, and considering how males might endeavour to spend the maximum time in relatively favoured positions, downwind or upwind, above or below, possible explanations for the dancelike motion of such swarms become apparent.

Acknowledgements

I wish to thank Dr I. Vine for the stimulus to publish this paper: his paper,[34] treating the influence of predation from a different point of view and showing one way in which gregariousness may be beneficial to the group as a whole, was sent to me in manuscript.

References and notes

1. F. Galton, *Macmillan's Magazine* **23**, 353 (1871).
2. G. C. Williams, *Michigan State University Museum Publications, Biological Series* **2**, 351 (1964).
3. G. P. Baerends and J. M. Baerends-van Roon, *Behaviour*, Suppl. **1**, 1 (1950).
4. E. Horstsmann, *Zoologischer Anzeiger*, Suppl. **17**, 153 (1952).
5. K. Lorenz, *On Aggression* (Methuen, London, 1966).

6. N. Tinbergen, *The Study of Instinct* (Oxford University Press, Oxford, 1951).
7. J. P. Scott, *Comparative Psychology Monographs* **18** (96), 1–29 (1945).
8. J. H. Crook, *Behaviour* **16**, 1 (1960).
9. F. F. Darling, *A Herd of Red Deer* (Oxford University Press, Oxford, 1937).
10. G. M. Breder, *Bulletin of the American Museum of Natural History* **117**, 395 (1959).
11. S. Springer, *Ecology* **38**, 166 (1957).
12. H. R. Bullis, Jr, *Ecology* **42**, 194 (1960).
13. G. C. Williams, *Adaptation and Natural Selection* (Princeton University Press, Princeton, NJ, 1966).
14. R. Hesse, W. C. Allee, and K. P. Schmidt, *Ecological Animal Geography* (Chapman and Hall, London/Wiley, New York, 1937).
15. V. M. Sdobnikov, *Trudy Arkticheskogo Nauchno-Issledovatel'skogo Instituta, Leningrad* **24**, 353 (1935).
16. V. C. Wynne-Edwards, *Animal Dispersion in Relation to Social Behaviour* (Oliver & Boyd, Edinburgh, 1962).
17. W. G. Moore, *Turtox News* **26**, 262 (1948).
18. H. Pryer, *Proceedings of the Zoological Society of London*, 532 (1884).
19. J. A. Hudleston, *Entomologist's Monthly Magazine* **94**, 210 (1958).
20. N. Tinbergen, *Social Behaviour in Animals* (Methuen, London, 1953).
21. Y. Espmark, *Zoologische Beiträge* **14**, 155 (1968).
22. E. E. Austin, in F. W. Edwards, H. Olroyd, and J. Smart (eds), *British Bloodsucking Flies*, pp. 149–52 (British Museum, London, 1939).
23. H. Kruuk, *Behaviour*, Suppl. **11**, 1 (1964).
24. D. Lack, *Ecological Adaptations for Breeding in Birds* (Methuen, London, 1968).
25. Cannibalism in gregarious species raises different problems and may help to explain why the nests of gulls do not become very closely aggregated (see 'the gull problem' in N. Tinbergen, *The Herring Gull's World* (Collins, London, 1953)). Insects have another kind of 'wolf in sheep's clothing': the predator mimicking its prey (W. Wickler, *Mimicry in Plants and Animals* (Weidenfeld & Nicolson, London, 1968)).
26. U. R. Evans. *Transactions of the Faraday Society* **41**, 365 (1945).
27. E. N. Gilbert, *Annals of Mathematical Statistics* **33**, 958 (1962).
28. J. L. Meijering, personal communication.
29. R. E. Miles (*Mathematics of Biosciences* **6**, 85 (1970)) has greatly expanded analysis of this case (footnote added to proof of original publication).
30. P. J. Clark and F. C. Evans, *Science, N.Y.* **121**, 397 (1955).
31. F. Galton, *Inquiries into Human Faculty and its Development* (Dutton, New York, 1928; first published 1883).
32. R. F. Hunter, *Advancement of Science* **21** (90), 29 (1964).
33. J. Fisher, in J. Huxley *et al.* (eds), *Evolution as a Process*, pp. 71–82 (Allen and Unwin, London, 1953).
34. I. Vine, *Journal of Theoretical Biology* **30**, 405 (1970).

CHAPTER 8

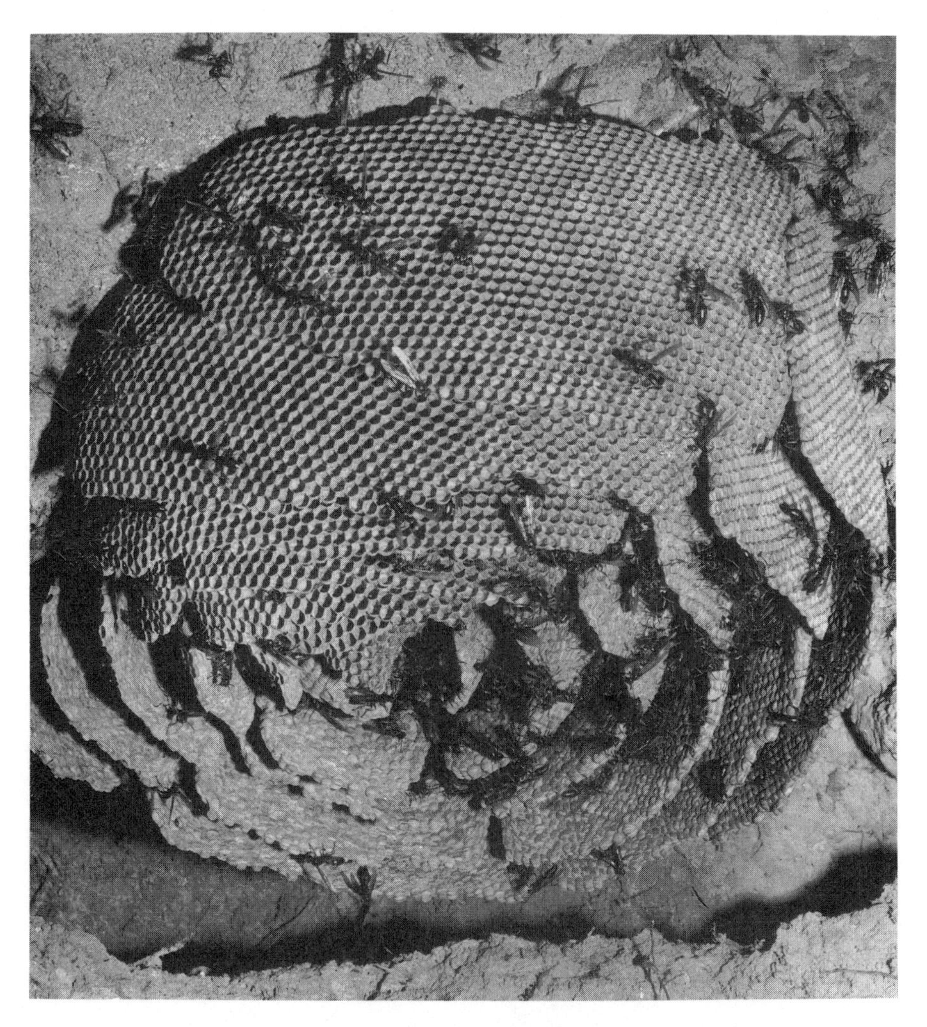

Nest of a thousand queens

This nest of the wasp *Agelaia angulata* was in the central chamber of a large disused terrestrial termite mound in Mato Grosso, Brazil. The mound had been previously excavated and its colony destroyed by an armadillo, whose burrow the wasps later adopted for their entry. To reach the nest I had to dig a 4-m trench (covered with tarpaulin for protection), followed by a 2-m tunnel that enabled me to break into the chamber from below and behind. The worker wasps are extremely fierce and leave their stings in human flesh like honey bees: even after losing them they continue to cling, buzz, and mimic the motions of stinging, so contributing to a general terror: where unable to sting due to a bee veil, they jet venom and this often reaches the eyes. Inspite of bee clothing (veil and gloves) I was incapacitated for two days after my first observation of this nest.

Hung by the edges of combs and almost perfectly circular, the nest is 42 cm in diameter. The large number of queens in such polybiine nests is an adaptation predominantly at the group level to the chance that the nest will be invaded by army ants (*Eciton*): small queens remain highly mobile and thus many escape with the workers to re-start the colony. But how can the much lowered average relatedness produced by multiple-queening be consistent with the high altruism, as evident in suicidal stinging? This is still under study (see ref. 42, page 269).

CHAPTER 8

SORORITY AVENUE

Altruism and Related Phenomena, Mainly in Social Insects

Que penser de la Nature, quand elle semble donner aux insectes la faculté de détruire leur propre espèce, lorsqu'elle permet aux Abeilles de tuer mâles, et qu'elle laisse aux Bourdons le droit et'envie de dévorer les oeufs nouvellement pondus?

Ne sembleroit-il pas naturel d'en conclure, qu'elle veut opérer une déstruction totale? Cependant les espèces se conservent, les familles se multiplient, les loix ne sont point changées; et il paroit au contraire, que c'est par le sacrifice de quelques individus que la conservation de l'espèce est assurée.

P. HUBER[1]

Perhaps, also, it may not be amiss to further suspect that at a future time some of the 'best minds' among annularis queens will learn that it is possible to deposit eggs in nests made by others, and leave the care and feeding of their young, in cuckoo-bird fashion, to the labour of others.

P. RAU[2]

IN the late 1960s I was invited by Gerhard H. Schmidt of the University of Würzburg to write a chapter for a proposed multi-author volume on social insects to be published in German. He wanted me to explain my ideas about how relatedness affects the evolution of social insects. I was to write an English version that would be translated. The request was attractive because what I had written on social insects in 1964 (see Chapter 2) was overdue for revision and extension, besides which I was delighted at the idea of a translation specially reaching German readers. I had learned immensely from German entomological and social insect papers in framing and checking ideas about altruism and this would be a kind of repayment. I had been following social insect work ever since I thought seriously about altruism in the early 1960s. I had, of course, a particular interest in any that brought me information on relatedness. In addition, there had been the two long trips to Brazil in 1963–64 and 1968–69, during which I studied social wasps and along the way learned much about bees from Warwick Kerr and his colleagues at Rio Claro and then, on my second trip, from the set who had followed him in his move to Ribeirão Preto. I was aware of various mistakes and inadequacies in my 1964 paper and here was a chance to rectify them. For example, I had not in the earlier paper attempted to cover inbreeding whereas now, thanks in part to my re-derivation of my main results under the stimulus from George Price, I saw fairly clearly how to do this. German, of course, wasn't the ideal language in which to publish a revised treatment but I had in mind that once it was all done and ready for translation I might seek a journal that could publish a similar version in English. This is roughly how it happened except that for some reason the German book was extremely slow and did not come out until 1974, 2 years after the paper in this chapter appeared in *Annual Review of Ecology and Systematics*.

The chapter in Schmidt's book[3] is covered in this paper in the sense that all its material is here plus a little more. One anecdotal comment about the German version seems worth directing to people who read it. The translation was done by my father-in-law. German by birth, be became a Lutheran minister, who, rather than tolerate Nazi interference in his pastoral work, emigrated to Britain shortly before the Second World War. In the late 1960s he was working for the Church of Ireland in County Mayo, Republic of Ireland. As his work on my paper progressed many letters and telephone calls had to be exchanged between him in Co. Mayo and my wife and me in Berkshire, these largely concerned with the ideas and the words to be used in the translation. In spite of these discussions it seems that not everything had been fixed quite satisfactorily by the time the resulting manuscript was dispatched to Germany because yet more exchanges then occurred between us and the editor, clarifying some 'unexpected phrases'. The Bible has, of course, quite a lot to say about such matters as relatedness, mating, incest, altruism, and the like, and it was inevitable that Canon Friess, proceeding literally and very carefully and using familiar idioms, should choose some expressions that were hardly the standard ones of German population genetical, ethological, or evolutionary writing. If any such oddities remain in the German version, this note explains why: I imagine their effect is just surprising, not seriously misleading.

Speaking of what may be the suitable words in this branch of science, whether they are German or English, possibly this is an appropriate place to digress for a few paragraphs on the subject of word use in science, and particularly concerning those words I hijacked out of common speech and applied in kin-selection theory. Was it appropriate, for example, to use 'altruism', 'selfishness', and 'spite' for the major behavioural categories discussed in my papers? A mother who runs into a burning building to save her own child and gets badly burned will receive sympathy, of course, but probably not the medal that might be given to a woman who had saved an unrelated child in similar circumstances. It is the act of the latter we think of as altruism in a high human sense, although we may need to be careful here about any implied restriction to humans when there are hints that dolphins, for example, may be capable of similar acts as I mentioned in the introduction to Chapter 1. Going to another kind of donor trait we certainly wouldn't say that an uncle who spends money dining influential friends in order to get his nephew off to a good start in industry is being 'altruistic' on behalf of the nephew: nepotism is the chosen word here. Nepotism would be also a moderately appropriate general word, if hardly the common usage, for the mother saving her own child from the burning building. Aiming explicitly to avoid common usage together with the emotions associated with it, some have suggested that very inert terms such as 'donorism' would be even better than nepotism. I take it that the risk foreseen here is of the biological usage debasing the higher human meaning—that is, through the theory's constant emphasis on a kind of cost-benefit analysis, encouraging us to overlook the truly unbiological cases where neither relatedness nor any expectable return is involved in the beneficent act. In general it seems to me, however, that the danger of unleashing a vogue either for cynical perception of altruism or laudatory perception of nepotism is slight, and that we risk frustrating the mnemonic and associative strengths of our language if we are too fussy.

There are, of course, dozens of words in other scientific disciplines borrowed from common usage and given precise new definitions: energy, work, power, stress, buffer, and current are just a few. Has an employer ever tried not paying an employee for the hour he spent 'holding something steady' on a building site on the grounds that, by the physics definition, he was not doing any 'work'—that is, making no movement against a force? I would not want to be the person explaining this physicist's point of view to the workman. Has any one suggested that the nepotistic uncle already mentioned should have a knighthood for his 'altruistic' enrichment of the echelons of industry with his relatives? The uncle may, of course, improve his chance of a knighthood by promoting grateful nephews along with others in sufficient numbers: all can be expected later to promote him. Obviously, however, such an egoistical strategy has nothing to do with human altruism and even not very much with kin selection. I suspect that the 'others' might actually be more important than the relatives for a man manoevering on these lines within a European culture although real relative support might prove more important in traditional societies.[4]

Regarding other common words bidding for usage in behavioural biology, the danger that use of a mundane word could do harm seems sometimes higher and the confusing notions or emotions engendered by the words may then outweigh the cost of the longer phrases or unmemorable inventions that could be used in their place. Thus use of the word 'rape' for the male-enforced copulation observed in beetles or in birds, for example, and the discussion of this activity as natural and adaptive—as it undoubtedly must be when it is found commonly enough in a species—could combine, conceivably, to give the impression that human rape is at worst sometimes justified and at second worst ineradicable because in some sense it is 'natural'. If so, this is a case where the more purely descriptive expression 'forced copulation' probably pays for its extra 13 letters. Nevertheless, use of the word 'rape' in animal contexts usefully provokes thought about whether human rape is ever or ever has been adaptive.[5–8] If it sometimes is now adaptive then quite clearly it is necessary to change laws, environment, and treatment of rapists so that it is no longer so. Even if it is not currently adaptive but instead atavistic (i.e. a legacy from what was once adaptive in our evolutionary past), then in order to have any hope of preventing rape today we need to understand how the past selection that created the behaviour worked—what, for example, is or was its trigger, whether now or in Pliocene Africa? All this may help us to treat and deter rapists or, at the very least, better advise potential victims on dangerous situations. Although the palaeolithic and primate background is by far the most relevant here, it needs to be remarked that a theme of forced copulations goes much further back, examples being scattered almost throughout the animal kingdom (see Chapter 13). If those other puzzles of sexuality have some faint bearing on rape, we need to understand them too.

Seeing deep correspondences in seemingly unrelated things is the essence of science and is vital in mathematics and philosophy as well. Newton saw that the fall of an apple is the motion of a planet; Lamarck, that the spiky scales of dogfish are the cousins to the teeth in his jaws. Darwin, adding to Lamarck, came closer to the emotional crux we are currently engaged with, realizing that the back-drawing of the lips that expose teeth in the baboon's snarl is, likewise, the country cousin of a human sneer. Let us forget dogfish scales for the present and try to grasp this slightly harder one: behaviours and emotions can also be homologous—that is, they can have a common evolutionary origin from which continuous threads of their modification lead to the examples we see. Practically none of our basic behaviour, perhaps only our linguistic behaviour and even that uncertainly,[9,10] is wholly unique to humans. Any scientist who saw a barnyard cock copulating with an unwilling hen and then tried to explain the act in terms of reflexes and the hormonal states of the animals would not be talking nonsense but would certainly be boring and missing the main point: he would be failing to see an adaptation. Boring likewise would be he who, while accepting to think in terms of an adaptive life programme for *Gallus gallus*, still refuses to anthropomorphize or to speculate on any possible connection between, say, the usual subordinate social status of the cock in question (a matter of fact) and the typical low social status of human rapists (likewise well attested[5–7]). Such a person who goes always straight to the level of 'how', missing the 'why' on the way, or who avoids the commonsense comparisons, may cripple his or her understanding of both the phenomena; he becomes like an economist trying to understand the economy of a country by concentrating on the manufacture of paper because he has noticed that most of the financial operations of the country are conducted using writing on paper,[11] or, equally absurd, like one who studiously avoids mentioning money just because money is printed on paper—paper, a material never (except in Japan where it is used to build houses) shown to have more than the most shadowy utilitarian value. It is those who bar obvious comparisons in these matters and descend pedantically to mechanism who should be called reductionist.

Speaking for myself, even though I am persuaded in the case of the word 'rape' that respect for other people's sensitivities may make it better to use such an expression 'forced copulation', it is certain that no thought police can stop me using an obvious word and idea in my own thinking. This applies wherever I see even a possible analogy, leave alone a homology, between animal and human phenomena; consistently I prefer the simplest, most vivid words available. I suppose the opponents of the use of the word 'rape' in biology should applaud Bowdler for eliminating 'cuckold' from Shakespeare: doubtless they, too, wish to see this word removed from the English language as a whole because, in parallel to their thinking on 'rape' in biology, its existence can only be giving men and women the idea that, because cuckoo birds do naturally something that vaguely resembles what the word means for humans, its acceptance implies human cuckoldry to be natural and right. To replace 'cuckold' a politically correct modern Shakespeare needs to contemplate. I suppose, some new and specific abbreviation such as MADEPC ('male-animal dupe of an extra-pair copulation'), while, in parallel, MHDEPC might be used for the human counterpart. Possibly even that resemblance between the 'animal' and a 'human' term is inadmissable, as is also any implication that animals might be 'dupes' in a human sense.

When two phenomena give me even a hint of similarity I try as a matter of course swapping modes of thought applied to them, forcing myself to contemplate each one in the light of the other. This playful effort continues until distinctions and special pleadings stick so much clay to my feet that the back and forth with ideas is no longer fun. Or else if new and unsuspected correspondences accumulate while distinctions prove rare or trivial (a situation that I am undoubtedly helped to notice through word usage bringing various other meanings and situations swiftly and repeatedly to mind) then I grow serious and skip ever faster, sensing a find that may be of beauty and use, a new tool with which to understand the world. Biological extensions of such common words as 'rape' and 'altruism' are not on equal footing with the 'power' and 'work' of mechanics yet, but that they aren't seems to me more connected with shibboleths and the 'let us hope' of the cartoon Victorian ladies than with any real distinction in our usage. The ladies' view of evolution, which was a joke even in their time, seems to have been laid gently to rest along with Bowdler's Shakespeare in the rubbish bin of history. It is my hope that they can soon be joined there by some of our present objections to supposedly 'emotive' usages in biology and a whole heap of 'political correctness' as well.

Returning to evolution and the paper to follow, it is partly because of Phil Rau's daring and his unscientific terminology (even his irony, as in 'best minds') that I have set the quotation from his paper at the head of this chapter. Combining cuckoos, selfish wasps, and humans into one sentence, it is a good illustration of the kind of laterality of thinking that I have been referring to. Rau was a shopkeeper, not a trained biologist, but in spite of that saw in most ways farther than his contemporary academic advisor and museum specialist, Joseph Bequaert, in rather the same way as Jane Goodall saw farther into chimpanzees than her academic contemporaries because of her emotional involvement and lack of training. Good in the museum and at times a field man too, Bequaert probably missed much through his academic soundness. For some time he denied, for example, Rau's tentative new species because he was unable to believe that nest traits and behaviour could be a guide. In the quotation, Rau in effect predicts the existence of cuckoo-like *Polistes*. Sure enough wasps of this kind were soon found and there were even species of such cuckoos as my paper mentions, although curiously the group seems to have only one origin and is confined solely to the Alps of Europe.[12–14] Short of such 'professionals', however, cuckoo-like behaviour and strife concerning who lays eggs for larvae that are to be reared is rife throughout the genus. Documentation of this kind of labour or property parasitism, which I first addressed in my papers of 1964

(Chapter 2) and later, as in Chapter 6 and in this one, endorsed as omnipresent among living things, continues strongly.[15,16] Whether a cell, a built home, a working offspring, or a tool, if it took time and energy to make it, then guard it and fight for it. Otherwise you'll lose it. Thieves, part time or professional (which in the biologese go as con- or allo-specific kleptoparasites), are evolving to grab it from you.

Returning to the more formal parts of my 1972 paper, besides correcting some of my previous errors (see also the Addendum to Chapter 2), the main novelty compared with Part II of my 1964 paper is the effort to cover inbreeding. The formulas for relatedness and to some extent the text are now sprinkled with Sewall Wright 'F statistics'. All the new and more general formulas that I give are correct for what they claim so far as I know, but they do not by any means complete the theoretical unravelling of inbreeding and inclusive fitness in diploids. A series of further refinements and generalizations or relatedness were to follow this paper, mostly at the hands of others. One that I contributed jointly with Richard Michod forms a chapter in the sequel volume; but that, too, was soon superseded. How far the extension of relatedness and inclusive fitness has gone may be suggested by mentioning versions that now invade the study of reciprocation and of interactions between members of different species.[17–20]

The present paper, of course, is far short of all such developments but as a start in their direction, in addition to the quantitative treatment of inbreeding, the paper does mention how reciprocation might play a part in the lives of social insects. By the time of its publication many sociobiologists (as we were about to be named) had recognized reciprocation to be the second main factor capable of building high co-operation in the living world. For me, serious thought on this issue had begun in 1969 on a seminar visit to Harvard just previous to the 'Man and Beast' symposium. There I met Robert Trivers, then a graduate student. He explained to me his ideas of 'reciprocal altruism'. I was, of course, already aware that the great mass of human co-operative behaviour cannot be ascribed to relatedness but I had generally excluded this kind of co-operation from the concepts I initially set out to discuss. Adam Smith, whose great work *Wealth of Nations* I had first read while at Cambridge, was my exemplar for 'all that other kind of co-ooperation' as I then saw it. His theory seemed to me enlightening but basically common sense and I was at the time little interested in what needed to be added to it. Robert Trivers changed this dismissive attitude. It had not occurred to me that if one focused carefully on specified trading relationships one might try to reason reciprocatory behaviour in detail, rather on the same lines as I had reasoned out kin-based altruism. My quest for biological altruism had carefully excluded anything I saw as possibly reciprocatory because it seemed that although behaviours of this category could mount a semblance of altruism, a donor always expected a benefit itself, at least in the long term: it was semblance only. I still believe the reciprocal altruism that Trivers explained to me and which he so named in his pioneer paper[21] was misnamed—even though from the start I was enthusiastic about his idea of the conditions for the phenomena he described. This was obviously the beginning of a useful line. The test for a truly altruistic element of behaviour, however, is whether one can imagine it being exaggerated by degrees until it becomes suicide and then, even in this extreme condition, having a possibility of being advantageous (as, for example, it obviously can be on the basis of any benefit per cost ratio greater than one when the donor and the recipient are identical twins). The activity I prefer to call reciprocation can never include suicidal acts. The issue is complicated by the fact that in the course of reciprocation some individuals may die while in seemingly altruistic grace if death chances to catch them before the expected return benefit; nevertheless, for the average of the acts of their type, reciprocatory donors expect a return. If human they generally know this quite well even if the tension of expectation is concealed most of the time. Even when concealment has been carried into the subconscious it usually abruptly re-emerges if the opportunity for the return arrives and is ignored by the other party: hurt and even vengeful feelings usually supervene immediately. Again, this is just etymology and psychology rather than useful theory. Name the ideas as he might, Trivers's emphasis on the predominance of reciprocation over kin co-operation in large areas of social behaviour, and on its special universality among humans, was obviously timely and important, as were his further points of how reciprocation is facilitated by long-term interaction and recognition; his points hold whether the interacting individuals form a single population or are from different species.

Reciprocation occupies only a small section at the end of my 1971 paper, so I'll leave further discussion of it until Volume 2, where it becomes central. Note, however, that this paper has on pages 301 and 302 the first remarks made by me (possibly the first ever made) on a theme of symbiosis that is receiving increasing attention today. First, mutualistic symbiosis virtually always involves consideration of kinship besides reciprocation—this quite contrary to unexplained theories that see symbiosis somehow as an alternative to the 'selfish genes'.[11,22–24] Second, the idea that may be original is that high mobility of either partner in a symbiosis is favourable to the evolution of antagonistic and damaging relations whereas viscosity, if present in both the populations, favours an approach to mutualism.[25–32] This second point, however, is less straightforward than the first. A recent counter[33,34] (but see also Queller,[35] and the introduction to Chapter 2) rightly says that we should not think of viscosity of a population as opening a completely free road to social benefits: the territorial battle of good with evil is more subtle. In effect, Good needs not merely a defence strategy—such as it may achieve through viscosity—but also an ability to penetrate, at least from time to time, along the open files and diagonals of its spatial world, to project distantly the fruits of its co-operation to places where they can wrest territory at least at the margins, better still deep in the heartlands, of Evil.[36] This is a fascinating theme that for me still has many unsolved puzzles both in theory and in interpretation. In the light of the ills of overexpansiveness of human populations, which are now all too clear, might we, for example, even want to reverse the terms Good and Evil for some issues? Leopards re-inhabiting the wastes that currently separate Israel and Syria might, for a start, see advantage at least to leopards in such a change.

On another issue, as in 1964 and 1971, I remain undecided whether in Hymenoptera (say) powerful flight in a species helps or hampers progress towards high sociality. The usually small sweat bees, the Halictidae, are not slow as insect fliers go but they are still far from being darting, hovering streaks in the air as are the brilliant orchid bees, the Euglossinae. It is the latter, undoubtedly among the most powerful of all insect fliers, that seem barely able to reach a level of eusociality[37] that Halictidae reach easily and perhaps this is because their species populations too easily mix all genes together. At the other extreme, however, it is also true that no wholly flightless insect is eusocial. Slow wingless *Cryptocercus* cockroaches of the decaying oak logs of the Appalachians should, if viscosity and the opportunities intrinsic to a lifestyle are the key, be eusocial already. Yet they are actually less social than some of the ultra-mobile euglossine orchid bees. If both extremes—powerful dispersal and none at all—are bad, where is the optimum? For kin selection to work best, whether mixed or not with reciprocation, what *is* optimal dispersal? Some progress on this question for cases of simple competition is discussed in Chapters 11 and 15 but there is obviously still a great deal to be worked out both in fact and in theory.[36]

As illustrated in the last paragraph, the writing of these introductions gives me a good opportunity to point out where I may have had the luck to squirrel ideas that happen to have re-emerged and are becoming trees: at the same time, of course, I can airily pass over all those nuts that I stored but which neither grew of themselves nor were reused by me. One more nut, however, that has grown at least to be a bush is worth pointing out. This is the brief expression in my 1971 paper of my awareness of the necessity for female hymenopterans to be working mainly on rearing of sisters, not brothers, and never to be equally devoted to both if the haplodiploid relatedness pattern is to play its special part in shaping eusocial trends. Although aware of the point I missed mentioning it in the 1964 paper and even in the 1971 paper do not give the matter anything like the attention it deserved.[38] On whether haplodiploidy helps shape in some degree the pattern of sociality the case seems clear: the Neodarwinian arguments of Chapters 2 and 5 show that it has to shape that pattern; for this to be wrong the whole argument must be flawed and no one has shown that it is nor provided any alternative rationale for how automatic altruism can arise. Elements of the actual pattern, such as the almost complete non-working of hymenopteran males in contrast to the 50 per cent of workers being male in termites, do in fact accord well with the theory. Whether other factors that must also apply to the evolution of sterility may be more potent forces, however, so that the special pattern imposed by haplodiploidy is swamped by them and therefore hardly detectable, is much less clear. On the side of haplodiploidy having relatively high power I take some tentative encouragement from the very recent discovery of a probably sterile soldier caste in thrips,[39] which is another haplodiploid group. However, in some thrips species males also develop into fighters and in at least one species have been observed attacking adventive enemies. Although males are apparently less active in this respect than females, the observation means, again, that biases of relatedness are clearly not the whole story. Unconnected with haplodiploidy, another long-desired ally on the relatedness front has also turned up, in this case in the 1970s, and now forms a well-researched topic. This is the discovery by Shigeyuki Aoki in Japan that sterile castes are well developed in some aphids and that many additional aphid species have sacrificial and workerlike tendencies even when they are not fully committed to sterility.[40,41] On the puzzle of the supposed 'absent' worker aphids, the recent discoveries show me the weakness of such *post hoc* explanations as I gave for absence in the 1964 paper. There into the soil went one glossy but maggot-doomed seed; however, out of the hole emerged, somehow, a stately palm—aphid soldiers!

Since the 1971 paper was written there has been striking advance in methods of measuring relatedness both theoretically and in terms of lab genetics, and this leads on to another major topic of the paper on which it seems worth giving one or two modern references: that of multiple-queen states in social insect colonies (see especially pages 288–299). My overall impression is that relatedness in colonies often turns out surprisingly low, lower than I would have guessed to be permissible in 1972.[42–44] Against this, however, first, selfish behaviour contingent on that low relatedness seems never to be absent. Second, there often seem to exist behaviours and breeding patterns that raise relatedness from time to time and thus check unlimited decline.[45–47,48]

References and notes

1. P. Huber, Observations on several species of the genus *Apis*, known by the name humble bees, and called Bombinatrices by Linnaeus, *Transactions of the Linnean Society of London* **6**, 214–98 (1802). Translation:

> What should one think of Nature when she seems to give to insects the ability to destroy their own species, permits honeybees to kill their males, and gives to bumblebees the right to devour newly laid eggs?

Does it not seem natural to conclude that she must want total destruction. Yet the species continue, families multiply, and these rules stay totally unchanged; it seems on the contrary that it is by the sacrifice of some individuals that the preservation of the species is assured.

2. P. Rau, Comparative nest-founding by *Polistes annularis* L., *Annals of the Entomological Society of America* **33**, 617–20 (1940).
3. G. H. Schmidt (ed.), *Sozialpolymorphismus bei Insekten* (Wissenschaftliche, Stuttgart, 1974).
4. A. L. Hughes, *Evolution and Human Kinship* (Oxford University Press, Oxford, 1988).
5. J. D. Wienrich, Human sociobiology: pair-bonding and resource predictability (effects of social class and race), *Behavioural Ecology and Sociobiology* **2**, 91–118 (1977).
6. R. Thornhill and N. W. Thornhill, The evolution of men's sexual coercion. (Target article), *Behavioural and Brain Sciences* **15**, 363–75 (1992).
7. R. Thornhill and N. W. Thornhill, The study of men's coercive sexuality: what form should it take? (Responses to 28 commentators), *Behavioural and Brain Sciences* **15**, 404–21 (1992).
8. B. B. Smuts and R. W. Smuts, Male aggression and sexual coercion of females in nonhuman primates and other mammals: evidence and theoretical implications, *Advances in the Study of Behaviour* **22**, 1–63 (1993).
9. S. Savage-Rumbaugh and R. Lewin, *Kanzi: The Ape at the Brink of the Human Mind* (Wiley/Doubleday, New York, 1994).
10. S. Pinker, Beyond folk psychology, *Nature* **373**, 205 (only) (1995).
11. L. Margulis and R. Fester (eds), *Symbiosis as a Source of Evolutionary Innovation, Speciation and Morphogenesis* (MIT Press, Cambridge, Mass, 1991). Margulis, for example, writes: 'The incorporation of "cost-benefit" analysis methods borrowed from insurance practices has led to the biologically puerile numerology that systematically ignores chemistry, biochemistry, molecular biology, and geology (sedimentology, palaeontology, and oceanography). Nevertheless such aseptic language dominates current evolutionary theory.' My reply is: of course, it depends on what interests you. Why *not* miss out those worthy disciplines when they are not germane? Who wants to hear about wood pulp when discussing economics, or chemistry when discussing altruism or sex-ratio strategy?
12. W. T. Wcislo, The roles of seasonality, host synchrony, and behaviour in the evolutions and distributions of nest parasites in the Hymenoptera (Insecta), with special references to bees (Apoidea), *Biological Reviews* **62**, 515–43 (1987).
13. J. Field, Intraspecific parasitism as an alternative reproductive tactic in nest-building wasps and bees, *Biological Reviews* **67**, 79–126 (1992).
14. M. Choudhary, J. E. Strasmann, D. C. Queller, S. Turilazzi, and R. Cervo, Social parasites in polistine wasps are monophyletic: implications for sympatric speciation, *Proceedings of the Royal Society of London* B **257**, 31–5 (1994).
15. C. J. Barnard (ed.), *Producers and Scroungers: Strategies of Exploitation and Parasitism* (Chapman and Hall, New York, 1984).
16. F. Vollrath, Kleptobiosis in spiders; in W. Nentwig (ed.) *Ecophysiology of Spiders*, pp. 274–86 (Springer, Berlin, 1987).

17. S. A. Frank, Genetics of mutualism: the evolution of altruism between species, *Journal of Theoretical Biology* **170**, 393–400 (1994).

18. D. C. Queller, Kinship, reciprocity and synergism in the evolution of social behaviour, *Nature* **318**, 366–7 (1985).

19. D. C. Queller, A general model for kin selection, *Evolution* **46**, 376–80 (1992).

20. D. C. Queller, Quantitative genetics, inclusive fitness and group selection, *American Naturalist* **139**, 540–58 (1992).

21. R. L. Trivers, The evolution of reciprocal altruism, *Quarterly Review of Biology* **46**, 35–57 (1971).

22. L. Margulis and D. Sagan, *Origins of Sex* (Yale University Press, New Haven, CT, 1986).

23. Margulis and Fester 1991 (see ref. 11).

24. T. Wakeford, Biology's grand unifying theory, *New Scientist* **144**, 61 (1994). Note: There are possible classes of social adaptation that are inexplicable as kin-selected altruism and also classes that are inexplicable by reciprocation. However, kinship and reciprocation overlap freely and whenever they do so relatedness cannot be ignored. Thus relatedness often resolves a 'Prisoner's Dilemma' game to a situation where co-operation is dominant, as explained in the paper of Chapter 6 (see also M. J. Wade and F. Breden, *Behaviour Ecology and Sociobiology* **7**, 167–72 (1980)). Relatedness always changes payoffs from apparent individualistic values. Even when relatedness is zero between potential reciprocators, as with partners in allospecific symbioses, relatedness commonly still enters when ever the return for a beneficent (or other) act from the other species comes to individuals different from the actor. Because of the usual disparity of size and/or longevity in allospecific partnerships such returns are frequent and it is then crucial whether the recipients of the return effects are relatives of the actor. If receivers are highly related to originators then mutualism is likely. This point appears not to be understood by Margulis (ref. 22). Oppositely, independent successful dispersal of partners leads to the return effects being to non-relatives. It seems to be still not generally realized how hostile this is to mutualism and how favourable to parasitism and pathogenicity (see, *inter alia*, Queller in ref. 18 above and Ewald and Frank in refs 25, 26, 31, and 32 below).

25. P. W. Ewald, Host-parasite relations, vectors and the evolution of disease severity, *Annual Reviews of Ecology and Systematics* **14**, 465–85 (1983).

26. P. W. Ewald, The evolution of virulence, *Scientific American* **268** (4), 56–62 (1993).

27. L. A. Dugatkin and D. S. Wilson, Rover: a strategy for exploiting cooperators in a patchy environment, *American Naturalist* **138**, 687–701 (1991).

28. M. Enquist and O. Leimar, The evolution of cooperation in mobile organisms, *Animal Behaviour* **45**, 747–57 (1993).

29. N. Yamamura, Vertical transmission and evolution of mutualism from parasitism, *Theoretical Populational Biology* **44**, 95–109 (1993).

30. C. C. Maley, A model of the effects of dispersal distances on the evolution of virulence in parasites; in R. Brooks and R. Maes (eds) *Artificial Life* IV, 152–9 (MIT Press, Cambridge, 1994).

31. S. A. Frank, A kin selection model for the evolution of virulence, *Proceedings of the Royal Society of London* B **250**, 195–7 (1993).

32. S. A. Frank, Kin selection and virulence in the evolution of protocells and parasites, *Proceedings of the Royal Society of London* B **258**, 153–61 (1994).

33. D. S. Wilson, G. B. Pollock, and L. A. Dugatkin, Can altruism evolve in purely viscous populations?, *Evolutionary Ecology* **6**, 331–41 (1992).

34. P. D. Taylor, Altruism in viscous populations—an inclusive fitness model, *Evolutionary Ecology* **6**, 352–6 (1992).

35. D. C. Queller, Genetic relatedness in viscous populations, *Evolutionary Ecology* **8**, 70–3 (1994).

36. M. A. Nowak and R. M. May, The spatial dilemmas of evolution, *International Journal of Bifurcation and Chaos* **3**, 35–79 (1993).

37. M. L. Santos and C. Garófalo, Nesting biology and nest re-use of *Eulaema nigrita* (Hymenoptera: Apidae, Euglossini), *Insectes Sociaux* **41**, 99–110 (1994).

38. R. L. Trivers and H. Hare, Haplodiploidy and the evolution of social insects, *Science* **191**, 249–63 (1976).

39. B. J. Crespi and L. A. Mound, Ecology and evolution of social behavior among Australian gall thrips and their allies; in J. Choe and B. Crespi (eds) *Ecology and Evolution of Social Behavior in Insects and Arachnids* (Princeton University Press, Princeton, 1996).

40. S. Aoki, A new species of *Colophina* (Homoptera, Aphidoidea) with soldiers, *Kontyu* **45**, 333–7 (1977).

41. D. L. Stern and W. A. Foster, The evolution of soldiers in aphids, *Biological Reviews* **71**, 27–79 (1996).

42. D. C. Queller, J. E. Strassmann, and C. R. Hughes, Genetic relatedness in colonies of tropical wasps with multiple queens, *Science* **242**, 1155–7 (1988).

43. J. E. Strassmann, Altruism and relatedness at colony foundation in social insects, *Trends in Ecology and Evolution* **4**, 371–4 (1989).

44. R. Gadagkar, On testing the role of genetic asymmetries created by haplodiploidy in the evolution of eusociality in the Hymenoptera, *Journal of Genetics* **70**, 1–31 (1991).

45. M. J. West-Eberhard, Temporary queens in Metapolybia wasps: non-reproductive helpers without altruism?, *Science* **200**, 441–3 (1978).

46. J. E. Strassmann, D. C. Queller, C. R. Solis, and C. R. Hughes, Relatedness and queen number in the Neotropical wasp, *Parachartergus colobopterus, Animal Behaviour* **42**, 461–70 (1991).

47. J. E. Strassmann, K. R. Gastreich, D. C. Queller, and C. R. Hughes, Demographic and genetic evidence for cyclical changes in queen number in a neotropical wasp *Polybia emaciata, American Naturalist* **140**, 363–72 (1992).

48. Three small textual corrections made to this chapter are as follows. On page 273 in the penultimate paragraph (page 196 in the original paper) 'the average of inclusive fitness of' is changed to 'of such effects in'. On page 275 the last fraction of the second series is change from 44/51 to the correct value 11/13. Finally, on page 289 between the two formulas, 'at least 1/3 times' is now replaced by 'at least 4/3 times'.

ALTRUISM AND RELATED PHENOMENA, MAINLY IN SOCIAL INSECTS†

W. D. HAMILTON

In what sense can the self-sacrificing sterile ant be considered to 'struggle for existence' or to endeavour to maximize the numbers of its descendants? Since the founding of the theory of evolution by natural selection, most biologists have evaded this question by focusing attention exclusively on the colony as the reproducing unit. There is a powerful precedent for this. Darwin himself took this course. He saw only a 'minor' difficulty in the evolution of sterility, and he passed over it in a few lines as he proceeded to discuss the 'great' difficulty of how the special aptitudes of the workers could be passed on in latent form by their fertile sisters.[1] A difficulty over sterility exists, nevertheless, and it is the more surprising that Darwin should have passed over it in that he discussed—but left unsolved—a parallel one raised by the social virtues (courage and self-sacrifice) in humans.[2] He saw that such qualities would be promoted in intergroup selection but counterselected within each group. Perhaps the possible avenues of indiscipline in social insects had been so little reported in Darwin's time that the problems they raised were easily overlooked.

Darwin's inadequate understanding of heredity may, likewise, have helped to keep the problem out of focus. With better knowledge of heredity and with more facts regarding the social insects to draw upon, Weismann[3] recognized the possible conflict between intergroup and intragroup selection in the evolution of worker attributes. He made the perceptive comment that, 'Obviously the workers must be more rapidly improved when all in a hive are progeny of one queen—i.e. they are all alike or almost alike.' But this comment was made in the course of discussion of another topic, and he did not pursue the matter. Soon Mendelian genetics resolved Darwin's difficulty of latency, apart from details of mechanism. But the disappearance of this problem does not seem to have given greater prominence to the other, and the question of how worker sterility comes to be selected continued to receive only occasional comment for

†*Annual Review of Ecology and Systematics* 3, 193–232 (1972).

a long time. Sturtevant[4] in 1938 again outlined it with admirable clarity and with special reference to multiqueened (polygynic) organization, which was by then well known. Rau[5] in 1940 also briefly touched on this crux when he noted how small might be the step separating worker-like behaviour in auxiliary *Polistes* queens from behaviour tending towards social parasitism. In 1951 G. C. and D. C. Williams[6] formulated the problem mathematically and were able to show the strong necessity of a genetical relationship between queen and prospective worker. More recent work on this theme has been that of Hamilton (see Chapters 1 and 2) and, less directly relevant, Levins[7] and Hamilton (Chapters 5 and 6).

From another point of view the existence of sterile workers expressing characteristic adaptations provides a strong argument in favour of Darwin's theory. Darwin himself pointed out that if worker-like adaptations were a difficulty for his theory, they were certainly a worse one for Lamarck's. Weismann[3] used this point against the Lamarckian concessions of Spencer (and indeed against those of Darwin himself) and in support of his own monolithic hypothesis of evolution by natural selection. The barbs on the sting of a worker bee, for example, which lead to their owner's death when she stings a vertebrate,[8] are not developed by use and are not passed on to offspring. Thus, the inheritance of acquired characters could not provide a universal explanation of adaptation. No similar bar to universality has been produced against natural selection. In a somewhat rhetorical reply to this argument Spencer[9] tried to drive a wedge into the crack created by the occurrence of laying workers. Weismann had already admitted the incomplete sterility of some workers to be a flaw in his argument, but he considered the occurrence too rare to be of importance. In any case he had been able to point to species in which the workers were completely sterile. On the Lamarckian issue the counterattack was ineffective.

Nevertheless, laying workers are more common than Weismann supposed,[10] and they remain an important issue. In the Hymenoptera, due to male haploidy, unfertilized worker-laid eggs generally produce males. Some biologists have been so impressed by the evidence for male production by unfertilized 'workers' that they have suggested that all males may arise in this way. This is certainly not true of some species (e.g. *Apis*), but it may be true of some others.[11,12] If it is true of any, we have a difficulty for the common view (supported, as we have seen, by Darwin) that selection in social insects is entirely an intercolony matter. If it is to the advantage of the colony to have one or a few individuals highly specialized as producers of eggs, how can we explain cases where the specialization is only in the production of female eggs, while males are produced by unspecialized females?

This problem may serve as a paradigm for the many other details of social behaviour and adaptation which seem to demand a more careful explanation than the usual passing reference to the benefit of the species or (less vacuous) to the benefit of a colony. This review tries to show how consideration of exact paths of gene propagation—using concepts of relatedness and inclusive fitness—can make intelligible many details that are otherwise obscure. Although the argument is potentially quantitative, social biology is still very far from providing the multiple measurements of fitness and the coefficients of relatedness that would permit exact tests of the theory. So, instead, relevant evidence is sought in the mass of mainly qualitative observations that are already stored in the literature. Even on the topics selected for discussion, however, this review aims to supplement and collate existing reviews rather than to provide a complete guide in itself. Certain recent reviews are particularly relevant and will enable the reader to judge the fairness of the present account: for bees;[11,13] for wasps;[14] for all social insects;[15] and briefly extending the field to vertebrates, especially primates.[16]

INCLUSIVE FITNESS

Replicas of a gene occur not only in descendants: a replica may be present in any relative. Of course, only one ancestor at a given level can actually have had the original copy; but, not knowing which ancester this is, one can associate a probability of possession with each ancestor. In a normal bisexual outbreeding system this probability is simply the reciprocal of the number of ancestors at the given level. Analogous probabilities can be associated with all other relatives by well-known methods of calculation based upon numbers of steps in pedigree diagrams.[17] To avoid complications when the relative might have two copies of the gene due to inbreeding, it is convenient to refer to a probability, r, that a gamete of the relative has the gene. As will appear, such a probability is well suited to the present needs.

A gene is being favoured in natural selection if the aggregate of its replicas forms an increasing fraction of the total gene pool. We are going to be concerned with genes supposed to affect the social behaviour of their bearers, so let us try to make the argument more vivid by attributing to the genes, temporarily, intelligence, and a certain freedom of choice. Imagine that a gene is considering the problem of increasing the number of its replicas and imagine that it can choose between causing purely self-interested behaviour by its bearer A (leading to more reproduction by A) and causing 'disinterested' behaviour that benefits in some way a relative, B. Specifically, let δA and δB represent the alternative increments to fitness that the gene in A is able to cause. Let the probability that our gene in A occurs as a replica in a random gamete of B be r_{AB}. Let F_A be the inbreeding coefficient of A (that is, the probability that the two homologous genes in A are replicas). Elementary considerations of probability show that the chance that our gene is passed on in a successful gamete of A is $\frac{1}{2}(1 + F_A)$. Our supposedly intelligent gene knows this as well as we do: it sees both this chance and (through supposed knowledge of the pedigree connections) the chance r_{AB} that a replica will be passed on through a successful gamete of B. Obviously it will decide what to do by weighing $\frac{1}{2}(1 + F_A)\delta A$ against $r_{AB}\delta B$.

The other gene in A will give exactly the same weightings. Thus, in this naive model, behaviour of A may be expected to reflect a 'unanimous decision' of its two genes: if b_{AB} denotes the fractional weighting which A gives to one unit of B's fitness compared to one unit of his own

$$b_{AB} = \frac{2r_{AB}}{1 + F_A}. \tag{1}$$

Such a weight can appropriately be called the *regression coefficient of relatedness*, or simply the *relatedness*: it measures the regression of B's additive genotype on that of A. (Strictly, the argument in terms of probabilities applies to selection in an infinite population. But the formula can be generalized to cover finite populations by treating r_{AB} and F_A as correlation coefficients—the usage of Wright.[18] It is then possible for b_{AB} to be negative with interesting implications, as was pointed out to me by G. R. Price (see Chapter 5). Note that usage of r in this article differs from usage in Chapters 1 and 2 of this volume.)

We can now abandon the fanciful viewpoint of individual genes. Likewise we can abandon our 'either ... or ... ' and suppose that the genotype of A simply gives rise to a fixed pattern of social behaviour and that this has fixed average effects on A and on relative B, and possibly on many other individuals as well. All the effects which A causes may be weighted by their appropriate bs and collected together in a quantity which may be named (see Chapter 2) the *inclusive fitness effect* of A. An approximate criterion for natural selection can then be given as follows: if the inclusive fitness effect of a genotype is above the average of such effects in the population, that genotype is currently favoured in natural selection. Allowing for slower rates of selection due to the 'dilution' (probabilistic) of the replica genes in relatives, the basic concepts of genetical selection theory, based on individual advantages, continue to hold. In stating this, however, a caution must be given that the social theory based on inclusive fitness is not yet so rigorous as the classical theory. For example, we do not yet know how to calculate r_{AB} and F_A precisely in a population that is undergoing selection, and difficulties in the interpretation of genotypic inclusive fitnesses also remain in cases of inbreeding. In short, the usefulness of the present tentative theory will have to be judged from the success of its predictions.

If an altruistic act by A greatly increases the fitness of B, A's inclusive fitness may be increased in spite of decrease in A's individual fitness. The criterion for a benefit to inclusive fitness depends on the ratio of gain to loss and on the relatedness. Suppose k units of fitness are gained by B for every unit sacrificed by A. Then A's act increases his inclusive fitness if $kb_{AB} - 1 > 0$; that is, if $k > (1 + F_A)/2r_{AB}$.

With inbreeding, F_A rises, but if B is a member of the same inbred deme, r_{AB} also rises above the value which is implied by connections through the most recent common ancestors. In a completely isolated colony not subject to mutation, genetic drift would eventually lead to $F_A = 1, r_{AB} = 1$, and hence $b_{AB} = 1$ for all individuals and pairs within the deme. This is genetic fixation: all the available loci carry replicas of a single allele. Mutation, multiplicity of demes, and some panmictic outcrossing normally prevent fixation and at least insure that the same allele is not fixed in all demes at once.

In outbred populations we can ignore F_A and calculate r_{AB} from immediate connections shown in the pedigree. The values of b for some important relationships are then as summarized below:

Self; member of same mitotic clone	1
Parent; offspring; full sib	$\frac{1}{2}$
Grandparent; grandchild; uncle; nephew; half sib; double first cousin	$\frac{1}{4}$
Greatgrandparents, etc.; first cousins, etc.	$\frac{1}{8}$

Even for outbred relationships the coefficients are not necessarily always of the form $(\frac{1}{2})^n$. For example, two children of a levirate marriage (same mother but different fathers who are brothers) have the relatedness $\frac{3}{8}$. Finally, a verbal illustration of criteria based on these coefficients may be useful. A sacrifice on behalf of half sibs is adaptive if the half sib receives more than four times as much reproductive potential as the altruist loses, but an altruistic act which is adaptive when directed to full sibs because $k > 2$ might be disadvantageous when directed to half sibs (if $k < 4$) or to a levirate sib (if $k < 2.6$).

TERMITE WORKERS AND OTHERS

The sacrifice of reproductive function by a worker social insect is a case of altruism to which such criteria can be applied. Some further elaboration of the coefficients will be necessary in the case of the male-haploid Hymenoptera, so let us first consider the termites, whose reproduction is normal.

As with all other social insects,[19] the co-operation and altruism within a colony of termites is in strong contrast to their hostility to members of other colonies.[20,21] Termites seem even more consistently hostile to strange queens in transference experiments than are ants, bees, and wasps. Evidently, relatedness is not a matter of indifference.

The basic system of monogamy ensures that relatedness within colonies will be high. Relationships with $b < \frac{1}{2}$ probably only occur when one colony has destroyed the reproductives of another and taken over its worker force.[21] On the other hand, inbreeding is evidently common and this, over successive generations, will rapidly increase the general level of relatedness within colonies. It is, indeed, tempting to suppose that the wood-eating habits of the cockroach-like ancestors of the Isoptera led to a claustral life with much inbreeding. But other groups in a similar niche and with even more reason to inbreed (e.g. Passalidae, if flightless) have attained only subsocial habits and have failed so far to plunge into social life like the termites. Thus, termite offspring must have found themselves able to increase the reproduction of their parents in some specially effective way, making the benefit to their inclusive fitness outweigh their sacrifice. Considering the extreme slowness with which primitive termites get their colony started and the evidently very high mortality of founding pairs, one can see roughly the nature of this opportunity. It has also been pointed out[22] that when the termites stood at the threshold of their wood-eating and scavenging niche they had taken on not only the help of protozoan symbionts, but also, because nymphs have to be re-infected after each ecdysis, a commitment to social contact. In such circumstances it is easy to see advantages of division of labour once the termites began to extend their burrows, to build, and to achieve homeostasis of their dark environment. These particular factors have no connection with genetical relatedness, but, of course, insisting on the necessity of relatedness in no way precludes other factors as necessary or contributory.

The mean inbreeding coefficient[23] of alates and the genetical correlation of mated alates are at present unknown, but they are becoming more easily measurable through techniques of biochemical genetics. At present, we only know that reproductives are commonly replaced. The substitute reproductive is probably an offspring of the individual replaced. Thus, it may be mating with its parent, or, if two reproductives have been replaced in quick succession, with a sib. If alates are inbred and yet themselves pair with non-relatives, their offspring will be highly heterozygous. But the offspring will be uniformly so and this will be reflected in a high b_{sibs} for this first generation ($b_{sibs} = 1$ in the extreme case where $r_{alates} = 0$ and $F_{alates} = 1$). However, if the following generation is produced by replacing both the primaries by two first-generation sibs, much segregation occurs and in the second generation the average relatedness falls. For example in the extreme case the coefficients run 1, 2/3, 5/6, ... After the third generation the coefficient rises steadily. If both reproductives are not replaced simultaneously and the surviving parent is always mated to an offspring, the hiatus in relatedness is to some extent smoothed over. In the extreme case the coefficients now run, 1, 5/6, 3/4, 11/13 ... The lowest value is not so low and occurs in the third generation instead of in the second. Relatedness of the mating alates also tends to smooth over the hiatus.

It is improbable that termites in any sense ‘know’ what generation they belong to; it is slightly less improbable that they might react to observed genetical heterogeneity in their fellows. Even so, it seems unlikely that termites of the second or third generation of multigeneration colonies will prove any less co-operative than those of the first, even if a minimum really occurs.

The overall effect of inbreeding, taking averages of all generations, is to increase relatedness within colonies. This reduces the selective scope for selfish behaviour within the colony. It does so, however, at the price of lowering the species ability to adapt quickly in a fluctuating environment. Viewing the colonies as superorganisms, we see this as implying a longer 'generation time' for colonies with many generations of internal inbreeding. In changing environments, this is a disadvantage. On the other hand, inbreeding makes the superorganism more nearly a reality. It should facilitate the peaceful coexistence of multiple reproductives (where advantageous) and facilitate the co-operation of their broods.

The most obvious manifestation of a genetical 'selfish' trait would be a raised threshold for response to a 'queen-substance' type of pheromone; selfish juveniles would be those quickest to start on the path to sexualization when any deficiency arose. However, since in termites the worker repertoire of behaviour includes the destruction of excess reproductives, a probable outcome of the spread of 'selfishness' of this kind is an increasing incidence of cannibalism. This is wasteful; any advantage in evolutionary flexibility possessed by the outbreeding species has to be weighed against present inefficiency from intra-colonial strife. Obviously, at present we cannot even speculate on what kind of a balance is struck.

That both sexes 'work' in the termites is consistent with the symmetry of the coefficients of relatedness with respect to sex. The situation with the Hymenoptera is in strong contrast. However, before passing to the Hymenoptera and their unusual system of reproduction, it is worth pausing to consider whether any other groups of animals with normal reproduction (like the termites) show even incipient tendencies to trophic or reproductive altruism apart from that involved in parental care.

Among groups related to the termites, Embioptera and Zoraptera perhaps deserve further study from this point of view, although it is already obvious that they have no mutual aid systems remotely comparable to those of termites. Earwigs (Dermaptera) have parental care and sometimes progressive provisioning; so do some crickets (Orthoptera: Gryllidae and Gryllotalpidae), and one even produces special trophic eggs. The mother in this species (*Anurogryllus muticus*) finally allows her own body to be eaten by the claustral brood, and there are hints that males may sometimes submit to the same fate.[24,25] The omnivorous diets, the manipulative ability (burrowing, entrance closing, food storing), and even the stridulation of crickets suggest pre-adaptations to social life. The wing polymorphism that is frequently present, involving a low proportion of macropters in both sexes, may also be pre-adaptive: local population viscosity permits the evolution of friendly relations between neighbours while some distant dispersion prevents gene fixation and preserves adaptive plasticity. Nevertheless, the concurrence of these various factors in crickets seems to have been less than enough, so far, to produce worker-like altruism. In many ways the social pattern achieved resembles the pattern of certain small rodents[26] more than that of the social insects. Of particular interest here is that, according to Alexander,[27] colonist males are much more aggressive and territorial than males in the crowded populations from which they came. Considering average relatedness of neighbouring males in the two situations, this is what we expect.

Among endopterygote insects apart from Hymenoptera, Coleoptera show the highest form of parental care. One case illustrates very clearly that parental care is not necessarily connected with altruism towards individuals other than the offspring. In *Necrophorus* several individuals of both sexes will combine in burying the corpse of a small mammal, but once it is buried they fight, sometimes fatally, until only one pair is left in possession. Later the 'nestling' larvae can be observed to compete and to importune the mother for food much in the manner of young birds.[28] Occasionally, in some species, the father also feeds them.

Among other normally reproducing arthropods, spiders seem a likely group for social life, with parallels to most of the supposed pre-adaptations mentioned for crickets. But beyond parental care and gregariousness, mutual aid in overcoming large prey and the grouping together of young of several mothers (either as eggs or as spiderlings) seem to be their highest social achievements.[29,30] By its concerted attacks *Agelena consociata* is said to be able to capture larger prey than individuals could overcome alone. This implies some danger to the first spiders that bite, and this in turn suggests an opening for parasitic behaviour, a habit of joining at the communal feast after hanging back at the kill. Thus, it is surprising to learn that the colonies show no hostility either to members of other colonies or to spiders of a different but congeneric species.[31] If the reason for this is that the experimental events are outside the range of normal adaptation, then it must be rare in nature for a spider to wander from one communal web to another. It would be interesting to know more about dispersal and how new colonies are formed. In many spiders the young disperse by ballooning on silk threads. Since this mixes young spiders with non-relatives, it is slightly more adverse to social co-operation than the macroptery of crickets and it seems very unlikely that *Agelena consociata* could maintain indifference to strangers if dispersal were by this means. In fact, up to a stage when they would be too heavy to balloon, the spiderlings do not leave their natal retreats or attack live prey: they feed on prey brought in by the adults.[32] Small swarms bud off as new colonies just outside the limits of an established one; but, as Krafft remarks, this does not readily explain the wide range of the species in the African forest.

Examples of trophic altruism are probably commoner in higher animals than is known: to mention recent reports we have the pooling of litters in mice[33] and regurgitation of food to non-filial young in carnivores such as hyenas, wolves, and dogs (e.g. *Lycaon*[34]) that hunt in groups. It is noteworthy that at least in the canids both sexes serve the young and that in mice males will perform such acts as retrieving young that have got out of the nest.[35] All these animals also show the strong contrast between co-operation within the clan and hostility to outsiders that is so characteristic of social insects.[19,36–39] But, again, there is no hint of non-reproductive 'workers', unless we take as a faint suggestion of this the rather obscure and disputable phenomenon in which social supernumeraries and outcasts put up less fight to better their lot and die of 'stress' more readily than seems reasonable from the point of individual selection. (The best evidence so far that social supernumeraries are physically capable of bettering their lot concerns not rodents but birds. Watson[40] implanted androgen into two non-territorial red grouse cocks (*Lagopus lagopus scoticus*) and found that they promptly resumed aggressive behaviour and won small territories. Neither bird bred in the first summer and one was ousted and died in the following fall. However, the other enlarged its territory and did breed the following year. Had these birds not been injected they would almost certainly have died within a few months along with other non-territorial males.)

In birds there are a few cases of females pooling their broods in a common nest and also a rapidly growing list of cases of young helping their parents to rear subsequent broods.[41,42] The rate of natural mortality in birds must ensure that some of these 'helpers' live and die as 'workers', but, so far as I know, there is no hint of a distinct caste. In some species both sexes are equally represented among the helpers, but in several cases the helpers are mainly or wholly males, and in one or two this seems to depend on a bias in the sex ratio which is either primary or appears very early through differential mortality. In the apparently unique hierarchical system of Welder brush turkeys, sibling males remain in life-long association and display and fight as a team. Normally only the dominant male of the locally dominant team can mate. The brothers' acquiescence in this has the appearance of altruism. Female turkeys dissociate from their siblings and show no such co-operation.[43] At first sight, these cases seem a radical departure from the pattern of both the termites and the Hymenoptera, but there is an intriguing possibility. The haplodiploid system of inheritance is strictly analogous to sex-linked inheritance. In the Hymenoptera the female is 'XX'; in birds the male is. Are genes for the behaviour of 'helper' males carried on their X chromosomes? The significance of this question will appear shortly. It seems rather unlikely that they are. Are genes on the sex chromosomes influencing the sex ratio (Chapter 4)? This seems more reasonable. In spite of the fairly high chromosome numbers characteristic of birds, the X chromosome is stated to form 10 per cent of the total genome in birds, compared to only 5 per cent in mammals.[44] Thus, the X is potentially more powerful.[45] Is there a connection between the statement[46] that the social-nesting cuckoos *Crotophaga* may have 'a bias in the primary sex ratio' and the suggestion[47] that the genes for the gentes of *Cuculus canonis* may be carried on the Y? Does the X chromosome of *Crotophaga* make the male into a helper while the Y makes the female into a cuckoo? Fortunately or unfortunately, the behavioural genetics of cuckoos provides a safe niche for such speculation at present.

HYMENOPTERA: THE MATRIFILIAL COLONY

In the Hymenoptera males normally arise from unfertilized eggs and are haploid. Excepting mutations, every sperm has exactly the genetic constitution of the male producing it. Regarded as a form of parthenogenesis, this system is called *arrhenotoky*; on the basis of its known cytology it is called *male-haploidy*. Strictly, arrhenotoky is a less-specific term implying that unfertilized eggs give males only, but since every case of arrhenotoky that has been investigated cytologically has been found to involve haploidy of the males, the terms are practically synonymous. Although it may be less stable than normal reproduction, arrhenotoky is not a degeneration of sexuality: males must occur in each generation if females are to be produced in the next. In this, arrhenotoky is very different from thelytoky (unfertilized eggs giving females), which allows male production to be abandoned. Thelytoky is common in some groups of Hymenoptera, but its scattered distribution suggests that it is everywhere derivative; and this is what we expect from consideration of the adaptive inflexibility of asexual reproduction in a fluctuating environment. Arrhenotoky alternated or interspersed with thelytoky seems, *a priori*, much safer than pure thelytoky. There are some evidently ancient systems of alternation—in the gall wasps for example.

Referring to pure arrhenotoky, let us consider the construction of coefficients of relatedness.[48] My previous account of this subject contained an error (Hamilton in ref. 49). Male-haploid sex determination corresponds formally to an XO system in which the X chromosome has become multiple and autosomes have disappeared. The calculation of correlations of genes in relatives under sex-linked inheritance is well explained by Li.[17] With caution as to the base population, which provides the variance for the correlation coefficient and the supposed population of non-replicas for the probability concept,[50] correlation coefficients for genes can be identified with probabilities that genes are replicas. Considering the probability that a particular gene in an adult has a replica in a particular gamete, we at once see an asymmetry in the situation of males and females: a gene in a male is certain to be passed on; for a gene in a female the chance is $\frac{1}{2}(1 + F)$, as before. However, if for every male we take $F = 1$, the same formula serves for both sexes and so does formula (1) for the coefficient of relatedness. For social interactions between individuals of like sex, these coefficients are what we need, but for interactants of opposite sex there remains another important asymmetry to be considered. It can be shown that in terms of contributions to the gene pool of a distant future generation $\frac{1}{3}$ of the current gene pool is vested in the aggregate of males and $\frac{2}{3}$ in the aggregate of females, irrespective of the sex ratio. Thus the reproductive values of individuals of each sex are proportional to $1/N_1$ and $2/N_2$ respectively, where N_1 and N_2 are the numbers of males and females. Figure 8.1 is an elementary illustration of this; it shows that when males and females are equally numerous and the population is constant a male expects one successful gamete and a female expects two. Hence, we arrive at what may be called *complete* or *life-for-life* coefficients of relatedness, B, which, at least for heterosexual pairs, are distinct from the *gamete-for-gamete* coefficients (b) discussed so far. Using $c = N_1/N_2$, Table 8.1 summarizes the formulas according to the sex of donor and the recipient of the action.

The most important specific relationships are those between parent and offspring and those between sibs. Making appropriate substitutions for the r in Table 8.1, we obtain the formulas in Table 8.2.

The only new symbol requiring definition is r_s. This is the correlation of sperms drawn randomly from the spermatheca of a female. The lower right-hand formula in each case is the one that holds generally under normal reproduction (autosomal as opposed to sex-linked inheritance). Strictly, this formula needs to be adjusted for sex ratio by the factors c and $1/c$ for heterosexual relations as Tables 8.1 and 8.2, but, as will be seen, there is less tendency for sex-ratio bias to occur with normal reproduction. Although the same formula, $B = (1 + 5F + 2r_s)/4(1 + F)$, will serve for any sibs under normal reproduction (ignoring sex ratio) and for sisters under male haploidy, there is an important difference in the range of values open to r_s in each case. If r_m denotes the

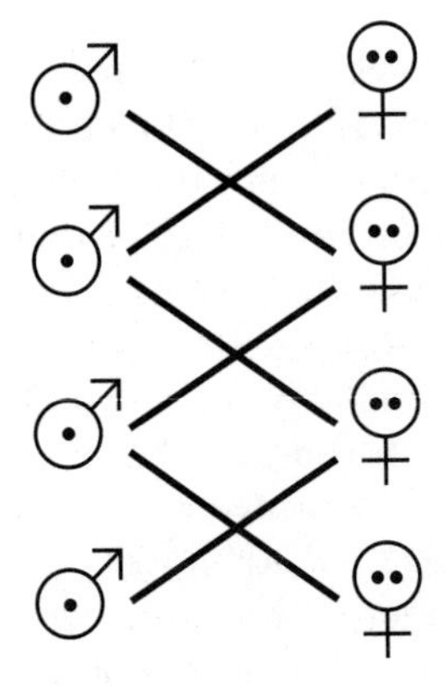

Figure 8.1 Male haploidy: the simplest population.

Table 8.1

		Recipient	
		(m)	(f)
Donor	**(M)**	$B_{Mm} = b_{Mm} = r_{Mm}$	$B_{Mf} = 2cb_{Mf} = 2cr_{Mf}$
	(F)	$B_{Fm} = \frac{1}{2c} b_{Fm} = \frac{r_{Fm}}{c(1+F)}$	$B_{Ff} = b_{Ff} = \frac{2r_{Ff}}{1+F}$

Table 8.2

	Parent-offspring; offspring parent		Sibs	
	♂	♀	♂	♀
♂	F	$c(1+F)$	$\frac{1}{2}(1+F)$	$\frac{1}{2}c(1+3F)$
♀	$\frac{1}{2c}$	$\frac{1+3F}{2(1+F)}$	$\frac{1+3F}{4c(1+F)}$	$\frac{1+5F+2r_s}{4(1+F)}$

correlation of random sperm from the whole population of males, the ranges are as follows:

	High multiple	insemination	Single insemination		
Male haploidy	r_m	$<$	r_s	$\leq$	1
Normal reproduction	r_m	$<$	r_s	$\leq$	$\frac{1}{2}(1+F)$

The value of r_m in a population in a steady state must be near to F. Mating would have to be extremely promiscuous for r_s to approach r_m closely, however, so F must be considered a hardly attainable lower limit for r_s. On the other hand, 1 and $\frac{1}{2}(1+F)$ are readily attained upper limits—they are attained whenever mating is single.

The heterosexual coefficients can only be compared to the others when the sex ratio is specified. Unfortunately, the factors affecting the sex ratio in social Hymenoptera are very complex. To anticipate a little, the queen may be inclined to produce more males than the sterile workers regard as ideal. On the other hand a laying worker may want more males than the queen does—provided the extra males are her own (the worker's) offspring. The factor of multiple insemination also affects these issues and so do differing roles of the two sexes in gene dispersion. In advanced swarming species, such as *Apis* and *Eciton*, the complete dependence of the queen upon workers becomes important. Outbreeding, multiple insemination, and helplessness of queens should all correlate with high male production. Regarding the first two factors there are indications that such correlation exists—for example, in the list of sex ratios of primitively social bees given by Michener,[13] *Lasioglossum marginatum* produces the highest percentage of males and also is the one bee in the list for which there is evidence of multiple insemination.[51] For present simplicity, let us assume that the sex ratio is wholly dependent on the behaviour of the queen. This, at least, is plausible for incipiently social species. In more highly social outbreeding species it may be hoped that some of the other factors cancel each other out. Under inbreeding, as is easily seen, the various preferences converge. Assuming determination of sex ratio by the queen, the sex ratio will be at equilibrium when $B_{mother\text{–}son} = B_{mother\text{–}daughter}$; that is, when $c = (1 + F)/(1 + 3F)$. Thus, in going from panmixia to complete inbreeding the ratio goes from 1:1 to 1 male:2 females. This tendency to sex-ratio bias is peculiar to male haploidy and sex-linked factors. (This source of sex-ratio bias due to inclusive fitness was overlooked in a previous paper (see Chapter 4) and supplies the explanation for the discrepancy mentioned on page 159. It is a much less striking effect than that due to differential dispersal, which was the main theme. Colony inbreeding plus lone-founding by females may be an important factor promoting female excess in some social insects: it certainly is so in the formicid social parasites which practice adelphogamy.[52])

Figure 8.2 shows B plotted against F on the assumption that $c = (1 + F)/(1 + 3F)$. Figure 8.2b shows that sister–sister relatedness may be either higher or lower than mother–daughter relatedness, depending upon the extent of multiple mating. Although the range above is equal to the range below, the range above is more readily attainable, as already pointed out. Data on the mating habits of social insects are still scanty and data on the effective multiplicity of insemination exists for only one or two species[53] (see also Part II of the paper in Chapter 2). But various facts suggest that the social groups of Hymenoptera originated from ancestors which were effectively monogamous in the female sex. Male Hymenoptera are readily polygamous (and expected to be so—see below) and are generally less common than sexual females (at least in solitary and incipiently social species), and a mature male usually has ready for use many more sperm than a female can store in her spermatheca (with exceptions only in some highly social species[11,54]). Assuming single mating in the socially inclined ancestor, an attractive explanation for the basic feature of hymenopteran sociality, worker-like attributes in females, is at once apparent. In the sense of inclusive fitness a female prefers sisters ($B = \frac{3}{4}$) to sons or daughters ($B = \frac{1}{2}$) and therefore easily evolves an inclination to work in the maternal nest rather than start her own.

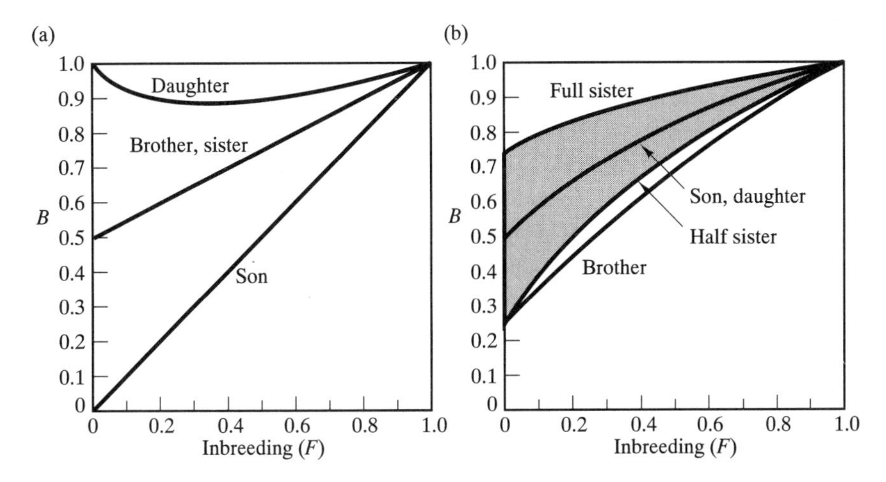

Figure 8.2 Complete coefficients of relatedness for males (a) and females (b) respectively, assuming the sex ratio to be $(1+F)$ males: $(1+3F)$ females, as expected if sex ratio is controlled by mothers. The average relatedness between sisters is affected by the degree of multiple insemination of the mother and may vary between the limits shown.

Such a bias latent in the male haploid system helps to explain the multiple origins of social life in the Hymenoptera, contrasted with one origin only for the Isoptera.[22] Nevertheless, the occurrence of this special relatedness to sisters must not be overemphasized. Male haploidy is certainly not the only prerequisite for evolving a sterile caste. Perhaps the preadaptations of solitary nesting Hymenoptera as porters and builders are equally important. We have to explain why other male-haploid groups have not evolved social life, and also why clonal aggregations like those of aphids have not done so. Equally troublesome, why is thelytoky not more common in social insects considering that it occurs at all?[10,55] Why is thelytokous *Ceratina dallatoreana* not social?[56] Why has *Apis mellifera capensis*, with its ready system of thelytoky, become a degenerate rather than a superefficient social insect?[57] Presumably, the catch lies in the danger of abandoning sexual reproduction, and this danger is still imperfectly understood. The aphids of a clone are related like the cells of any metazoan body, and in theory there is no reason why they should not develop an equal diversity of function. But what co-operation there is in aphids is slight and of a very passive kind, and with such sedentary sap-feeders it is difficult to see how this could be otherwise. Perhaps it is no more surprising that workers have failed to appear in aphids than that they have disappeared in many lines of parasitic ants, the reason being, in both cases, that there is very little that they can do. Also, the simpler neural and sensory equipment of aphids is less adequate for communication and may not permit them to recognize the clonal siblings to whom alone they should direct altruism. (Brain size increases in higher social Hymenoptera, but diminishes in higher termites.[11,58] The explanation of this odd contrast is unknown.) Similar inadequacies apply to most other male-haploid groups (in Rotifera, Coccoidea, Aleyrodidae, Acarina), but there are two groups which seem more promising. Arrhenotoky[59] and male haploidy[60] have been demonstrated recently in ambrosia beetles of the genus *Xyleborus*, and evidence from sex ratios[61] or arrhenotoky[62] suggests that several tribes may be male haphoid. Elaborate gallery systems have been described in *Xyleborus*.[63] In the Thysanoptera, colonial fungus-feeding and gall-forming[66] species in the Phlaeothripinae deserve further study. It is too soon to be sure that these groups have no examples of worker-like adaptation. However, *Xyleborus* has close inbreeding. For the present, we defer discussing the very varied implications of this extra factor and pass to other contrasts shown in Fig. 8.2.

BEHAVIOUR OF HAPLOID MALES

From Fig. 8.2a we see that a male is always more related to a daughter than to brothers and sisters. Thus, a male is not expected to evolve worker instincts, and correspondingly, in the Hymenoptera he has hardly done so. That this is not due to any deep-seated inability on the part of the male to perform constructive tasks is shown by the sphecid genus *Tachysphex*. The males dig short tunnels in the sand. They rest in these and defend territories around them.[67] The greater relatedness of males to daughters than to sibs declines with inbreeding, and so it may be suspected that those ants in which long-lived males do some nursing and food-distributing prior to reaching sexual maturity,[68] are inclined to inbreed.[69] The species of *Cardiocondyla* in which Santschi[70] saw ergatoid males carrying larvae almost certainly do inbreed.

At first sight the male's high relatedness to daughters suggests that parental care by males is especially likely. But if the male cannot distinguish his mate's female eggs from her male eggs, any parental care would be spent on individuals of average relationship $\frac{1}{2}(1+0) = \frac{1}{2}$; so the male would do better to devote himself to polygamy. Moreover, with inbreeding, leading to female excess, polygamy may be easy to achieve. All the same, there are many nest-making hymenopteras in which an unbroken series of female cells is followed by an unbroken series of male cells—or, more rarely, vice versa. Such a system encourages outbreeding and it also creates a situation in which a male could 'know', roughly, which sex a cell was intended for, and so contrive to help only the daughters. But in one group in which the male does associate with the new nest, the subgenus *Trypargillum* of *Trypoxylon*, we begin to have evidence (Medler[71] and personal observations) that the female randomizes the sequence of sexes. Perhaps this is not a coincidence. The male's main interest seems to be in mating the female when she comes in;[71,72] but if his presence does protect the nest, it seems that he has been tricked into guarding his unrelated sons as well as his daughters! This interpretation is, of course, somewhat speculative.

Houston[73] described a single nest of a bee in the genus *Chilalictus* in which, besides unremarkable females, he found one ordinary male and 10 others of very bizarre form. They had short wings and greatly enlarged heads and mandibles. One ordinary and three of the strange males were also identified among the pupae. Referring to an even briefer previous report on this phenomenon by Rayment and to a mention of entrance guarding by Halictine males, also by Rayment, Houston suggests that the strange males may function like soldier ants and defend the colony. The outward resemblance to dinergates is certainly striking. But the one strange male that was dissected had normal gonads and sperm, and this suggests an alternative comparison to the armed and ornamented males that appear so widely in the animal kingdom due to sexual selection. From Michener's failure to find any workers in closely related Australian halictines[74] and from the estimate of 40 cells in use, it may be suspected that Houston's nest is the work of several independent females sharing an entrance. Such suspicion is fortified by an earlier report from Japan of monstrous-headed males, winged this time, in *Evylaeus ohei*, which is closely related to *Chilalictus*.[75] *E. ohei* is nonsocial and at least sometimes an entrance sharer. One large-headed male was seen to fly to a nest and enter, and adult male bees found in nests were always of the large-headed type. The Japanese authors also reviewed thoroughly other cases of large-headed males then known. Apart from Rayment's case most are andrenine bees. Andrenines are also non-social but in some instances are known to share entrances. Altogether, there seems ground for the conjecture that these males are parallel developments initiated when subterranean mating occurs in an outbreeding species; selection derives from fierce competition for such mating among usually unrelated males. In *Chilalictus* the strange males are unable to fly, so unrelatedness within the colony must depend on unrelatedness among the mothers sharing the nest system. In *Evylaeus ohei* the males can presumably fight for possession of any nest, but entrance-sharing (as found) is still theoretically probable in that, whether most mating is before or after nest founding, it sets up the females in ready-made harems, thus providing the strongest basis for Darwinian sexual selection. Conspicuous male weapons are rare in Hymenoptera, but male combat is by no means unknown.

A common characteristic of the enlarged heads of these various bees is extension of the genae. This is certainly not entirely for muscle attachments,[75] and the forward placement of the large falcate mandibles suggests head-on combat in burrows with the genal processes shielding the neck. Thus, I suspect that even if, as in *Trypargillum*, these males do sometimes defend burrows against parasites, their adaptations are intended primarily for aggression and defence against other males.

A parallel male head polymorphism is known in a sphecid, the crabronine wasp *Ectemnius martjanowii*, which makes 'linear or branched linear' burrows in dead wood.[76] Unfortunately, nothing is known of the habits of the males. Going further afield, the same can be said of the even stranger males of the scolytid beetle, *Ozopemon brownei*.[61] The wingless, blind larviform male with falcate mandibles is reminiscent of the fighting larval forms in 'solitary' hymenopterous parasitoids.[77,78] Comparisons to Platygasteridae are suggestive concerning the role of relatedness which is certainly important[79] (for a similar behaviour contrast in adult Scelionidae see ref. 80). Colonizing females are said to make their burrows in close proximity, so one may guess that galleries meet under the bark and so bring unrelated males together. To judge from the sex ratio, these males may be haploid.

In the mite genus *Caloglyphus*, also probably male haploid, an armed and aggressive male morph arises when population density is low.[81] It is pheromonally inhibited at high density. Here, it is known that the armed male easily kills many that are not so armed (but never females) and that fighting increases if mites from a different culture are introduced. Again these various details suggest that the fighting morph is adapted to succeed in the mating competition that arises when weakly related socks meet in new habitats.

In another mite that should have some competing unrelated males (*Dicrocheles phalaenodectes*, see below), there is no report of antagonism. The case of the hymenopterous parasite *Melittobia acasta* is similarly unexpected but in the opposite sense. Here, the extremely biased sex ratio indicates close inbreeding: it seems likely that a host is usually colonized by one female only and that the few males on a host are normally brothers. Thus, the fierce and mortal fighting between them is a surprise. But really the ecology of these cases is not well enough known for us to say that they present a serious anomaly. For example, the emphasis on mortal combat in *Melittobia* varies from one account to another,[82–85] while from illustrations[82,86] there is an interesting hint that there may be in this case also some polymorphism, or race variation, in male head size.

LAYING WORKERS AND RIVAL QUEENS

Referring again to Figure 8.2b, a female is more related to her offspring than to her brothers. Thus, workers are expected to be comparatively reluctant to 'work' on the rearing of brothers and, if circumstances allow, inclined to replace the queen's male eggs with their own. Not all social Hymenoptera have laying workers but, as mentioned earlier, the phenomenon is very widespread.

Montagner[87] found evidence for species of *Vespula* that queens which changed from laying female eggs to laying only male eggs were unable to maintain their dominance: laying workers attacked such queens and killed them or drove them off. The laying workers also fought among themselves, sometimes fatally. On the other hand, queens which laid only female eggs or which laid some male eggs interspersed with their female eggs were maintained, although laying workers added male eggs simultaneously. Pooling results from his 14 captive colonies, Montagner found that the workers produced far more males than the queens did. These results have to be applied cautiously for two reasons. First, in view of the variability in the queens' performances, the number of cases is small. Second, the colonies had all been upset by separation of their queens for 4 days while they received radioactive food to mark their eggs.

Spradbery,[88] after analysing 89 wild *Vespula* colonies, considered that laying workers contributed insignificantly to the production of males. His evidence was (1) that the highest proportion of male eggs occurred in the lowest combs where the queen spends most of her time, (2) that queenless colonies were rare, mostly young, and almost without males, and (3) that less than 4 per cent of workers had developed ovaries. Although Spradbery's colonies have more claim to be considered natural, Montagner's results suggest tensions in the normal discipline which became manifest under stress and which follow the lines that the theory suggests. In *Vespa orientalis* egg-laying by dominant workers on the upper combs may be more normal.[89,90]

Laying by unfertilized females occurs in most social halictines. In *Evylaeus malachurus* workers are consistently smaller than queens, never fertilized, and produce most of the males.[91]

In Brazil (as reported by W. E. Kerr[11]), Zucchi has found that most males of *Bombus atratus* come from dominant egg-laying workers, and Beig has found the same for *Trigona postica* and two other meliponines. As always, the predicted preference for sons over brothers decreases with inbreeding, and it is probably significant that the cases just mentioned come from generally outbreeding groups. Regarding Meliponini, Kerr has genetical evidence of outbreeding in *Melipona marginata*, although the existence of worker-laying in this particular species has not been ascertained. In another species with genetically proven outbreeding, *Apis mellifera*, worker-laying certainly contributes little to male production (10 per cent in *A. mellifera adansoni*, hardly any in *A. m. mellifera*.[11] Although it is far from clear why only some outbreeding species should have male production from laying workers, detailed considerations of inclusive fitness such as those outlined here seem to offer the best prospect of explanation.

Social Hymenoptera which do not normally have laying workers may produce them abundantly when the queen is removed. The laying workers are mutually hostile and may develop territories on the combs.[89,92] Hostility is expected because a female is always more related to her own sons than to the sons of sisters. Equally, of course, a fertilized female prefers her own daughters to those of any other fertilized female; hence, the animosity of queens. As already remarked, it is difficult to believe that it is efficient for the colony to have a specialized reproductive to lay female eggs and unspecialized reproductives to lay male eggs. It is likewise difficult to believe that all the fighting, the reciprocal egg-eating (in *Lasioglossum*,[93] *Bombus*,[94] and *Polistes*[95]), and all the time wasted in milder dominance interactions really serve colony efficiency. The phenomenon in *Bombus* has been a continuing source of wonder and speculation ever since Pierre Huber remarked on it in 1802.[96] For me, another feature of female rivalry in bees is more puzzling. This is that the queen of *Apis*, beyond mere hostility to sisters, appears to possess a structural adaptation for killing them. This is her curved sting,[97] a feature shared with such sinister relatives as *Psithyrus, Sulcopolistes*, and the parasitic Vespinae,[98] all of which use their weapons against other social Hymenoptera rather than against vertebrate enemies. But, if the function of the curved sting has been correctly interpreted, the existence of an adaptation for killing sisters is so unusual as to suggest that the fighting queens are sometimes not sisters; do *Apis* queens sometimes return from their nuptial flight to strange colonies?

ASSOCIATION, POLYGYNY, AND PARASITISM

Michener[99] has suggested that in the Halictinae the sociality may have arrived without direct family relationship between the bee behaving as a worker and the brood she helps to rear. The suggested sequence is roughly as follows: first, nest burrows are aggregated; then an entrance burrow is shared while bees provision and oviposit in independent branch burrows; then cells are provisioned co-operatively while dominance behaviour or nutritional advantage determines which bee lays the eggs. This stage may be said to involve 'association',[100] and the helpers who do not lay eggs (or do not lay eggs that survive[95] are called 'auxiliaries'. In theory, there is no reason why such a colony should go on to develop filial workers (Michener's final stage), but in fact almost all halictine colonies that start with mated auxiliaries have been found to employ filial workers later and some have filial workers without having had auxiliaries. This is a point against Michener's hypothesis since it suggests that the matrifilial colony comes first. There are other adverse points. One of Michener's reasons for proposing the idea has weakened since it was put forward. He drew attention to the rarity of confirmed examples of the stage which seems the plausible antecedent to the matrifilial colony: daughters returning and independently extending the maternal nest. Likewise, however, there is a lack of evidence that the conglomerated nests classed by Michener as 'communal groups'[13] are not generally built up by groups of sisters. A careful study by Zucchi *et al.*[101] showed just this situation for a nest of a euglossine bee, *Eulaema nigrita*, and this finding makes it seem more likely that other conglomerate nests (*Euglossa*,[102] *Xylocopa*, etc.) are the work of sisters. In the Vespoidea such a stage has long been known, for example—Ducke's *Zethus lobatus* (now *Zethusculus*). (On the basis of his case in *Zethusculus lobatus*, Ducke seems to have guessed correctly about the large nests of *Eulaema nigrita*.[103]) Of course, cases of daughters nesting near the maternal nest are extremely common.

It is impossible for worker-like characters to evolve on the basis of benefits given to unrelated bees. The case is very different if the bee benefited is a sister, but the minimum ratio benefit–sacrifice that has to be achieved is still forbidding. This ratio must exceed

$$B_{daughter}/B_{niece} = 4(1 + 3F)/(1 + 13F + 2r_s).$$

For example, under outbreeding and with high multiple mating the benefit must be about 4 times the sacrifice, and under outbreeding and monogamy it still must be at least 4/3 times the sacrifice. Inbreeding brings these limiting factors down towards 1 and so makes altruism more easily adaptive. If the mother is still egg-laying when a young bee makes her 'decision' whether or not to become a worker, she should always prefer working on behalf of sister larvae rather than daughter larvae. For sisters we have

$$B_{daughter}/B_{sister} = 2(1 + 3F)/(1 + 5F + 2r_s),$$

which under outbreeding goes from 2 for multiple mating to 2/3 for monogamy. The last limiting ratio is the remarkable one. To state its implication again, a bee may be content if she manages to rear a number of sexual sisters smaller than the number of daughters she could have produced on her own. Socialization can proceed in spite of slight reproductive inefficiency of the colonial mode of life.

We arrive here at the seeming paradox that a species might be injured by the selection of 'altruism' among its members: in a monogamous outbreeding species the preference for sisters over daughters could assume the role of a vice. Thus, while the high relatedness of sisters may help to get socialization started, it may be that new social species succeed best if they subsequently take up mating habits that slightly increase the relatedness of daughters relative to sisters. According to whether such adjustment comes from inbreeding or from multiple insemination, colony structure would be channelled towards polygyny (and association) or towards monogyny (and lone-founding).

Figure 8.3 shows how sisters and nieces should be valued relative to daughters. It plots the straight line contours implied by the two formulas above. Contour height shows how many sisters or nieces must be produced for the sacrifice of one daughter to be worthwhile in terms of inclusive fitness. In a species represented near the lower-left corner, if selection is favouring the production of about two sisters in the place of three daughters (dotted contours) it is clear that the population must suffer numerically. On the other

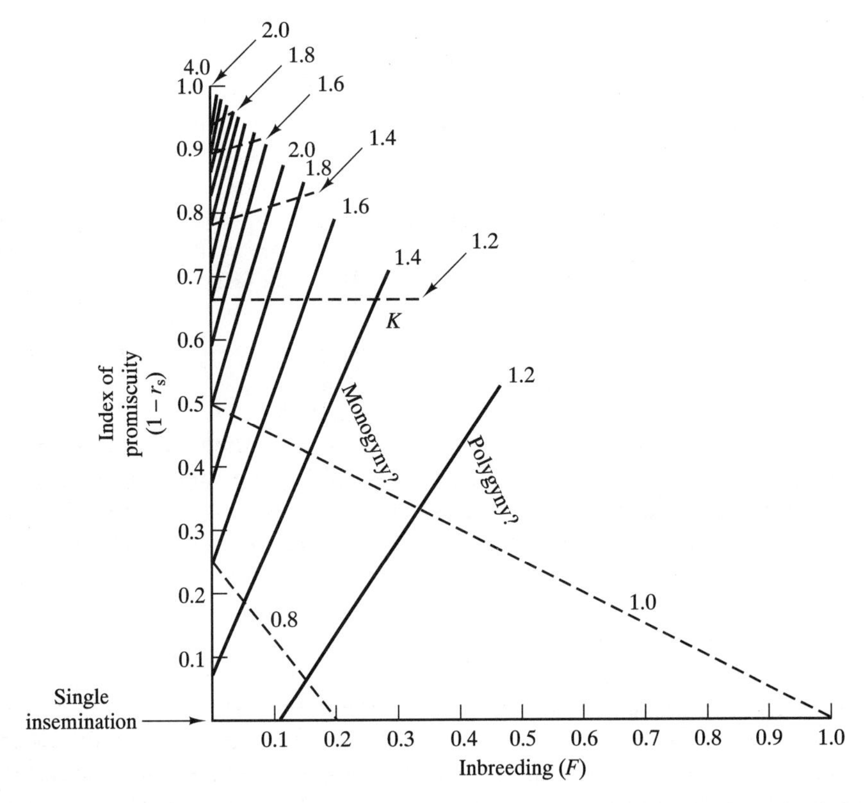

Figure 8.3 The relatedness of nieces (solid contours) and of sisters (dashed contours) relative to relatedness of daughters, showing dependence on parameters of inbreeding and multiple insemination. All contours are at intervals of 0.2.

hand, in a species at K the conditions that would make it advantageous for a reproductive female to transform into a worker must imply a net gain to the reproduction of a supposed two-female colony. For any social species represented near the upper-left corner of the diagram, the efficiency of colonial compared to solitary reproduction should be considerable. Such species are less stable in the sense that conditions favour incipient parasitism. Strife between the sisters, or between mother and daughters, is likely to break out if efficiency falls. Of course, once caste differentiation has begun, the worker no longer has the real alternative of nesting alone; efficiency of a kind is then guaranteed and the species can change its breeding habits with less danger. Rather than complete breakdown of the eusocial system, we expect then merely more or less production of males by the workers.

What do these theoretical considerations imply for Michener's hypothesis? It is certainly not impossible for worker-like behaviour to evolve in a group of sisters if the advantage to the colony is high enough. On the other hand, high advantage is unnecessary to arrive at the matrifilial colony: male haploid animals gravitate naturally towards this condition provided that the sex ratio or some ability to discriminate enables the worker to work mainly in rearing sisters. Therefore it seems likely that the worker-like attributes involved in association—submission, ovary inhibition, etc.—arise during a matrifilial phase and that these attributes subsequently permit association between foundresses when certain additional conditions are satisfied. The following conditions are suggested: (1) all adult females have latent potentiality to behave as workers, (2) homing instincts tend to keep groups of sisters together, (3) inbreeding further increases the relatedness of sisters and also relates the occasional pairs of non-sister associates, and (4) an associated group of n females has, at the very least, more than n times the reproductive potential of a nest-foundress working alone.

The halictine group which comes nearest to having association without subsequent filial workers is, not surprisingly, that which originally stimulated Michener's hypothesis, namely the Augochlorini (as now defined by Eickwort[104]). Association, with filial workers later, is known in many species,[13] and in most of these there are indications that the cofounding females are usually sisters. For bivoltine species of *Pseudaugochloropsis*, however, Michener and Kerfoot[105] found that the first-generation bees were lone founders whilst the second-generation bees became associates after the death of the short-lived mother. While admitting that the evidence was less than conclusive, the authors thought that these bees illustrated the 'semisocial' link in Michener's system. But whether or not the mother now lives long enough to receive help from her daughters, from sequences in the other known Augochlorini it seems quite as likely that her ancestors once did so, as that the described pattern is itself ancestral to the others.

Similar reservations apply in the Xylocopinae to the 'allodapine' bees.[13] Some have not quite crossed the threshold to eusocial life, some linger just the other side with cycles rather similar to *Pseudoaugochloropsis*, apart from a short overlap between mother and daughters, and some may even have crossed the threshold and then withdrawn.[106] Meanwhile, an unusual number of independent lines seem to have taken advantage of the confusion by becoming parasitic.[107]

The social patterns common in other halictine groups contrast interestingly with those in the Augochlorini.[108] In the *Agapostemon* group,[104,109] entrance-sharing is known, but not association, and on the evidence it is unlikely that the sharing bees are closely related. Workers have never been found. In the *Halictus–Lasioglossum* group, on the other hand, there are some cases of association.[110] A report on one case suggests Michener's semisocial stage. Batra[111] found that the females of *H. rubicundus* associated in nest-founding and that the daughters did not remain in the nests where their mothers were still alive. However, the data concern only three nests in artificial rearing rooms. As a whole, the *Halictus–Lasioglossum* group has less association than the Augochlorini. Instead, the social species show greater emphasis on the matrifilial colony. In some species such colonies reach the highest social levels known in the Halictinae, being of elaborate architecture, populous, with marked caste differentiation, and practising brood care. Social advances achieved in this group suggest that association may be a diversion on the road to populous, complex colonies. Moreover, theory suggests that if vagile usurpers follow on the heels of the auxiliaries[110] (see Chapter 2, Part II, page 68), association may be a permanent check or even a step on the return road towards the solitary condition.

Regarding condition (4) above, Michener has pointed out that it seems never to be achieved on the basis of colony growth rates alone, so that the presence of workers must be particularly important for the survival of colonies.[112] His observation suggested the importance of tunnel-guarding in ground-nesting bees. In theory, co-operative tunnel-guarding does not seem especially likely to lead to worker-like reproductive differentiation and could exist (fulfilling condition 4) even if the nest-sharers were not related; nevertheless, guarding is evidently a very important function of workers in all social insects, incipient or otherwise.

Data showing the effect of foundress association on colony growth rate in *Mischocyttarus* and *Polistes* show, like the data examined by Michener, productivity increasing too slowly to justify reproductive altruism on this basis alone, although the latest (and best) data were the least adverse.[5,113–115] Censuses exclude the failures, and when we take into account the high mortality of young nests, which certainly falls most heavily on the lone-founded ones,[116] the association is more understandable. A further point, relevant at least to *Polistes*, is that the associating queens differ in reserves and other endowments so that one cannot assume that all would have been equally successful in founding nests alone. The less endowed females should thus readily become the auxiliaries.[117]

If males as well as females usually remain near and mate near their nest of origin, we have an 'island' or 'stepping-stone' population structure,[118] which ensures at least some relatedness between all members of an aggregation. This could help the association. Apart from homing, feeble flight by itself will tend to bring about local genetic homogeneity. It is worth suggesting that the very powerful flight characteristic of Sphecoidea, Euglossini, Anthophorini, etc.,[119] may partly explain why these groups have failed to cross the bridge into social life—with few exceptions (*Microstigmus comes*,[120] *Exomalopsis*,[121,122] and, most dubiously, *Eulaema*.[101,102,123,124] (Tsuneki[125] has observed what seems a puzzling mixture of communistic and aggressive behaviour in a gregarious sphecoid. He found the flight range to be very short, although Fabre for the same genus (*Cerceris*) had recorded homing from 2 and 3 km; for another subsocial sphecoid see Eberhard.[126])

Suppose there is no tendency within an aggregation for associating females to be sisters. How much inbreeding is necessary, how restricted must migration between aggregations be, for a given level of general relatedness to be achieved? As a step towards understanding the possibilities for an 'island' population structure, I have considered a model in which males only are allowed to migrate between islands. Figure 8.4 shows for each population size N the fractions s of males that must mate within their natal island in order to maintain the average relatedness of females at the rather high level of 0.75. N is the total number of sexual forms, and it is assumed, rather artificially, that the two sexes are produced in equal numbers. It can be seen that the permitted fraction of outbreeding which allows such a high level of relatedness to be maintained falls rapidly at first and then more slowly. With island populations of five (2.5 females), only 24 per cent of matings can be with outside males; with islands twice this size (five females), only 9 per cent of matings can be outbred; with 50 females per island, the limit is just less than 1 per cent. Corresponding restrictions in a stepping-stone model would certainly be less severe, but no analysis has been worked out.

It should be noted that, although all the states represented in Fig. 8.3 have associating females of the same generation as closely related as full sisters under outbreeding, mother–daughter relatedness is rising along with F as N increases ($B_{mother\text{–}daughter}$ approaches the limit 7/8 as F approaches the limit 3/5).

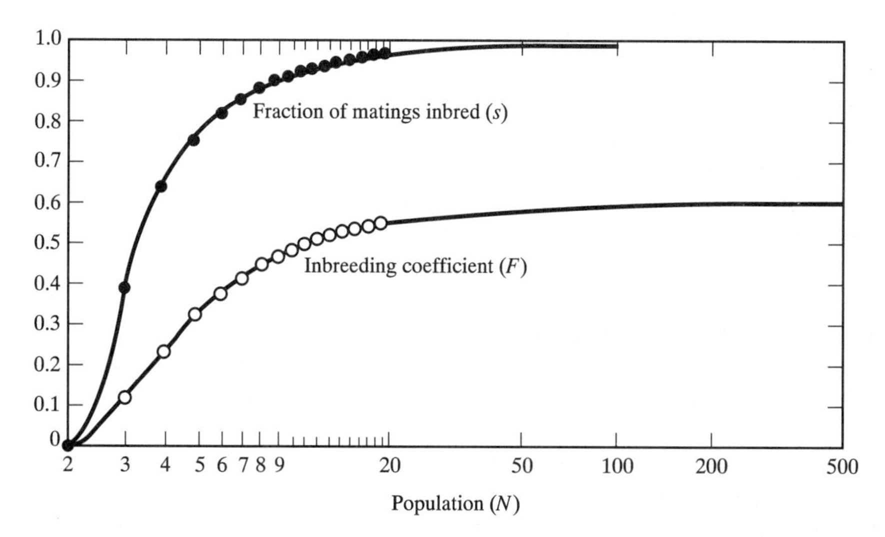

Figure 8.4 The fraction of inbred matings necessary to maintain the average relatedness of females at 0.75 in an 'island' population structure. There are $N/2$ males and $N/2$ females per island deme. Only males migrate; in each deme $(1-s)N/2$ males are replaced by random immigrants in each generation. Within demes mating is random. The mean coefficient of inbreeding F (open circles) is also shown. Population is shown on a logarithmic scale

It only happens that females are more related to their female contemporaries than they are to daughters in groups having fewer than three females. Thus, inbreeding may help harmony and altruism in the colony, especially as regards the behaviour of (and the treatment accorded to) the males, but it does not especially encourage worker sterility. Unfortunately, the question of whether colonies of parasocial bees do ever approximate the degrees of endogamy indicated in Fig. 8.4 cannot be answered with present data.

Association and the briefly outlined model lead to the problem of permanent polygyny. Polygyny (multiple queens) occurs very widely in wasps and ants. In ants[55] (and even more in bees[11,13]) its occurrence is sporadic, but in the major group of social wasps, Polistinae, it is almost universal. The populations of small open nests in the genera *Polistes, Mischocyttarus, Ropalidia*, and *Belonogaster* are usually monogynic. (In *Belonogaster*[14] and in Stenogastrinae[127] it is still unsettled whether there are females which are unattractive and cannot mate; but some certainly function as workers.) Apart from these, all nest populations analysed have shown multiple egg-layers and most egg-layers carry sperm. In my opinion the polygyny in Polybiini (for example), and its contrast with organization of *Polistes* and Vespinae, provides the most testing difficulty for the interpretation of the social insect pattern which is offered in this review.

Unless there is a very high degree of inbreeding, why does not intracolony selection for queen-like behaviour break down the system? Why do workers work so willingly and by what device are the fierce struggles for dominance that occur, for example, in queenless *Apis*[92] and *Vespa*[89] colonies prevented? If inbreeding is the answer, would we not expect more genetic diversity between colonies, relative to uniformity within colonies, than we actually observe? These questions cannot be answered yet. Richards and Richards[100] suggested that the frequency of attacks by ants gave special advantage to polygyny in tropical wasps. When rebuilding after an ant attack, it will add greatly to the productivity of a colony if it can be quickly restocked with young larvae so as to give full employment to the worker force. Both queens and workers easily escape from the ants by flying, but, of course, a single physogastric queen could not do so and would be destroyed. It is easy to see the advantage of polygyny to the colony in this situation, but difficult to see how it is regulated so as to prevent strife and parasitic tendencies. Richards and Richards did not discuss this problem. To add to it, in nests that I have been able to watch, dominance behaviour is less in evidence than it is in nests of *Polistes*. As yet, moreover, no parasite species has been found in the Polybiini, whereas parasites have evolved in European *Polistes* and several times in northern Vespinae (the evidence supporting the claim that many parasite species exist in tropical *Mischocyttarus*[128] seems too tenuous to be accepted at present).

There seems to be some connection between social parasitism and latitude. This applies to ants[12,129,130] as well as to bees and wasps, but one has to repeat the caution that the social insects of the tropics have been much less studied. Richards[131] outlined a plausible course for the divergence of the non-social cuckoo bees, *Psithyrus*, from a social industrious group ancestral to their present bumblebee hosts. He used as illustration the sporadic parasitism practised by the more southerly *Bombus terrestis* on its more northerly sibling *B. lucorum* in Britain. A parallel illustration, with apparently greater dependence on the host, has since been found in a different subgenus of *Bombus* in the high arctic.[132] It seems that the most northerly of all social insects has a social parasite! The migratory habits which tend to be favoured in severe climates must help to bring together different subspecies. Also, in a single continuous population which is mixed by migration, the gene causing parasitic behaviour will have a better chance of establishment since it will not, as in a more viscous population, tend to destroy its own replicas in relatives. There are, of course, also the usual difficulties regarding the full emergence of a parasitic *species* in a single interbreeding population. (The existence of many quasi-isolated populations together with the factor of altitude providing discordance in the seasonal timing of reproductive activities (as required in Richards's hypothesis) may help to explain why the Alps are so prolific in social parasites. There seems to be yet another example in *Bombus*;[133] parasitic *Sulcopolistes* is centered here;[134] and there are many independent examples in ants.[135]) Definite migrations have been recorded in *Polistes fuscatus*[136] and in *Vespula rufa*,[137] and there are occasional records of migrating queens of other wasps and of *Bombus*.[138] These facts hint at a connection between migration and both (1) the mutual intolerance of queens which precludes polygyny[139] and (2) the emergence of social parasites. However, the evidence from *Polistes* is not so clearly favourable on this score as when I attempted to review it in 1963: the evidence still favours a correlation between associative nest founding and low latitude in *Polistes*, but there are certainly exceptions. *P. subsericeus* and *P. niger* in Brazil appear to found nests alone, whereas *P. fuscatus* has some association even in the northern part of its range in the USA. Moreover, *P. fuscatus* sometimes migrates and has no known social parasite. However, probably most species of *Polistes* are variable in the occurrence of foundress associations.[140,141] Nevertheless, the evidence that when *Polistes* queens associate they are usually sisters is now strong.[115,142]

The known facts of association in ants[143–145] do not contradict the view that relatedness, or lack of it, is an important factor (see also Chapter 2, pp 65 and 72). For example, Poldi, as reported by Brian,[144] found for *Tetramorium caespitum* that only sister queens would associate without fighting. These queens remained amicable even after workers had been produced. Whether association ever passes on into permanent polygyny when the queens are derived from a massive mating flight seems debatable. A *Polistes*-like phenomenon in which one queen dominates her associates and causes regression of their ovaries is known in *Leptothorax gredleri*. It would be interesting to know how the queens normally come together (and if possible their relatedness in specific instances) in this and other species of *Leptothorax*, and also in *Myrmecina graminicola*, where the phenomenon is more variable and may overlap with permanent polygyny.[143,145]

Colony reproduction by swarming or fission is known for many polygynous species. This mode tends to result in loosely connected supercolonies,[146,147] and the existence of these can, perhaps, account for the odd cases in which colonies are observed to 'adopt' strange queens after a mating flight.[148] Perhaps these queens are usually returning to their own tribe. Combining fact and plausible speculation, Table 8.3 shows features of life history and behaviour which seem to show intelligible correlation with polygyny in ants.

Some examples of incipient divergence in *Myrmica*, with at least some characters as shown in Table 8.3, are given by Brian,[149] who also suggests a connection with type of habitat (open, fugitive; closed, climax). *Monomorium pharaonis*, another polygynic myrmicine, also fits in well.[150] The female-biased sex ratio[152] confirms that inbreeding is usual in this species. However, many other well-known ants fit less well than these. Probably the least-supported correlation concerns the source of males. *Myrmica* fails here,[153] as does the *rufa* group of *Formica*, and there are probably many monogynic and oligogynic genera like *Atta* and *Solenopsis* that have completely sterile workers. For the extensive, polydomous, polygynic colonies of *Iridomyrmex humilis*, one report[154] records such puzzling features as: (1) a mating flight for which males are produced in great excess; (2) in spite of this flight, queens added to the colony are mainly ones mated and dealated within the nest; and (3) queen massacres at certain periods. The first feature suggests vigorous outbreeding with males serving as the main vehicle for gene dispersion. The third suggests that the polygyny is not fully stable, which is understandable if the species does not inbreed—if, for example, some of the flown queens are readopted and some retained queens outcrossed to adventive males entering the nests. An equally puzzling situation holds for *Rhytidoponera*. Here Haskins and Whelden[155] found that queens are readopted. At the same time, they give evidence not only of habits that would normally lead to outbreeding, but also of genetical intolerance to close inbreeding. The queens showed no reluctance to mate twice

Table 8.3

	Monogyny	**Polygyny**
Mating flight	Massive, synchronous	Local, or mating on or in the nest
Founding	Lone queens or temporary association	Swarming or temporary parasitism
Hostility	All colonies hostile; strange queens killed[151]	Hostility only between supercolonies; strange queens sometimes accepted
Males	Ex-worker eggs, abundant	Ex-queen eggs, less numerous

and their spermathecas were usually less than completely filled. These facts (as the authors say) make it unlikely that the average relatedness of females within a colony is as high as that of normal sibs. Yet the colonies are cohesive and the species abundant. The swarming of polygynic species is, of course, against gene dispersion, but the dispersion of the winged male is presumably more important in fixing the level of inbreeding.

Haskins (personal communication) has observed two budded colonies of *R. metallica* derived from a flourishing captive colony. Each consisted of just seven females, of which only one was able to produce workers. If typical, such initial monogyny will help greatly to keep up the level of relatedness within colonies. If there are several queens in a swarm, it helps in the same direction (1) if they are sisters, (2) if they or their descendant queens differentiate their roles in such a way that most of them produce only worker brood. Of course, if (2) holds, (1) tends to follow. There is little evidence, but a hint of (2) appeared unexpectedly in the course of Brian's intensive studies on reproduction in *Myrmica*.[156] Cases where some apparent queens are actually non-functional have already been mentioned.

As in many other evolutionary lines in ants, the road to polygyny in *Rhytidoponera* lies through the sexualization of ergatoid forms accompanied by gradual failure to produce the perfect alate female. Early stages on this road are suggested by the occasional examples of wingless microgynes in normal colonies. Buschinger[145] saw a fight between the queen and a large 'worker' in a captive colony of *Myrmecina graminicola*. The following day this 'worker' was dragged out of the nest, apparently injured, by two workers. When dissected she proved to have developed ovaries and had been inseminated. Kutter[157] found a crop of microgynes in one nest of *Leptothorax acervorum*. Apart from their lack of wings, their morphology was very similar to that of a winged, workerless parasite species which he had discovered in *L. acervorum* nests in the same area. These facts suggest that winged microgynes, with their ability to infect unrelated colonies, may evolve into obligate parasites; whereas wingless microgynes, whilst originally 'selfish' and detrimental to colonies, may come to form the basis of a new and efficient polygynic system. Nevertheless, the stem of formicid evolution seems to lie through forms with winged reproductives in both sexes, and from this point of view the trend in genera like *Rhytidoponera* is degenerate and is probably headed for ultimate extinction.

The idea of stabilization of an originally 'selfish', sexualized worker suggests a possible modification of Kerr's explanation of the evolution of genetic caste determination in *Melipona*. Kerr[158] suggested a succession of heterotic mutations which made the heterozygote more sexual than either homozygote. But if originally the mutation sexualized its homozygotes as well, if the workers so sexualized could pass on their genes by producing sons (as is now known to occur in Meliponini), and if the population was outbreeding (known by genetic evidence for *Melipona marginata*[11]), the mutant gene would spread and reach an equilibrium even though its spread reduced the mean fitness of colonies. The check to its spread would come from the inefficiency or outright failure of colonies producing too many sexualized workers. As a queen, a mutant homozygote would be liable to the disaster of producing no proper workers, so presumably there would be selection of modifiers diminishing the sexual attractiveness and 'ambition' of the homozygous mutant. All this involves little change from Kerr's original hypothesis. The main difference is the implication that each episode of selection might reduce mean colony fitness and hence reduce, at least temporarily, the density of the species. A genetic implication, not easily testable, is that the males and the genetic workers will have only one of Kerr's postulated alleles at each locus, the old allele being retained only in queens.

Apart from the recent discovery of the importance of laying workers in male production, the extraordinary ritual that surrounds oviposition by the queen in the Meliponini[159] seemingly records an ancient struggle over egg-laying and oophagy. One is tempted to suggest that it stands in the same relation to the genuine struggle in *Bombus*[122,160] as does a civilized wedding ceremony to 'marriage by capture', although, of course, the human ritual has a quite different kind of hereditary basis. Another curious feature in this tribe is the parasite-like modification of the queen.[161] It is extremely unlikely that all Meliponini are descended from a robber-parasite social group like *Lestrimelitta*, so probably these characters are also connected with the aforementioned struggle and have the same sort of explanation as the curved sting of the *Apis* queen.

Bumblebees in South America are evidently newcomers to the tropical environment, compared to most of the Polybiini. Correspondingly, they have retained lone founding of nests, but their colonies are potentially perennial by supersedure of queens, and in the process of supersedure there is a phase during which sister queens cohabit after the death of their mother.[11] Perhaps this is the beginning of polygyny. The genus *Vespa*, thought by Van der Vecht[162] to have originated in subtropical Asia, seems similarly to have retained lone-founding in tropical Malaysia and Indonesia.[163] Complexities of the geographic situation in *Polistes* have already been touched on. The history of these genus may be like that of *Bombus*, but with an earlier and more tropical origin. A disproportionate amount of speciation has occurred in South America, even though the most primitive species are in Southeast Asia.[162] Again, true polygyny hardly occurs, if at all.

Returning to ants, there is one good morphological indicator of inbreeding. This is brachyptery or aptery of males. The phenomenon is rare in ants, rarer still in bees, and absent, so far as we know, in wasps. If after mating with such flightless males in or on the nest some females remain at home and become egg layers, the colony develops or maintains a polygyny with high intrarelatedness of its gynes, and such polygyny should be very harmonious. I am not able to review the known facts on wingless-male species in the Formicidae. However, a review of rare ant species by Wilson[164] indicates that there may be a correlation between polygyny and inbreeding, with wingless males often involved. Great interest attaches to genera such as *Cardiocondyla*,[165] *Plagiolepis*,[166] and *Hypoponera*,[167,168] in which species with winged males can be compared with closely related species with wingless males. Ant ancestors are thought to have become social while living as predators in rotting wood.[169] Inbreeding is common in this habitat and thus it is easier to believe for ants than it is for bees that the polygyny of primitive species in Myrmeciinae and Ponerinae is itself primitive. There is no evidence that ancestral males were wingless. Bethylidae and other primitive vespoids seem to suggest the opposite—that if one sex alone was wingless it was the female. But the point is equivocal, for these are the groups that failed to become social.

Parasitic ants with wingless males such as *Formicoxenus nitidulus, Plagiolepis xene*, and *P. grassei*, which are polygynic, point toward more extreme workerless forms such as *Wheeleriella, Anergates*, etc. Although some other extreme parasites have winged males, aptery is certainly far more common among these parasites than in ants as a whole, and also much more common than in less extreme parasites. In wing condition, mating habits, sex ratio, and dependant mode of life, such ants have come to resemble Hymenoptera Parasitica more than other ants. Apart from dangers of the dispersion flight (which would be much greater for a rare species if the flight was also a mating flight as in other ants), such species lead a sheltered life in a very stable environment, so the sacrifice of outbreeding is not immediately detrimental. Nevertheless, the rareness of such species may illustrate the long-term peril of such sacrifice.

TOLERATION AND CO-OPERATION IN THE ABSENCE OF KINSHIP

Intraspecific cases

Anergates and *Wheeleriella* eliminate the host queen; thus it is surprising if their colonies have time to proceed into polygyny. There is no evidence that *Wheeleriella* does so, but *Anergates* has been found with up to four physogastric queens in a single nest[176] and, although it seems to be more usual for there to be only one,[171] this raises a problem. One would expect this 'polygyny' to be tolerated neither by the workers nor by the parasites themselves. Although their situation is probably less exigent than that of larvae of 'solitary' Hymenopterous parasitoids which always fight to the death when deposited together in a host,[77,78] associating queens of *Anergates* are unlikely to be related and each certainly expects a lower fertility in the presence of the other. Presumably, the advantage of hybrid vigour and recombination in descendants outweighs this disadvantage. The same excuse serves for the same difficulty with *Melittobia* and *Sclerodermus*, parasitoids with rather ant-like habits which also associate peaceably on their hosts. (*Sclerodermus* females are said to help their sisters to eclose, but perhaps they just want to eat the coccoons.[172] *Melittobia chalybii* has a dimorphism in the female sex, although neither form is worker-like.[84] As already mentioned, *Melittobia* males are strikingly unsocial.) The obscure benefits of outbreeding may also account for the almost peaceable association of females of the mite *Dicrocheles phalaenodectes* in the ears of noctuid moths.[173] Young fertilized female mites are dispersed from moth to moth by way of the flowers visited. Thus, females colonizing any particular moth are very likely to be unrelated. The sex ratio[174] suggests that association may be uncommon in nature.

In *Dicrocheles* a much more striking and positive social habit also deserves mention. This is the refusal of colonizing females to parasitize both ears of a moth. After the first mite has selected and entered an ear, subsequent colonists always go to the same one. It appears that if that ear is already overcrowded the latecomers will abandon the host they have laboriously boarded and explored rather than set up home in the vacant ear. The reason is obviously connected with the fact that an ear cannot function for the moth and house mites at the same time. A moth's ears are tuned to the sonar pulses by which bats find their prey; a deaf moth cannot take evasive action. The restrained parasitism of the mites suggests that piercing one ear puts the moth at only a mild disadvantage, whereas piercing the second ear brings almost immediate nemesis to moth, piercer, and all other inhabitants.

That predation by bats can impose this remarkable discipline on the behaviour of unrelated mites strengthens the suggestion that predation by ants on polybiine wasps can, as the Richards believed, enforce a mix of queens and workers which enables an ant-shattered colony to start again with near optimal efficiency. It likewise strengthens Michener's case that parasite pressure on ground-nesting bees can induce at least some co-operative guarding by unrelated females and perhaps also co-operative provisioning if this reduces the time that cells are left unattended.

The situation in the last case might be roughly expressed: 'If you guard my nest while I forage, I'll guard yours while you forage.' Taken at all literally this certainly supposes too much of the intelligence of bees. Indeed, if the situation really required a threat of change in behaviour if the partner is observed not to reciprocate, guard behaviour probably could not have arisen in any insect. Actually, in blocking the shared entrance of a system the bees guard their own cells as well as those of the others, which must greatly aid the evolution of the habit. Moreover, the guard behaviour observed does not suggest strict alternation of duties.[74] However, the above anthropomorphism (while still fanciful of course) would not suppose too much of the intelligence of some higher animals and it serves to introduce the subject of reciprocal altruism.

Reciprocal altruism, as defined by Trivers,[175] has no need of relatedness. It requires instead that interactants should remain together long enough for their roles as donor and recipient to reverse several times, and also that they should be endowed with flexible behaviour that curtails further benefits to individuals which are observed not to reciprocate. Evidence for reciprocal altruism is covered in Trivers's review. Regarding the intraspecific cases it here suffices to make two points. First, since 'remaining together' often implies population viscosity, it may be difficult to find cases that do not involve elements due to inclusive fitness as well. Second, there is most scope for this kind of selection (and for the ramifying complications which Trivers has outlined) in perceptive and intelligent animals. This points to primates and especially to humans.

Interspecific cases: symbiosis

For interactions between different species the factor of kinship is excluded. Trivers has discussed some examples, notably cleaning symbioses, in which an exploitative act analogous to non-reciprocation could give immediate benefit. He argues that, as with the intraspecific cases, only when the same pair of individuals normally associate for some time does selection establish an inhibition against the exploitative act (e.g. an inhibition on the part of a predatory fish against eating its cleaner at the end of servicing). This argument leads to an interesting point concerning those more intimate symbioses where benefit (or possibly harm) is traded almost continuously.

Drawing an analogy between a clone of parasitic bacteria and a long-lived organism, the argument suggests that bacteria should most readily evolve benign relations with their hosts where it is members of the same clone that suffer most if the host (or local host population) is killed by the infection (or epidemic). Obviously the mixing of clones will depend, among other things, on the mobility and mixing of the hosts. Mobile hosts help a virulent pathogen to keep on the move, away from areas it has already depleted, and also help to emphasize competition between (rather than within) strains of the pathogen. Thus, in a fluid host population we expect pathogens to remain virulent and symbiosis to be unstable. In a viscous one, on the contrary, pathogens may evolve into symbionts. It would be inappropriate to try to survey evidence in general for this proposed correlation; instead, let us briefly consider the arthropod 'symphiles' of social insects as possible pathogens or symbionts of their host colonies.

We need to consider the dispersive ability of both the symphile and its host. We expect to find that *Apis*, with highly mobile colonies, has a smaller proportion of its symphiles harmless or beneficial than have most termites and ants, especially those in which colony reproduction is mainly by budding. And we expect to find that harmless and beneficial symphiles are more often flightless than those which are harmful. The subject is very complicated, but broadly these expectations may hold. The brief list of symphiles of *Apis*, for example, includes very few that could even be suspected of benefiting the host colony. With ants and termites on the other hand there are some definitely benign relations and many that are undecided. In some it has become clear that the hosts are dupes of chemical and behavioural trickery by very injurious parasites.[15] Often, however, there is no evidence that substances proffered by the symphiles to their hosts are injurious—they could as well be vitamins as addictive drugs. A brenthid beetle living with ants has been shown both to solicit and to return liquid food.[176] It is tempting to link this with another discovery, that adult wasps are dependant on their larvae for the digestion of proteins into sugars.[177] If adult wasps shut down parts of their enzymatic machinery on such basis as this, and if the secretions of certain symphiles are available almost as constantly as those of larvae, it seems possible that a social insect species may sometimes have entered an arrangement with a symphile similar to that existing between the wasp and its larva.

With the homopteran associates of ants, it was probably the ants that moved from a hunting and exploitative to a dairying and conservative emphasis,[144] rather as with humans and their cattle. There is no doubt as to mutual benefits here. A change of emphasis of the same sort must have occurred as ants moved into the ant-plants. Now some ants are entirely dependent on them, and in one case at least it has been proved that the plants are almost equally dependant on protection by the ants.[178]

A puzzling problem of myrmecophily is raised by the wasp–ant associations involving the dolichoderine genus *Azteca*. A tree bearing a large colony of the ants usually has several nests of polybiine wasps as well, and as many as eight species of wasp have been found in a single tree. Some of these species (e.g. *Synoeca virginea, Polybia sulcata*, and a species of *Clypearia*) I have never found nesting away from the ants. Some of the wasps (e.g. *Polybia rejecta*) are very fierce and the combination with the ants has obvious advantages for defence. However, on the whole, the wasps associating with *Azteca* are not notably fierce and some are timid. Some (e.g. *Mischocytarrus ater*) may use an ant repellent for protection,[179] but with others the ants are often seen running on the associated nests. *Polybia rejecta* workers seem to regard ants on the nest with apprehension; they try to knock them off with sharp pecks, but fall back if the ants appear in numbers. Several scraps of evidence suggests that the ants allow these various wasps to use their protection, but occasionally turn on them and treat them as a reserve food supply. For example, whereas a species of *Stelopolybia* (close to *S. fulvofasciata*) is often found peacefully nesting in a cavity in the hanging carton nest of an *Azteca*, I have seen fierce fighting between the two species around such a cavity, with dismembered wasps being carried by the ants into their nest. Perhaps similar relations may hold for some inter-ant associations.

Predictions based on inclusive fitness alone have also proved inadequate in cases of mixed colonies and of trophallactic exchanges between related species, especially in *Formica*.[180–182] These cases must represent either very subtle forms of parasitism or forms of symbiosis. Two remote analogies suggest themselves here. One connects this problem with another, equally unsolved, concerning mixed colonies of cellular slime molds (Acrasiales),[183] and this in turn links with that of the evolutionary step which is fundamental to all eukaryotes: the disciplined co-operation of the two genomes of the diploid cell. The other analogy suggests that we might expect to find elements resembling those which have liberated humans from a purely tribal outlook and state of society. Of these, some form of barter seems the most possible for social insects. But this analogy also suggests that the above 'either ... or ... ' may be too simple.

References and notes

1. C. Darwin, *The Origin of Species by Natural Selection*, 6th edn (Murray, London, 1882).
2. C. Darwin, *The Descent of Man and Selection in Relation to Sex*, 2nd edn (Murray, London, 1874).
3. A. Weismann, The all-sufficiency of natural selection, *Contemporary Reviews* **64**, 309–38 (1893).
4. A. H. Sturtevant, Essays on evolution II. On the effects of selection on social insects, *Quarterly Review of Biology* **13**, 74–6 (1938).
5. P. Rau, Cooperative nest-founding by the wasp, *Polistes annularis, Annals of the Entomological Society of America* **33**, 617–20 (1940).
6. G. C. Williams and D. C. Williams, Natural selection of individually social adaptations among sibs with special reference to social insects, *Evolution* **11**, 32–9 (1957).
7. R. Levins, Extinction, in *Some Mathematical Questions in Biology*, pp. 77–107 (American Mathematical Society, Providence, Rhode Island, 1970).
8. H. R. Hermann, Sting autonomy, a defensive mechanism in certain social Hymenoptera, *Insectes Sociaux* **18**, 111–20 (1971).
9. H. Spencer, A rejoinder to Professor Weismann, *Contemporary Reviews* **64**, 893–912 (1893).
10. C. Plateaux-Quénu, Un nouveau type de société d'insectes: *Halictus marginatus Brullé* (Hymenoptera, Apoidea), *Année de Biologie* **35**, 327–444 (1959).
11. W. E. Kerr, Some aspects of the evolution of social bees, *Evolutionary Biology* **3**, 119–75 (1969).
12. M. W. Wheeler, *The Social Insects* (Harcourt Brace, New York, 1928).
13. C. D. Michener, Comparative social behaviour of bees, *Annual Review of Entomology* **14**, 299–342 (1969).
14. O. W. Richards, The biology of the social wasps (Hymenoptera, Vespidae), *Biological Reviews* **46**, 483–528 (1971).

15. E. O. Wilson, *The Insect Societies* (Belknap Press/Harvard University Press, Cambridge, MA, 1971).
16. H. Markl, Vom Eigennutz des Uneigennützigen: die Evolution hochentwickelter Sozialsysteme in Tierreich, *Naturwissenschaften* **24**, 281–9 (1971).
17. C. C. Li, *Population Genetics* (University of Chicago Press, Chicago, 1955).
18. S. Wright, The interpretation of population structure by F-statistics with special regard to systems of mating, *Evolution* **19**, 395–420 (1965).
19. C. R. Ribbands, The role of the recognition of comrades in the defence of social insect communities, *Symposia of the Zoological Society of London* **14**, 159–68 (1965).
20. J. J. C. Nel, Aggressive behaviour of the harvester termites *Hodotermes mossambicus* (Hagen) and *Trinervitermes trinervoides* (Sjostedt), *Insectes Sociaux* **15**, 145–56 (1968).
21. A. L. Pickens, in C. A. Kofoid (ed.), *Termites and Termite Control*, Ch. 14, 15 (University of California Press, Berkeley, 1946).
22. E. O. Wilson, Behaviour of social insects, in P. T. Haskell (ed.), *Insect Behaviour*, pp. 81–96 (Royal Entomological Society, London, 1966).
23. To measure F, the mean coefficient of inbreeding or fixation index, a sample of the population classified for neutral genotypes is required. If A, H, and B are the numbers of genotypes *aa*, *ab*, and *bb*, then an estimate is

$$F = \frac{4AB - H^2}{(2A + H)(H + 2B)}.$$

Li (ref. 17) and Wright (ref. 18) may be consulted for the theory of this and more refined measures of inbreeding.
24. R. Alexander and D. Otte, The evolution of genitalia and mating behaviour in crickets (Gryllidae) and other Orthoptera, *Miscellaneous Publications of the Museum of Zoology, University of Michigan* **133**, 1–62 (1967).
25. M. J. West and R. D. Alexander, Sub-social behaviour in a burrowing cricket *Anurogyllus muticus* (De Geer). Orthoptera: Gryllidae, *Ohio Journal of Science* **63**, 10–24 (1963).
26. P. K. Anderson, Ecological structure and gene flow in small mammals, *Symposia of the Zoological Society of London* **26**, 299–325 (1970).
27. R. D. Anderson, Aggressiveness, territoriality and sexual behaviour in field crickets (Orthoptera: Gryllidae), *Behaviour* **17**, 130–223 (1961).
28. H. v. Lengerken, *Die Brutsfursorge- und Brutspflegeinstinkte der Käfer*, 2nd edn (Akademische Verlagsgesellschaft, Leipzig, 1954).
29. V. E. Kullmann, Social Phaenomène bei Spinnen, *Insectes Sociaux* **15**, 289–97 (1968).
30. W. A. Shear, The evolution of social phenomena in spiders, *Bulletin of the British Arachnological Society* **1**, 65–76 (1970).
31. B. Krafft, Contribution à la biologie et à l'éthologie d'*Agelena consociata* Denis (araignée sociale du Gabon), Part 3, *Biologica Gabonica* **7**, 3–56 (1971).
32. Ibid., Parts 1 and 2, **6**, 179–369 (1970).

33. A. Sayler and S. Salmon, An ethological analysis of communal nursing by the house mouse (*Mus musculus*), *Behaviour* **40**, 62–85 (1971).

34. W. Kühme, Freilandstudien zur Sociologie des Hyanenhundes (*Lycaon pictus lupinus* Thomas 1902), *Zeitschrift für Tierpsychologie* **22**, 495–541 (1964).

35. F. A. Beach, Maternal behaviour in males of various species, *Science* **157**, 1591 (1967).

36. H. Kruuk, *The Spotted Hyena* (Chicago University Press, Chicago 1972).

37. H. v. Lawick-Goodall and J. v. Lawick-Goodall, *Innocent Killers* (Collins, London, 1971).

38. K. Lorenz, *On Aggression* (Methuen, London, 1966).

39. L. D. Mech, *The Wolf: The Ecology and Behavior of an Endangered Species* (The Natural History Press, New York, 1972).

40. A. Watson, Territorial and reproductive behaviour of red grouse, *Journal of Reproduction and Fertility*, Suppl. **11**, 3–14 (1970).

41. J. H. Crook, The adaptive significance of avian social organisations, *Symposia of the Zoological Society of London* **14**, 181–218 (1965).

42. C. H. Fry, The social organisation of bee-eaters (Meropidae) and co-operative breeding in hot-climate birds, *Ibis* **114**, 1–14 (1972).

43. C. R. Watts and A. W. Stokes, The social order of turkeys, *Scientific American* **224** (6), 112–18 (1971).

44. S. Ohno, *Sex Chromosomes and Sex-linked Genes* (Springer-Verlag, Berlin, Heidelberg, New York, 1967).

45. Other evidence of the power of sex chromosomes to cause special social and sex-ratio effects might be searched for in spiders. Most spiders have an X_1X_2O sex mechanism and a fairly low haploid number of autosomes (about 12). But as in Hymenoptera the habits of the sexes tend to be very different even in solitary species.

46. D. Lack, *Ecological Adaptations for Breeding in Birds* (Methuen, London, 1968).

47. R. C. Punnett, Inheritance of egg colour in the 'parasitic' cuckoo. *Nature* **132**, 892–3 (1933).

48. R. H. Crozier, Coefficients of relationship and the identity of genes by descent in the Hymenoptera, *American Naturalist*, **104**, 216–17 (1970).

49. G. C. Williams, *Group Selection* (Aldine-Atherton, Chicago, 1971).

50. See page 273 and Wright 1965 (ref. 18).

51. C. Plateaux-Quénu, Un nouveau type de société d'insectes: *Halictus marginatus* Brullé (Hymenoptera, Apoidea), *Année de Biologie* **35**, 327–444 (1959).

52. F. Bernard, *Les Fourmis (Hymenoptera, Formicidae) d'Europe occidentale et septentrionale* (Masson, Paris, 1968); H. St J. K. Donisthorpe, *British Ants, their Life-history and Classification* (Routledge, London, 1927); L. Passera, Interactions et fecondité des reines de *Plagiolepis pygmaea* Latr. et de ses parasites sociaux *P. grassei* Le Masne et Passera, et *P. xene* St. (Hymenoptera, Formicidae), *Insectes Sociaux* **16**, 179–93 (1969).

53. W. E. Kerr, R. Zucchi, J. T. Nakadaira, and J. E. Butola, Reproduction in the social bees (Hymenoptera: Apidae), *Journal of the NY Entomological Society* **70**, 265–76 (1962).

54. For *Apis* and insects in general see G. Parker, Sperm competition and its evolutionary consequences in the insects, *Biological Reviews* **45**, 525–68 (1970).

55. E. O. Wilson, The social biology of ants, *Annual Review of Entomology* **8**, 345–68 (1963).

56. H. V. Daly, Biological studies on *Ceratina dallatorreana*, an alien bee in California which reproduces by parthenogenesis (Hymenoptera: Apoidea), *Annals of the Entomological Society of America* **59**, 1138–54 (1966).

57. R. H. Anderson, The laying worker in the Cape honeybee, *Apis mellifera capensis*, *Journal of Agricultural Research* **2**, 85–92 (1963).

58. P. E. Howse, *Termites: A Study in Social Behaviour* (Hutchinson, London, 1970).

59. P. F. Entwistle, Inbreeding and arrhenotoky in the ambrosia beetle *Xyleborus compactus* (Eichh.) (Coleoptera: Scolytidae), *Proceedings of the Royal Entomological Society of London* A, **39**, 83–8 (1964).

60. K. Takagi and T. Kaneko, Biology of some scolytid ambrosia beetles attacking tea plants. V. Chromosome numbers and sex determination of tea root borer *Xyleborus germanus* Blandford, *Applied Entomological Zoology* **1**, 29–31 (1966).

61. F. G. Browne, The biology of the Malayan Scolytidae and Platypodidae, *Malayan Forestry Record* **22**, 1–255 (1961).

62. P. Buchner, Endosymbiosestudien an Ipiden. I. Die Gattung *Coccotrypes*, *Zeitschrift für Morphologie und Okologie der Tiere* **50**, 1–80 (1961).

63. E. F. Iton and G. R. Conway, Studies on a wilt disease of cacao at River Estate. III. Some aspects of the biology and habits of *Xyleborus* spp. and their relation to disease transmission, *Cacao Research 1959–1960* (Imperial College of Tropical Agriculture, Trinidad, 1961).

64. R. C. Graves and A. C. F. Graves, The insects and other inhabitants of shelf fungi of the Southern Blue Ridge region of Western North Carolina. IV. The Thysanoptera, *Annals of the Entomological Society of America* **63**, 96–8 (1970).

65. G. D. Morison, Thysanoptera of the London area, *Nature*, Reprint No. 59, 1–131 (1949).

66. L. A. Mound, Gall-forming thrips and allied species (Thysanoptera: Phlaeothripinae) from *Acacia* trees in Australia. *Bulletin of the British Museum (Natural History), Entomology* **25**, 387–466 (1971).

67. F. E. Kurzewski, Comparative behaviour of male digger wasps of the genus *Tachysphex* (Hymenoptera: Sphecidae, Larrinae), *Journal of the Kansas Entomological Society* **39**, 436–53 (1966).

68. B. Hölldobler, Futterverteilung durch Männchen im Ameisenstaat, *Zeitschrift für vergleichende Physiologie* **52**, 430–55 (1966).

69. B. Hölldobler, Zur Frage der Oligogynie bei *Camponotus ligniperda* Latr. und *Camponotus herculeanus* L. (Hymenoptera, Formicidae), *Zeitschrift für angewandte Entomologie* **49**, 337–52 (1962).

70. F. Santschi, Fourmis de Tunisie capturées en 1906, *Revue suisse de Zoologie* **15**, 305–34 (1907).
71. J. T. Medler, Biology of *Trypoxylon* in trap nests in Wisconsin (Hymenoptera, Sphecidae), *American Midland Naturalist* **78**, 344–58 (1967).
72. G. W. Peckham and E. G. Peckham, On the instincts and habits of solitary wasps, *Wisconsin Geology and Natural History Survey Bulletin* **2**, 1–245 (1898).
73. T. F. Houston, Discovery of an apparent male soldier caste in a nest of a halictine bee (Hymenoptera: Halictidae) with notes on the nest, *Australian Journal of Zoology* **18**, 345–51 (1970).
74. C. D. Michener, Notes on the biology and supposed parthenogenesis of halictine bees from the Australian region, *Journal of the Kansas Entomological Society* **33**, 185–96 (1960).
75. S. F. Sakagami, Y. Hirashima, and Y. Ohe, Bionomics of two new Japanese halictine bees, *Journal of the Faculty of Agriculture, Kyushu University* **13**, 673–703 (1966).
76. K. Tsuneki, The biology of East-Asiatic *Cerceris* with special reference to the peculiar social relationship and return to the nest in *Cerceris hortivaga* Kohl., *Etizenia, Fukui* **9**, 1–46 (1965).
77. G. Salt, Competition among insect parasitoids, *Symposia of the Society for Experimental Biology* **15**, 96–119 (1961).
78. 'Solitary' means that a female deposits only one egg per host. When several larvae of a 'solitary' species occur in a host they are normally progeny of unrelated mothers and we expect them to fight, but when a batch arises by polyembryony from a single egg we expect even less agonistic adaptation than in species where such groups are normal siblings.
79. C. P. Clausen, *Entomophagous Insects* (McGraw-Hill, New York, 1940).
80. N. Hokyo, K. Kiritani, F. Nakasugi, and M. Shiga, Comparative biology of two scelionid egg parasites of *Nezara viridula* L. (Hemiptera: Pentatomidae), *Applied Entomological Zoology* **1**, 94–102 (1966).
81. J. P. Woodring, Environmental regulation of andropolymorphism in tyroglyphids (Acari), *Proceedings of the 2nd International Congress of Acarology*, 433–40 (1969).
82. G. S. Graham-Smith, Further observations on the habits and parasites of common flies, *Parasitology* **11**, 347–84 (1919).
83. G. A. Hobbs and M. D. Krunić, Comparative behaviour of three chalcidoid (Hymenoptera) parasites of the alfalfa leaf-cutter bee, *Megachile rotundata, Canadian Entomologist* **103**, 674–85 (1971).
84. R. G. Schmeider, The polymorphic forms of *Melittobia chalybii* Ashm. and the determining factors involved in their production, *Biological Bulletin, Marine Biological Laboratory, Woods Hole* **65**, 338–54 (1933).
85. R. G. Schmeider, The sex ratio in *Melittobia chalybii* Ashmead, gametogenesis and cleavage in females and in haploid males (Hymenoptera; Chalcidoidea), *Biological Bulletin, Marine Biological Laboratory, Woods Hole* **74**, 256–66 (1939).
86. L. Berland, *Atlas des Hyménoptères de France, Belgique, Suisse*, Vol. 1 (N. Boubee, Paris, 1958).

87. H. Montagner, Sur l'origine des mâles dans les sociétés de guêpes du genre *Vespa, Comptes Rendus de l'Académie des Sciences* **263D**, 785–87 (1966).

88. J. P. Spradbery, Seasonal changes in the population structure of wasp colonies (Hymenoptera: Vespidae), *Journal of Animal Ecology* **40**, 501–23 (1971).

89. J. Ishay, Observations sur la biologie de la guêpe orientale *Vespa orientalis* F., *Insectes Sociaux* **11**, 194–206 (1964).

90. In *Vespa crabro* groups of workers in summer leave the maternal nest and build small nests nearby (often in dense grass) where they rear males. My own observations on this are reinforced by those of G. E. J. Nixon and J. Ishay (personal communications).

91. C. Plateaux-Quénu, Tendences évolutives et degré de socialisation chez les Halictinae, *Annales de la Société Entomologique de France* NS **3**, 859–66 (1967).

92. S. F. Sakagami, Occurrence of an aggressive behaviour in queenless hives, with considerations on the social organisation of honeybees, *Insectes Sociaux* **1**, 331–45 (1954).

93. S. W. T. Batra, Behaviour of the social bee *Lasioglossum zephyrum* within the nest (Hymenoptera; Halictidae), *Insectes Sociaux* **11**, 159–85 (1964).

94. S. F. Sakagami and R. Zucchi, Winterverhalten einer neotropischen Hummel, *Bombus atratus*, innerhalb des Beobachtungskastens, *Journal of the Faculty of Science, Hokkaido University, Series 6, Zoology* **15**, 712–62 (1965).

95. G. Gervet, Le comportement d'oophagie différentielle chez *Polistes gallicus* L. (Hymenoptera, Vespidae), *Insectes Sociaux* **11**, 343–82 (1964).

96. P. Huber, Observations on several species of the genus *Apis*, known by the name of humble-bees, and called Bombinatrices by Linnaeus, *Transactions of the Linnean Society of London* **6**, 214–98 (1802).

97. F. W. L. Sladen, *The Humble Bee* (Macmillan, London, 1912).

98. J. de Beaumont, Le parasitisme social chez le guêpes et les bourdons, *Mitteilungen der Schweizerischen Entomologischen Gesellschaft* **31**, 168–76 (1958).

99. C. D. Michener, The evolution of social behaviour in bees, *Proceedings of the 10th International Congress on Entomology, Montreal, 1956* **2**, 441–7 (1958).

100. O. W. Richards and M. J. Richards, Observations on the social wasps of South America, *Transactions of the Royal Entomological Society of London* **102**, 1–170 (1951).

101. R. Zucchi, S. F. Sakagami, J. M. F. de Camargo, Biological observations on a neotropical parasocial bee, *Eulaema nigrita*, with a review of the biology of Euglossinae (Hymenoptera, Apidae): a comparative study, *Journal of the Faculty of Science, Hokkaido University, Series 6, Zoology* **17**, 271–380 (1969).

102. R. B. Richards and C. H. Dodson, Nesting biology of two communal bees, *Euglossa imperialis* and *Euglossa ignita* (Hymenoptera: Apidae), including description of larvae, *Annals of the Entomological Society of America* **60**, 1007–14 (1967).

103. A. Ducke, Uber Phylogenie und Klassifikation der sozialen Vespiden. *Zoologisches Jahrbücher, Abteilungen für Systematik, Ökologie und Geographie der Tiere* **36**, 303–30 (1914).

104. G. C. Eickwort, Tribal positions of Western Hemisphere green sweat bees, with comments on their nest architecture (Hymenoptera: Halictidae), *Annals of the Entomological Society of America* **62**, 652–60 (1969).

105. C. D. Michener and W. B. Kerfoot, Nests and social behaviour of three species *Pseudoaugochloropsis* (Hymenoptera: Halictidae), *Journal of the Kansas Entomological Society* **40**, 214–32 (1967).

106. C. D. Michener, The bionomics of *Exoneurella,* a solitary relative of *Exoneura* (Hymenoptera, Apoidea, Ceratinini), *Pacific Insects* **6**, 411–26 (1964).

107. C. D. Michener, Social parasites among African allodapine bees (Hymenoptera, Anthophoridae, Ceratinini), *Zoological Journal of the Linnean Society* **49**, 199–215 (1970).

108. I thank Dr. G. C. Eickwort for pointing out to me the different behavioural tendencies of the suprageneric group.

109. G. C. Eickwort and K. R. Eickwort, Aspects of the biology of Costa Rican halictine bees. I. *Agapostemon nasutus* (Hymenoptera: Halictidae), *Journal of the Kansas Entomological Society* **42**, 421–52 (1969).

110. See for example, G. Knerer and C. Plateaux-Quénu, Usurpation de nids étrangers et parasitisme facultatif chez *Halictus scabiosae* (Rossi), *Insectes Sociaux* **14**, 47–50 (1967).

111. S. W. T. Batra, Behaviour of some social and solitary bees within their nests: a comparative study (Hymenoptera: Halictidae), *Journal of the Kansas Entomological Society* **41**, 120–33 (1968).

112. C. D. Michener, Reproductive efficiency in relation to colony size in hymenopterous societies. *Insectes Sociaux* **11**, 317–42 (1964).

113. P. Rau, The nesting habits of *Polistes rubiginosus* with special reference to pleometrosis in this and other species of *Polistes* wasps, *Psyche* **38**, 129–44 (1931).

114. P. Rau, The nesting habits of several species of Mexican social wasps, *Annals of the Entomological Society of America* **35**, 81–95 (1940).

115. M. J. West Eberhard, The social biology of polistine wasps, *Miscellaneous Publications of the Museum of Zoology, University of Michigan* **140**, 1–101 (1969).

116. P. Rau, Ecological and behaviour notes on the wasp *Polistes pallipes, Canadian Entomologist* **62**, 43–7 (1930); and personal observation.

117. M. J. West, Foundress associations in polistine wasps: dominance hierarchies and the evolution of social behaviour, *Science* **157**, 1584–5 (1967).

118. M. Kimura and G. H. Weiss, The stepping stone model of population structure and the decrease of genetic correlation with distance, *Genetics* **49**, 561–76 (1964).

119. D. H. Janzen, Coevolution of mutalism between ants and acacias in central America, *Evolution* **20**, 249–75 (1966).

120. R. W. Matthews, *Microstigmus comes:* sociality in a sphecid wasp, *Science* **160**, 787–8 (1968).

121. C. D. Michener, Evidence of cooperative provisioning of cells in *Exomalopsis* (Hymenoptera: Anthophoridae), *Journal of the Kansas Entomological Society* **39**, 315–17 (1966).

122. S. F. Sakagami and R. Zucchi, Estudo comparativo do comportamento de varias especies de abelhas sem ferrao, com especial referência ao processo de aprovisionamento e postura das cellulas, *Ciência e Cultura, São Paulo* **18**, 283–96 (1966).

123. C. H. Dodson, Ethology of some bees of the tribe Euglossini, *Journal of the Kansas Entomological Society* **39**, 607–29 (1966).

124. R. Zucchi, B. L. Oliveira and J. M. F. de Camargo, Notas bionomicas sôbre *Euglossa (Glossura) intersecta* Latreille 1838 e descrição de suas larvas e pupa (Euglossini, Apidae), *Boletim da Universidade Federal do Paraná, Zoologia* **3**, 203–24 (1969).

125. K. Tsuneki, The biology of East-Asiatic *Cerceris* with special reference to the peculiar social relationship and return to the nest in *Cerceris hortivaga* Kohl., *Etizenia, Fukui* **9**, 1–46 (1965).

126. W. G. Eberhard, Altruistic behavior in a sphecid wasp: support for kin selection theory, *Science* **172**, 1390–1 (1972).

127. K. Yoshikawa, R. Ohgushi, and S. F. Sakagami, Preliminary report on entomology of the Osaka City University 5th Scientific Expedition to Southeast Asia, 1966, *Nature and Life in Southeast Asia* **6**, 153–82 (1969).

128. J. F. Zikan, O genero *Mischocyttarus* Saussure (Hymenoptera, Vespidae) com a descrição de 82 especies novas, *Boletim da Parque Nacional de Itatiaia, Brazil* **1**, 1–251 (1949).

129. W. L. Brown, The first social parasite in the ant tribe Dacetini, *Insectes Sociaux* **2**, 181–6 (1955).

130. R. Stumper, Les associations complexes des fourmis. Commensalisme, symbiose et parasitisme, *Bulletin Biologique de la France et de la Belgique* **24**, 376–99 (1950).

131. O. W. Richards. The specific characters of the British bumble-bees, *Transactions of the Royal Entomological Society of London* **75**, 233–68 (1927).

132. H. E. Milliron and D. R. Oliver, Bumblebees from North Ellesmere Island, with observations on usurpation by *Megabombus hyperboreus* (Schonh.) (Hymenoptera: Apidae), *Canadian Entomologist* **98**, 207–13 (1966).

133. I. H. H. Yarrow, Is *Bombus inexpectatus* (Tkalcu) a workerless obligate parasite, *Insectes Sociaux* **17**, 95–112 (1970).

134. J. Scheven, Beitrag zur Biologie der Schmarotzerfeldwespen *Sulcopolistes atrimandibularis* Zimm., *S. semenowi* F. Morawitz und *S. sulcifer* Zimm., *Insectes Sociaux* **5**, 409–37 (1958).

135. H. Kutter, Die sozialparasitischen Ameisen der Schweiz, *Neujahrsblätt der Naturforschende Gesellschaft in Zürich* **171**, 1–62 (1969).

136. G. Beall, Mass movement of the wasp, *Polistes fuscatus* var. *pallipes* Le P., *Canadian Field Naturalist* **56**, 64–7 (1942).

137. G. Rudebeck, On a migratory movement of wasps, mainly *Vespula rufa* (L.) at Falsterbo, Sweden, *Proceedings of the Royal Entomological Society of London* A **40**, 1–8 (1965).

138. C. G. Johnson, *Migration and Dispersal of Insects by Flight* (Methuen, London, 1969).

139. K. R. Eickwort, Separation of the castes of *Polistes exclamans* and notes on its biology (Hym.: Vespidae), *Insectes Sociaux* **16**, 67–72 (1969).

140. M. J. West, Range extension and solitary nest founding in *Polistes exclamans* (Hymenoptera: Vespidae), *Psyche* **75**, 118–23 (1968).

141. K. Yoshikawa, Predatory hunting wasps as the natural enemies of insect pests in Thailand, *Nature and Life in Southeast Asia* **3**, 391–7 (1964).

142. V. M. Rodrigues, Estudo sôbre vespas sociais do Brasil (Hymenoptera-Vespidae), DS thesis, Universidade de Campinas (1968).

143. C. Baroni-Urbani, Ueber die funktionelle monogynie der ameise *Myrmecina graminicola* (Latr.), *Insectes Sociaux* **17**, 219–22 (1970).

144. M. V. Brian, *Social Insect Populations* (Academic, London and New York, 1965).

145. A. Buschinger, Zur Frage der Monogynie oder Polygynie bei *Myrmecina graminicola* (Latr.) (Hym., Form.), *Insectes Sociaux* **17**, 177–82 (1970).

146. P. B. Kannowski, The flight activities and colony founding behavior of bog ants in southeastern Michigan, *Insectes Sociaux* **6**, 115–62 (1959).

147. P. I. Marikovsky, Colonies of yellow ants (*Lasius flavus* de Geer) as theatre of struggle between nest colonies of other ant species, *Insectes Sociaux* **12**, 63–70 (1965).

148. See, for example, P. I. Marikovsky, Material on sexual biology of the ant *Formica rufa* L., *Insectes Sociaux* **8**, 23–30 (1961).

149. Brian 1965 (ref. 144), pp. 88–90.

150. A. D. Peacock, I. C. Smith, D. W. Hall, and A. T. Baxter, Studies in Pharoah's Ant, *Monomorium pharaonis* (L.), (8) Male production by parthenogenesis, *Entomologist's Monthly Magazine* **90**, 154–8 (1954).

151. A. J. Pontin, Field experiments on colony foundation by *Lasius niger* and *L. flavus*, *Insectes Sociaux* **7**, 227–30 (1960).

152. M. Petersen and A. Buschinger, Das Begattungs-verhalten der Pharaoameise, *Monomorium pharaonis* (L.), *Zeitschrift für angewandte Entomologie* **68**, 168–75 (1971).

153. M. V. Brian, Male production in the ant *Myrmica rubra* L. (Hymenoptera, Formicidae), *Insectes Sociaux* **16**, 249–68 (1969).

154. G. P. Markin, The seasonal life cycle of the Argentine ant, *Iridomyrmex humilis* (Hymenoptera: Formicidae), in Southern California, *Annals of the Entomological Society of America* **63**, 1238–42 (1970).

155. C. P. Haskins and R. M. Whelden, 'Queenlessness', worker sibship, and colony versus population structure in the formicid genus *Rhytidoponera*, *Psyche* **72**, 87–112 (1965).

156. M. V. Brian and J. Hibble, Studies of caste differentiation in *Myrmica rubra* L. 7—Caste bias, queen age and influence, *Insectes Sociaux* **11**, 223–38 (1964).

157. H. Kutter, Ueber *Doronomyrmex* und verwandte Ameisen, *Mitteilungen der Schweizerischen Entomologischen Gesellschaft* **23**, 347–53 (1950).

158. W. E. Kerr, Evolution of the mechanism of caste-determination in the genus *Melipona*, *Evolution* **4**, 7–13 (1950).

159. First described by Sakagami and Zucchi (ref. 122); for other reviews see refs 11 and 13.

160. J. B. Free, I. Weinberg, and A. Whiten, The egg-eating behaviour of *Bombus lapidarius* L., *Behaviour* **35**, 313–17 (1969).

161. J. S. Moure, Sobre a origem do Meliponinae parasitas (Hymenoptera, Apoidea), *Ciência e Cultura, São Paulo* **15**, 183–4 (1963).

162. J. van der Vecht, The geographical distribution of the social wasps (Hymenoptera, Vespidae), *Proceedings of the 12th International Congress of Entomology, London 1964*, 440–1 (1965).

163. J. van der Vecht, The Vespinae of the Indo-Malayan and Papuan areas (Hymenoptera, Vespidae), *Zoologische Verhandelingen, Leiden* **34**, 1–83 (1957).

164. E. O. Wilson, Social modifications related to rareness in ant species, *Evolution, Lancaster, Pa* **17**, 249–53 (1963).

165. A. Forel, Le mâle des *Cardiocondyla* et le reproduction consanguine perpétuée, *Annales de la Société Entomologique de Belgique* **36**, 485–62 (1892).

166. Passera 1969 (see ref. 52).

167. G. Le Masne, La signification des reproducteurs aptères chez la fourmi *Ponera eduardi* (Forel), *Insectes Sociaux* **3**, 39–259 (1959).

168. R. W. Taylor, A monographic revision of the ant genus *Ponera* Latreille (Hymenoptera: Formicidae), *Pacific Insects Monographs* **13**, 1–112 (1967).

169. S. I. Malyshev, *Genesis of the Hymenoptera and the Phases of their Evolution* (Methuen, London, 1969; transl. from the Russian).

170. K. Gösswald, Ökologische Studien uber die Ameisenfauna des mittleren Maingebietes, *Zeitschrift für Wissenschaftliche Zoologie* A, **142**, 1–156 (1932).

171. Donisthorpe 1927 (see ref. 52).

172. J. C. Bridwell, Some notes on Hawaiian and other Bethylidae, with description of a new genus, *Proceedings of the Hawaiian Entomological Society* **4**, 291–314 (1920).

173. A. E. Treat, Social organization in the moth ear mite (*Myrmonyssus phalaeonodectes*), *Proceedings of the 10th International Congress of Entomology, Montreal, 1956* **2**, 475–80 (1958).

174. A. E. Treat, Behavioural aspects of the association of mites with noctuid moths, *Proceedings of the 2nd International Congress of Acarology, Budapest*, 275–86 (1969).

175. R. L. Trivers, The evolution of reciprocal altruism, *Quarterly Review of Biology* **46**, 35–57 (1971).

176. G. Le Masne and C. Torossian, Observations sur le comportement du Coléoptère myrmécophile *Amorphocephalus coronatus* Germar (Brenthidae) hôte des *Camponotus*, *Insectes Sociaux* **12**, 185–94 (1965).

177. J. Ishay and R. Ikan, Gluconeogenesis in the oriental *Vespa orientalis* F., *Ecology* **49**, 169–71 (1968).

178. D. H. Janzen, Coevolution of mutualism between ants and acacias in Central America, *Evolution* **20**, 249–75 (1966).

179. R. L. Jeanne, Chemical defense of brood by a social wasp, *Science* **168**, 1465–6 (1970).

180. R. Chauvin, G. Courtois, and J. Lecompt, Sur la transmission d'isotopes radioactifs entre deux fourmilières d'espèces differentes (*Formica rufa* et *Formica polyctena*), *Insectes Sociaux* **8**, 99–107 (1961).

181. R. L. King and R. M. Sallee, *Formica fossaceps* Buren and *Formica obscuriventris clivia* Creighton as slaves of *Formica rubicunda* Emery, *Proceedings of the Iowa Academy of Science* **66**, 472–3 (1959).

182. G. Scherba, Analysis of inter-nest movement by workers of the ant *Formica opaciventris* Emery (Hymenoptera: Formicidae), *Animal Behaviour* **12**, 508–12 (1964).

183. M. F. Filosa, Heterocytosis in cellular slime molds, *American Naturalist* **96**, 79–91 (1962).

Altruism and a group ambush

A *Paratemnus* pseudoscorpion of Brazil seizes the antenna of a large ant (*Odontomachus*) and pulls backwards into a narrowing crevice, hoping to avoid the ant's gin-trap mandibles. Other adults, sensing action, orient to the ant and will join in holding and subduing it when any appendages come within their reach. The reality of altruism and danger for first strikers is shown by the adult male towards upper left who has lost a chela during some previous encounter. After the ant is paralysed (by neurotoxic poison injected from the chelae) juveniles (not shown) will move up and be the first to feed. All individuals in the group reproduce. In the upper right, a non-feeding gravid female has built a silken tent in which to hatch her eggs. During the event of a capture she is undisturbed. (In the observation chamber a sheet of glass replaces the tree trunk to which the bark flakes concealing the colony are normally attached — thus the view is as if from within a transparent tree trunk.)

CHAPTER 9

FRIENDS, ROMANS, GROUPS . . .

Innate Social Aptitudes of Man: an Approach from Evolutionary Genetics

Then to the rolling Heav'n itself I cried,
Asking, 'What Lamp had Destiny to guide
Her little Children stumbling in the Dark?'
And—'A blind Understanding!' Heav'n replied.

OMAR KHAYYÁM[1]

In 1972 for a second time I found myself asked to discuss at an international symposium the implications of kinship theory for human evolution. In this case the invitation came from Robin Fox, a social anthropologist (now at Rutgers University), and from Irven De Vore, a Harvard primatologist. The symposium was to be in Oxford in July 1973. Compared to the Smithsonian symposium it was to be as small as the other was large and slanted much more to the side of Man, less to Beast. Fox, whom I first met at the Washington conference, was at a quite different intellectual pole of his profession from such as Sir Edmund Leach, who, as mentioned in Chapter 1, had turned down my intended subsidiary study at Cambridge. Fox and I at different times had both worked at the London School of Economics. Perhaps repressed deep in the soul of that institution lies a streak of heterodoxy that has played a part in our convergence. Or there could be another factor. Fox had long been interested in evolution and in animal parallels to human phenomena, as uncovered, for example, by Konrad Lorenz and his followers in the field of ethology. He had a close collaborator Lionel Tiger, whom I had also first met at the Smithsonian

'Man and Beast' symposium. If, as one sometimes sees signs, names are formative in personality development, this zygotic (and perhaps here tetraploid) combination of the two men could be argued to have shaped their work. If so, it could be cited as an example of a beneficial cultural mutation, recombination, or whatever. As illustrated in the papers and books they wrote together, both were busy importing much good 'animal' common sense into a discipline that has often appeared to me progressively losing itself in bye-ways of purely human tradition and verbiage.

A year later the topic of the meeting might have been named 'human sociobiology'. Even the title it had, 'Biosocial Anthropology' (also given to the symposium volume), could doubtless have gained its organizers a share of the hot coals that, by the year of publication, had begun to shower on E. O. Wilson's brave head over his book *Sociobiology*. But such few coals as did shower, as mentioned below, fell on me and on the whole all was quiet. Probably this is because the meeting's and the book's intention was not of asserting any sweeping change in the world view of behavioural science, nor of predicting imminent extinctions and mergers of disciplines. More it was to record the steady but rapid extension of an existing paradigm, Neodarwinism, which was now in course of application to behaviour and to humans. The new ideas underlying the papers of the meeting were, however, largely the same as underlay the offending human chapters of E. O. Wilson's massive work. Like Wilson, Robin Fox had noticed the rapid accumulation of changes both in the facts and the theories in his area; he wanted a meeting to pull various new lines together and to help establish a new perspective. Thus Fox and Wilson went at their tasks in complete independence but drew on much of the same material. Simultaneously, and even more independently, Richard Dawkins was working on his book *The Selfish Gene*, destined to come out in 1976. Our book was a very small brother to the other two yet, nevertheless, it was another mark of rapidly changing times and the emergence of a new view of evolution and of ourselves.

Robert Trivers was later to refer to the article that I contributed to Fox's volume as my 'Fascist paper'. I believe he was referring not mainly to his own impression but rather to what others were saying about it and particularly to one strong response by a noted anthropologist, S. L. Washburn, in which, singling my paper out of the whole volume, he called it 'reductionist, racist, and ridiculous'.[2] Washburn obviously didn't like the whole thing but focused his objection on some of my more speculative paragraphs concerning the warlike propensities of pastoral peoples and the possible involvement of these in recurrent trends in the histories of Old World civilizations. Very possibly on account of my having used an introductory quotation from a paper of his own he felt a need to distance himself from what he saw as wrong views. Having admired Washburn as an anthropologist of a seemingly brave and biological orientation (indeed in a way similar to Fox) I had naively expected him to be pleased by the quotation I intended to be a dedication and also to be pleased by my paper as a whole. Fox may have been surprised, too, and disappointed at the reaction.[3] It had been he who suggested to the meeting that we contributors as a group might dedicate the whole of the volume to Sherwood Washburn and Niko Tinbergen jointly, as was done. I wonder if people who struggle to extend the frontiers of a discipline against a current of peer disapproval sometimes need to convince themselves and others that they are not quite the heretics and outlaws everyone thinks and this need is expressed through an extra militancy against further extension in the direction they themselves have been taking.

Be that as it may, whether my paper really is racist or absurd I will leave to the reader's judgement and just comment that I have hardly changed my opinions and certainly haven't with regard to what Washburn indicated as the most offending passage. All I allow is that the matter is still, of course, speculative, and I add the thought that I could, and perhaps should, have left out one word, 'genetic', from the sentence preceding that which has reference to Eshel on page 346. I had indicated earlier that the effects I had been talking about could be partly cultural and there was no good reason only to mention the genetic possibility when referring to the same idea below. However, I have never seen any evidence that a genetic interpretation or a genetic component is out of court and the idea in general continues to be justified to my mind. I do not believe that it poses any threat or insult to any races; multiracial mixture almost throughout our species population is involved in the pattern I suggested. Thus in the most obvious case I am talking about a racial interaction involving the two ends of Asia. In general, I think it would be sad to have to accept a culturist's conclusion that current minority and traditional cultures have nothing that they can usefully contribute genetically to the systems that dominate them. Few I think would deny a contribution of minority genes to America's basketball excellence for example. If such a contribution is accepted for a physical trait, why not look for others? The writer Douglas Adams is very unlikely to have read or been influenced by the ideas in this paper yet he expressed an idea singularly close to mine in his successful and popular book *The Hitch Hiker's Guide to the Galaxy*. So far as I know his passages on the thoughts of the bulldozer driver who had a touch of the Golden Horde in his ancestry aroused no protest. In all it is a pity to see scientists struggling to tie each other's hands in respect of some kinds of understanding and in effect crippling themselves in this way relative to poets and novelists. I think that people who do not agree with speculations should simply say why hints that suggested them are worthless, what contrary facts support their own view, and so on. Washburn did none of this.

Independent of these and the paper's other human ideas, I am proud to have included the first presentation of Price's natural selection formalism as applied to group-level processes. I wish George Price had been alive to see it published. He did see a draft and expressed approval of the formal part (i.e. the part he had helped to inspire). As to the rest he spoke guardedly, so perhaps his feelings were akin to Washburn's. I remember something like: 'There's a lot I wasn't expecting—interesting but I need to read it again. We must talk about it later.' Well, again, be that as it may. As to the formal part I am fairly sure it does lay out what he had meant when he asked me so pointedly whether I had seen how his formula applied to group selection. He himself published one application of the formula to groups but I think it less explicit and general than mine, indeed almost as if he was trying still to conceal his formula's full significance.[4] For myself, I consider the format of analysis I was able to achieve through his idea brilliantly illuminating. But why hadn't he explained it more explicitly, why didn't he publish it all himself? To answer this I have to say a little more about his life.

Up to the time when he came to London and read my paper he had been, he told me, an almost militant atheist. In contrast, his wife was a practising Roman Catholic. It is easy to imagine how, when his two daughters reached independence, such difference might begin to exacerbate and push his marriage into difficulties. There had been other factors upsetting his life at about the same time, notably a tumour of his thyroid, the surgery of which had damaged a nerve and left one arm and shoulder slightly paralysed. As regards the tumour, the operation was a complete success but in addition to the damaged nerve it had, of course, left him with a need to take thyroxine for the rest of his life. There had been a positive side to the operation also, however. Somehow out of it all he received a big insurance benefit and after his divorce he used this money, as he told me, to finance his transfer to London. He had betted on ill health and won.

Always an extremely clear thinker he was also very explicit, never a man to compromise or to pretend to opinions he didn't hold. This certainly applied in both his anti-religious and his religious periods. The only paper that I know of his atheist and 'science journalism' period was a long article in *Science*, stating that no claims of paranormal phenomena could yet be considered acceptable.[5,6] He highlighted especially the possibility of fraud, which he said had never yet been taken seriously enough: all supposedly positive tests needed to be done again with a care that would exclude every conceivable way of cheating. Scientists by their nature were inclined to accept evidence at face value and to assume others to be as honest and objective as they tried to be themselves. This made them very vulnerable to fraud. He believed the general public to be even more gullible, being less trained in logic and in the pitfalls of probability. He laid out in detail the sort of experiment that could provide convincing evidence of, for example, clairvoyance: no one, not even the experimenter, was to know what object was placed in the welded metal container that the clairvoyant was to penetrate by his powers; even the welds of the box were to bear seals; and so on. This long article, as can be imagined, brought him a huge correspondence, and no matter how naive or bizarre the letters George appears to have tried to answer them all, as I came to see as I attempted to put his papers into some sort of order before taking them to his brother in New York in 1976. The main point for the present is simply that, following his operation and his divorce, he arrived in London as an ardent and outspoken atheist plus a general sceptic. He read my main papers on kin selection, I think he told me, in the Senate House Library. By suggesting to him that life evolved purely by natural selection would not be nearly as benign as he had previously supposed, the papers seem to have had a profound effect on him. Could this limited nepotistic altruism be the best, the most 'humane', that evolution could achieve? If so the prospect for humanity seemed stark. He set to work to try to understand genetics and to verify what I had done. Once he had convinced himself that something at least close to what I claimed was true, he became very depressed. Was it the limit or was there something else?

It was fairly soon after, I think only a week or so, that he had a religious experience. He never described it to me in detail, remarking that he could tell that I was practically as he had been in a former life, and not open to anything that would seem intrinsically supernatural. He described himself running through the streets of London in the neighbourhood of his flat in Marylebone (north-east of Oxford Circus—land of rich doctors, offices, and the BBC). He was looking for a church, he said, and entered the first he came to and prayed for guidance. The immediate result was his complete dedication to Christianity. He told me which church but I have forgotten other than it was Anglican; but even that is a hint that could, I suppose, facilitate finding the clergyman who became George's mentor in the first weeks of his conversion. However, it seems likely that the clergyman soon realized that he had an unusual and a potentially difficult convert on his hands (I understood from George that before long the two were not on very close terms). George believed that everyone who claimed to be Christian should follow the teaching in the New Testament literally and absolutely. There being no shortage of the unloved poor in the streets near Oxford Circus, George started to try to be a Samaritan to them all. This rapidly drained the remainder of the funds and possessions he had brought from America. He brought derelicts to his flat, where, as he admitted, they stole, smashed things up, and made a mess to the point where his landlord eventually told him he must leave. By then he had virtually no money and, therefore, with the permission of Professor Cedric Smith (my old University College supervisor), he started sleeping on the floor of the small office he had been given the use of in the Galton Laboratory (by now the Galton had moved from where I had first known it in the main building of University College to the present site in Stephenson's Way, a hundred yards or so to the north). From this base George divided his time between three main projects. First was trying to help the alcoholics and destitutes he found in the streets, initially near to Oxford Circus but later more around London's three great north-facing railway stations, Euston, St Pancras, and King's Cross. Where I had sympathized with flowers and bracken fern perhaps partly because there wasn't much I could be expected to do for them, making my sympathy cheap, he gave himself wholeheartedly to the crying children and homeless humans wandering the same streets. He helped the people he found with money, tried to find them jobs, spoke up for their characters in court. Second, he worked on his new formalism of natural selection, this being the line that brought him into contact with Cedric Smith and me. Third, he conducted an exact and scrupulous exegetical reading of the Bible, especially the New Testament.

Charitable activity was his highest priority but all three endeavours were evidently treated as interwoven. This led to some odd results. One of his early discoveries from his reading of the Bible and scholarly texts that he consulted, was that Christianity as founded by Jesus had to be considered, similar to many other sectarian beliefs of its time, a mystery religion. This implied that not all the intentions and ideas of its creator or of the testament writers were intended to be easily understood through casual reading. Their divine origin ensured there would be no ultimate contradiction or mendacity, but that was about all. Short of outright untruths, meanings could be concealed so as to confuse any superficial reader: but if instead one dedicated one's whole intelligence, better still one's life, to the task of reading, clarity would slowly dawn. George's view would have Omar Khayyám wrong in the passage I quote at the start of this chapter. Had Omar listened more carefully he would have heard Heaven's reply: 'Study more carefully.'

George claimed to discover from his reading that the terse style of the gospel writers where it concerned the events of Holy Week—

that is, the traditional 7-day period preceding the crucifixion of Jesus—had given rise to a confusion. He maintained that if one followed all clues to chronology, Holy Week as a week simply could not work; instead, he concluded the period covered by this 'week' had to be 12 days. He wrote up all his reasoning for this conclusion in a long article entitled 'The twelve days of Easter' and had hopes of publishing it in a major Sunday paper around the time of some Easter festival. He tried this in 1972 but his hopes were dashed after his article was sent to theologians for review and he had further tried it personally on, I think, one other theologian. From his account biblical scholars are much like the rest. New ideas from an unknown source (even if much more carefully supported than are my speculations in anthropology and history as in this paper) are received the reverse of enthusiastically and often not even read. To get serious attention in any field ideas do well to have not just factual support but political and human support as well. For my part I found George's reconstruction of the 12 days both believable and a gripping tale, even in spite of the intrusion of one or two minor miracles that he believed in but I couldn't (actually they hardly affected the general idea). It all reminded me of George's best work in evolutionary theory. There was the same Sherlock Holmes mind, the same acute attention to detail. I thought it would indeed be good reading in an intellectual Sunday paper at Easter and offered to try to combine with him for some Mark II of our Trojan Horse in order to get it into *The Sunday Times* or *The Observer* for Easter of 1975. Sadly, it was impossible. Living now completely a life that was almost identical to that of the derelicts he had tried to help, apart from their addictions, and accommodated as I described in the introduction to Chapter 5, Christmas 1974 was his last great Christian festival and beyond it he lived only a few days.

What has all this to do, if anything, with the paper of this chapter? Apart from his contributing his bi-level (and therefore potentially multilevel) covariance formula there is, I believe, still quite a significant connection. George believed that the discovery he had made in evolutionary theory was truly a miracle. God had given him this insight where he had no reason to expect it. It was ludicrous, he told me, to suppose that he, a person who had never understood or used statistics and had hardly known previously a covariance from a coconut, could have discovered the simple formula that should prove to be the most transparent yet found to partition and interpret the working of natural selection. Because it was a formula missed by the world's best population geneticists throughout the past 60 years, it was clear to him that he had, somehow, been chosen to pass on a truth about evolution to a world that was, somehow, just now deemed ready to receive it. How was he supposed to do it? How much was he expected to tell and how? He did not say exactly this to me but he dropped various hints that have led me rather strongly to believe he decided it was right to treat the matter in just the way he saw divine Truth being handled in the Testaments—that is, as partly a mystery, wrapped in parable, perceived only slowly by disciples and those who very much wanted to know. Through such first interpreters, through such glass, so darkly, along with religious truth, evolutionary truth was supposed to filter outwards. In this process I believe I was chosen to be his first initiate. I am flattered, of course, if this is right but do not agree much with his philosophy of gradualism or with being obscure in any matter of science (for that matter, I might add, neither did he in the little he wrote for publication); nor do I agree with him about an inevitable underlying truth or unique inspiration in the Bible. The Bible is a very great book and with much great teaching but inevitably as a compilation of both fact and myth it has its contradictions and absurdities. Still, in my own way I do subscribe to a 'mystery' belief that most is perhaps not too different from George's. It does seem to me that amazing clues are made gifts to us in the book of nature around us. Clues to almost everything we can want to know seem to be there. Hence I believe that enough effort will give the answer to almost any question. Thanks to a path that seems to have been paved (for us in particular?) with fragments of a giant Rosetta Stone that when reconstructed may eventually unlock for us the text of our nature and of the whole universe, there seems to be nothing that is at all interesting that we cannot hope to find out. This optimism may eventually prove to apply to morals as well;[7] if all this is true, Omar Khayyám's haunting, pessimistic quatrain is wrong in my review also, my comment being: 'Not so; listen again.'

I was not thinking of any of this, however, when I started to write the paper of this chapter, just fulfilling Fox's request and fitting out a text that I hoped would stir people's imagination—a frigate as I vaguely thought of it, armed partly with Price's guns and partly with my own, and to be sent to stir up trouble in some Spanish Main. I think I even liked the idea of people being torn between citing my paper for its first-ever just formalism for group selection and not citing it because of the various 'politically incorrect' notions I had packed in on the other side (perhaps, one might say, the frigate's more buccaneering side). Although I was surprised by Washburn's reaction because of the dedication, I was not surprised that the paper drew fire elsewhere.[8] As to Pricean hierarchical selection analysis as part of some great unfolding, I don't know quite why the public was supposed ready to receive this genuinely illuminating tautology[9] so precisely in the mid-1970s. I retain an impression that at least the scientific section of the public may not be ready even yet because the paper is little or unappreciatively cited even in works that discuss exactly the multilevel-selection problem that it was so well adapted to deal with.[10] I hope that the present reprinting of the paper will give Price's idea renewed attention and just here make readers tolerant of a few final details of the paper's invisible second author and his contribution.

George's move to live in his office at the Galton Lab certainly did not reduce the chaos of his life that had been begun by his divorce and increased through his conversion. I believe that some at the Galton from the start regarded his sleeping there as dubiously justified. Gradually other difficulties accumulated. At first he tried to keep his new address secret from his alcoholic friends but some found him and tried to visit him. There came a big man of violent temper, a woman beater whose wife (or partner) George had assisted to hide in a secret place. On detecting George's part the man not surprisingly became abusive and resentful. He visited the lab several times, first to request and later to try to compel George to divulge the location. George refused steadfastly and the man became increasingly violent. He was prevented from entering the building and so took to standing outside in the street below and shouting up threats and demands towards the window of George's office. This, of course, disrupted work in the whole lab and Cedric Smith, like George's previous landlord, had to ask George to leave. It was after this that George started living in third-rate bedsitters and squatter communes in various disused buildings of the area. After several such moves he came to the building in Tolmers Square.

A factor almost certainly contributing to George's erratic behaviour during this period was his need for continuous thyroxine medication. He told me that sometimes when depressed about the deepening troubles of his life, including the seeming counterproductiveness of his efforts to help people in Christian fashion, he stopped taking the tablets. He did so because he believed that if God wanted him to continue with his work then He would somehow show his approval of what George was doing by supplying the regulator chemical through some miracle: otherwise, if He chose not to help, it would be right for George to simply accept whatever followed. Twice, I think he told me, he took this step and the intervention he hoped for occurred. Once his account (completely honest I'm sure) included a fall downstairs and the breaking of a wrist. The accident brought him to hospital where, on discovery of his hypoglycaemia, the doctor or anaesthetist provided thyroxine without George even being told what he was receiving. He concluded he was supposed to live on and, following the doctor's instructions, started taking the tablets again. Another time was a month or so before he died, but I forget what happened then. At his inquest evidence was presented that yet again, just before his death, he had been neglecting his tablets. His very brief suicide note said that he had been getting depressed because of increasing difficulty he was encountering in the aims of his life, especially in the social work he had chosen, while at the same time he felt himself to be becoming a burden on his friends; he had therefore decided the time had come for him to go to his Maker. This time, clearly, the miracle hadn't come. There was an inquest and later I was notified of a funeral service to be held in Camden Town just north of the rail stations. Suicide isn't yet a way of dying approved by orthodox Christianity or Camden Town and this is probably why the service was not held in a church but instead in a bare room with desks like a school room, my memory failing me again as to exactly what and where the room was. Besides myself and John Maynard Smith only some half-dozen people, mainly those who had known him in his social work, were present. I vividly remember two or three red faces framed in shaggy hair, the hair in turn touching the shoulders of old dark coats: these I guessed to belong to some of the people George had tried to help. While their appearance suggested his lack of tangible success, their presence and their obvious sorrow about his death were moving to see.

So it was that when the paper was finally published George Price never saw it and what had begun simply as the commission from the two kindly, ethology- and evolution-oriented anthropologists became in part a valediction for a man I have always felt it a great privilege to have known. I am very glad that he at least did see the draft before he died. If what he thought to be his key idea of his London period was right then I suppose from where he is now he sees clearly the whole matter that I am trying to delineate. He knows exactly what I made out of his second idea using the papers and hints he had given me; but then I personally don't believe this is where he is and consider his second idea actually to have been more important than the first.

To complete the present introduction I now need to say that George Price is not the only person for whom I feel the paper to be valedictory. From much earlier in its development I had planned that it should include a kind of farewell to my father and in fact it is unique among my papers in having found an occasion to refer to my father's sole important publication, his account of the 4 years he spent building a road through a land of mountains and brigands, his *Road through Kurdistan*. Not long after the invitation to the Oxford symposium came, my father's own slow and uncomplaining 'reversion to type', the minor aspects of which he had seemed surprised that I, an evolutionist, would be unaware (Chapter 3), took the plunge that eventually carries all of us far below the level of even the lowest of mere primate forbears, to that state that underlies the lowest roots of the whole evolutionary tree. My father died, merged into those organic and inorganic molecules out of which life once came, the 'clays of our cold star'.[11] By his own wish his cremation ashes were to be scattered on his native Cashmere Hills near Christchurch, New Zealand.[12]

I have never seen my father's road. He hoped it would bring peace and order and prosperity to the mountain people he came to love; but even by the time he died there was enough disaster looming and in progress in the Middle East to show that these hopes were vain.

Beginning in his lifetime the road he built has since been fought over many times (Iran versus Kurds; Iraq versus Kurds; perhaps also Iran versus Iraq—certainly that fight occurred not far off; and now Kurds versus Kurds). Early on my father wrote letters to various ministers and administrators trying to smooth a path of peace; and yet I doubt whether now a single bridge that he built on the road can have survived bombing at one time or another. First Xenophon the Athenian mercenary with his book, then my father, and last of all palaeoanthropologist Ralph Solecki, who wrote to him asking about caves and finally presented him with a copy of his own book on the Neanderthals of the famous Shanidar rockshelter in Iraq—it was these that combined to suggest to me the image of the recording angel of Shanidar cave. Fortunately on the time scale of that angel, neither my father's nor George Price's lost causes really matter. The effort that they made (which, it will be noted, was not in the least on behalf of kin or expecting any tangible return) nevertheless made changes. Faith in me says that, while it is always right that we look to see how effort might have been better spent towards goals intended or towards others related but less illusory, the effect of true altruism like this is never lost completely.

References and notes

1. O. Khayyám, *Rubáiyát* (Macmillan, London, 1905); quatrain XXXIII of 1st edition.
2. S. L. Washburn, Sociobiology, *Anthropology Newsletter* **18** (3), 3 (1976).
3. N. Yamamura, Vertical transmission and evolution of mutualism from parasitism, *Theoretical Populational Biology* **44**, 95–109 (1993).
4. G. R. Price, Extension of covariance selection mathematics, *Annals of Human Genetics* **35**, 485–90 (1972).
5. G. R. Price, Science and the supernatural, *Science* **122**, 359–67 (1955).
6. G. R. Price, Where is the definitive experiment?, *Science* **123**, 17–18 (1956).
7. R. D. Alexander, *The Biology of Moral Systems* (Aldine de Gruyter, New York, 1987).
8. S. M. Boorstein and P. W. Ewald, Costs and benefits of behavioural fever in *Melanoplus sanguinipes* infected by *Nosema acridophagus*, *Physiological Zoology* **60**, 586–95 (1977).
9. S. A. Frank, George Price's contributions to evolutionary genetics, *Journal of Theoretical Biology* **175**, 373–88 (1995).
10. D. S. Wilson and E. Sober, Re-introducing group selection to the human and behavioral sciences, *Behavioral and Brain Sciences* **17**, 585–654 (1994).

11. W. Owen, *The Collected Poems of Wilfred Owen* (Chatto & Windus, London, 1963); *Futility*, first published 1920.
12. Actually, they weren't: as I discovered during a recent visit to New Zealand; his brother arranged their burial beside my Hamilton grandparents' grave in Waimate.

INNATE SOCIAL APTITUDES OF MAN: AN APPROACH FROM EVOLUTIONARY GENETICS†

W. D. HAMILTON

It has become clear that, although learning has great importance in the normal development of nearly all phases of primate behaviour, it is not a generalized ability; animals are able to learn some things with great ease and other things only with the greatest difficulty. Learning is part of the adaptive pattern of a species and can be understood only when it is seen as a process of acquiring skills and attitudes that are of evolutionary significance to a species when living in the environment to which it is adapted.

WASHBURN, JAY, and LANCASTER[1]

SURVIVAL OF THE FITTEST

The phrase directs attention to differential survival. Darwin accepted it from Herbert Spencer as adequately expressing the idea of natural selection. While accusations of tautology seem hardly fair on this small phrase itself, it must be admitted that some descendent ideas in the theory of natural selection are open to attack. For example, the idea of measuring ability to survive and reproduce—biological 'fitness'—has undoubtedly been useful, but a slight haziness still lingers, a lack of precise and general definition: we do not know exactly what qualities natural selection is after. I think it is doubt like this rather than doubt about the reality and effectiveness of natural selection that inspires a present spirit of caution in evolutionary biology, including caution and distrust towards Spencer's ideogram whenever it renews aspirations to become a slogan.

A part of the difficulty, and the part I am mainly concerned with now, is that of saying exactly what are the things that natural selection is supposed to select.

†In Robin Fox (ed.), *ASA Studies 4: Biosocial Anthropology*, pp. 133–53 (Malaby Press, London, 1975).

The fittest what? Is it a trait, an individual, a set of individuals bearing a trait, or bearing its determinants expressed or latent? Can it be a population, a whole species, perhaps even an ecosystem? In such a confusion of possibilities (and of fervent opinions either way) the individual organism stands out as one clear and obvious choice, with the number of its offspring as the measure of its fitness. But, beyond the problem of when to count and how to weight offspring for their ages, there is the problem that in sexual species the individual is really a physical composite of contributions from two parents and it may be composite in slightly different ways for different parts. Moreover, Mendel's principles concerning the fair distribution of genes to gametes and fair competition of these in fertilization do not always hold, so that the set of offspring of a given individual may carry a biased sample from the composite. Does this matter? For safe conclusions, do we have to descend to the level of the individual gene, perhaps ultimately to that of changed or added parts of the replicating molecule? Or can we, on the contrary, confidently follow the consensus of biologists to a higher level, in believing that the generally significant selection is at the level of competing groups and species? I shall argue that lower levels of selection are inherently more powerful than higher levels, and that careful thought and factual checks are always needed before lower levels are neglected. In this I follow a recent critical trend in evolutionary thought[2,3] (see also Chapter 8 and references therein). Incidentally, to a biologist, a rather similar critique seems to be invited by the supposition that cultural evolution is independent of evolution in its biological substratum: to come to our notice cultures, too, have to survive and will hardly do so when by their nature they undermine the viability of their bearers.[4] Thus we would expect the genetic system to have various inbuilt safeguards and to provide not a blank sheet for individual cultural development but a sheet at least lightly scrawled with certain tentative outlines. The problem facing a humane civilization may be how to complete a sketch suggesting some massive and brutal edifice—say the outlines of an Aztec pyramid—so that it reappears as a Parthenon or a Taj Mahal. These ideas concerning cultural evolution will not be expanded in what follows, but I hope to produce evidence that some things which are often treated as purely cultural in humans—say racial discrimination—have deep roots in our animal past and thus are quite likely to rest on direct genetic foundations. To be more specific, it is suggested that the ease and accuracy with which an idea like xenophobia strikes the next replica of itself on the template of human memory may depend on the preparation made for it there by selection—selection acting, ultimately, at the level of replicating molecules.

Returning to the problem of units of selection, Darwin himself, vague about the process of heredity, based most of his arguments on considerations of the fitness of individuals. He made occasional exceptions, as for the social insects where he treated the 'family group' as the unit of selection. I believe even these limited concessions were incautious (Chapter 8), and value his judgement more where, discussing the evolution of courage and self-sacrifice in man, he left a difficulty apparent and unresolved. He saw that such traits would naturally be counterselected *within* a social group whereas in competition *between* groups the groups with the most of such qualities would be the ones best fitted to survive and increase. This open problem which Darwin left is really the starting-point of my own argument, but it is historically interesting to note that after some initial wavering between the calls of Spencer, Kropotkin, and others, almost the whole field of biology stampeded in the direction where Darwin had gone circumspectly or not at all.

Until the advent of Mendelism uncritical acceptance of group selection could be understood partly on grounds of vagueness about the hereditary process. For example, courage and self-sacrifice could spread by cultural contagion and, in so spreading, modify heredity as well. But in the event neither the rediscovery of Mendel's work nor the fairly brisk incorporation of Mendelism into evolutionary theory had much effect. From about 1920 to about 1960 a curious situation developed where the models of 'Neodarwinism' were all concerned with selection at levels no higher than that of competing individuals, whereas the biological literature as a whole increasingly proclaimed faith in Neodarwinism, and at the same time stated almost all its interpretations of adaptation in terms of 'benefit to the species'. The leading theorists did occasionally point out the weakness of this position but on the whole concerned themselves with it surprisingly little (references in Chapters 2, 6, and 8).

With facts mostly neutral and theory silent it seems that we must look to the events and the 'isms' of recent human history to understand how such a situation arose. Marxism, trade unionism, fears of 'social darwinism', and vicissitudes of thought during two world wars seem likely influences. Confronted with common social exhortations, natural selection is easily accused of divisive and reactionary implications unless 'fittest' means the fittest species (man) and 'struggle' means struggle against nature (anything but man). 'Benefit-of-the-species' arguments, so freely used during the period in question, are seen in this light as euphemisms for natural selection. They provide for the reader (and evidently often for the writer as well) an escape from inner conflict, exacting nothing emotionally beyond what most of us learn to accept in childhood, that most forms of life exploit and prey on one another.

LEVELS OF SELECTION

Often the problem is not acute. There are many traits like resistance to disease, good eyesight, dexterity which are clearly beneficial to individual, group, and species. But with most traits that can be called social in a general sense there is some question. For example, as language becomes more sophisticated there is also more opportunity to pervert its use for selfish ends: fluency is an aid to persuasive lying as well as to conveying complex truths that are socially useful. Consider also the selective value of having a conscience. The more consciences are lacking in a group as a whole, the more energy the group will need to divert to enforcing otherwise tacit rules or else face dissolution. Thus considering one step (individual vs. group) in a hierarchical population structure, having a conscience is an 'altruistic' character. But for the next step—group vs. supergroup—it might be selfish, in the sense that the groups with high levels of conscience and orderly behaviour may grow too fast and threaten to over-exploit the resources on which the whole supergroup depends. As a more biological instance similar considerations apply to sex ratio, and here a considerable amount of data has accumulated for arthropods (Chapter 4).

A recent reformulation of natural selection can be adapted to show how two successive levels of the subdivision of a population contribute separately to the overall natural selection.[5] The approach is not limited to Mendelian inheritance but its usefulness in other directions (e.g. cultural evolution) has not yet been explored.

Consider a population consisting of a mixture of particles, and suppose we are interested in the frequency of a certain kind of particle G. Suppose the particles are grouped: let the subscript s denote the sth subpopulation. For subpopulation and for the whole we define parameters relevant to natural selection as follows:

	Subpopulation	Whole population
Number of particles	n_s	$N = \sum n_s$
Frequency of G	q_s	$q = \sum n_s q_s / N$
Mean fitness	w_s	$w = \sum n_s w_s / N$

Fitness measures the amount of successful replication of particles in one 'generation'. Thus the total population of the next generation will be $N' = \sum n_s w_s$. The symbol $'$ (denoting 'next generation') is used again in the same sense in the following further addition to notation:

	Subpopulation	Whole population
Change in frequency of G in one generation	$\Delta q_s = q_s' - q_s$	$\Delta q = q' - q$

With such notation it is easy to derive:

$$\begin{aligned} w\,\Delta q &= \sum n_s w_s (q_s - q)/N + \sum n_s w_s\,\Delta q_s/N \\ &= \mathit{Covariance}\,(w_s,\ q_s) + \mathit{Expectation}\,(w_s\,\Delta q_s), \end{aligned} \tag{1}$$

where *Covariance* and *Expectation* are understood to involve weighting by the n_s as indicated. This is Price's form.[6,7] The covariance term represents the contribution of *intergroup* selection, so quantifying the intuitive notion that high q_s must cause high w_s for selective change to occur. The expectation term represents the contribution of *intragroup* selection. It is possible to apply the formula within itself, to expand $w_s\ \Delta q_s$. For example, if the next level is that of diploid individuals and *si* indexes the *i*th individual of the *s*th group we have $n_s = \sum 2$, $q_s = \sum 2q_{si}/n_s$ and $w_s = \sum 2w_{si}/n_s$ where these summations are understood to cover all i instead of all s as previously. Then $w_s\ \Delta q_s$ decomposes into two terms, one of which represents ordinary diploid selection with strictly Mendelian inheritance, while the second represents the effects of genetic 'drift' (random sampling effects), and 'drive' (non-Mendelian ratios). Even this latter term can be reformulated using equation (1), but then our 'groups' are the fundamental particles themselves, which, neglecting mutation, must give $\Delta q_{particle} = 0$, so that here finally the second term goes out.

An often useful rearrangement of (1), which shows the dependence of selection on the variability in its units, introduces the regression coefficient of w_s on q_s. If β_1 is this coefficient:

$$w\Delta q = \beta_1\ \mathit{Variance}\,(q_s) + \mathit{Expectation}\,(w_s\ \Delta q_s). \qquad (2)$$

Conceptual simplicity, recursiveness, and formal separation of levels of selection are attractive features of these equations. But, of course, being able to point to a relevant and generally non-zero part of selective change is far from showing that group selection can override individual selection when the two are in conflict. Moreover, even the possibility of devising model circumstances in which a positive group-selection term (first term) outweighs a negative individual selection one (second term, assuming no further levels), gives no guarantee that 'altruism' can evolve by group selection: we have to consider whether the population can get into the specified state, and, if it can, whether its present trend will continue. For example, if we suppose persistent groups with no extinction and no intergroup migration it is easy to arrange that the group-beneficial effect (β_1), of frequent altruism in a certain group is so large that the rapid expansion of the group with the highest frequency of G (q_m say) draws the population q rapidly upwards. But q will never reach or pass q_m, and must eventually fall, remaining below the ever-falling value of q_m. Admittedly, all this is reasonably obvious without the equation; but the equation does emphasize that natural selection depends on a certain variance which in this model must at last die away as the best group increasingly predominates. This is the essential objection to an algebraic model of Haldane[8] for selection of altruism, which other writers have wrongly treated as the first successful analytical model for altruism. In verbal discussion Haldane himself admitted the necessity of a device to maintain diversity. He suggested that if groups split on reaching a certain size, random assortment of altruists and egotists would raise the frequency of altruists in some daughter moieties, and if the critical size was low enough and the group advantage of altruism high enough, a process having endless overall enrichment in altruism might be devised. Increasing the inter-group variance by random (or, better, associative) division of existing tribes leaves less variance within groups, which, as a development of the equation will shortly make plain, weakens the power of individual selection, and this further improves the case. But Price's equation does not seem to lend itself to a detailed analysis of Haldane's suggestion—indeed the lack of analysis by Haldane himself suggests that it is not easy. The value of the covariance approach lies not so much in analytical penetration as in clarifying the approach to a problem.

Therefore, noting hopeful auguries in Haldane's tribe-splitting no-migration idea, let us now turn to a model at the opposite extreme in which groups break up completely and re-form in each generation. Suppose that on reaching maturity the young animals take off to form a migrant pool, from which groups of n are randomly selected to be the group of the next generation. Assume completely asexual reproduction (or perfect matriclinal or patriclinal inheritance of a cultural trait in an ordinary population), and assume that an altruist gives up k units of his own fitness in order to add K units to the joint fitness of his $(n-1)$ companions. These companions are a random selection from the gene pool and therefore, in a supposed infinite population, have the expected gene frequency of the gene pool. Thus compared to a non-altruist, the altruist is putting into the next gene pool fewer of his own genes plus a random handful from the pool of the last generation. Obviously his trait is not enriching the population with genes that cause the trait. The specification of grouping has been a mere gesture. Nevertheless it is instructive to see how equation (2) handles the matter.

With asexuality individuals are basic particles, so, as already explained, the recursive use of (2) to expand its second term gives simply:

$$w\,\Delta q = \beta_1\ Var(q_s) + E\{\beta_0\ Var_s(q_{si})\}.$$

All units are now of the same size, so *Var(iance)* and *E(xpectation)* can have their conventional meanings. Since β_0 does not vary with group constitution,

$$w\,\Delta q = \beta_1\ Var(q_s) + \beta_0\ E\{Var_s(q_{si})\} \qquad (3)$$

and the expectation is what is commonly called the within-group variance.

With random grouping, the distribution of the different compositions of groups will be binomial with parameters (q, n). The variance of q_s is $\frac{1}{n}pq$. Likewise it is easily shown that $E\{Var_s(q_{si})\} = \frac{n-1}{n}pq$, so that

$$w\,\Delta q = \frac{1}{n}pq\{\beta_1 + (n-1)\beta_0\}. \qquad (4)$$

This already shows the characteristically greater power of the lower level of selection as dictated by a ratio of variances that is bound to hold when grouping is random or nearly so.

In a group with v of its n members altruistic fitnesses are as follows:

Group mean	Selfish member	Altruistic member
$1+\frac{v}{n}(K-k)$	$1+v\frac{K}{n-1}$	$1-k+(v-1)\frac{K}{n-1}$

Thus by inspection and by subtraction, respectively

$$\beta_1 = K - k \text{ and } \beta_0 = -k - \frac{K}{n-1}. \tag{5}$$

Substituting in (4) we find:

$$w\,\Delta q = -kpq. \tag{6}$$

This confirms the earlier argument that altruism cannot progress in such a model. It seems at a first glance that the benefits dispensed by altruists have been entirely null in the working of the model, but they affect it through their involvement in mean fitness:

$$w = p + q(K-k).$$

This being the only involvement of K, we see that the most that altruism can achieve in the model is a slowing of the rate at which natural selection reduces its frequency—an effect which I explained earlier as altruism diluting each new gene mixture by adding, as it were, handfuls taken randomly from the previous one. Apart from this minor effect the model, like Haldane's algebraic one, is a failure, in spite of having shifted to the opposite extreme in respect of migration. It reveals a group-selection component which is not zero but which is bound in an unchanging subordination to the individual selection component. However, the relation between the two variances in this case suggests how we must change the model to make altruism succeed: $Var(q_s)$ must be increased relative to $E\{Var_s(q_{si})\}$. As already mentioned, this can be done by making G assort positively with its own type in settling from the migrant cloud. Suppose it assorts to such a degree that the correlation of two separate randomly selected members of a group is F. If this correlation is achieved by having a fraction F of groups made pure for each type and then the remainder again formed randomly, then it is easily shown that the between-and within-group variances are respectively:

$$\frac{1}{n}pq(1-F+nF) \text{ and } \frac{1}{n}pq(n-1)(1-F)$$

Putting these results and those of (5) into equation (3) we find as the generalization of (6):

$$w\,\Delta q = pq(FK - k) \tag{7}$$

so that the criterion for positive selection of altruism is

$$\frac{K}{k} > \frac{1}{F}.$$

Now the model can be made to work. Moreover, the simple form and the independence of group size suggest that the criterion may hold beyond the limits of the rather artificial model discussed here. Careful thought confirms that this is indeed the case: the criterion is completely general for asexual models with non-overlapping generations, and also holds for sexual diploid models when the coefficient F is suitably redefined (Chapters 2 and 5). The easiest way to see the basis of generality is to notice that the benefits of altruism do not now fall on a random section of the population and therefore do not simply enlarge the existing gene pool; instead they fall on individuals more likely to be altruists than are random members of the population. Indeed, the existence of the positive correlation F could be interpreted as implying in this case that there is a chance F that the K units of fitness are definitely given to a fellow altruist, while with chance $(1 - F)$ they are given (as they always were in the previous version) to a random member of the population.

The redefinition necessary for diploid organisms involves specifying a regression coefficient, b_{AB}, representing the regression of the genotype of recipient B on genotype of donor A. Often this is the same as the correlation coefficient of such genotypes (it always is so in the haploid case), but where they differ it is the regression coefficient that gives the prediction of gene content that we need. To get the form like (7) which applies to diploid selection other changes are obviously necessary, notably dividing pq by two to get the variance of gene frequency between pairs instead of that between individuals and other more complex changes connected with dominance and details of the assortative process. However, it is striking that a criterion like $(FK - k) > 0$ can be shown to determine positive selection of each genotype, and can be generalized to cover cases where A distributes various effects, positive or negative, to numerous individuals B, C, D, ... all having different regressions on A. Including A himself in the list of recipients we arrive at the idea of A's 'inclusive fitness' his basic non-social fitness, plus all the effects caused by his action when each has been devalued by a regression coefficient.

The usefulness of the 'inclusive fitness' approach to social behaviour (i.e. an approach using criteria like $(b_{AB}K - k) > 0$) is that it is more general than the 'group selection', 'kin selection', or 'reciprocal altruism' approaches and so provides an overview even where regression coefficients and fitness effects are not easy to estimate or specify. As against 'group selection' it provides a useful conceptual tool where no grouping is apparent—for example, it can deal with an ungrouped viscous population where, owing to restricted migration, an individual's normal neighbours and interactants tend to be his genetical kindred.

Because of the way it was first explained, the approach using inclusive fitness has often been identified with 'kin selection' and presented strictly as an alternative to 'group selection' as a way of establishing altruistic social behaviour by natural selection.[9] But the foregoing discussion shows that kinship should be considered just one way of getting positive regression of genotype in the recipient, and that it is this positive regression that is vitally necessary for altruism. Thus the inclusive-fitness concept is more general than 'kin selection'. Haldane's suggestion about tribe-splitting can be seen in one light as a way of increasing intergroup variance and in another as a way of getting positive regression in the population as a whole by having the groups which happen to have most altruists divide most frequently. In this case the altruists are helping true relatives. But in the assortative-settling model it obviously makes no difference if altruists settle with altruists because they are related (perhaps never having parted from them) or because they recognize fellow altruists as such, or settle together because of some pleiotropic effect of the gene on habitat preference. If we insist that group selection is different from kin selection the term should be restricted to situations of assortation definitely not involving kin. But it seems on the whole preferable to retain a more flexible use of terms; to use group selection where groups are clearly in evidence and to qualify with mention of 'kin' (as in the 'kin-group' selection referred to by Brown[10]), 'relatedness' or 'low migration' (which is often the cause of relatedness in groups), or else 'assortation', as appropriate. The term 'kin selection' appeals most where pedigrees tend to be unbounded and interwoven, as is so often the case with humans.

Although correlation between interactants is necessary if altruism is to receive positive selection, it may well be that trying to find regression coefficients is not the best analytical approach to a particular model. Indeed, the problem of formulating them exactly for sexual models proves difficult (Chapter 2). One recent model that makes more frequent group extinction the penalty for selfishness (or lack of altruism) has achieved rigorous and striking conclusions without reference to regression or relatedness.[11] But reassuringly the conclusions of both this and another similar model (more general but less thorough and much less well explained[12]) are of the general kind that consideration of regression leads us to expect. The regression is due to relatedness in these cases, but classified by approach these were the first working models of group selection.

TRIBAL FACIES OF SOCIAL BEHAVIOUR

One of the conclusions of the models just mentioned is that with a grouped population the migration between groups is crucially important in determining the general level which altruism can reach within a group. This is something which should now seem fairly obvious but which has been surprisingly overlooked in most discussions of group selection previous to Eshel's. The less migration there is the more relatedness will build up within groups. This will permit selection of acts with low gain ratio (i.e. ratios like K/k) but the gain ratios must always exceed one, and this means that the act must actually aid group fitness in some way—reduce its chance of sudden extinction,[11,12] or increase its rate of emission of migrants.[13] With the last eventuality it is better for altruism if the migrants get together in small groups to found new colonies than if they all enter existing groups, since entering undermines the assumption of low migration—in other words reduces intergroup variance. If groups of founder migrants are assortative so much the better, although if they are so by coming all from the same parent group this could be treated as fission. Likewise if migrant acceptance is the established mode, so much the better if groups selectively accept altruists. The ability of animals to exercise such discrimination may seem dubious when behaviour even in humans is rather indefinite in this respect, but it is noticeable that with many of the tight-knit groups of social carnivores and primates the would-be immigrant does go through a probationary period of hostile treatment and low status, which sometimes terminates his attempt to join.[14] Similar phenomena of possibly similar significance are certainly not lacking in people, witness the harsh requirements of achievement and service for an aspiring Amerindian brave (or neophyte British doctor for that matter) and the general suspicion, hostility, and low position accorded to wealthless immigrants. I should add here that the idea that such human behaviour is natural does not mean that it is right or even sensible under modern conditions. For example, the immigrants may bring new skills and aptitudes, a point to which I return later. And as regards 'altruism', recent tribal immigrants are likely to be net importers of this precious stuff—themselves the losers when they expose their natural communistic generosity to civilized exploitation. On the other hand, when experience of ambient guile and cupidity has taught them better, such immigrants may learn to confine this generosity again among themselves and to turn outwards a contemptuous and unsympathetic attitude which is also typically tribal; but such expected ambivalence in tribal feeling is another matter to which I must return.

I have carefully spoken of 'migration' rather than 'migration rate' so far, and in doing so intend to emphasize that it is the number of acts of successful migration that is important for mean intragroup relatedness. The size of demes may matter surprisingly little. An indigenous villager may know some of his many connections with other villagers and be aware of a plexus of relationship through the misty past. What might surprise him (as it surprised me) is that relatedness as measured here (and as manifested in physical similarity) builds up just as much eventually in a large unit, say a remote town, as it does in a village, if the same actual numbers enter and leave each generation. In other words, connections which the remote townsman does not so easily know of make up in multiplicity what they lack in close degree. Of course, a large unit usually does have more migration, and consequently less intrarelatedness, but the important thing is that it is the number of migrants rather than the size of colonies that determines this. For Wright's simple island model where migrants go anywhere among infinitely many colonies the approximate formula for mean intragroup relatedness (after migration has occurred) is very simple, $b = 1/(2M + 1)$ where M is the number of migrants (assumed small) per subpopulation per generation. So with one migrant exchanged every other generation we find $b = \frac{1}{2}$, the same as for siblings in a panmictic population, and we therefore expect the degree of amicability that is normally expressed between siblings. If three migrants go (and three come) every generation we get $b = \frac{1}{7}$. This is slightly more than the relatedness of outbred cousins ($b = \frac{1}{8}$), so such colonies should be slightly more intra-amicable than groups of cousins would be. If, as normally happens, migrants tend to go mainly to neighbouring populations, then emigrant and immigrant genotypes will tend to correlate and so a given level of relatedness can be maintained with more migration. However, recent achievements with the analysis of the harder stepping-stone model,[15,16] which covers the island model as a special case, show that the difference is not very great as regards own-deme relatedness. Consider the case where the colonies are supposed spread on the plane in a square lattice. Suppose that an act of migration is either 'distant', with the migrant going to any deme among the infinitely many, as in the island model, or 'close', with the migrant going to one of the four neighbouring demes, and suppose that the odds on events of the two kinds are specified. With odds 100:1 for 'distant' to 'close' $M = 0.5$ leads to $b = 0.5$ within colonies to a good approximation: in other words, the change produced by such a small amount of local migration is negligible. If the odds are reversed to 1:100, implying local migration much more probable than distant migration, the Kimura–Weiss solutions show that the relatedness only rises to 0.68. With 10 times as much migration (i.e. about five exchanged per deme per generation) the corresponding relatednesses are 0.09 and 0.17, so relatedness still only doubles when migration is local rather than distant. A much greater contrast is apparent in the relatedness of individuals of neighbouring demes: when distant migration preponderates this relatedness tends to be extremely small, but when close migration preponderates members of a neighbour deme can easily have more than half the relatedness that applies to an own-deme member. And up to a point increasing migration reduces the contrast between own and neighbour deme, so that there are genetical as well as cultural reasons why, in humans, intergroup migration and marriage should decrease intergroup hostility.

Two other points seem worth making about the stepping-stone model. One is that, in the one-dimensional version of this model which could apply to demes in a linear habitat such as a coastline or river, relatedness holds up much more strongly as local migration is increased, and relatedness to neighbour–deme member more strongly still (relative to the two-dimensional case). This means that hardly any extra hostility is expected to members of neighbouring demes. From this point of view, a seashore phase of hominid evolution, if it occurred, should have been particularly harmonious. The other point concerns the distribution of gene frequencies. The apparent variability of colonies is expected to change rather sharply at certain critical levels of migration. These are $M = 0.5$ for the island model and $M = 1$ for the two-dimensional stepping-stone model with close migration predominant. This means that at about the point where the colony members are related to each other like outbred sibs it should become relatively easy for individuals to detect a fairly clear difference in appearance when comparing fellow colony members with outsiders. Actually, in the stepping-stone model the possibilities with regard to patchiness and cline-like effects are complex, but, considering simultaneously several traits which are independently inherited and at most weakly selected, the complex overlap of patterns should make possible fairly accurate separation of 'us' and 'them' at the level of colonies. We shall shortly see why natural selection might favour motivation and ability so to discriminate.

What is happening to the ordinary families embedded in these supposedly endogamous colonies? Siblings, parents, and offspring will still be the individual's closest relatives. Owing to the inbreeding, their relatednesses will be above the value of $\frac{1}{2}$ that applies under outbreeding. Thus an individual should be more altruistic than usual to his immediate kin. But other neighbours who are not immediate kin are now also closely related, and it is this reduced contrast between neighbours and close kin that will give what is probably the most striking effect: we expect less nepotistic discrimination and more genuine communism of behaviour. At the boundary of the local group, however, there is usually a sharp drop in relatedness. If migrants (or whole groups) are very mobile, leading to an 'island' rather than a 'stepping-stone' situation, this drop may be such as to promote active hostility between neighbouring groups.[17] Even though these groups have some relatedness, as practical limitations to distant migration naturally ensure, the contrast is still such that a minor benefit from taking the life of an outsider would make the act adaptive. Recent studies on hunting dogs[18,19] and hyenas,[20] show strangers sometimes being killed, while within-the-group relations are usually amicable and even communistic. The most serious wounding which Lawick-Goodall[21] recorded in her study of chimpanzees occurred when two males combined to attack a male of another group. Bygott[22] witnessed a fierce attack by a group of male chimpanzees on a stranger female. The female escaped but the males caught and ate her infant. Trespassers may sometimes be killed in wolves,[23] and in rats.[24] In lions,[25] langurs,[26] and probably in rats and mice there is also killing of strange young, but this is probably in a rather different category because it is done only by males whose aim seems to be to sire new offspring on the mothers they bereave.

These phenomena are reminiscent of the intercolony hostility so often observed in social insects, where again actual killing may be frequent along the frontiers. With regard to relatedness, the situation is the same except that intrarelatedness in groups is usually due to all colony members being descendants of a single queen. But polygynic ants (e.g. the common red ants of the genus *Myrmica*) may approximate the breeding structure of group hunting carnivores rather well, and this tempts one to apply the superorganism concept often used for social insects to the co-operative social mammals. Such a view would compare the killing of occasional trespassers to the occasional minor wound with death of cells which occurs in the restrained fighting between, say, two individual dogs.

The basis for thinking that group-hunting carnivores are highly related within groups is the known low rate of migration and the reluctance with which migrants are accepted. Why such reluctance? The probable reason has already been touched on: the group has a co-operative job to do, necessary for its survival. This job is the hunting and killing of prey which are too large for one individual to tackle alone. The more work is invested in a task prior to its fruition the more worthwhile a parasitic option of behaviour becomes—at least, until parasites are too numerous (Chapter 6). And the more co-operation is involved in any endeavour the more scope there is for the inconspicuous idler. What is to stop a hunting dog from watching the hunt from afar and trotting up, by all possible short cuts, just after the prey has been killed? Probably this has happened and probably groups over full of the offspring of such idlers have found themselves unable to kill prey and have died out. This would give a slow selection for features of pack behaviour (either cultural or genetic) that make infiltration progressively more difficult. Simply cutting down on immigration would have the desired effect through raising relatedness but, as mentioned earlier, a selective entrance requirement, with the applicant's behaviour watched through a probationary period, would be even better. It should be mentioned here that a development which closed a group's frontiers completely would probably also fail in the long run for reasons of general adaptation: complete inbreeding abandons the obviously important advantages of sexual reproduction, whatever these are.

Roughly, as we currently see it, a cunning ape-like creature once pushed boldly out from near niches now held by baboons and chimpanzees. Whether or not (as one quite plausible view holds) it first left its less enterprising cousins to take a holiday on the seashore, eventually it reappears on the African savannah participating in the Pleistocene wildlife bonanza as a group-hunting carnivore. In spite of the now-not-so-prehensile foot which it kept all the while in the door of an omnivorous diet, it seems likely that this creature would have needed the same population structure as the other group-hunting carnivores and for the same but more urgent reasons. It is difficult to see what was the first factor in the escalation in cunning of this particular primate line, but the choice seems mainly between tools and language. The great benefits that these could confer to a co-operative hunter through improved technique and organization would ensure rapid selection for their development. But they would also affect the social situation in significant ways and indirectly this might escalate their selection: (1) both would provide extra cultural clues to group identification; (2) tools (and later other valued artefacts) would give further scope for parasitical behaviour, first intragroup but later between groups as well. Tools and possessions could be appropriated instead of made. With language in rapid evolution, learning, experience, and even intelligence would become increasingly open to parasitism. Meanwhile, increasing intelligence would make possible a very plastic approach to parasitical and altruistic behaviour, which in turn would increase the complexity of the semi-serious deception and coalition games which are so characteristic of behaviour within primate groups. Real rewards in food and mating are the incentive to this activity and thus escalate the selection for skill in play. The main point is that intelligence, plus (1) and (2), plus what has been explained about the real differentiation of genetical relatedness suggest the development of an explosive situation. Close frontiers to migrants a little more, or slightly increase group mobility, and it is possible to imagine the sudden success of a policy which makes any frontier incident an occasion for an attempt at violent incursion by the more populous group with losers killed, enslaved, or driven off. Successful occupation of the captured territory would soon bring the victors into contact with still less related 'stones' of the 'stepping-stone' lattice, which they could attack adaptively with even less reason for restraint. Increasing foresight would mean that a group would not necessarily wait until large enough to need neighbouring territory, if attacking a weak group while it is weak helped to ensure that space could be occupied as needed. Increasing ability to abstract and generalize would enable groups to reanimate their intragroup coalition games in the more serious intergroup context. The usual and firmest coalitions would be between related groups, as is the case with coalitions of individuals (usually males) in wild turkeys,[27] lions,[25] and chimpanzees.[21] Are group fights necessarily more serious for the species if, on the analogy of the superorganism, we are allowed to equate a few deaths to a minor wound? Perhaps not, and of course it is possible that making groups more aggressive would not 'melt' the lattice structure to the extent suggested. Moreover, groups might be units in supergroups that are themselves in a 'stepping-stone' lattice. In such cases warfare—for that is the behaviour we now survey—might carry over from intragroup behaviour (or itself spontaneously develop) quite orderly restrained procedures involving little loss of life.

In developing this admittedly speculative outline of certain cultural and genetic processes in tribal evolution, I confess a bias towards discovering the patterns of coalitions, warfare, language, contempt, and so on that are documented in certain remote peoples of the present day—for example, the Yanomamo[28] and various New Guinea highlanders.[29,30] Admittedly in these cases there is agriculture; it is possible to claim that most hunter–gatherers are more peaceable. For example, why not aim to derive the customs of the Kalahari Bushmen (San)? But most hunter–gatherers are certainly less peaceable than Bushmen. The record of human violence goes back far indeed, even if the earliest attributions (Dart's cases in *Australopithecus*[31]) are doubtful. A trace of homology with the sporadic violence of chimpanzees seems not impossible. Probably with hominids, as with chimpanzees, actual violence towards outsiders would contrast with restrained violence, or mere threats, used within the group, while within the group too there would be much sharing and co-operation. One Neanderthal skeleton of Shanidar Cave had bone damage suggesting a stab wound.[32] Another skeleton also had bone defects but of quite different implication: one forearm was lacking, perhaps from birth, certainly for a long time, and a healed injury to the skull showed that one eye was blind. Goodall's chimpanzees were part hostile, part sympathetic, and part indifferent to comrades suddenly crippled with polio: they did nothing positive to support them. In contrast, the Neanderthals of Shanidar evidently supported a cripple, and on his death they buried him in the cave where, in other graves, they also sometimes buried their dead with flowers. These hints of violence and loyalty and (perhaps most purely human) of incipient love of things for themselves evoke a startlingly familiar and sympathetic portrait. Considering only the same affectional attributes in the present-day tribal and pastoral Kurds (as opposed to attributes connected to the ever-accelerating change in material culture), a recording angel perhaps notes today much the same events in Shanidar Cave as he noted an ice age ago.[32,33]

Probable instances of cannibalism in *Homo erectus* and Neanderthals have been plausibly compared to similar recent cannibalism in New Guinea. In New Guinea, it is interesting to note, this practice acted as a kind of population control, since by eating the brain of his victim, it was believed, the headhunter won a name for a child of his own—in effect, won a birthright. Other usually less drastic beliefs and practices affecting fertility that are widespread in human cultures may help to explain how they manage to be as peaceful as they are. In so far as the practices amount to effective birth control, they cut warfare at its demographic root. Unfortunately, it is possible that in doing so they also cut an important link that has escalated the selection for intelligence.[34–37] No hunt needs quite so much forethought or ability to communicate complex instructions as does a war, nor do such drastic demographic consequences hinge on the outcome.

The rewards of the victors in warfare obviously increase for peoples past the neolithic revolution. There are tools, livestock, stores of food, luxury goods to be seized, and even a possibility for the victors to impose themselves for a long period as a parasitical upper class. Hunter–gatherers, on the other hand, at most win only mates and land. It might seem that these things would not repay the expected cost of the fighting, but it has to be remembered that to raise mean fitness in a group either new territory or outside mates have to be obtained somehow. The occurrence of quasi-warlike group interactions in various higher primates[26,38] (and references in Bigelow[35]) strongly suggests that something like warfare may have become adaptive far down in the hominid stock. These primate examples suggest the prototype war party as an all-male group, brothers and kin, practised as a team in successful hunting and at last redirecting its skill towards usurping the females or territory of another group. Out of such cells can be built the somewhat less stable organism of the post-neolithic army. The Homeric *Iliad* gives a vivid inside view of the process of coalition, while the siege it describes emphasizes the existence of economic surpluses supporting the warriors on both sides (something hunter–gatherer warriors would never have). If the male war party has been adaptive for as long as is surmised here, it is hardly surprising that a similar grouping often reappears spontaneously even in circumstances where its present adaptive value is low or negative, as in modern teenage gangs.[39]

Whether or not the neolithic revolution brought an increase in the *per capita* incidence of violence it does seem that from then on warfare looms larger in the affairs of men. The situation seems reflected in the fact that only one of various series of pre-neolithic cave paintings depicts warfare,[40] whereas for most early civilizations the earliest known written records of warfare, booty, captives, and the like.

It has been argued that warfare must be a pathological development in humans, continually countered by natural selection, and this claim is sometimes based on a sweeping *a priori* view that habits of mortal intraspecific fighting must always endanger the survival of a species.[24] While endorsing such a view as regards wars between the few frightfully armed superpowers of today, I see no likelihood for it as regards fighting of individuals or of groups up to the level of small nations. Of course, for the species as a whole, and in the short term, war is detrimental from the biological demographic point of view, but, as shown above and elsewhere, detriment to the species does not mean that a genetical proclivity will not spread. Anyway, what is bad at one level may be good at another and the cost to the species may be paid in the long run. The gross inefficiency of warfare may be just what is necessary, or at least an alternative to birth control and infanticide, in order to spare a population's less resilient resources from dangerous exploitation. Maybe if the mammoth-hunters had attacked each other more and the mammoths less they could be mammoth-hunters still. And the rich ice-age fauna of the Americas might have had time to adapt to the human predator as it adapted in Africa if fighting had induced man to draw his curtain of overkill across the continent less rapidly. Many examples in the living world show that a population can be very successful in spite of a surprising diversion of time and energy into aggressive displays, squabbling, and outright fights. The examples range from bumble bees to European nations. In case all this reads like a paean for fascism let me add one caution from the geological record. Arms and armour seem to weigh one down in the end: it is hard in the modern human world to see warfare as a stabilizing influence.

The relatively peaceable Bushmen may tell us something valuable about the aetiology of wars, but I am doubtful if they tell us much about the role of this factor in the main stream of human evolution. However, it is noteworthy that the Bantu who replaced the khoisaniform races in much of the rest of Africa were warlike in ways that evoke comparisons from the dark ages of Europe. The Bantu were, of course, mainly pastoralists and agriculturalists, for whom, as stated above, booty would be an important additional incentive to warfare. Pastoralists tend to be particularly warlike and the histories of civilization are punctuated by their inroads. Pastoral tribes have to be mobile in following or driving their herds and this mixes tribes and reduces relatedness of neighbours. Viewed as booty, the mobility of stock is a great convenience. Both factors must contribute to the warlike propensity.

The incursions of barbaric pastoralists seem to do civilizations less harm in the long run than one might expect. Indeed, two dark ages and renaissances in Europe suggest a recurring pattern in which a renaissance follows an incursion by about 800 years. It may even be suggested that certain genes or traditions of the pastoralists revitalize the conquered people with an ingredient of progress which tends to die out in a large panmictic population for the reasons already discussed. I have in mind altruism itself, or the part of the altruism which is perhaps better described as self-sacrificial daring. By the time of the Renaissance it may be that the mixing of genes and cultures (or of cultures alone if these are the only vehicles, which I doubt) has continued long enough to bring the old mercantile thoughtfulness and the infused daring into conjunction in a few individuals who then find courage for all kinds of inventive innovation against the resistance of established thought and practice. Often, however, the cost in fitness of such altruism and sublimated pugnacity to the individuals concerned is by no means metaphorical, and the benefits to fitness, such as they are, go to a mass of individuals whose genetic correlation with the innovator must be slight indeed. Thus civilization probably slowly reduces its altruism of all kinds, including the kinds needed for cultural creativity (see also Eshel[11]).

RECIPROCATION AND SOCIAL ENFORCEMENT

The last suggestion is rather different from saying, as has sometimes been said, that civilization selects against all kinds of creative intelligence. It seems to me that there are some aspects of innate intelligence that civilization steadily promotes. Mercantile operations, for example, are an inseparable part of Old World civilizations and need complex models in the minds of their operators, just as military ventures do. The main difference is in more emphasis on prudence and less on daring. It is probable that civilization has given steady selection for the intelligence needed for this mercantile kind of preparatory modelling. The intelligence that gives a good appreciation of the real principles involved in a new technology, as opposed to seeing it as a kind of magic, is probably also constantly favoured, since improvers of a technology avoid the arrows of contempt and penury that face pioneers and can do very well. However, my main reason for turning to the subject of trade is to introduce the idea of another kind of positive social arrangement which thrives in a mercantile and technological atmosphere, for which intelligence is more necessary and relatedness much less so. This refers, of course, to reciprocation.

Starting perhaps with something like the meat-sharing of chimpanzees ('feed me while you have plenty and I'll feed you when I have plenty'), proceeding through barter (where differing aptitudes may begin to be important), reciprocative activity branches out into all the various business-like arrangements of modern humans. The key words are 'client' and 'partner' as opposed to 'kinsman' and 'friend'.

Establishing a basis for reciprocating has problems of natural selection closely similar to those of altruism as discussed so far. It is very frequently necessary for one party to execute his half of a bargain without any way of being certain that the other party will later stick to his. The best response if the other does not reciprocate is to cut off any further benefits to him.[41] Unfortunately, this leaves the selfish non-reciprocator better off than the 'altruistic' initiator and unless the two are related this is against the habit of reciprocation, at least when the trait is rare (see discussion of Prisoner's Dilemma in Chapter 6; also Boorman and Levitt[42]). However, this initial barrier to selection is a slight one if the rewards of the interaction are high, as they would be when an advantageous exchange can be repeated many times.[41,42] Once the barrier is passed by genetic drift or the like, non-reciprocation finds itself in the category of maladaptive spite—harming the self to harm others more. Nepotistic altruism, of course, also has an initial barrier to pass when it first occurs by mutation. But once positive selection supervenes the resemblance between the two situations fades: reciprocal altruism of the kind described is less purely altruistic. Indeed, the term altruism may be a misnomer: there is an expectation of benefit of the initiating individual, not just an expectation of benefit to the genotype. To put the matter another way, reciprocal altruism can never be suicidal, whereas suicidal nepotistic altruism can and has evolved—it is apparent, for example, in worker sacrifices in the social insects.

Whether reciprocation involves altruism or not, we see that in so far as it involves repeated acts between the same two individuals this useful and immensely variegating type of interaction can spread genetically, given only an ability to remember individual faces of those who have helped and those who have cheated in the past. Unfortunately, by the very aid it gives to the growth and diversification of social systems, reciprocation tends to undermine the basis of its success. Situations demanding reciprocation just once between individuals destined never to meet again naturally become more common and it becomes easier for cheaters to specialize in these and to hide from retribution. Cheating can also become more subtle, especially along lines which make it hard for victims to be sure just who has cheated them.

In considering this problem I think there may be reason to be glad that human life is a 'many-person game' and not just a disjoined collection of 'two-person games'. Admittedly, it may not seem so at first. At first reading the theory of many-person games may seem to stand to that of two-person games in the relation of sea-sickness to a headache. But given also a little real intragroup altruism endowed from the tribal past, it may turn out that the one is at least a partial cure for the other (see Fig. 9.1). The idea here is that for pairs in isolation the problem of cheating in a single exchange may be insoluble, and that therefore we have all evolved, more or less in proportion to our exposure to civilized (i.e. relatively panmictic) conditions, into potential cheaters. But at the same time we also have every reason to agree as to the parasitical nature of cheating as it affects the welfare of the community as a whole and to deplore its successful practice by others. So detection of cheating arouses indignation in everyone except the accused, and everyone sees a benefit to both group and self in trying to punish the cheater and in forcing restitution (some part of which, as an added incentive, may be diverted as a fee for those who administer the collective justice). The reason I believe that a little real intragroup altruism is also necessary for the evolution of efficient justice on these lines is that individuals must feel the difference between the usefulness of this behaviour and the futility of using collective power arbitrarily in ways profitless to the group. A healthy society *should* feel sea-sick when confronted with the endless internal instabilities of the 'solutions', 'coalition sets', etc., which the theory of many-person games has had to describe. One hears that game theorists, trying to persuade people to play even two-person games like 'Prisoner's Dilemma', often encounter exasperated remarks like: 'There ought

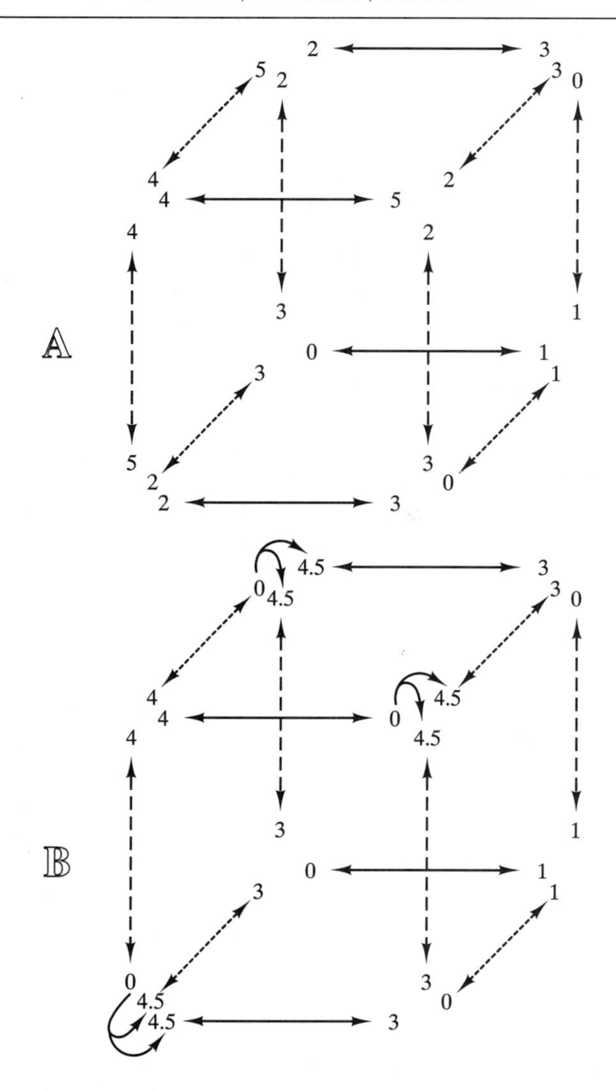

Figure 9.1 Justice and the 'Prisoner's Dilemma'.[43] Case A shows a 'three-person Prisoner's Dilemma'; case B shows the expected modification of this when justice can be enforced by a numerical majority. In every case a player has two strategies, 'co-operative' and 'selfish', and decisions between them are indicated by double-ended arrows, which are solid, dashed, or dotted to correspond to the identities of the three players. Arrow ends tending towards the near upper left corner of the cube represent a decision to be co-operative; arrow ends tending to the far lower right corner of the cube represent a decision to be selfish. Numbers at the ends of arrows show the players' payoffs. Thus, naming the players in the obvious way, if 'solid' plays co-operatively but 'dash' and 'dot' both play selfishly the outcome is represented at the far lower left corner of the cube: 'solid' gets zero, while 'dash' and 'dot' get three each.

In case A, payoffs are arranged so that every face of the cube presents a 'Prisoner's Dilemma' to each of the three possible pairs of players. Thus whatever the other pair is

Figure 9.1 *(continued)*
doing a player does best by playing selfishly; as in Prisoner's Dilemma the rational yet 'paradoxical' result of this is that the players all play selfishly and get one unit each, whereas had they all played co-operatively they could have four units each.

In case B, it is assumed that if two players play co-operatively while a third is selfish, the two use their united strength to 'punish' the third. His selfishly-gained five units are taken from him and divided between the two co-operators. This creates a situation where no player sees an advantage in departing from triple co-operation. Likewise no player has an advantage in departing unilaterally from the triply selfish corner; but now any two players can decide to depart from this corner by a coalition which is not only advantageous (as it was also in case A) but also secure in that neither member can benefit by defaulting.

to be a law against such games!' Some of the main points of this paper can be summarized as an answer to this comment: that often, in real life, there is a law, and we can see why, and that sadly we also see the protean nature of this Dilemma, which, when suppressed at one level, gathers its strength at another.

References and notes

1. S. L. Washburn, P. C. Jay, and B. Lancaster, Field studies of Old World monkeys and apes, *Science* **150**, 1541–7 (1965).
2. G. C. Williams, *Adaptation and Natural Selection: a Critique of Some Current Evolutionary Thought* (Princeton University Press, Princeton, NJ, 1966).
3. R. C. Lewontin, The units of selection, *Annual Review of Ecology and Systematics* **1**, 1–18 (1970).
4. R. Fox, in *Kinship and Marriage* (Penguin, London, 1967), also emphasized this in discussing human kinship systems and why some which are easily conceivable never actually occur. His discussion of the incest taboo is also very pertinent to the idea that follows.
5. G. R. Price, Extension of covariance selection mathematics, *Annals of Human Genetics* **35**, 485–90 (1972).
6. G. R. Price, Selection and covariance, *Nature* **227**, 520–1 (1970).
7. Although Price first pointed out the generality and usefulness of this relation, earlier partial recognition of it seems to be due to Alan Robertson. See A. Robertson, A mathematical model of the culling process in dairy cattle, *Animal Production* **8**, 95–108 (1966).
8. J. B. S. Haldane, *The Causes of Evolution* (Cornell University Press, New York, 1966; first published 1932).
9. For example, J. Maynard Smith, Group selection and kin selection, *Nature* **201**, 1145–7 (1964) and Lewontin 1970 (ref. 3).
10. J. L. Brown, Alternative routes to sociality in jays, *American Zoologist* **12**, 642 (1973).
11. I. Eshel, On the neighbour effect and the evolution of altruistic traits, *Theoretical Population Biology* **3**, 258–77 (1972).

12. R. Levins, Extinction, in M. Gerstenhaber (ed.), *Some Mathematical Questions in Biology* **2**, pp. 77–107 (American Mathematical Society, Providence, Rhode Island, 1970).
13. S. Wright, Tempo and mode in evolution: a critical review, *Ecology* **26**, 415–19 (1945); Factor interaction and linkage in evolution, *Proceedings of the Royal Society of London* B, **162**, 88–104 (1965); and other papers 1945–65.
14. S. A. Altmann and J. Altmann, *Baboon Ecology, African Field Research* (S. Karger, New York, 1972).
15. M. Kimura and G. H. Weiss, The stepping stone model of population structure and the decrease of genetic correlation with distance, *Genetics* **49**, 561–76 (1964).
16. T. Maruyama, Analysis of population structure, II. Two-dimensional stepping stone models of finite length and other geographically structured populations. *Annals of Human Genetics* **35**, 182–96 (1972).
17. As pointed out elsewhere (see Chapters 5 and 6), even spiteful behaviour, harming oneself in order to harm another more, is a theoretical possibility. The mean relatedness to the entire species population other than self is $-1/N - 1)$ where N is the population. If by inbreeding or otherwise a colony has grouped n identical genotypes together, then relatedness to the average outsider is $-1/(N - n)$. Thus with only a few large long-isolated groups spite is more possible.
18. W. Kuhme, Freilandstudien zur Soziologie des Hyänenhundes (*Lycaon pictus lupinus* Thomas 1902), *Zeitschrift für Tierpsychologie* **22**, 495–541 (1965).
19. H. van Lawick and J. van Lawick-Goodall, *Innocent Killers* (Collins, London, 1971).
20. H. Kruuk, *The Spotted Hyena* (University of Chicago Press, Chicago, 1972).
21. J. van Lawick-Goodall, *In the Shadow of Man* (Collins, London, 1971).
22. J. D. Bygott, Cannibalism among wild chimpanzees, *Nature* **238**, 410–11 (1972).
23. L. D. Mech, *The Wolf: the Ecology and Behaviour of an Endangered Species* (The Natural History Press, New York, 1972).
24. K. Lorenz, *On Aggression* (Methuen, London, 1966).
25. G. Schaller, *The Serengeti Lion: a Study of Predator–Prey Relations* (Chicago, Chicago University Press, 1972).
26. Y. Sugiyama and M. D. Parthasanathy, A brief account of the social life of Hanuman langurs, *Proceedings of the National Institute of Sciences, India* **35B**, 306–19 (1969).
27. C. R. Watts and A. W. Stokes, The social order of turkeys, *Scientific American* **224** (6), 112–18 (1971).
28. N. A. Chagnon, *Yanomamo: the Fierce People* (Holt, Rinehart & Winston, New York, 1968).
29. P. Matthiessen, *Under the Mountain Wall* (Heinemann, London, 1962).
30. R. A. Rappaport, The flow of energy in an agricultural society, *Scientific American* **224** (3), 116–32 (1971).
31. R. Ardrey, *African Genesis* (Collins, London, 1961).
32. R. S. Solecki, *Shanidar: the Humanity of Neanderthal Man* (Allen Lane, London, 1971).

33. A. M. Hamilton, *Road through Kurdistan* (Faber, London, 1937).
34. A. Keith, *Essays on Human Evolution* (Watts, London, 1946).
35. R. Bigelow, *The Dawn Warriors* (Hutchinson, London, 1969).
36. R. D. Alexander, The search for an evolutionary philosophy of man, *Proceedings of the Royal Society of Victoria* **84**, 99–120 (1971).
37. S. Andrewski, The case for war, *Science Journal* **7** (1), 89–92 (1971).
38. H. Kummer, *Social Organization of Hamadryas Baboons* (Chicago University Press, Chicago, 1972).
39. J. Patrick, *A Glasgow Gang Observed* (Eyre Methuen, London, 1973).
40. L. Pericot, The social life of Spanish palaeolithic hunters as shown by levantine art, in S. L. Washburn (ed.), *The Social Life of Early Man* (Methuen, London, 1961).
41. R. L. Trivers, The evolution of reciprocal altruism, *Quarterly Review of Biology* **46**, 35–57 (1971).
42. S. A. Boorman and P. R. Levitt, A frequency dependent natural selection model for the evolution of social cooperation networks, *Proceedings of the National Academy of Sciences, USA* **70**, 187–9 (1973).
43. The game-theoretic situation known as Prisoner's Dilemma has as paradigm the dilemma of a criminal in custody who is offered the inducement of a light sentence if he will give state evidence to clear up an important crime, while knowing that a confederate, also in custody, has the same offer and that the pay-off situation which both face is as follows, arranged from lightest to heaviest sentence:

    (c) I confess, he doesn't > (a) We neither confess > (d) We both confess > (b) He confesses, I don't.

    So that (c) can be better than (a) it is assumed that they can certainly be convicted of some minor offence whereas at least one confession is needed in order to settle the major crime. For further information on this 'game' see A. Rapoport and A. M. Chammah, *Prisoner's Dilemma* (University of Michigan Press, Michigan, 1965).

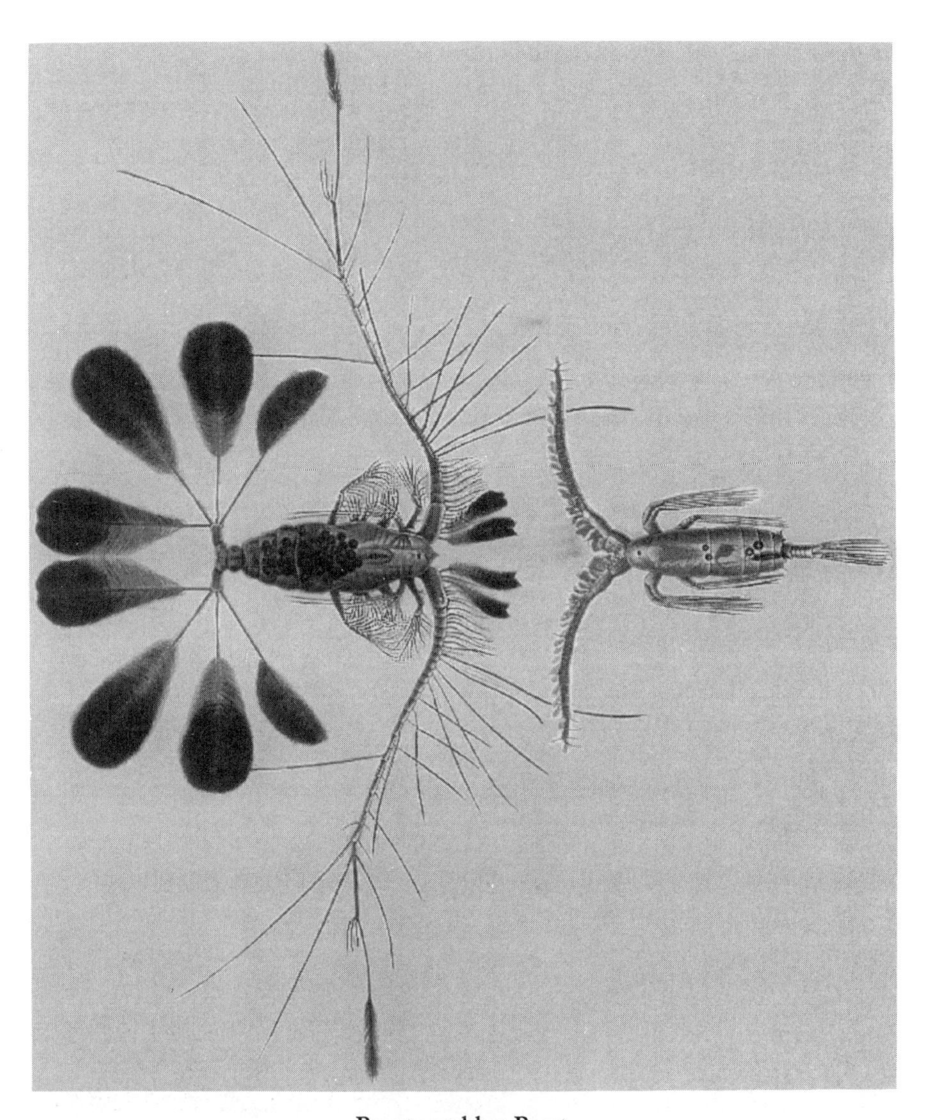

Beauty and her Beast

A spectacular pelagic female zooplankter of the Mediterranean, *Calocalanus pavo*, is shown facing her less-ornamental male. Little seems known of this copepod but conduct in others suggests the male may provide a substantial present of nutriment surrounding his sperm, attaching his package externally to the female's body from which she later harvests it. Thus, in a reversal of the usual situation (although not necessarily the human one), it is expected that the female's high ornamentation is a necessity to dazzle and court the male before his gift will be transferred.

CHAPTER 10

VENUS TOO KIND

Gamblers Since Life Began: Barnacles, Aphids, Elms

Sexual reproduction is the chef d'oeuvre, the masterpiece of nature.
ERASMUS DARWIN[1]

THIS book does not intend to republish book reviews but I include this one partly because it is long and I gave it unusual care, and partly also because it marks a turning point in my interests away from social behaviour and towards the problem of sexuality. Just as altruism and its related topics were the first major interest of my life, the study of sexuality has been, roughly, the second. My present research concerns states of parasitism and disease, and counteradaptations of hosts, as all of these relate to sexuality. Papers mainly about how sex can be supported by parasitism will make up Volume 2 of these collected works. At the stage when I wrote the review, although I had not seen the particular relevance that parasitism might have, I had for many years seen sex looming ahead and had reached the stage of being excited by the possible primary role of biotic interaction.[2] I had decided that it was in aspects of the interspecies struggle, and not survival in the inanimate environment, that I had to search for the main factor. Adaptation to new physical habitats might be helped by novelties made possible through sexuality but these adaptations could not be the main reason for its existence.

In the light of those new interests, I was mining the libraries again but with a new lure drawing me on. It was not in any sense sex on my own behalf, not prurient interest, nor yet focus on single species. It was now variation in the widest sense, sex *en masse*—of entire flowering meadows, sex of tropical forest, of crowded deep-sea floors, a multitroilism and multibestiality, if you like, where always many species besides that of my focus of the moment were participating. If my own body came into it, it was to wonder about what sex meant for the worms (of two phyla at least) that could lie entwined in love within my own veins or capillaries. I almost wanted to be infected by those worms so that I might 'feel' their strategies; indeed, I did for a while imagine, wrongly, that I might have acquired one of them (*Onchocerca*) during my time in Brazil. After such admissions I hope it will not offend the two authors of the excellent books I review in the chapter to say that in general my thoughts on all of this massive theme seem to have run on fairly parallel lines to theirs, especially to those of George Williams, whose evolutionary ideas have always seemed convergent like those of a twin or older brother from whom I was parted at birth.

In my own case two developments had been helping to bring sex into focus. One was the publication by J. F. Crow and M. Kimura in the mid-1960s of two papers challenging a certain long-accepted orthodoxy, which I will call the Weissmann–Muller–Fisher theory. Essentially, Crow and Kimura had pointed out the uncharacteristically 'groupish' natural selection that had been invoked for the case of sex by these three normally very 'individualist' originators. Although allergic virtually since my school days to ideas of 'group benefit' and equally to 'species benefit' reasoning about evolution, I had somehow let my awe of R. A. Fisher slip this one argument, or lack of argument, past me without serious question. Never had I given it the careful thought that Crow and Kimura now applied. The second development arose from my own reasoning on sex ratios, which had shown how, under conditions of extreme inbreeding, where any advantages of sexuality via its consequences for gene recombination (or heterozygosity) are nearly eliminated, reproduction could and did evolve to be much more demographically efficient. Such efficiency came through a drastic reduction of male production, raising very acutely the problem of why males were ever there in the first place. Both my first strongly upward-arching graph of my simulation in my 'Extraordinary sex ratios' (Chapter 4, Fig. 4.1) and of all that paper's small inbreeders with their male-deficient sex ratios, seemed now to be joining in chorus to force me to attend to the issue. In human life it has always been mainly females shouting at me that I don't understand something; now from inside my own breeding vials the shout was the same. If from outside, from the human world, the shout was mainly 'Do something that matters!', from inside the vials it was deeper and worse: 'Why are males bothering us, why are you there?' In so far as such hubbub had come via my sex-ratio paper, of course, it had to be from females: the males were hardly to be seen, which was one of the paper's main points. But more directly serious was that many of the most vociferous species were extremely successful and some indeed so successful as to be pests.

These facts, and the not infrequent phenomenon of parthenogenesis found close to or among these species, turned my attention to the question of why evolution, which was generally such a thrift-loving process and so attentive to detail and to purpose, had ever brought males into existence at all. How did they persist? Or, as a question slightly less acute because much more distant (it concerns events of the Precambrian era), how did they first arise? All British gardeners know the military efficiency with which purely female aphids invade their rose bushes or broad beans and with what perceptible daily increase squadrons of bulbous bodies sweep onwards along a stem. I knew from my reading that parthenogenesis such as these aphids have is present sporadically throughout both animal and plant kingdoms. If suitable mutations to parthenogenesis can happen, as these cases prove, and if parthenogenesis is so efficient, why weren't waves of such self-sufficient females overwhelming more than just the stems of roses—in short, replacing sexual organisms on all sides? How in the long run has this crazy whim maleness proved itself the opposite of what it should have been—a fleeting disaster and a long-abandoned experiment?

When the second volume of this work comes out, you will know more, if you read it, about these questions, and about sex as seen from my perverse, unsexy, yet fundamental angle than you probably ever wanted to. For now, however, the double book review that follows hopefully introduces some of the questions that will inspire the chapters of Volume 2.[3] In so doing it can speak for itself: I tried hard to make it as readable as possible. How much and how little of what I now see as the true solution I was understanding in the early 1970s is I think revealed in the second half of that paragraph towards the end, where I speak of the possible sources of periodic selection including the biotic.

Reading it makes me wish I could have been more far-sighted; on the other hand, the naivety that I frown at now is, I suppose, the sign that I have made progress.

References and notes

1. E. Darwin, *Phytologia, or the Philosophy of Agriculture and Gardening* (1799); see p. 103.
2. C. M. Lively, Rapid evolution by biological enemies, *Trends in Ecology and Evolution* **8**, 345–6 (1993).
3. If eager for this theme, a good start can be made through Matt Ridley's *The Red Queen* (Viking, London, 1993).

GAMBLERS SINCE LIFE BEGAN: BARNACLES, APHIDS, ELMS[†]

W. D. HAMILTON

A REVIEW OF

The Economy of Nature and the Evolution of Sex
BY MICHAEL T. GHISELIN
University of California Press, Berkeley, Los Angeles, and London
xii + 346 pp.; index. 1974

Sex and Evolution
BY GEORGE C. WILLIAMS
Princeton University Press, Princeton, NJ
x + 200 pp.; index. 1975

When Kostoglotov read in *Pravda* (in Solzhenitsyn's *Cancer Ward*) of a complete change of the personnel of the Supreme Court of the Soviet Union, 'Beethoven's four muffled chords of fate' thundered above his head. When he read two days later that prime minister G. M. Malenkov had resigned, he heard those chords again. I think Kostoglotov's sensations were mine exactly when I opened first one and then the other of these two books sent for review. Like those issues of *Pravda*, these books also push to one side some old idols, once almost as grim enemies of mine as were stalinists for Kostoglotov. Yet this is not really the cause for excitement: those idols, 'Evolution for the benefit of the species' and related notions, have been dealt with in earlier books by G. C. Williams. The chords of fate sound for me because these books intend, each in its own way, to bring under the full searching gaze of science what seems to me biology's most outstanding problem. Even to see clearly that such a problem exists is partly to solve it, and one book (Ghiselin's) appears almost to claim that it is solved already.

[†]*The Quarterly Review of Biology* **50**, 175–80 (1975).

As may be guessed from the two titles, this problem is simply that of—well, how state it most vividly?—why all this silly rigmarole of *sex*? Why this gavotte of chromosomes? Why all these useless males, this striving and wasteful bloodshed, these grotesque horns, colours, ... and why, in the end, novels, like *Cancer Ward*, about love?

Both books evoke an urgent sense of change. The reader feels near to evolutionary ideas of a generality hardly envisioned since the idea of natural selection. Even chapter headings suggest fearsome generality: 'The elm–oyster model', 'The strawberry–coral model' (Williams); 'The loves of the plants', 'A new theory of moral sentiments' (Ghiselin). And beyond biology, what a distant influence a 'solution' of sex may have—for example, on how we understand love. Perhaps Ghiselin chose chapter headings mainly with sales of his book in mind, but I think few will deny that there is really a deep and serious theme here, or at least a threat of a theme, as doubters of the omnipotence of natural selection may prefer to see it.

Besides attention to the same central theme, the books have much else in common both in matter and approach. Most striking, an emphasis on the overwhelming predominance of selection at the individual level leads both authors to insist that sex must be shown to be advantageous to the individual sexist, not just to population or species as a whole. Further, both see the advantage of sex as connected with environmental uncertainty and concentrate, in effect, on the problem of why sex does not give place to apomixis rather than on the derivation of fertilization and meiosis in the first place. Both (let W stand for Williams and G for Ghiselin) start at much the same level and invite a very general readership by using little (W) or no (G) mathematical formalisms. They are effectively open to anyone who has grasped Darwin, fundamental genetics, and a little about the diversity of the plant and animal kingdoms. Nevertheless, the reader who is determined to be more than superficial may, if he has read no other recent work in evolutionary theory, find the books hard going, hard in the one case (W), perhaps, because rather deeper in theory, and hard in the other (G) because so encyclopaedic in its references to species and taxa—and also, perhaps, because hard to give needful resistance to the writer's literary skill.

In spite of such differences, I think the books will be liked and disliked by the same broad groups. Both writers are hopeful of finding sweeping evolutionary convergences in phylogenetically distant groups. There are biologists who reject such hopes. The books will also not appeal to those who dislike speculation. Still less will they appeal (as I hope is hardly relevant for biologists) to those who feel that even valid evolutionary interpretation must debase a more pristine and joyful vision of the living world. I hear this said, follow syntax, try to believe it sincere: for my own part the case could hardly be more different. As when first told about evolution it was an intense delight to see the star-like flowers of cabbage and turnip shine on the twigs of some much vaster tree, so now when I am shown a realistic vision of spiders as barnacles of the air, their effete males dwarfed by similar causes (G), or, still more abstract, just some tentative new linkage between old puzzles—say, between the physiques, the mortalities, and even the fidelities of man and woman (W).

Now more as to differences. *The Economy of Nature and the Evolution of Sex* is a much longer book and ranges over a wider field, including sexual selection (which Williams also treats but more briefly), social behaviour, and even something of the philosophy of biology and of scientific investigation in general. Ghiselin aims to give a rather full survey of the phenomena relevant to each of his biological topics and marshalls a very impressive list of references. This organized coverage should make the book valuable to all who are interested in the problems it surveys, whether or not its interpretations are accepted.

The author (G) is particularly good on early literature, including that of the last century. This reflects, however, an odd and contentious streak. He seems to see less a recent acceleration in evolutionary biology leading up to the present interest than a 100-year hiatus of thought. All that has gone on since Darwin, he seems to say, is mere rediscovery of, or ornament to, the ideas of the master. It is stimulating to be forced to consider such a point of view, but it seems to me that, brave and perspicacious as Darwin's reviews undoubtedly were and timid and feeble as has been much of the writing on similar issues since then, Ghiselin sometimes draws too much out of rather vague and tentative statements by Darwin, and also overlooks far too much in later work. The neglect of recent work is especially marked in the realm of more formal theory. It is striking that J. B. S. Haldane, R. A. Fisher and S. Wright get just one reference each in the list of 1500. (But since they hardly do better in Williams's book perhaps this just reflects our new focus: anyway, I quickly found out that it is not necessarily a compliment to be in that 1500!) I do not recall reading praise for any recent contribution to theory whereas all manner of theoretical utterances are actively criticized. A conspicuous talent for epigram and swingeing metaphor is very entertaining, providing one is not oneself in the line of fire. Fortunately I agreed with most of his attacks on anti-selection and species-benefit thinking in the early chapters. One sentence will show the style: 'The idea that natural economies evolve in the direction of efficiency is just as much of a metaphysical delusion as the notion that good will inevitably result if only government will let capitalists do as they will.'

Such broadsides enliven an already well-written story. Metaphors shine and no tables, algebra, or graphs bar the reader's path—characteristically, the only diagrams in the whole book are put there for our amusement rather than for information. But what of the book's own theme? Are the writer's opinions to be taken any more seriously than all those he so ably attacks, or than the ideas of other almost too readable books of recent years in the field of ethology and evolution? It seems to me that there are indeed germinal insights in this book but also a need for much caution and further refinement.

Ghiselin's main theme is that many adaptations of obscure function, including sex itself, would be more readily interpretable if we were to apply, both in an intraspecies and an interspecies context, a model of a *laissez faire* human economy. In other words, organisms adapt on much the same principles as men are supposed to act in the pages of Adam Smith, towards the maximization of individual advantage in every case. 'The economy of nature is altogether individualistic, and altruism is a metaphysical delusion' (p. 25). Such a model is indeed much better than one of a wholly co-operative economy would be, and I agree that what amounts to tacit acceptance of the latter has confused the study of evolution for a long time. But just as the *laissez faire* model has shown inadequacies in economics, so even more obviously the idea of total individualism is inadequate for the rest of the living world. To support his statement about altruism Ghiselin has to say, in effect, that a 'family' or 'breeding stock' is the equivalent of an individual. Maybe in some sense it can be *almost* equivalent; nevertheless, it seems to me both more exact and less 'metaphysical' to stick to common usage and say that worker bees are the individuals and that they behave altruistically when, like the historic Burghers of Calais, they are seen to offer their lives to save their colony. The economic model is certainly a useful guide to insight and has been recognized as such by various biologists since Darwin—by Fisher for example—but no analogy can fully substitute for careful thought about the particulars of a problem. Throughout the book the lack of more specific models led to a sense of enlightenment deferred: from remarks about firms, markets, advertisements, and the like, which seemed a mere preliminary to the main revelation, one would find oneself hastening on into new snow drifts of fascinating facts, new battles against erring opinions, as though this was another load that had to be appreciated before the whole could be made clear—but it never was.

Possibly it is on issues of energetic adaptation that Ghiselin's approach is most revealing: I liked, for example, his discussion of why large-colony social bees are polylectic while unsocial bees tend to be specialists. But it is to be noted that this issue about social bees is much less intrinsically social than many of the others treated, like sexual dimorphism or sex ratio. On these, Ghiselin's failure to consider relatedness and the implications of breeding structure give him a most serious handicap. To use his own kind of analogy, it is as if he discounted entirely the possibility that the different directors of a firm could ever fall out with one another, or as if he supposed that firms would lose nothing in efficiency if laws forced them to assemble a multinational staff without regard for common language.

It is odd, yet in a way characteristic, that the extreme proponents of both group selection and individual selection should equally neglect the problems of breeding structure and grades of relatedness. I even find it encouraging as to the significance of my own interest in this field; at least, I feel it easier and safer to try to hold a territory between Ghiselin and V.C. Wynne-Edwards than to camp, as I once saw it, in the neutral strip between Fisher and Sewall Wright. Most of Ghiselin's criticisms of my own views on social insects are in effect answered in a paper [see Chapter 8] which is only briefly mentioned in this book (perhaps having been seen too late). Where he tries to sustain his view that the social insect colony is in some way entitled to treatment as an individual, he seems to be saying, in rough paraphrase, that he agrees that 1 is different from 0 but he cannot see any appreciable difference between $\frac{3}{4}$ and $\frac{1}{4}$. The common male-biased sex ratio among reproductives, which he regards as 'perhaps the most serious difficulty' to my thesis, does indeed suggestively fulfil expectations on the superorganism view in some cases, yet still I see the matter rather differently. In a species whose colonies reproduce by swarms the 'female' unit of reproduction is the swarm itself—queen plus the workers necessary to make a viable unit—so if by Fisher's Principle we make the mass of males equal to the mass of emitted swarms we see that males are expected to be much more abundant than queens in genera like *Apis* and *Eciton*. However, I believe that Fisher's Principle is only *exactly* applicable under a most unlikely set of conditions: workers completely sterile, mother queen in full control of sex ratio and of larval nutrition, population panmictic, males with no greater dispersive power than the swarms, and doubtless others. The matter is obviously complex and a careful review is needed before we can say whether predictions based on relatedness are upheld. Meanwhile, it can only be said that the situation in *Apis* and *Eciton* gives striking support to Fisher's general idea on the equilibrium when the sexes are differently expensive to produce. At another point it seems worth recording a counterobjection both to Ghiselin and to those he joins in emphasizing the role of parental manipulation of progeny in the development of the worker caste. In the Hymenoptera why don't the ruthless mothers manipulate *sons* to form the worker caste? Being wholly derived from the mothers they should be much easier to produce—a female need not even wait to get mated. Proponents of parental manipulation seem to be at pains to point out that males can do some useful acts of work, which tends to forestall any claims that they are simply not supplied with suitable instincts. Manipulation by parents undoubtedly *is* a factor in the case and I have to admit that enthusiasm for the 'altruistic' factor has led me to underemphasize the mother's position of power. Yet I am certain that an 'altruistic' willingness to be manipulated is also a factor and I still see hymenopteran daughters as natural masochists in this respect.

With characteristic omniscience Ghiselin cites papers unknown to me on male acts of work in *Polistes* and meliponines. But one of these and two out of three others that I did look up on his theme of dwarf males suggest his citations may sometimes lack much relevance to his themes. If we accept as 'dwarf' the male referred to in a mite, then so must the female of *Homo sapiens* be called dwarf; and the 'dwarf male' of the gastropod *Capulus ungaricus* turned out to be a hermaphrodite.

Much more strikingly dwarfed males could easily have been selected in mites, but whether they would have been any more relevant seems questionable, since the whole discussion of the 'dwarf male' theme in terrestial arthropods seemed confused, in contrast to the discussion of aquatic dwarf males, which I found fascinating. For the aquatic ones I know less of the facts, but since the author is a marine biologist and started the idea of the book out of this theme it is probable as good as it seems. On land Ghiselin's dwarf-male biofacies does link up well with similar conditions in spiders, but it also links fairly well with a 'wingless female' biofacies in insects, which surprisingly Ghiselin fails to notice. This covers coccids, psychid moths, gypsy moth and winter moth (and relatives), some embioptera, psocoptera, thrips, and various others. It can hardly be coincidence that all these insects are either known to have some 'planktonic' larval dispersal or have some possibility of it through small size, silk production, or general windiness of habitat. The adult male is typically much less bulky than the female. Besides these cases, there is another insect assemblage which tends to have small males: these could hardly be more different in general adaptation from the 'wingless female' lot, being for a start usually winged in the female and wingless in the male. Ghiselin notes a few examples and does see that they don't fit with a male dispersal hypothesis, but he fails to note the impressive range of their other convergences. They disperse as adult females, never as larvae, and have to my knowledge no parallel in the sea. Additional differences from all male-dispersing cases are in their habitual inbreeding, female-biased sex ratios and specialized diets (wingless female species tend to polyphagy: blown larvae must find food where they land, or else perish).

Similar superficiality seems probable elsewhere. The discussion of sexual selection showed little if any understanding of Fisher's suggestion about positive feedback from female preference, and virtually no recognition of Peter O'Donald's important work on this subject. Yet this is still the best hypothesis on hand to explain the exaggerated nature of so much sexual ornament.

I have said little on Ghiselin's views on the great problem of sex itself that I emphasized at the outset. This is because most of what he does say is said again and more clearly by Williams. This is not to imply, of course, that his thoughts on the subject are derived or that what he writes is redundant if one has read Williams. I found his review of the circumstances and phenomena of sex very exciting. But it is like reading Inspector Lestrade's report on a crime rather than Sherlock Holmes's, or, a closer analogy, like reading Erasmus Darwin's *Zoonomia* rather than *The Origin*. The crucial snag for more facile theories of sex is that which John Maynard Smith first brought fully to our attention: what Williams refers to as the *twofold cost of meiosis*. Although in one place he emphasizes the wastefulness of sex (p. 52), Ghiselin shows no clear recognition of this problem. I suspect that this is connected with failure to appreciate the issue of relatedness: he does not keep constantly in mind that a sexual offspring is only a half offspring to its parent.

Sex and Evolution is a much shorter book, and, in keeping with the tradition of its series, more tightly reasoned around a limited theme. It has the almost ruthless objectivity one has come to associate with its author and also much of his characteristic knack of seeing a problem from a totally fresh point of view. This has a bracing effect even if one doesn't agree (as happens, in my case, when he argues that there is no evidence that parthenogenetic taxa are evolutionary dead ends). He is much more willing than Ghiselin to suggest specific models but sometimes seems to have difficulty in suiting them to his intuitions, or at least in writing it all down. I remain very hazy about the details of some of his models, although for most I have come to see the outlines of some valid idea. On the other hand many passages are admirably clear throughout, and, to return to the main theme, the first chapter is one of these. After finishing it the reader will have no doubt about the seriousness of the cost of meiosis and will have lost, if he had not done so already, most of what he thought he knew about the reasons for sex—Müller, Fisher, good mutations being quickly combined into a single stock, and so on. The flaming sword which Williams dares to set at the close of the first chapter seems worth quoting in full: 'Maynard Smith's ... analysis convinces me of the unlikelihood of anyone ever finding a sufficiently powerful advantage in sexual reproduction with broadly applicable models that use only such general properties as mutation rates, population sizes, selection coefficients, etc.'

Yet a few chapters later it may appear that Eden is regained. So interesting and successful have been the models presented that the reader may feel that Williams has evaded his claimed 'unlikehood', and his final assessment of his own achievement in the very last chapter may seem like false modesty. Nevertheless, I think he is objective in this too. He has indeed achieved much in reviewing the problem and in providing special-case solutions. Yet something more general has yet to be shown if we are to understand the near universality and homogeneous properties of sex. Why the enthusiasm for pairs, for example, or for forming chiasmata?

As in the other book the idea I liked best was one which I suspect was germinal in this chain of models. This is the idea of genotypes as lottery tickets whose prizes are sites that are won totally if won at all. The idea is that in competition for such local prizes, asexual reproduction simply duplicates a ticket number, and so gives no extra chance of winning.

Another important concept that appears early on is that of sexual and asexual reproduction achieving equilibrium within a single species. This has no particular novelty as regards the vegetative reproduction of plants but when carried towards cases where identical propagules are involved I found it increasingly counterintuitive and felt correspondingly enlightened when I saw how, on the assumptions made at least, it does work. Morning mists of definition and explanation in the aphid–rotifer model left me confused on this point (p. 18), but in subsequent models these cleared and now instead I am unsure why Williams places as many restrictions as he does. For example, equilibrium easily occurs without the restriction to local dispersal.

Consider a site reached by s sexual and t asexual propagules. All those t are supposed to have identical 'tickets', so each sexual propagule has a chance

$$\frac{1}{s+1}$$

of winning the site while each asexual one has a chance of only

$$\frac{1}{s+1} \cdot \frac{1}{t}.$$

However, if an asexual does win it can express the greater productivity that results from being asexual—a constant factor $(1 + d)$ say. At $t = 1$ any $d > 0$ gives a definite advantage to the asexual, but at $t = 2$ even $d = 1$ (which implies fairly perfect efficiency of the asexual process) does no more than just cancel the disadvantage of genotype redundancy. So counterselection of the asexual strain is certain (assuming $d \leq 1$) in all groups that receive three or more of its propagules ($t \geq 3$). Thus very generally we expect asexual strains to persist at densities which give them on average only one or two propagules landing in each site opened to competitive colonization. To be more specific, if a and b are the mean numbers of sexual and asexual propagules per site in a random settling model (Poisson distributions) we find

$$\Delta q = q \frac{\mathrm{d}/b - f(a) - f(b)}{1/b + (1+d)/a + f(a) + f(b)},$$

where $q = b/(a+b)$ and $f(r) = \frac{1}{2}(\coth r/2 - 2/r)$. If a and b are considered constrained ecologically by an equation like

$$a + b = k[1 - \mathrm{e}^{-(a+b)} + (1 - \mathrm{e}^{-a})(1 - \mathrm{e}^{-b})d/a],$$

where k is a constant relating propagule density on the sites to the level of production from the previous site-holding adults, then it is easy to show that only for very low densities of colonization (low k) can the asexual strain win outright. Weedy apomictic or cleistogamic plants (often with wind-blown or bird-dispersed seeds like *Taraxacum, Heiracium*, and *Rubus*) would seem to fit here, while, as emphasized by reference to elm and oyster, large individuals with huge production of unmobile propagules are at an opposite extreme and should be very secure against inroads of parthenogenesis.

In so far as aphid and rotifer clones do competitively oust one another in the course of one or a few seasons, almost the same arguments can apply to the maintenance of their periodic sexual generation. Sexuals of some tree aphids (perhaps especially elm aphids, oddly enough) seem almost to be trying to pretend that they are like elm gynoecia on the twigs of their clone: the female sexuale lays just one egg, a brood that would be an absurdity for any species that did not have asexual generations behind it and to follow.

Doubtless some will see in the specification of sites, colonist groups and abundant clone extinctions within these a basis for a claim that Williams has made sexual reproduction stable only by admitting his old enemy, group selection, through a back door. Maybe so, but this is merely a semantic issue. It is a very far cry from the specifications and limited fields of success of these models to the panglossian all-pervading scope of 'group selection' as it held sway recently. I see this book as the culmination of a lonely crusade against the paradigm of species-benefiting adaptation which Williams began almost 20 years ago. Some rearguard actions of the crusade are still fought in this book (as against R. Levins on p. 157 and even against Ghiselin—accused of being insufficiently Darwinian!—on p. 122) but on the whole his cause has been so successful that he can afford to be less critical now and—which is a pleasure of this book—consequently has leisure to be more creative. Even so, perhaps he still over-restricts the circumstances where the term group selection can be useful. In one place he says: 'A high frequency of ... extinctions is necessary to a theory of group selection.' In my opinion a situation with strong differentials in group productions of colonists can give effects which justify the term equally well; indeed it may be best, using something like the Price formulation of selection, to think of most selection process as having components at group, individual, and haploid levels and to talk about the relative magnitudes of these components rather than try to force all effects to submit to any one term.

Other important ideas of this book can be no more than touched on here. Williams's concept of a 'Sisyphean' genotype is perhaps sufficiently illustrated if I say that in Britain, and probably in the USA too, the present crop of elm seedlings is going to need some of these if the elm population is to recover from present ravages of Dutch elm disease. The common elm of Britain, incidentally, could as well fit into the strawberry–coral model of this book as into the elm–oyster model. The view from my window here will, in a year or two, show whether a horde of young elms can clonally perpetuate the genotype of the stricken mother whose tattered outline shows in the sky above them. And if none of these root offshoots survives and yet a few seedlings grow up resistant to the disease (while being more susceptible to, say, plagues of elm aphids), this elm can help to exemplify Williams's 'cod-starfish model' too.

So by the end of *Sex and Evolution* it begins to seem easy to see how sexuality can be stable if environments are sufficiently heterogeneous in space and time. It helps if the species has a very high fecundity. Williams himself seems to have despaired of showing advantage for sexuality for low fecundity organisms and concludes, in effect, that most practise sex because they haven't found suitable tricks for eliminating it yet (can he really believe this for so many vertebrates?). This conclusion changes the focus to what we mean by heterogeneity and randomness in the environment. Nature's lottery is a continuum whose rule must be to pay prizes to certain phenotypes which are *nearest*, or simply near, to a more or less random target point. Only if the target point is bound to occur within a circumscribed range is a sexual phenotype likely to be so much nearer to it that nearness outweighs the doubled fecundity of the asexual. In other words, it seems to me that we need environmental fluctuations around a trend line of change. For the source of these we may look to fluctuations and periodicities inherent in our solar system, and also to the possibility of others generated by life itself. The latter line of thought tends to carry us back from the egg of sex to the hen of a multi-species system. To get the best possibilities of limit cycles and pseudo-random fluctuations offered by present ecological theory it seems that we might need to give preeminence to the annual life cycle—and so back again to the solar system.

Apart from all this there also remains the problem of pairs. What do these do that simple mixtures of haploids would not? Here perhaps something generalized from that slight hysteresis between the curves of forward and backward selection when selection coefficients reverse (e.g. figure 2.2 in D. S. Falconer, *Introduction to Quantitative Genetics*) may be effective to preserve variability that would disappear too quickly from the haploid mixture. And triploid sex may fail to develop because it gives too much power of disruption to selection at the haploid level.

Haldane once said something—I forget where and exactly what—to the effect that evolution could roll on fairly efficiently without sex but that such a world would be a dull one to live in. I agree with this and at several levels. Without sex there might be no multicellular organisms for a start, as seems to follow from an expected tendency for diploids to procrastinate the 'costly' event of meiosis. Given multicellularity we would still have to be content with autumnal tints on flowerless trees and the harsh patterns of warning coloration. Moreover, sexual selection probably plays a part in making us all so over-intellectualized as to love our bright and capricious world as we do, so far beyond any utility to us. Ghiselin considers the problem of our intellectual addictions—humour, music, science, and the like—in his last chapter, but I confess that I hardly expected to feel persuaded by what he said because earlier he had failed to attach importance to Fisher's idea on the escalation of sexual selection, and this excludes my own pet theory. Williams, too, seems to admit having grown averse to Fisher's idea because 'this would require an unrealistically high heritability of fitness' (p. 130). This of course fits with his present theme. But a preference guided by genes is guaranteed some persistance: there may be something like fashions here, ephemeral in the long run yet none-the-less urgently heeded by couturiers for the present. And for us the timescale is long, what I see as a genetic fashion for intellectual antlers having lasted since the Aurignacian cave paintings at least.

So magical are these antlers (if I am right) that before they pull us down in the bog they may even offer to disappear—that is, offer us an escape into a nirvana of parthenogenesis. Williams seems to have proved in his book that man's extremely low fecundity gives him no evolutionary strength to object to or withstand this offer when it comes. Personally, like Haldane, I seem to lack spirit for the thing . . . and this is not just male chauvinism (as I see it we are like reindeer anyway—I think Fisher's idea might work for both sexes at once, although no one has proved this yet). It is more like the feeling of J. E. Flecker when he wrote: 'All that calm Sunday that goes on and on: When even lovers find their peace at last, and Earth is but a star that once had shone.' Well, there is still time for all restless lovers of mental challenge, all nature's ardent analysts, to read these two books.

Cinderella seeds

A flower head of bristly ox-tongue, *Picris echioides*, loses its seeds to the wind—but not all. A few outer fruits of a distinct type stay firmly clasped against the maternal bracts. Paler and slightly larger in the photo, and with defective 'parachutes', these fruits will fall to the ground with the decaying mother in autumn. There along with siblings, similar seeds from neighbours, and yet others arrived by wind, they will compete to replace the mother in the following year. Ox-tongue is believed largely parthenogenetic: if this is true, no conflict occurs between seeds (or with the mother) over assumption of disperser or stay-at-home roles; if it is not true, then conflict can exist.

CHAPTER 11

ELM AND AUSTRALIAN

Dispersal in Stable Habitats

It works in me like madness, dear, to bid me say goodbye . . .
GERALD GOULD[1]

IT would be a gross understatement to say that sexuality was the only evolutionary problem worrying me in the mid-1970s. Generally at any one time I have a half dozen problems, some in the foreground of my mind, some like quiescent volcanoes at the back, some completely laid aside but not forgotten. But in those years one in particular besides sex had become major and was proving to be the special successor to one from the past. More than sex itself, it was proposing very clear-cut and even quantitative questions, similar to those I had had concerning optimal sex ratios under the various groupings and types of control that I had treated in the early 1960s. The new problem was that of dispersal.

I had become very interested in variations in wing development of insects. In the countryside around Silwood Park I had been finding many species with mixtures of winged and wingless adults. My examples came at first especially in the grass of old meadows but eventually I found as many or more in what was my new favourite habitat, rotting trees. I already knew of some such cases as an aspect of sex differences; as is seen in Chapter 4 (e.g. Table 4.1) they had been not uncommon in the set of my 'spanandrous' (scarce male) sex-ratio examples. Among these cases commonly the males are wingless but the females winged and this correlates strongly with the brother–sister mating occurring before dispersal. But even while busy with the sex-ratio paper I had also noted cases where the female sex alone might be polymorphic for wings, or else the males might be so, alone or in addition to their females. Evidently wingedness wasn't simply a trait linked with sex. It wasn't that males were sometimes built to disperse more and sometimes less than females, rather it seemed that the state of being winged and more able to disperse is subject to its own special pressures and that these sometimes correlate with sex and sometimes not. I had seen this much long ago but now here the problem was again right before my eyes. Far from libraries, a stream of new examples were coming in—or more accurately were crawling or flying according to wing endowment out of the depths of the sweep net I was using in the meadow. Or, if instead in the woods, they were there running on my dead trees and under the bark either as fully winged adults or as the wingless mock juveniles, the latter easily fooling you to be actual juveniles until you discovered them mating. Obviously all these were species that as adults sometimes needed and had wings and sometimes equally advantageously saved themselves the encumbrance and the expense. That the environments of grass tussocks and fallen trees would sometimes call for ability to emigrate and sometimes not seemed fairly obvious but the dynamics over generations of pressures to fly and how these pressures would lead to polymorphism was obscure.[2]

When I went to Brazil for my third long visit in 1975 and found myself working on the pollinator and other wasps of the wild figs (see Chapter 13) I discovered in the males of parasitic wasp species such an array of strange and deeply differing degrees of wing development within single species, as well as other modifications, as I had never imagined. It became apparent that the issue of being preformed as a disperser or as a stay-at-home was pervasive and that it applied probably to all organisms, not just to insects. Other groups were showing dichotomy over the same issue and the forms of dispersal involved were proving various and often totally unrelated to wings. Dispersal polymorphism indeed seemed second only to sex in its ability to create simultaneous different body forms within species, forms often so different they must have been slowly modified and perfected through very long periods. Sometimes the morphs differed so much that they had been originally described as different species or even as different genera. More of the biology connected with this matter will be treated in later papers. The point here is that I had realized that whether to disperse from the locality of your birth or not do so, and the parallel but not identical problem of whether as a parent you should to preform offspring to disperse or to stay at home, must be omnipresent dilemmas. The dilemmas must be there even when the evolutionary results do not display in such a neat away as in the possession or lack of wings. Invisible musculature or traits of behaviour may take the place of wings as the varying character. The dilemma applies even to humans. Let us think of son Howard in a Victorian family, how he would wander out of the garden even as a toddler, how he went seafaring at seventeen, founded a line of our family in Australia (or, perhaps, as another worthwhile human outcome, returned later bringing great wealth). Everyone knows of examples. Think also of Howard's brother Frank who hardly, even in adult life, showed inclination to stray beyond the county where he was born. Why such great differences? Was it just that Frank was the elder and custom dictated that he would inherit the family farm? I am sure there are cases where he wasn't the elder. Whether the differences are a pre-endowment as with the insect wings or derive from parental harshness or expectations, they are often dramatic and they manifest in very diverse ways and spread over very diverse groups. There are plants of the daisy family, for example, where some species have all their seeds 'winged', as dandelions (*Taraxacum*) do, whereas others have no winged seeds, as is the case with the common daisy (*Bellis perennis*) whose flowers star the green of our English lawns. Between these extremes are some daisy species with a mixture of 'winged' and 'wingless' within a single flowerhead. It was easy to imagine the 'wingless' seeds (or to use the proper botanical term, those lacking a pappus) falling directly below their mother and there waiting, like Frank, to take over the family home when mother dies. Meanwhile many winged sisters, 'Howard' daisy seeds, blow far away and with luck establish on distant sites. Ideally, how many should a mother programme to be like Frank, how many like Howard? How many should want of their own choice to be like the one or the other? The answering of these questions can be left to the paper itself.

The pappus example is perhaps closest to my original image. The dilemma I had of how many winged seeds it would be best to make is represented at its simplest in the text figure on page 378. A little thought makes clear that, as with the prototype of the sex-ratio problem in which I imagined females colonizing breeding sites in pairs and mated female offspring dispersing from the sites to colonize others, there must exist a safest and unbeatable course to follow, a disperse:stay ratio, which, when used, will minimize or eliminate the chance of a takeover by any other ratio producer. What is the ideal ratio? Does it differ according to what agent (e.g. parent or offspring) as was the case with sex ratios, or with what genetic element (e.g. on Y or X or autosome) determines it? And so on.

The problem looks simple and in the form presented in the first diagram does indeed have a simple solution. I am a little ashamed not to have seen this solution for myself. But one day after watching for about the third or fourth time tentative outlines of my algebraic attack plan perplex and fade behind the ragged skyline of my view from the Silwood office where I sat, I thought of an analytical wizard newly present in the Field Station, a man already busy cutting a broad swathe through the longstanding theoretical problems of my ecological colleagues. I decided to take my problem to him: dumping it in his hands, I thought, I would break loose, go to that skyline my window looked on, and participate in the rites of a grand funeral I knew to be there in the making. That seemed more fun . . .

At this point I should explain that it was a year when a particularly enormous scythe of death was sweeping the English countryside. It was affecting a conspicuous species in a way that had hardly happened since the great plagues attacked humans or, in the recent 1950s, myxomatosis the rabbit. Fortunately this new plague wasn't killing humans, just elm trees; nevertheless these were the trees that largely made our lowland landscape and they were going. In short, the fungus *Ceratocystus ulmi*—'Dutch elm disease'—was in full epidemic. The mighty elms of my skyline had first 'shown the flag' some years back in the form of those first yellowed and then wilted twigs that became so dreary to see, the equivalent of the first sneeze of a plague-struck victim. They were now withered in whole huge branches and were completely doomed. I had ceased to think of them as still living in their own right, as habitats and homes for those insects that required leaves (the white-letter hairstreak butterflies, for example); I now thought more of the grand funeral feast that was well on its way and of the insects that would take part in that. Who would be there? *Scolytus* for certain. This is both the part-murderer and part-undertaker of the trees; it was still hard for me to know what to think of the beautiful, chunky, slanted beetle, professional engraver of the straight vertical groove on the trunk but also the deadly nibbler of the twigs where the fungus is first introduced. *Ptinella, Heteropeza*, of course, they'd be there—coming and already, at least in the worst patches, moving out. Would *Xylocoris* with its pitchy and homosexual rapist male, who is said to reach some ova of females by homosexual proxy, injecting his sperm into the body cavities of other males? Would *Xyleborus*, bringer of another fungus, borer and stainer of the wood?

I forget what I actually found among the elms that day but remember well this as the general background of my activities when Robert May, present on one of his summer visits to the Field Station from Princeton University, became the originator of first solutions for the optimal dispersal problem. As we stared at my diagram and Bob muttered slowly 'It would seem so, I'll give it some thought', two things were already plain. First, there had to be some intermediate optimum between all stay-at-homes and all wanderers. Second, that the solution had to be another instance of the more general concept, the evolutionarily stable strategy (ESS), that was currently so popular. Newly named by John Maynard Smith,[3,4] the idea of the ESS was epidemic in the evolutionary world as much as *Ceratocystis* was among the elm trees, although fortunately the ESS disease was more benign. The search for these 'final' strategies that are to be expected in defined microevolutionary contexts via successions of mutations had become like a popular game for puzzle-minded ecologists and evolutionists. People were picking out new solutions everywhere or else busy challenging the 'defined contexts' of others as unrealistic, showing instead more subtle ESSs applying in the alternatives that they thought were better.

I like to think that I had contributed to the vogue myself when, in the 1960s, I had written about 'unbeatable' sex ratios in my paper of 1967 (Chapter 4). It was essentially the same idea as the ESS except that I had used a stronger but more restrictive criterion. This required that, relative to a type I designated 'unbeatable', no other type, no matter in what starting frequency in a mixture, would be able to increase. Maynard Smith's ESS only required that it be shown that if almost all individuals in the population are using the ESS, no other strategy starting from a *low* frequency can invade that population. Thus the ESS does not address the problem of what might happen if half the population, say, suddenly became practitioners of some differing strategy, as could conceivably happen through a mass immigration. Once present commonly, a mutant strategy that had failed to progress at low frequency might forge ahead, move to an equilibrium with the existing strategy, or the like. In cases where I could find it my 'unbeatable' strategy showed no equilibria to be possible. In short, an unbeatable strategy, if found, always implied an ESS but the reverse is not true. However, for most simple cases so far studied the ideas were effectively equivalent and the way of approaching the evaluation of an ESS that Maynard Smith had outlined had the advantage of being elegantly simple. He also drew attention to the generality of the idea of such endpoints for microevolutionary trends. Undoubtedly it was the combination simplicity and generality that brought about the vogue; new problems needing answers of the kind were ever coming to light, my sex-ratio case having been just one, and Maynard Smith's approach made results quite easy to get. In the case of the dispersal problem it may have been because it was still my instinct to attempt the more generally 'unbeatable' ratios, which involves looking at all frequencies at once, at every stage of the game, that I bogged down and in the end turned to Bob May for help.

With the flair he had brought to so many ecological problems May solved my particular simple case over the weekend. Besides some generality he had already added unasked, early next week he was suggesting to me more variables that we might include in further work with an aim of greater realism at every stage. All the analysis that followed in this line, whether realistic or simply fun to do, was due to Bob: I contributed merely notational suggestions, issued some of the new challenges, but in general merely enthused and marvelled at the results. In recognition of his single-handed role in the analysis I still believe Bob should have been the first author; he, however, would not allow this, insisting that I had originated the problem and had understood from the start the major points that an ESS must exist, and that selection for dispersal needed neither any continuous instability of habitat nor any flickering existence of extant and annihilated sites. As to our desired 'greater realism at every stage', the reader will shortly see that realism to an ESS modeller is still far from the realism of dying elms on the Silwood slopes. If it is the real reality that you are after, you had better be happy as I tend to be with these graphs that, like shoulders of Cairngorm mountains, jut towards you the first hints of some things you should be looking for; having appreciated those hints you may then spend the rest of your time contemplating not any more of the model Braeriachs (graphed, actually, after Bob had returned to Princeton) but instead being out among the teeming, dying trees, watching the insect families come and go, or bent over your dissecting microscope observing your cultures. I am glad to say that facts and theory concerning dispersal seem to have flourished since we published this paper,[5–7] in addition to our own relatively minor further effort, which was simply to enthuse a second superanalyst, Hugh Comins, with the problem (see Chapter 15).

The correctness in spirit of the answers we gave has never been challenged but typos and other slips in the formulas caused considerable justified complaint: between us Bob and I did a very bad job of the proofreading. These various small errors have been corrected in the version that follows.

References and notes

1. G. Gould, 'Wander-thirst', in L. Chisholm (ed.), *The Golden Staircase*, p. 267 (T. C. and E. C. Jack, London, 1928).
2. One advantage of the economy of not having wings to a female, which occurs in many wing-polymorphic species, is that she is able to lay more eggs. It is also worth pointing out that once the developmental decision in an insect to have wings or not has been made there is no going back because the moult into the adult form (which alone is ever winged) is quite irreversible; however, often a half step back is made after the dispersal flight in the sense of a resorption of the wing muscles and transferral of material so remobilized into the eggs (or possibly into the gonads in the case of a male). These details, however, are fairly unimportant for the present theme.
3. J. Maynard Smith and G. R. Price, The logic of animal conflict, *Nature* **246**, 15–18 (1973).
4. J. Maynard Smith, *Evolution and the Theory of Games* (Cambridge University Press, Cambridge, 1982).
5. D. A. Roff, The evolution of wing dimorphism in insects, *Evolution* **40**, 1009–20 (1986).

6. D. A. Roff, Habitat persistence and the evolution of wing dimorphism in insects, *American Naturalist* **144**, 733–98 (1994).
7. D. L. Wagner and J. K. Liebherr, Flightlessness in insects, *Trends in Ecology and Evolution* **7**, 216–20 (1992).

DISPERSAL IN STABLE HABITATS†

W. D. HAMILTON and ROBERT M. MAY

Simple mathematical models show that adaptations for achieving dispersal retain great importance even in uniform and predictable environments. A parent organism is expected to try to enter a high fraction of its propagules into competition for sites away from its own immediate locality even when mortality to such dispersing propagules is extremely high. The models incidentally provide a case where the evolutionarily stable dispersal strategy for individuals is suboptimal for the population as a whole.

In nature, adaptations for dispersal are ubiquitous and are often applied in almost suicidal ventures or in ventures which seem too feeble to be worthwhile (such as dehiscent seeds that fall only a few feet from the parent plant). This behaviour is clearly advantageous if the habitat is unstable or offers many empty, if transient, patches. We discuss here some simple models that help to explain why such dispersal is also advantageous even in stable and saturated habitats.

To begin, we consider a wholly parthenogenetic species in an environment that provides a fixed number of sites, at each of which just one adult can live. In a fixed season at the end of its life each adult produces a certain constant number of offspring, m. A fraction v of these are programmed to be migrants (e.g. insects provided with wings and appropriate instincts, seeds with a pappus for wind dispersal). The remaining fraction $(1 - v)$ are destined to be sedentary competitors for the home site. The mother's genotype, by means of some maternal influence on each ovum (or testa or fruit), determines the fraction v. When ready, at about the time of the mother's death, the migrant offspring take off and after mixing with all other migrants, and suffering a mortality such that only a fraction p survive, they are distributed equally to all the sites. There they compete on equal terms with the resident young which, it is supposed, up to this stage suffer no mortality (alternatively, if they do, p is the relative survival of the migrants). In effect, one offspring is chosen at random from

†*Nature* **269**, 578–81 (1977).

among the young present on a site to become the adult at that site in the new generation.

We acknowledge that this simple model probably has few close parallels in the real world. Nevertheless it may usefully force a re-examination of some widely held ideas about migration.

For example, it has been claimed as 'intuitively obvious' that in a saturated and time-invariant environment 'organisms can never gain any advantage by changing their locations'.[1] The schematic illustration below presents a counterexample based on the above model.

			x			x		
	o	o	o		o	o	o	o
Offspring	o	o	o		o	o	o	o
(after dispersion)	o	o	o		o	o	o	o
	o	o	o		o	o	o	o
	o	o	o	x	o	o	o	o
Adults	O	O	O	X	O	O	O	O
Site labels	*a*	*b*	*c*	*d*	*e*	*f*	*g*	*h*

The environment (represented by the eight sites labelled a to h) here is saturated and time-invariant. In the absence of any migrating mutant like X, and with full survival of stay-at-home propagules, the majority genotype O which keeps all propagules at its own site can perfectly maintain the population. But the 'O' strategy is not evolutionarily stable. The genotype O will be replaced by the mutant X, which keeps only one propagule at home, even though it loses two of its remaining four propagules due to mortality in migration (assumed to be 50 per cent; $p = 0.5$): obviously X has a chance of 1/6 of winning each of sites c and f; meanwhile, at least against so ill-advised a genotype as O, it certainly retains its base at d. Hence X is certain to become the established type. Of course the particular migration probability of X ($v = 4.5$) illustrated here will not itself prove to be the evolutionarily stable strategy, or ESS,[2] except for some special value of the survival factor p. Normally other mutations would supervene, after the spread of X, until finally a migration probability that was evolutionarily stable was approximated.

In general, the one or more ESS migration probabilities which the model might have can be determined as follows. The population is imagined to contain two types, using migration fractions v and v'. An expression is written for the fitness, w', of one adult of type v'. This fitness, or expected number of sites to be gained by offspring, will consist of the chance of retaining the home site plus the expectation of sites to be gained elsewhere by migrant offspring. From this we find the value of v which has the property that $w' \leq 1$ for all $v' \neq v$ (mean fitness in the model is unity, so that $w = 1$). The value of v so found, symbolized v^*, is unbeatable (Chapter 4) in the sense that any genotype with strategy $v' \neq v^*$ will have a diminishing frequency in any mixture.

For the simple limiting case where the number of sites and the number of propagules per parent are large, we find by this method that the ESS or unbeatable migration probability is

$$v^* = 1/(2 - p). \tag{1}$$

Because this formula shows no dependence on the composition of the mixture, stable mixtures are not possible, and the ESS can be considered safe against both rare mutations and any massive invasion by a different genotype.

A striking conclusion to be drawn from equation (1) is that even when migrant mortality is extremely high (small p), and the environment offers no vacant sites for colonization, it is still advantageous to commit slightly more than half of the offspring to migration.

Pre-eminent among the artificialities of this model are: (1) insistence on death and replacement of every parent in each generation; (2) absence of vacant sites (stemming from the deterministic description of the propagation processes); (3) pure parthenogenesis. We now indicate how the model may be extended to encompass such effects, paying particular attention to (2).

DEATH AND REPLACEMENT OF PARENTS

There is one simple and often realistic assumption that allows for perennation of the parents, while preserving the result (1). The assumption is that all parents, irrespective of age, have some constant probability q to survive and retain their sites into the next season. This corresponds to the so-called 'Type II survivorship curve', which is known to be not far from true for many organisms. For such organisms, both stay-at-home and migrant offspring have their expectations of inheriting a vacant site devalued by exactly the same factor, namely $1 - q$, which consequently cancels itself out of the analysis. (On the other hand, for organisms whose chance of surviving from one year to the next diminishes with age, the ESS will be age-dependent, if this is biologically feasible. A likely outcome is a 'bang-bang' strategy,[3] with all propagules dispersed until their parent reaches a certain age, whereupon all propagules stay at home.)

VACANT SITES

In our simple model, the evolutionary pressure that commits at least half the offspring to migration, no matter how risky migration is, results from an advantage in arranging competitive interactions as far as possible with unlike genotypes. It can be imagined that in nature such a pressure to take risks could cause a wastage that was damaging to the population's chance of survival. This cannot happen in the simple model because its deterministic assumptions keep all sites occupied; to investigate wastage we need some version that relaxes this assumption (2). Such a version may have some bearing on our confidence in models of theoretical ecology in general, because these often take it for granted that the species discussed already have, or will evolve towards, maximal efficiency of resource utilization.

To this end, we stochasticize the earlier model by assuming the number of propagules produced at any given site is given by a Poisson distribution, with the mean equal to the previously constant value m. It follows that the numbers for emigrant and for stay-at-home offspring, and for immigrants, are all generated by Poisson distributions for which the same symbols and expressions as occurred in the simple model are now to be taken as the means. Note that m is the mean brood size and also the variance in brood size, so that the relative magnitude of statistical fluctuations goes as $m^{-\frac{1}{2}}$. The parameter m thus comes to have an important influence, especially when relatively small; when m is large, the present stochastic model tends to revert to the earlier deterministic one.

Now sites can become vacant. Vacancies begin or continue whenever a site happens to have no surviving stay-at-home offspring and receives no immigrants. Obviously, some migration is essential if extinction is to be avoided, and from the population point of view there will be some level of migration that keeps extinction at furthest reach.[4]

For any specified combination of m, v, and p, the system settles to some stable level of site occupancy f (that is f = fraction of sites occupied), with a zero level signifying that extinction is inevitable. This stable level f is given (for $f > 0$) by the implicit relation

$$f/(1-f) = \mathrm{e}^{m(1-v)}[\mathrm{e}^{mvpf} - 1]. \tag{2}$$

For given m, the site occupancy function f defined by equation (2) may be mapped over the unit square of admissable values of v and p; $f(v, p)$ is found to have the form of a smooth promontory (Fig. 11.1 for $m = 3$), or, for larger m, a squarish headland (Fig. 11.2 for $m = 10$). In any such figure, for fixed v the occupancy $f(p)$ falls convexly with decreasing p, and clearly must always hit extinction somewhere short of $p = 0$. The smallest p that still allows the population to exist may be shown to occur when $v = 1/m$, that is when just one migrant is dispatched from each site. The concomitant least value of p is given by $p_{\min} = \exp(1 - m)$, which is very small for moderate values of m; slightly larger values of p carry f up to a 'plateau' at a level of nearly complete occupancy. Conversely, for fixed p, the occupancy f as a function of v at first rises (albeit very slightly for large m) to a maximum as v decreases; then, as v decreases further, this is followed by a steepening fall in f, which goes to zero short of $v = 0$.

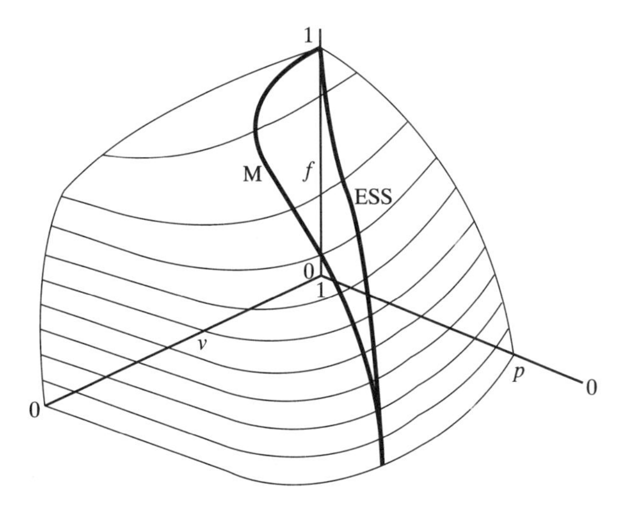

Figure 11.1 The surface depicted here gives f (the equilibrium fraction of sites occupied) as a function of v (the fraction of offspring that migrate) and p (the relative survival probability for migrants), under the assumptions delineated in the text, for a mean 'brood size' $m = 3$. The contour lines are for constant values of f, spaced at intervals 0.1 apart (that is $\Delta f = 0.1$). On this surface, the trajectory labelled ESS shows the dispersal strategy that is evolutionarily stable, $v^*(p)$, as a function of p. The ESS trajectory is to be contrasted with that labelled M, which shows the strategy which maximizes site occupancy for given

$$p, \hat{v}(p).$$

Thus we may trace out the locus of the strategy, $\hat{v}$, that maximizes the site occupancy f for given p; this locus moves up a 'cliff', and across a 'plateau' to $p = v = 1$ (Figs 11.1 and 11.2). This strategy $\hat{v}$ is that which is 'best for the population'.

This locus of the strategy for maximal occupancy, $\hat{v}$, is to be contrasted with the locus of the ESS v^*, which is determined by the method described above. The ESS migration probability, $v = v^*$, is given by

$$\exp[-m(u - vpf)] = \frac{v^2pf - vpf(u - vpf)}{v^2pf - u(u - vpf)(1 - mv)}. \tag{3}$$

Here $u = 1 - v$, and f has the value implicitly fixed by equation (2). The derivation of the results (2) and (3) will be set out elsewhere (our work with H. N. Comins); unlike the earlier simple result (1), these equations are made more complicated if age-independent or 'Type II' perennation of the parents is introduced.

The locus of the ESS migration fraction v^*, so determined, is shown in Figs 11.1 and 11.2. We note that ESS migration is always higher than maximum-occupancy migration, although they converge at the two extremes (at perfect

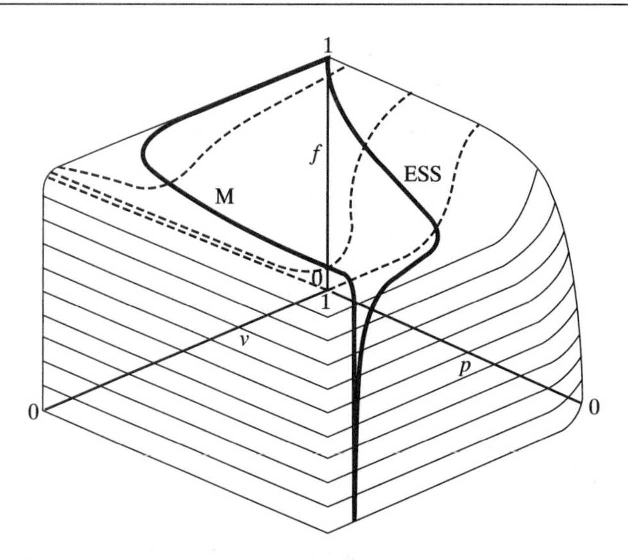

Figure 11.2 As for Fig. 11.1, except that now the mean brood size is $m = 10$. Again the solid contour lines depict constant values of f, from $f = 0$ to 0.9 at intervals of 0.1; the increasingly broken contour lines on the 'plateau' are for $f = 0.99$, 0.999, and 0.9999, respectively. For a more full discussion of Figs 11.1 and 11.2, see text.

and at least-possible survival, $p = 1$ and $p = p_{\min}$). This is true for all brood sizes m. As m is increased towards more usual biological values (from $m = 3$ in Fig. 11.1 to $m = 10$ in Fig. 11.2), the separation between the ESS and maximum-occupancy loci with respect to v also increases strikingly. This separation results mainly from the way the changing shape of the site-occupancy surface $f(v, p)$ controls the course of $\hat{v}$; that is $\hat{v}(p)$ is much more affected by changing m than is $v^*(p)$. The separation between the two loci with respect to f, however, remains very small, even for populations that are on the 'cliff face' of the f surface and hence in some danger of extinction.

In other words, the ESS v^* can demand far more migration than is 'best for the population' ($\hat{v}$), but such excess does not, in this model at least, substantially affect population safety nor create many extra vacant sites that would give other species opportunity to evolve as competitors for the niche. Notwithstanding the small difference in occupancy levels, there is usually strong selection to establish v^* rather than $\hat{v}$. One incidental consequence of the relatively abundant supply of migrants is that some must overflow from the system and occasionally have the luck to found new populations elsewhere. It therefore seems that our models show, after all, a case where a trait is positively adaptive at more than one level: the ESS is best for the genotype, suboptimal at the level of an isolated population, and advantageous in carrying the population to new areas.

SEXUAL ORGANISMS

If the organisms are taken to be sexual rather than purely parthenogenetic, with migration probability depending on the genotype of the offspring, lower values of v^* are definitely expected.

This follows from the observation that our models directly treat the genetic success of a mother. A parthenogenetic offspring has an identical genotype, so that whatever maximizes inclusive fitness (Chapter 2, Part I) for the mother maximizes it for the offspring. In contrast, the inclusive fitness of a sexual offspring normally differs from that of its mother, and, in some direct relation to the hazards of migration, will find its optimum at a lower migration rate. Hence there is a conflict of interest between parent and offspring (ref. 5 and Chapter 2, Part II). Its outcome will depend on the biological circumstances: as regards applying a pappus to a seed to make it blow away it is clear that the parent has a strong position; as regards making wing buds become wings in an insect it is much more likely that the offspring can have its preference.

In nature, less migration generally means more inbreeding; such inbreeding reduces the genetic contrast between parent and offspring, and thus tends to diminish the difference between parthenogenetic models and sexual models with offspring-determined migration. Various situations can arise. If males mate before migration in a one-per-site situation, inbreeding is total and the model is again effectively as under parthenogenesis. Unfortunately, the small arthropods which almost meet this extreme of inbreeding and which have a dispersal polymorphism (e.g. some mites of the genus *Pygmephorus*; ref. 6 and unpublished results of W.D.H.) are far from meeting the assumption of fixed unitary sites, being rather exploiters of patchy, ephemeral resources. If inbreeding occurs because males sometimes survive along with a sister, then our one-per-site assumption does not hold.

But if male offspring are assumed always to migrate, then outbred sex can be brought in and the one-per-site feature retained; moreover, this version may approach realism for some of the many insect species that have free-flying males while females are either wholly flightless or polymorphic for flight. For the simple, non-stochastic version of this model with allele semi-dominance, the ESS migration probability for the female offspring is indeed lower than for the parthenogenetic model: in place of equation (1), we now have

$$\begin{aligned} v^* &= 0 && \text{for } \tfrac{1}{2} > p > 0 \\ v^* &= (2p-1)/(4p-1-2p^2) && \text{for } 1 > p > \tfrac{1}{2}. \end{aligned} \qquad (4)$$

The fact that $v^* = 0$ for $p < \frac{1}{2}$ suggests that when this model is made stochastic (along the lines followed previously) the ESS may imply situations where the population and the species is dangerously liable to extinction. We are pursuing this open question.

(Roff[7] has used computer simulations to study migration in models which not only have sexual reproduction and multiple adults per site with a density-dependent migration structure, but also have temporal variability in site suitability. It is obvious from the facts of nature,[8] and is confirmed in both Roff's model and several others,[9–11] that erratic habitat suitability (including local extinction) is an extremely important factor in favouring adaptions for dispersal and migration. More relevant to the present discussion, however, is Roff's demonstration that his ESS migration probabilities are considerably lower than those which would keep populations highest and most secure from extinction. Although partly attributable to the workings of sexuality and Mendelian inheritance in Roff's models, this effect arises primarily because there are many adults per site, which changes the calculation of inclusive fitness in such a way as to favour the short-term, selfish option of not dispersing.)

It would be satisfying to conclude by listing a few biological illustrations of principles suggested by the models. Unfortunately most of the examples which have come to mind are more in the region of jokes than of reality: they range from *Caulobacter crescentus*,[12] whose asexual fission always produces one stalked sessile cell and one motile one that swims away, through the inevitable lemmings and on to the non-first sons in Victorian families who joined the army or went to Australia. Both our simplest model, and the refined versions (1) and (2), support what these examples might illustrate, namely a minimum rule that 'at least one must migrate whatever the odds'. All three examples, however, fail to have fixed sites and, except for *Caulobacter*, fail also in regard to asexuality. There are flight dimorphisms in non-sexual insects (e.g. aphids in summer: the tiny subcortical beetle *Ptinella errabunda*[13]) and probably parallels in some weeds (e.g. Compositae)[14,15] but in being colonist species these fit badly with the assumptions of fixed sites and of no extinction or recreation of habitats on a large scale. But the fact that colonist plants in arid environments, where perennation of the whole plant (as opposed to survival of its seed) may be hardest to achieve, seem especially inclined to produce actually dimorphic or differently shed propagules with one class buried or otherwise retained beside or under the parent[16] (*Sieglingia decumbens* is the best example in Britain) does at least suggest that a plant's perennating body (or its tubers, corms or the like) should be considered its bid to retain the home site. We expect the investment in such aids to perennation, as opposed to investment in flowers and dispersing seed, to show positive correlation with the chance that seed if produced would end up destroyed or in sites that never could support an adult. For example, the more successful a specialized epiphyte or parasite is at getting its propagules to suitable sites, even when these are objects of intense competition, the more it should produce them. A mistletoe should flower more than an epiphytic orchid; and we should not be surprised to find that weedy perennial species put more net assimilation into seed production than do climax perennial species.[17] This may be so but is not particularly exciting; other explanations are available. It is not easy to test the quantitative conclusion that, because $v^* > \frac{1}{2}$ for most values of p, we expect to find at least one-half of net assimilation going into some form of growth or propagule production that extends competition beyond the space that the organism already occupies.

The value of our models lies primarily in their approach and method, in their well-determined conclusions, and in the novelty of having made plain a new reason for the ubiquity of dispersal. We have shown that the habitat does not have to be patchy and of erratic suitability; substantial dispersal is to be expected even when the habitat is uniform, constant, and occupied completely.

Acknowledgements

We thank H. N. Comins, H. S. Horn, T. R. E. Southwood, and D. M. Waller for helpful comments. This work was supported in part by the NSF.

References

1. D. Cohen, *American Naturalist* **101**, 5 (1967).
2. J. Maynard Smith and G. R. Price, *Nature* **246**, 15 (1973).
3. S. Macevicz and G. Oster, *Behaviourial Ecology and Sociobiology* **1**, 265 (1976).
4. R. M. May, *Stability and Complexity in Model Ecosystems*, 2nd edn (Princeton University Press, Princeton, NJ, 1974).
5. R. L. Trivers, *American Zoologist* **14**, 249 (1974).
6. J. C. Moser and E. A. Cross, *Annals of the Entomological Society of America* **68**, 820 (1975).
7. D. A. Roff, *Oecologica, Berlin* **19**, 217 (1975).
8. T. R. E. Southwood, *Biological Reviews* **37**, 171 (1962).
9. M. Gadgil, *Ecology* **52**, 253 (1971).
10. L. Van Valen, *Evolution* **25**, 5 (1971).
11. R. Levins, in M. Gerstenhaber (ed.), *Some Mathematical Questions in Biology* **2**, pp. 77–107 (American Mathematical Society, Providence, Rhode Island, 1970).
12. N. Kurn and L. Shapiro, *Proceedings of the National Academy of Sciences, USA* **73**, 3303 (1976).
13. V. A. Taylor, thesis, University of London (1975).
14. J. L. Harper, P. H. Lovell, and K. G. Moore, *Annual Review of Ecology and Systematics* **1**, 327 (1970).
15. G. L. Stebbins, *Annual Review of Ecology and Systematics* **2**, 237 (1971).
16. M. Zohary, *Plant Life of Palestine, Israel and Jordan* (Ronald, New York, 1962).
17. J. L. Harper and J. A. White, *Annual Review of Ecology and Systematics* **5**, 419 (1974).

Whales and mice of rotting wood

The beetle *Titanus giganteus* (above) breeding in rotting palm trunks in Amazonia, is possibly the world's most massive insect. Below it, greatly enlarged (a millionth smaller in mass), four specimens of *Ptinella errabunda* from beech logs in Britain, are shown contrasted to a human hair (mine). As can be seen these almost smallest of all beetles exist in two forms. One is blind, unpigmented, flightless, and adapted for rapid expansion of a local underbark colony; the other is winged, tanned, with compound eyes, and possessing instincts for dispersal. The *Ptinella* shown are all female and their species parthenogenetic, but this is unusual: in most species of the genus, males exist, deliver gigantic sperm (longer than the beetles), and are bimorphic like the females. When present the flying wings are feather-like if expanded but mostly, as here, are seen only as dark bars under the horny outer wings where they lie folded in three sections. The vast order of beetles, the most speciose group of the living world, almost certainly originated in the habitat of rotting trees: its first representatives are found in the Permian period and closely resemble particular primitive rotten-wood species of today.

CHAPTER 12

FUNERAL FEASTS

Evolution and Diversity Under Bark

The angels keep their ancient places,
Turn but a stone and start a wing!
'Tis ye, 'tis your estrangèd faces,
That miss the many-splendoured thing.
FRANCIS THOMPSON[1]

MY interest in the odd sex ratios of arthropods had tangled to some degree with my older interest in the potential role of haplodiploidy in shaping the 'eusocial' pattern of insect social life. Most of the animals with inbreeding and male-deficient sex ratios were small. Many were male-haploid. Many were also social but in a minor not eusocial way. Taken all together, these associations sent me out looking for colonial thrips and finding these re-opened a door to old childhood excitements that I could now see with new eyes.

If you happen to read this book outdoors on a summer's day and you see a hyphen crawl across the page it is hopefully not the beginning of delirium tremens, more likely just a thrips. In the grassy and cultivated North Temperate zone the hyphen is most probably the minute insect *Limothrips denticornis*, a species that breeds and mates in grass leaf sheaths, including those of our cereal crops—the very thrips that is mentioned in the table of my paper on extreme female-biased sex ratios (Chapter 4, Table 4.1). Needless to say, it can be very abundant. Outside of its leaf-sheath natal home *L. denticornis* is a fairly normal plant thrips in appearance and habits and dispersers are seen singly. Even when still with its family inside the sheath it cannot really be described as social; the main oddity of its life there is the smallish, yellowish, wingless, haploid brother who matures first and mates with all his sisters before they fly out—your moving hyphen, as you probably were never interested to know, is unlikely to be virgin. But as I learned from the literature, there were other thrips equally haplodiploid that formed much larger colonies and some of these were reputed to have strange polymorphisms in both sexes: these species were mostly in rotting wood. Hence I was led at length to my first colonial thrips strolling by herself on the dried-apricot-like surface of a *Stereum* fungus fruiting on a rotten tree stump near Sevenoaks in Kent. This animal quickly introduced me to a throng of sisters, cousins, and colourful younger relatives clustered in dozens beneath the same fruiting crust. All of them together then tumbled me headlong back into the already half-familiar world that was to become the subject of the paper that follows. As a child I had always loved to turn stones to look at the life beneath (see my paper for the Insectarium in Volume 2) and it was the same with the slabs of loose bark on standing or downed trees. Having found my first *Hoplothrips* colony and beginning to know of various other targets of interest in the same habitat, I started pulling and chiselling bark in Silwood Park and at my old wood-surrounded home in Kent much more with a purpose and also with an eye for much smaller forms of life. As a child I had not had a binocular microscope.

A British tabloid newspaper used to have an advertisement that said simply: 'All human life is there.' In a similar spirit of exaggeration and of advertisement I will state of the under-bark habitat that: 'All arthropod life is there.' I am semi-serious—all of it is there except for representatives of the ancient aquatic branches. I am claiming that amongst all the bizarre life that is tensely and densely clustered in this habitat, most of terrestrial arthropod future is in the making and keys to much of its past can also be found. Here the world's smallest beetles fold within their half-millimetre bodies sperms longer than themselves (likewise fold the delicate feather wings that experts say they should not be able to fly with, but they do); here *Xylocoris*, the prototype bed bug, lives with its psychopath male crimes all long forgiven and forgotten, its astonishing act accepted throughout an entire descendant superfamily; here lives *Pygmephorus*, the *Stereum*-sucking mite that balloons its body to be a gigantic and watery play pen for sex-wise incestuous infants; here *Cephalonomia* the proto-ant, fitted with more castes-in-prospect than most of the real ants, moves weasel-like through tiny tunnels after doomed beetle-larva hosts; here *Heteropeza* nearly forgets the midge identity and open air of its ancestors and teems by bursting its stiff, fibre-like maggot offspring out of the larger mothers (when grown, these are due to explode themselves in birth and so onward in an endless progression); here beetle larvae lie in soft cradles quilted with living food, no more than turning their heads to feed all their lives through 'kindness' of an organism from another kingdom;[2] here fight syncytial amoebae; here luminous wood chambers still brighter-glowing lampyrid firefly larvae within it; here are polymorphisms for wings, harems sequestered by armoured blue-beards, horned beetles fighting, guarding, or in copula . . . A family style of life was so common that I came to expect ever new groups to reveal it. Always I would find many ages mixed together, all within small cavities or self-formed patches of other kinds. Worms, mites, woodlice, and insects of varied types, were among my examples.

Knowing of the existence under the bark and in old wood of these families and extended families reinforced my belief that something more than merely the overlap of generations is needed before such situations can evolve towards an outbreak of that extreme sociality and sterility such as is found in the ants and termites, both of which groups, incidentally—and illustrating the present thesis—are very likely to have originated in this environment. What the extra factors might be remains a subject of very active discussion. Haplodiploidy is probably just one of them. This itself, as I indicate in the paper, may through microbe interference be especially apt to become originated in the seething under-bark world. Other courses towards social life arise, somewhat paradoxically, from the fact that life in families is not necessarily peaceful;[3,4] others arise from the temporal limitation of the dead tree's supply of food and shelter, leading to the split incentives of the preceding chapter to be a stay-at-home breeder or a disperser;[5,6] yet others again arise from patterns of relatedness (see Chapter 8). Extremely rarely, as it appears, enough of these factors may concur and synergize so as to launch, without the help of haplodiploid asymmetry or superrelatedness, a new and thoroughly eusocial group.[7]

In general, the novelties present in arthropods under the bark appeared endless to me and the paper that follows touches only the edges. Many of the examples were brought directly to my attention as I worked with chisel, hammer, glass vials, pooter,[8] and inevitably black-rimmed fingernails in Silwood Park and Windsor Forest. My paper became for me like a funeral oration for trees I had seen die and in particular a reminder of the joyous feasts of the funerals I had attended. Imperial College—Imperial estates where I trespassed and worked! Feasts for and of those lordly beeches that had been such favourites of returned nabobs and of the Widow Queen. In fields beyond the ha-has, as at Silwood, on the estates of such builders, my now fallen trees had been fostered. Here, step by step, their growth, grandeur, and fall had mimed that of the short-lived Empire that planted them.

More seriously my attendance at the feasts had highlighted for me various curious correlations in insect phylogeny and habitat that still seem to me to deserve more attention than they get. As it has turned out I think my paper, perhaps just because of its 'oration' style, is little read and most evolutionists don't seem to see the interest of the dead-tree habitat for insect phylogeny as I do. However, it is also possible that I overdo, as usual, my advocacy. Thoroughbred entomologists are the only biologists likely to appreciate the rather dense associations of genera and groups that I instance in the paper; but these also may be just the biologists least likely to be sympathetic to synthetic overviews brought forward in a whimsical style that is almost as faded as Fabre and as obsolete as the Empire. Anyway, every entomologist must have his or her favourite group and habitat and even I admit I can imagine claims closely parallel to mine being made instead for the habitat of litter and dead leaves, or, to a lesser extent, for grass tussocks and other dense detrital habitats of various kinds. The claim that I regard as relatively certain, however, is that the open realms of leaves, living twigs, and waving grass, all that is up there in the wind and sun, is quite different in its style of evolution from any of these murky underworlds. Up there are most of our pest species as well as of many (and sometimes extremely many) of our fancy spectaculars such as butterflies, gay leaf beetles, grasshoppers, shield and leaf-legged bugs, and all the rest. These generally more widely outbred and mostly herbivorous insects tend to belong to species-rich lineages; and yet in spite of numbers they are lineages that are rather trivially differentiated in terms of body construction. This fact is a part of the puzzle I address.

The paper was 'read' (meaning, of course, spoken in a shorter version) at a meeting in South Kensington that my Silwood colleague, Nadia Waloff, organized jointly with Laurence Mound from the British Museum (Natural History). Mound is one of the world's few experts in thrips. This fact, however, is largely coincidental to my own interests in the group and also to the invitation for me to contribute this paper to a Royal Entomological Society Symposium. (Mound had, however, been kindly identifying the thrips species I was finding under bark.)

Besides again thanking Mound and Nadia Waloff who sponsored the paper, I would like to re-state here my gratitude to the Royal Entomological Society, which throughout my time at Imperial College gave me almost unfettered use of its library even though I never quite brought myself to become a member of the society. Use of the library was a marvellous freedom. Right to hand on the shelves (if sometimes dauntingly high up in the old book-lined Victorian rooms and only reached from the top of a creaky stepladder) was all the most ancient wisdom of entomology stretching back even before Bates and Darwin. It is characteristic in a field taxonomically as vast as entomology that the latest paper on a given group that you urgently want to learn about, need, in fact, today, turns out to be a paper written a century ago. A glance at the reference list here shows me papers of 1880 and 1901 but perhaps a better example is the reference in the next paper to Fritz Müller, Darwin's correspondent from Brazil, publishing on fig wasps in the 1880s (see the Addendum after the acknowledgements, page 474).When I worked in the Beit Building of Imperial College beside the Albert Hall (as I didn't normally, being at the Field Station, but I often gave lectures there) it was only necessary to walk a hundred yards to 41 Queen's Gate to try in the 'Royal Ent Soc' library for a paper I wanted. Failing to find it there (for example, if the work had found publication in some upstart journal that the RES didn't yet take) then a few steps down the same road brought me to the Commonwealth (now International) Institute of Entomology where I could try again. Finally my last resort could be opposite to the CIE, where loomed that great building always referred to by entomologists simply as 'The BM' (British Museum). In giving this title entomologists choose to ignore a puffed-up and supposedly more mighty branch known for its marbles, mummies, and Magna Carta 2 miles away in Russell Square. Most Londoners, of course, know the entomologists' 'BM' as the Natural History Museum. This is in fact now its proper title and a great improvement on the formerly official British Museum (Natural History), or BM(NH).

If ever I needed sheer professional insect taxonomy, as I rather seldom did, this museum would be my first destination rather than my last: it is, of course, for most groups, Britain's ultimate authority. Although most people know this museum better for its dinosaurs and birds and other large animals, it should be realized that it houses at least a thousand times more species of insect than it ever has had of birds, and maybe 10 000 times more than it has of dinosaurs. Several floors of a large wing are devoted to insects alone and I imagine only vague estimates can be made of the number of individual specimens stored here. It includes (or did then, before government grew mean to it) its own excellent entomological library. For the bibliographies of my paper on extraordinary sex ratios (Chapter 4), the paper of this chapter, and of the one that follows on fig wasps, I extend both my thanks and my nostalgia towards all three of these South Kensington libraries: they were the temples where, for many years, in respect of so much of my theorizing, I worshipped amidst a shadowy but uplifting congregation of ancestors and angels.

References and notes

1. F. Thompson, 'In no strange land', in *Selected Poems of Francis Thompson* (Methuen, London, 1909).
2. Darwin's prediction that no adaptation can be such as to benefit exclusively another species is not falsified, however, because the 'ambrosia' fungi associated with this beetle depend on it entirely for transmission from tree to tree. In feeding the beetle larvae the fungus creates vehicles for the dispersal of its own offspring, the case being similar to that of our own cultivated plants that rely on us for propagation and dispersal and that sometimes also devolve to having defective propagules (e.g. banana, sugar cane). Also, like domestic plants, ambrosia fungi have reduced sexuality even where not completely dependent for reproduction, as was well explained by V. Hutson and R. Law in *Proceedings of the Royal Society of London* B **213**, 345–451 (1982). As to other means of dispersal, most ambrosia fungi encounter no situations in the dead trees that could allow wind dispersal and accordingly create no wind-borne spores. Probably much other fungus transmission

between dead trees occurs by insect agency in a less regular manner. In any case, no example of mutualism yet shows either partner giving benefit for zero or negative return.

3. N. Yamamura and M. Higashi, An evolutionary theory of conflict resolution between relatives: altruism, manipulation, compromise, *Evolution* **46**, 1236–9 (1992).
4. N. Yamamura and N. Tsuji, Parental care as a game, *Journal of Evolutionary Biology* **6**, 103–27 (1993).
5. V. A. Taylor, A winged élite in a subcortical beetle as a model for a prototermite, *Nature* **276**, 72–5 (1978).
6. V. A. Taylor, The adaptive and evolutionary significance of wing polymorphism and parthenogenesis in *Ptinella* Motschulsky (Coleoptera: Ptiliidae), *Ecological Entomology* **6**, 89–8 (1981).
7. Y. Roisin, Intragroup conflicts and the evolution of sterile caste in termites, *American Naturalist* **143**, 751–65 (1994).
8. A pooter is a device for collecting small insects. Sharp inhalation on a rubber tube draws them off the surface where they are running and into a vial. In the USA, the device is called an aspirator: the etymology of the more evocative 'pooter' (plus its verb 'to poot') is obscure but I have heard the origin ascribed to Nadia Waloff and thus located in Silwood Park itself.

EVOLUTION AND DIVERSITY UNDER BARK†

W. D. HAMILTON

A dying tree opens a wide variety of habitats for colonization by insects. The basic anatomy of the tree remains: there are heartwood, sapwood, phloem, cambiums, and bark. The various layers differentiate further by size, regimes of dryness and temperature, height above ground, and in other ways. Underground the tree supplies rotting roots and rootlets and in the soil surface a litter of fallen branches and twigs; below ground, on its surface, and in the tree's standing or fallen hulk come patchy invasions and the fruit bodies of the major fungi.

Such a complex of habitats would be expected to have a rich fauna of insects and this is the case. The richness most peculiar to it, however, and that on which I will concentrate this account, is not the most obvious kind. Compared to certain other broadly defined habitats the dead-tree complex is not outstanding in its total species. A more remarkable diversity lies in the host of genera, families, and higher taxa, which, themselves tending to be few in species or even monotypic, are to found only in this complex,[1] or which, if found elsewhere, extend only into other kinds of decaying plant litter.

In trying to understand this 'phytonecrophily' and perhaps specific 'arbornecrophily' of so many of the odd and 'primitive' taxa of entomology, it is necessary to bring into view another peculiar and under-commented richness of the dead-tree arthropod fauna. Besides the many cases of deep phyletic *divergence*, illuminating common ancestry with outside groups, dead trees are rich in examples of functional *convergence*. Examples of such convergence will also provide ongoing threads in the theme of this essay and interweave with the problem of phylogeny. Four groups are outlined in the following paragraphs. Jointly they refer to certain similarities in *breeding structure*. Effects of breeding structure will form the essay's tentative thread of explanation.

†In L. A. Mound and N. Waloff (eds), *Diversity of Insect Faunas*, Symposia of the Royal Entomological Society of London No. 9, pp. 154–75 (Blackwell Scientific, Oxford, 1978).

1. Dead trees, and especially their habitats that lie immediately beneath the bark, are very productive of examples of *wing polymorphism* (list in Appendix A, Tables 12.2 and 12.3). Such polymorphisms seem to be almost never of the type that is simply switched by alternative genotypes; they are switched instead by environmental cues to habitat change (personal observations and refs 2–4). *Ptinella errabunda* is at once polymorphic for wings and wholly thelytokous.[2] Similar exceptions from sexuality are known in Psocoptera[5] and, of course, within summer clones, are ubiquitous in aphids. Convergences to paedogenesis, combined or not with thelytoky (*Plastosciara, Micromalthus, Heteropeza* etc.), have rather similar significance and are an extension of the adaptive pattern. All such cases index the occurrence of colonies that inbreed to multiply and then disperse. More trivially, convergences to phoresy and to caduescence of wings in adults, both characteristic in the habitat, are linked to the same pattern.
2. The occurrence and probable origin in this habitat of about four of the six or so inventions of *male haploidy* by insects (Hymenoptera, Thysanoptera, *Micromalthus*, certain Scolytidae; excluded are iceryine coccids and Aleyrodidae) is likely also to reflect a structure of local breeding. The situation for male-haploid Acarina may well turn out to be similar, while certain parahaplo-diploid groups (in Sciaridae and Cecidomyiidae) may be arrested steps in a similar direction.[6]
3. At least two of the origins of advanced insect *social life* (termites and ants) have their place in this type of habitat.[7] (Patterns of both solitary and social aculeate wasps are also conceivably derivable from a biofacies of roughly bethylid type in rotten wood.[7] In social bees wing polymorphism in a hypothetical ancestor of this type could have left an impress which is now renascent in the juvenoid-controlled caste differences between queen and worker.[8]) Here again special breeding structure and/or male haploidy are implicated (Chapter 8 and ref. 9). Cases of subsocial life in the dead tree complex are legion and without doubt many still await discovery. One example which seems novel enough to deserve mention here, even though somewhat marginal to our habitat and concerning a non-insect, is a certain social group-hunting pseudoscorpion which Dr L. A. O. Campos pointed out to me in Brazil. By co-operatively securing legs or antennae, bands of these pseudoscorpions are able to capture ants (their principal prey) many times their own size and far larger than would otherwise be possible. Although these atemnids are found most commonly under bark flakes of living *Eucalyptus* (where they waylay mainly *Atta*), I had also noticed similar bands in the habitat more typical for the group, that is, under the main slabs of bark of dead trees. Here ants of a similar size (e.g. *Camponotus*) were common. A seemingly very similar gregarious and myrmecophilous species was described by Turk[10] from Argentina.

4. Finally, as a less basic convergence, although one which is no less striking, we may note the very numerous developments of certain types of *sex dimorphism*. Frequently the cases are connected with subsociality and reflect particular roles of the male (including as usual, with regard to offspring care itself, a frequent lightness or lack of role). In and around dead trees and recently evolved out of them are found a great number of the wingless, or dwarf, or outsize, or bizarrely armed insect males that are known (Chapter 13).

Following the hint from these four groups of examples, that habitats of the dead-tree complex tend to force certain kinds of breeding structure, it will be argued that, besides calling forth specifically the convergences mentioned, such breeding structure is favourable to rapid evolution in a most general way: several interconnected reasons for this will be given. As to what may have actually been achieved by such postulated rapid evolution in dead trees, the strongest claim—and, as I judge from preliminary reactions from entomologists, the most difficult of acceptance—will be that many of the major insect groups have diverged there. More precisely this claim is that the phyletic divergences which now give systematists their separations at generic, family, and ordinal levels were initiated with disproportionate frequency among insects living in dead trunks and branches. Such a claim seems to be almost new: so far as I know only one author in a very brief note has argued for a special importance of dead trees in insect evolution.[11] A weaker and doubtless more acceptable version of the claim will be simply the same with the proposed crucial habitat complex widened to refer to vegetation detritus generally, not just to the hulks of woody plants. This version is much less novel (see, for example, refs 12–14) but even if the phenomenon is not more special than this it is still perhaps useful to look over a theory of Ghilarov's (or Mamaev's) type with the focus on breeding structure instead of on the more physical factors that have usually been emphasized.

BREEDING STRUCTURE

The argument will refer primarily to the *cavernous* quality of the insect living spaces which dead trees offer. Besides this an important auxiliary factor will be the uneven and *scattered* distribution of dead trees. This second factor, however, is less peculiar to dead trees since living trees tend to show it as well. Moreover, its consequences for evolution are already appreciated.

Sewall Wright has long claimed that species are best able to make rapid evolutionary advance if their populations are divided into many small quasi-isolated demes. Obviously rotting trees tend to impose subdivision. Tree bodies are large enough and often die in a sufficiently piecemeal way to provide food for several generations: yet in the end each is eaten up and dispersal enforced.

Returning to the cavernous quality already mentioned, dead-tree insects tend to live *in* dead trees and at least for those whose colonies can run several generations, without being forced to emerge, this affects the rigour of their isolation. Of course, in theory, colonies on the outside of live trees could perennate much longer still but in practice their isolation is much less certain; predators can make such insects fly or jump off and winds can mix them.

Support for Wright's thesis has come in recent years from demonstrations that evolutionary rates, as measured by morphology and chromosome re-arrangements, can differ very markedly from group to group, and from recent surveys of such differences[15,16] that strongly implicate an effect from breeding structures. This is seen to work partly on the lines of Wright's prediction and partly on the basis of a founder effect through local isolation which aids the spread of chromosome rearrangements in a process that amounts to a kind of sympatric speciation.[15,17] In so far as such speciation goes on in dead trees it may multiply species numbers less than it would elsewhere since the new karyotype may be more apt to take over the whole habitat from the old (see below); but if karyotype and morphology change are correlated in insects as they are in mammals the process is still important from the present point of view and one of the remarkable convergences of dead-tree fauna—that to male haploidy (also see below)—seems to attest this.

Wright's plasticity through drift and Bush's 'instant speciation' by no means exhaust the evolutionary facilitations that can arise from local isolation. Some others are briefly indicated in the following summary:

1. Evolutionary change facilitated through drift plus recombination.[18]
2. Social evolution promoted through kinship. Local kinship groups are an inseparable concomitant of genetic drift.[19]
3. Karyotype plasticity.[15,20] Via male haploidy[6,9] this synergizes (2).
4. Subdivision resists wide-scale disaster from drive (Chapter 4 and ref. 21). This point connects closely with (1), (2), and (3) but emphasizes deme extinction.
5. Interspecies reciprocation and symbiosis. The Trivers argument for individuals can be extended to relations of multigeneration local stocks (Chapter 8 and refs 22 and 23).
6. Polymorphism promoted by disruptive or alternating selection for sedentary breeders and dispersers.[24] Such polymorphism has synergized (2) in ants and termites.

DIVERSITY IN DEAD TREES

Compared to living oak trees[25] dead oaks do not have an impressively long species list of insects.[26,27] Inclusion of Acarina would reduce but not eliminate this contrast. Extension of the survey to other tree species on the other hand would accentuate the difference: considering the fauna of a whole British wood, and still more for that of a tropical forest, the part associated with dead trees will prove far less species-diverse than that associated with live vegetation.

This relative poverty in species extends, of course, to other detritus habitats and the reason for it certainly has to do with the chemical and other special defenses of living green plants.[28–30] It is well known that the stage of scolytid attack through the bark to the phloem is normally about the last where host specificity is at all marked. Subsequently arriving insects (perhaps no worse than most humans faced with debris of a tree) are poor botanists: in Britain, for example, they very roughly separate into degraders of conifers of hard woods, but in general they tend to ignore species and go for convergently similar conditions of bark and wood and state of invasion by fungi. Thus faunal lists for dead trees overlap very widely.[31]

If important evolutionary advance has an equal chance of occurring in all species (as is suggested by Mayr[32]) the diversity of species on living plants would be expected to give these groups a great advantage with respect to initiating major new groups. Since direct phytophages and also parasitoids contribute disproportionately to the immense total species list of the Insecta, this should apply particularly to them. Yet it can be said with near certainty that the main forks of the tree of insect evolution have not occurred in phytophages or parasitoids. Even for the predominantly phytophagous groups Lepidoptera and Hemiptera there is plenty of room for doubt that the common ancestor was phytophagous.[33] For the Coleoptera such an ancestor is hardly considered.[34] The scattered distribution of predation in generally phytophagous groups suggests that a transition from phytophagy to predation is fairly easy. It probably occurs most often via cannibalism in circumstances of severe competition.[35] However, possible cases of transition to detritophagy or mycetophagy seem much less common. This topic will be reconsidered below. As one immediate example, the recently preferred phylogeny of the Thysanoptera which I used to view as providing the most probable case of a fairly important reversion from green plants to fungi[36] has been thrown into doubt by Mound and O'Neill:[37] the group which they highlight (Merothripidae) raises a new possible image of the ancestral thysanopteran.

As might be expected the evidence for important reversions from parasitoidal existence is also weak. The most that can be claimed is that the ancestry of the aculeate Hymenoptera probably includes forms with at least ectoparasitoid habits.[7] Even this reversion can be partly turned to advantage for the present thesis in that the distribution of primitivity in Hymenoptera suggests that the group's origin lies occurred among parasitoids, very possibly polymorphic for wings, living in dead trees, rather than in parasitoids whose hosts were in more aerial habitats.[7,38]

If reversions from phytophagy and parasitism are few and special, the weight of species numbers associated with living vegetation and the smallness of the numbers associated with detritus can largely be dismissed from our argument.

PRIMITIVITY IN DEAD TREES

Paradox

Much more than in species numbers and superficial diversity, a serious, almost paradoxical, difficulty concerning the role claimed for dead trees arises when we consider how it is proper to interpret the present-day taxonomic distribution of 'primitivity'. Dead-tree insects are especially apt to be categorized as primitive by taxonomists (see Table 12.1). It has to be admitted that the most obvious interpretation of this is not at all that dead trees are the sites of radical innovations but rather that they are quiet backwaters serving as a kind of refuge for insect forms that have failed and vanished in the course of faster evolution going on elsewhere. The primitivity of dead-tree insects has indeed been referred to in such terms[39] and no doubt there are examples, particularly in those insects which have adapted to the difficult and nutritionally poor diets in the tree, to which the concept of a refuge and genetic stagnation really applies. But often the claims of systematists are confusing: they present the idea of a refuge and at the same time suggest that the habit of mycetophagy, for example, is more primitive in a group than that of feeding on the living parts of plants. This applies to Mamaev's discussion of Cecidomyiidae. The first impression on pulling slabs of bark from a dead tree trunk and seeing the dense and varied community, very rich in carnivores, that is present at a certain stage of decay of phloem is certainly not that of a quiet refuge.[40] Yet, at the same time, to mention one group in particular, the beetle species in this subcortical community often present to the viewer a kind of synopsis of the whole classification of Coleoptera. Further, combining both the fossil and modern evidence on this group, Crowson[34] has concluded that this is indeed the most probable habitat for the ancestral beetle.

Such a paradox may be partly resolved in the following way. Suppose a detritophagous stock evolves a phytophagous branch. Suppose that the lines diverge and attain, say, subfamily distinction, but all the time the phytophagous line speciates far more rapidly. A taxonomist trying to classify the whole group will have to heed and use far more characters in the phytophagous branch. The characters are needed to define numerous species and genera.

Table 12.1 Primitivity in insects connected with dead trees

Insect group / Order / Suborder	In dead or dying trees		On or in live land plants (examples not mentioned)	Comments
	Dead tree association exists in:	Primitive with respect to:		
Subclass **APTERYGOTA**				
Collembola				
Arthropleona	*Neanura*, etc.	Arthropleona	None	c.f. *Rhyniella*
Symphypleona	*Allacma*, etc.	—	Some	
Protura	Some	Insecta	None	
Diplura	Some			
Archaeognatha	Some			
Thysanura	Some			
Subclass **PTERYGOTA**				
Division Palaeoptera: two orders and five suborders, all aquatic unconnected with terrestrial plants.				
Division Neoptera				
Section Polyneoptera				
Dictyoptera				
Blattaria	Many	Dictyo-Isoptera	Some	Note *Cryptocercus*
Mantodea	A few	Neoptera	Most	
Isoptera	Most	Neoptera	None	Primitive termites in wood
Zoraptera	All	Neoptera	None	
Plecoptera	Some Archiperlaria?	Plecoptera	None	Aquatic
Grylloblattoidea	None?	Neoptera	None	Montane
Phasmida	*Eurycantha*, etc.	Phasmida	Almost all	
Orthoptera				
Ensifera	*Deinacridia*, etc.	Ensifera	Most	
Coelifera	None?	—	Most	
Embioptera	Some	Neoptera	Some (bark)	Soil and litter typical
Dermaptera	Many	Neoptera	A few	Note flattened subcortical Apachyoidea
Section Paraneoptera				
Psocoptera				
Trogiomorpha	*Lepinotus*, etc.	Psocoptera	?	Principally under or on bark
Troctomorpha	*Embidopsocus*, etc.	Psocoptera	None?	
Eupsocida	None?		Most	

	Thysanoptera				
	Terebrantia	Merothripidae	Terebrantia	Most	
	Tubulifera	Many		Many	
	Hemiptera				
	Coleorrhyncha	None?	Hemiptera	All?	Note also *Newsteadia*, *Orthezia* (see text)
	Homoptera	Some Fulgoroidea	Homoptera	Most	
	Heteroptera	Dipsocorimorpha	Heteroptera	Most	
Section	Oligoneopotera				
	Neuroptera				
	Megaloptera	Raphidioidea, *Chauliodes*	Neuroptera	None	Mainly aquatic
	Planipennia	†	—	Most	
	Mecoptera				
	Protomecoptera	Unknown	Mecoptera	?	
	Eumecoptera	None	Oligoneoptera	Many	
	Trichoptera	Xiphocentron	(not primitive)	None	Almost all aquatic
	Lepidoptera*				
	Zeugloptera	None?	Lepidoptera	All?	On bryophytes
	Monotrysia	Some Hepialidae, etc.	Monotrysia	Most	
	Ditrysia	Various	—	Most	
	Diptera				
	Nematocera	Some, in about $\frac{1}{2}$ of families	Diptera	Some	Aquatic species also common
	Brachycera	Some, in majority of families	—	Some	
	Cyclorrhapha	Some	—	Some	Leaf miners occur
	Hymenoptera				
	Symphyta	Siricoidea, Orussoidea	Hymenoptera	Most	‡
	Apocrita	Various	Apocrita	Many	

This table shows a conservative classification of insects to the level of suborders (or orders, if suborders are not defined), but with all wholly epizootic groups excluded. Column two indicates the degree of association of species with dead trees, and column three the extent to which these species are regarded as 'primitive' with respect their own suborder or to some more inclusive taxon (a dash indicates that primitivity is indecisive). Column four indicates the extent of attachment of each group to living plants: no attempt is made to mention specific groups and status within these, since this would greatly lengthen and complicate the table to reveal only a minor extent of primitivity.

* The earliest known lepidopteran fossil is a larval head capsule in Cretaceous amber embedded with webbing, frass, and crumpled plant remains; this suggests a feeding site on or under bark or else in litter at ground level.[33]

† Larvae of Psychopsidae and Berothidae under bark flakes—only on live trees? *Megalithone* (Ithonidae)?

‡ Species from living plants are mostly parasitoids of plant feeders; some are seed chalcids, gall wasps, etc.

Only a few characters will be needed in the detritophagous branch. Further, possibly it is just the characters that are most radical and pregnant with possibility of major evolutionary change which, in the few species of this detritophagous group, the taxonomist will neglect to emphasize because these characters are at once superfluous for definition and out of line with what are considered 'useful' kinds of character in the rest of his task. In other words, the primitivity of dead-tree insects could be partly an illusion arising out of the need to define species and create keys. For example, a horny quality of the adult forewings (e.g. *Issus*, *Merope*), or paedogenesis (e.g. Heteropezini), might be relatively overlooked as an apomorphous character of a dead-tree group because other characters had already sufficed to define it. Systematists have actually produced a name for such characters, autapomorphous, yet still seem to neglect them.[41]

Consider *Plastosciara perniciosa*. This is a parallel to *Heteropeza* in that, besides being a dead-tree insect in the wild,[42] it is also a pest of protected cultivation[43] and shows a kind of neoteny. Specifically, this species has a wingless and rather worm-like morph (genuinely adult and represented in both sexes) as an alternative to the normal alate morph.[42] The place of the species in sciarine taxonomy is established, however, mainly on characters of the normal adult and Steffan's discovery and the strange form of the claustral adults are unlikely to affect this. If the alate form were abandoned due to success with another method of dispersal (e.g. endoparasitic like a *Deladenus* nematode[44]) the worm-like form could conceivably become the ancestor of a future 'order' of vermiform soil-dwelling insects, even with potentiality to become a new class.

Direction of transition: into or out of dead trees?

The example of *P. perniciosa* was chosen because the special evolutionary potential of neoteny is widely recognized, but, apart from this, the argument of the preceding section would apply equally well if the hypothesized original stock had been non-detritophage. In general, whether a stock has emerged from dead trees (or simply from detritus) or has gone into it has to be settled from the particular evidence of the case, including the hints as to adaptation that appear in the systematist's reconstruction of the original type of a group.

Thus the fact that, so far as I know, the only place where *Heteropeza pygmaea* can be reliably found in the wild is under dead bark, where, in Britain, it is an abundant and characteristic pioneer in the decay of phloem, strongly suggests that the occurrence of this species in mushroom houses[43] is secondary. The case of *P. perniciosa* is actually less certain because although known from rotting logs[42] I do not know that this habitat is most typical in its native land, wherever this may be. Similarly, another sciarine, *Bradysia paupera*, may give a good illustration of a tentative switch of oviposition and larval feeding out of the usual habitat and on to green leaves;[43] but this is more likely to be a switch out of soil than out of dead trees. Nevertheless, dead trees are indeed rich in sciarine genera.[45] A more direct transfer to phytophagy from a subcortical habitat is likely in the mite *Rhizoglyphus echinopus*,[27,46] and such transfer is virtually certain in the case of the bug *Aradus cinnamomeus*.[47,48] At generic or higher levels Crowson[34] instances other cases in Coleoptera.

Brief mention in this text of three other examples of probable emergence from dead trees must suffice.

First, consider the beetle family Rhipiphoridae; this may be treated as a possible model for the unknown history of the parasitic order Strepsiptera. Females of the parasitic genus *Metoecus* rather unexpectedly lay their eggs on dead wood, although the larvae are parasitic on *Vespula*. Other rhipiphorids have larvae which are free-living in dead trees.[49] So this one family spans the lifestyles of a normal beetle and a stylops, and the oviposition site mentioned suggests dead trees as the likely ancestral habitat. Second, consider the path to another type of parasitism, that of Cimicidae. *Cimex* is flattened and also wingless and we would expect some sign of these characters in a bug proposed as its closest non-parasitic relative. Flattening suggests a subcortical insect and sure enough both characters are to be found under bark in representatives of the predatory and wing polymorphic genus *Xylocoris*. Usinger[50] finds *Xylocoris* to be the anthocorid genus with certain characters most suggestive of a primitive cimicid. Here flattening and winglessness suggest the direction of evolution. The genus *Anthocoris* itself suggests a bridge from subcortical life to phytophagy: in summer *Anthocoris* species are important predators of aphids, and feeding on the body fluids of aphids cannot be very different from imbibing phloem sap directly.

I do not know of any actually phytophagous Anthocoridae, but perhaps a parallel transition can be identified—my third example—in the Coccinellidae, where, along with the aphid-feeders, plant-feeding species also occur. Here, the divergence is more advanced and the family Endomychidae is the proposed parallel to genus *Xylocoris* (although whether larval endomychids are predatory or mycetophagous I have not ascertained). Coccinellids have evolved preference for drier and quieter hibernacula than the sites under bark where one finds *Endomychus*; on the other hand *Anthocoris nemorum* can be found in winter under bark mixed with *Xylocoris* colonies and at such times is conceivably predatory there—in what is here suggested to be its ancestral home.

Examples of insects that have most probably entered dead trees from elsewhere appear to me much harder to find. This may be partly due to my current bias and I will be glad to be informed of counter-cases overlooked. Perhaps the best case noted so far is that of *Forcipomyia*. Larvae of this genus are common under dead bark in Britain. They possess a closed tracheal system normally characteristic of aquatic larvae and in fact many related genera of the same family of flies (Ceratopogonidae) are aquatic as larvae.[51] It could be that the closed tracheal system was evolved for survival in species dwelling in bark habitats subject to frequent flooding and then later proved pre-adaptive for permanent life in water; but the opposite course of evolution seems at least equally likely. Outside the Insecta a rather similar example is provided by woodlice. These are isopod crustaceans (Oniscoidea) and the majority of isopods are aquatic. On land some woodlice are, as the British common name implies, abundant under loose bark, but all are found widely in litter, compost, and similar habitats as well. In Britain *Oniscus asellus* is, perhaps, the most constant bark species and its slightly flattened form, so similar to that of some subcortical roaches, suggests bark as its ecological 'headquarters' (*sensu* Elton). It seems most likely that the ancestor of terrestrial Oniscoidea came ashore (probably in at least two invasions[52]) via litter and soil, as some isopods and amphipods and other primarily aquatic crustacean groups seem to be doing at the present day. However, isopods in *Limnoria* and *Chelura* suggest that a course from water directly into dead trees is at least possible.

Rather as woodlice seem recently to have gate-crashed the land fauna and, in consequence of a lack of groups adapted to exploit them, may obtain a kind of freedom of diverse habitats into which they can begin to radiate, so some *Drosophila* colonists which had the fortune to enter the Hawaiian Islands may have found a similar freedom, in this case probably mainly freedom from competitors. Superabundant speciation in the *Drosophila* in question gives their case a different complexion from the isopods, but the parallel to be pointed out here is that the colonist *Drosophila* stocks seem to have been particularly successful in invading the subcortical habitat.[53] But again, an alternative possibility, that the invasion took place through species arriving in driftwood, should be borne in mind.

There are various enigmatic cases in Hemiptera which might be transferrals from phytophagy to mycetophagy. Most of them concern soil,[54] but some refer to rotting logs (e.g. most of the Achilidae) or to rotting tree ferns (e.g. *Oliarus* in Cixiidae[55]). I do not know about the primitivity status of these log- and soil-associated groups within Fulgoroidea but this group itself is treated by Goodchild[56] and others as the basal branch of Heteroptera. With Sternorrhyncha also the most primitive groups tend to be those most associated with soil and claustral habitats. Such a distribution of primitivity makes the mycetophagy, which is, for example, at least sometimes indulged by *Orthezia*,[57] lose weight as a counterexample. (A like argument also applies to what may be reversions to mycetophagy in scarabaeids[58,59].)

A more serious implication in Hemiptera comes from the Peloridiidae, usually considered the basal branch for the whole order. If the common ancestor which this family has with the rest did indeed suck mesophyll, as peloridiids are thought to do, then, in Goodchild's phylogeny, there is an important reversion in passing from this ancestor to the supposed litter-dwelling non-phytophagous ancestor of the Heteroptera. The change in the articulation of the rostrum to the front of the head is certainly suggestive of a move under bark. Predatory enicocephalids and anthocorids make good use of forwardly directable mouthparts in this situation; yet achilids and mycetophagous thrips manage without them. The uncertain reversion indicated by peloridiids and also the general implication of their moss habitat will be touched on again below.

Rotten wood is favoured by very diverse groups as a site for hibernation[31] but transition into rotting wood in this sense is usually a trivial event from the point of view of breeding structure. Whether it carries an implication of 'ancestral familiarity' with dead wood is an intriguing question in view of the use of this site by *Anthocoris*, coccinellids, ichneumonids, etc.[26] Perhaps an equal list of cases with no such likely ancestral connection could be cited. *Vespa crabro*, a 'primitive' social vespine, uses dead wood for its hibernation site, nest site, and nest material; on the other hand *Vespula* queens seem to use equally readily any other hibernation site that offers suitable cover and microclimate. The same applies to *Bombus*. As regards wood-nesting in *Xylocopa*, Hurd[60] claims that soil is the likely ancestral nest site, and Barrows[61] equally claims this for log-nesting Halictinae. Wood or soil should make little differences to breeding structure in these very free-flying flower-visiting insects.

Some other cases of transition which might repay study are indicated in Table 12.1 and in the following additional list: Sminthuridae (Collembola); Liposcelidae (Psocoptera); Derbidae (Hemiptera); Chrysomeloidea, Scarabaeoidea (e.g. *Oryctes rhinocerus*), Curculionoidea, Scolytidae (e.g. *Hypothenemus*), Elateridae, Nitidulidae (Coleoptera); Hepialidae (Lepidoptera); Tipulidae, Lonchopteridae, Syrphidae, Phoridae (e.g. *Megaselia*), Dolichopodidae (Diptera). Possible trends in Scolytidae have already been outlined by Schedl.[62]

The fossil record and comparative evidence

The record of arborescent land plants extends back as far as the record of insects but evidence directly connecting the two in the Palaeozoic is extremely scanty.[63] This applies almost as much to evidence of attack on living plants, arborescent or not, where damage that could be attributable to insects should be relatively easy to recognize, as it does to evidence of inroads on dying and dead remains. As regards indirect evidence, the point can at once be made that the earliest of all known insects (using a broad view of the class), the collembolan *Rhyniella*, is remarkably like *Neanura muscorum*,[64] a species which occurs today under dead bark and in rotting wood. It is also to be noted that some forms of attack on tree detritus, especially the habit of feeding on the associated fungi, would leave little trace of any kind that could be fossilized, and that wet conditions such as prevailed where Palaeozoic plant fossils were formed are hostile even to the preservation of arthropod coprolites.

Standing dead lepidodendroid trunks of Carboniferous age on Arran, described by Williamson[65]—hollow, and containing lodged within them twigs and 'leaves' and other fragments of various plants (and all finally buried in volcanic ash)—represent not only the type of habitat where one would confidently search for *Neanura* today, but also just that kind of food object, discrete, bulky yet cavernous, which our hypothesis has required. These trunks hint that we need not take too seriously the lack of evidence of bored insect galleries in pre-Permian plant remains. In truth such evidence of boring in the Palaeozoic seems to be scanty and unconvincing: sinous engravings on the cortex of *Sigillaria* in Germany[66] look to me as attributable to, say, impressions left by fungal rhizophores as to galleries of insects; and, in some other fragments, holes indubitably bored by some animal are of a size too minute to be the probable work of insects.[65]

Borings, however, appear in fossil wood of the Permian and so does bark.[34] As Hinton[67] has noted, there is a basic conflict between burrowing and the possession of wings and the most striking solution of this has been holometabolous development. Perhaps a move by arborescent plants to protect phloem beneath bark[68] was the principal stimulus to holometaboly. In this essay we are mainly concerned with the bonanza of dead phloem but no doubt in all ages there have been saprophagous insects that pressed their attacks earlier or brought in parasitic fungi to prepare ground ahead, much as cerambycids, scolytids, and siricids do today; bark needs to protect against these too. In this connection it is interesting to note the somewhat unexpected appearance of ovipositors in certain Palaeozoic groups.[69] Among their many other uses ovipositors serve to put eggs in deep crevices and under bark: ovipositors occur, for example, in various stages of evolution, in many present-day insects whose larvae live under bark or in wood (*Helops* and *Lonchaea* show cases of incipience). Species that lay eggs from the outside of bark are usually outbreeders, and this, according to our thesis, counterindicates the ancient groups with such appendages as being quite perfect images of the ancestors of major groups of the present day; correspondingly, of course, a connection of ovipositors with apparent evolutionary stagnation (as in raphidians, cupesids, siricids, orussids) is not surprising. Adults going in under bark, going in further through flattening or by development of horny forewings (as in roaches, beetles, and some psocoptera—and perhaps also in *Issus, Merope* and others) or by dealation (as in some Embioptera, Isoptera, Zoraptera, Psocoptera, Thysanoptera, and Hymenoptera), going perhaps further still through an apterous adult morph or total wing-loss (Appendix A, *Cryptocercus*, etc.), sending on unencumbered larvae ahead as specialized burrowers, and, finally, as the ultimate development in a few lines, allowing those larvae to breed for themselves without any need for more space (*Heteropeza, Micromalthus*)—this whole sketched evolutionary sequence for the increasingly rapid penetration and utilization of dead phloem hidden beneath bark seems to me eminently possible. One step is perhaps currently reillustrated by the sluggish 'pupal' stage of subcortical tubuliferan thrips. In the main stock formation of the pupa produced the great insectan advance to holometaboly.

Bast and wood

Bast—that is, phloem plus cambiums—offers a rich and well-balanced diet. On the death of a tree these layers are the most speedily consumed. Sapwood is consumed more slowly and by fewer species. Lignified heartwood and suberized bark, where these are present, go more slowly still.

The wide variety of taxa that can be found directly under the bark has already been emphasized, and the same habitat has had repeated mention in connection with other phenomena. Besides consumers of dead plant tissues there are many species that feed primarily on the bacteria, yeasts, and fungi that are soon abundantly present, including on parts of those fungi that are primarily concerned with the decay of wood. There is also a surprisingly high diversity of predators. All these insects tend to be small, with a size range overlapping (e.g. *Ptinella*, 0.6 mm) that of typical Acarina, which group is itself abundantly and diversely represented. They also tend to have short generation times which, more than offsetting rather low total fecundities, gives them potential for high rates of increase. As our hypothesis requires, many species do indeed readily mate within the habitat and the inbreeding which this implies is reflected in biased sex ratios which further improve the potential for colonization and increase. Polymorphism in colonial species of this habitat is summarized in Appendix A. It might be thought that winglessness in a morph would always be a further device for greater effective fecundity and increase, as it is in all other winged/wingless morph comparisons that have ever been made in non-social insects. No doubt this is often the case with the subcortical polymorphisms but one notable exception has recently appeared.

Studying the very marked polymorphism found in both bisexual *Ptinella aptera* and thelytokous *P. errabunda*, Taylor[2] found that winged females were longer lived, were more fecund, and had larger spermathecas than their wingless counterparts. This is suggestive of the beginning of a termite-like social development. In fact, on the basis of this example plus the general absence of wing polymorphisms in xylophagous insects, it can be suggested[24] that social termites arose from their roach-like ancestors in the habitat of dead phloem, and that *Cryptocercus* is consequently connected to them as a parallel invasion of the wood, rather than as a 'wax-work' image of their ancient way of life as is often implied.[70]

In contrast to the subcortical insects, those of dead wood are usually larger (and include the largest of all, such as *Titanus giganteus*, up to 200 mm), and tend to develop much more slowly; generation time of 2 or 3 years is common in temperate latitudes. Consequently wood feeders are rather slow to increase in a habitat and for reasons perhaps connected with this but not fully understood they are also much less inclined to endogamy and to claustral continuance of the colony even if the wood provides bulk for it (primitive termites, *Micromalthus*, and perhaps passalids are exceptions here). As expected if mating is outside the log, sex ratios are on the whole normal; both parthenogenesis and wing polymorphism are almost unknown (termites and *Micromalthus* again excepted).

Bearing in mind reduced faunal competition and slower generation turnover and bearing in mind also the more outbred breeding structure, it is particularly here that we would expect to find true relict insects. And possibly the concept can apply, for example, to *Cryptocercus*, Cupesidae, and Siricidae.

At the same time the wood has many obviously advanced invaders, for example Cerambycidae. These are suspected to have moved inward from the phloem relatively recently. Some lines in Scolytidae appear to be so evolving at the present time.[62] Where beetles move in as 'ambrosia' feeders the diet remains rich and rapid endogamous breeding is sometimes retained. Here might be mentioned the large genus *Xyleborus* which also illustrates various cases of transition towards attack on the green parts of living plants. *Micromalthus*, combining arrhenotoky, thelytoky, and paedogenesis, is much less successful but confined within rotten wood its trend is similar but more extreme. Certain cecidomyiids, moving in through shrinkage cracks without ambrosia but again feeding on fungi and not on wood, offer rough parallels to the ambrosia beetles—*Pezomyia* (or, probably better, the little-known *Micropteromyia*) parallel to *Xyleborus*, and *Heteropeza* parallel to *Micromalthus*.

PRIMITIVITY IN OTHER HABITATS

Apart from habitats of the dead-tree complex, others especially frequently mentioned in connection with 'primitivity' are litter and soil, moss (and to some extent other primitive green plants), and finally fresh water. The living parts of higher plants tend to carry insects that taxonomists rate as relatively advanced, the main exception to this being, perhaps, an undue frequency of 'primitive' insects feeding on pollen and the pollen-producing organs of plants (see below). Seed insects, on the contrary, and also those of 'stored-product' type environments, tend to be classed as advanced and so do parasites (probable routes to these various lifestyles can be drawn out of dead trees via the hollows in which birds, lizards, and mammals nest and keep their stores[71]).

Dung and carrion tend to carry types of rather intermediate primitivity. Carrion of large animals might at first thought seem very like the rich phloem masses of dead trees in size and nutritive value. Perhaps it is partly the quality of thinness that tends to reserve dead phloem for insects. Besides this carrion is even more scattered, even more nutritive, and less resistant to entry and destruction. Thus larger animals take a far larger share and tend to leave only dispersed fragments and bones—the latter a worse resource for insects even than the heartwood of trees. Altogether the attributes of carrion left available to insects almost necessitate that those that breed in it be good fliers and fly in every generation. This encourages panmixia. Excepting perhaps dried carcasses, carrion is too ephemeral to support colonies.

Moss

Plant litter and soil are habitats with no sharp separation from rotting tree trunks and branches. Moss, too, is intimately associated with all these, and like them provides sheltered and hidden spaces. The spaces in dense moss cushions may, like cavities in soil and wood (and like the similar spaces in dense grass tussocks), tend to encourage local inbred colonies. Yet, I believe—admittedly on no easily presented evidence—that feeding in living moss is more often derived from mycetophagy and detritophagy than vice versa. One example, that of Peloridiidae again, must serve to illustrate both moss primitivity and this bias of the writer. The family was largely lost as an illustration of my dead-tree theme when I read that some members had been shown to exist in *Sphagnum* and other mosses distant from trees;[72] however, this has not been shown for all species and I preserve a small hope that at least *Peloridium hammoniorum* may prove mycetophagous in rotting wood: this species was first found under a rotten log, and it is perhaps significant that this is also the one species known to show wing polymorphism.[73]

Litter and soil

Almost every group mentioned in connection with dead trees in Table 12.1 also has representatives in other kinds of plant litter and in soil. Soil and litter insects likewise tend to be more 'primitive' than collateral groups that feed (not necessarily as herbivores) on aerial parts of living plants. The special case made for dead trees is based partly on the *a priori* considerations already mentioned and partly on an impression that evolutionary novelty is really more common in dead-tree insects than in those of soil. What has soil to offer quite so odd as, for example, the intra-haemocoelic insemination of Cimicoidea and related bugs? There are, of course, abundant examples of ingenious adaptation for life in the soil and doubtless many routes to subaerial phytophagy do actually lie through soil-dwelling forms. But where such a path seems apparent the evidence is often somewhat equivocal. In the case of *Sminthurus viridis*, for example, derivation from the commoner litter habitat of other *Sminthurus* species is likely, but we note also the existence of the 'primitive' relative *Allacma fusca* which is associated with rotting wood and fungi. Similarly two rather unexpected pests of vines, *Lethrus* and *Vesperus*,[74] seem to have more connection with soil than rotting wood, but on surveying slightly more distant relatives this case too becomes more doubtful.

In a less-dismissive approach to the competing claims of soil and litter a stand can be taken on the already mentioned weaker version of our thesis; that is, on the claim that, like dead trees, soil and litter force subdivision and inbreeding on their inhabitants. Certainly a great deal of what has been said about necessities and adaptive responses of insects in dead trees applies to soil insects as well, and the evolutionary potential of these can be equally contrasted to that of the insects in more panmixial habitats on plants. It has to be insisted that neither in tree trunks nor in patchily distributed resources in the soil do we expect specially high rates of evolution unless some local inbreeding really occurs; if the resource, whatever its location, is used up in one generation and all offspring have to disperse and mix, then the species is certainly worse off than is, say, a coccid or a spider mite which is capable of local differentiation by drift through the perennation of its colonies and its restricted mobility. (Coccids have, in fact, very high rates of evolution at least for some traits;[75] for tetranychid breeding structure see McEnroe.[76]) Apart from soil insects this caution applies with force, as already mentioned, to a lot of dead-wood flies and beetles which have annual or longer life cycles. In being a more continuous habitat one might well expect soil and litter to enforce even less long-range dispersal than dead trees do; and the total loss of flying wings in many soil and litter insects of permanent habitats tends to confirm that this is so.[77] But unfortunately, apart from the vague portent of such flightlessness, we have little factual information about dispersal in environments presumed to be stable and uniform, either soil or any other. In the realm of theory too the problem remains little explored but models already developed make it clear that we cannot assume that existence of a stable ongoing resource will imply minimal dispersal (Chapter 11). (Comins has found it possible to extend the models cited in such a way that ESS dispersal probabilities can be given for some of the classical breeding structures of population genetics;[78] so far, however, this development is more relevant to expected inbreeding and other properties for the 'stepping-stone'-type distributions of dead tree insects than it is for the more purely 'viscous'-type distributions here supposed typical of soil insects.)

Rather as in the field of sociobiology the factor of relatedness has been added to others previously adduced in order to improve our understanding of, for example, evolution of social insects (e.g. Chapter 8 and refs 79 and 80), so the present considerations of breeding structure must be considered additions to a continuously improving, general picture of insect evolution. I see no conflict between the present emphasis on dead trees and Ghilarov's[12] emphasis on soil as mediating transition to drier habitats, or with Hinton's[14] deduction that conditions of alternate dryness and flooding were important in the stem of endopterygote evolution. Dead trees in a showery climate, indeed, almost idealize the concepts of these writers—a wick through which insects can evaporate to the air.

Water

Equally, trees fallen into water could encourage transition to fully aquatic life. The idea that insects as a whole originate in aquatic habitats is now almost abandoned; consequently even the modern palaeopteran orders have to be regarded as secondarily aquatic. I am not aware that these orders show any hint of a significant connection with dead trees or even with litter. (The terrestrial litter-dwelling larve of *Megalagrion oahuensis* in forest litter is obviously a re-transference to land.) Inability to flex the wings flat over the abdomen in these orders also strongly contraindicates any connection with bark (contrast the habitat and presence of this ability in Liposcelidae and Phylloxeridae with lack of it in related families); but then, of course, the arborescent plants had no substantial kind of bark at the time these groups originated.

For most of the rest of the numerous invasions of water by insects, a route through dead trees seems more possible but is not strongly indicated against routes through other kinds of wet litter. This particularly is true for the Diptera, and, as already indicated for Ceratopogonidae, there may be some good cases for transition from water into rotting trees on land rather than the reverse. However, a phyletic closeness of 'wood' and 'water' is also indicated by a surprising number of families and these are mostly at the 'primitive' end of the dipteran range; examples are Tipulidae, Chironomidae, Ceratopogonidae, Syrphidae, Stratiomyiidae, Tabanidae, and Dolichopodidae.

In Trichoptera the species with exceptional terrestrial larvae are not considered primitive. The larva of *Xiphocentron*, for example, lives on the outside of damp rotting trunks and that of *Enoicycla* lives in moss and litter. But both occur in woodland and so faintly suggest that tree trunks partly in water could have provided a bridge—to land in these cases, but also, perhaps, in an ancestor, to water. According to Ross[81] the most primitive genera of Trichoptera are strongly convergent to a habitat of 'cool moderately rapid small streams running through shaded woodland'. Trunks partly in water are particularly common in such places. With regard to the Megaloptera, such trunks at about the water level do in fact provide the headquarters of *Chauliodes*,[82] a genus which well connects terrestrial subcortical raphidians with the more fully aquatic genera such as *Corydalus* and *Sialis*. Notwithstanding some claims (e.g. for Coleoptera, instancing *Corydalus*[83]) there seems no need to suppose that any major group of insects, excepting possibly Diptera, has arisen via an ancestral aquatic larva. However, the distribution of fresh water itself often imposes a pattern of small demes so that the evolutionary characteristics of this habitat may not be so very different from those of dead trees. In fact, ponds and streams[84–86] are indeed, like dead trees and grass tussocks,[87] sites where wing polymorphism is common. And, corresponding to taxa of dead trees, aquatic taxa evidently also survive long, are often dubbed 'primitive' and at the same time manifest many very original adaptations: mention may be made here of labial masks of Odonata, subimagos of Ephemeroptera, and the cases and feeding devices of Trichoptera. Paedogenesis has been achieved by Chironomidae. Male haploidy has not evolved in aquatic insects but Plecoptera seem to approach it, and outside the Arthropoda its only other full attainment is in an aquatic phylum, Rotifera.

Pollen

Examples of primitivity in this habitat are few but quite striking. Here examples from three groups must suffice. Adult Micropterygidae feed on pollen but their larvae apparently on bryophytes—perhaps also on rotting materials. This family is regarded as the basal branch of the Lepidoptera, and by some as perhaps sister group to the Trichoptera. Three primitive genera of the Curculionoidea—*Cimberis* (Nemonychidae), *Allocorynus* (Oxycorynidae), and *Bruchela* (Anthribidae)—breed, respectively, in male cones of *Pinus*, of *Zamia* and in the seed-capsules of *Reseda*. The last beetle and site merit mention here because the carpels of *Reseda* show an unusual primitive gymnospermous character—they are not closed. Such 'primitive' genera connected with sporangia of 'primitive' plants convey an obvious suggestion. However, they are but three genera in an immense group; and even with attention confined to these families plus the others that are primitive for the superfamily (e.g. Aglyceridae and Brenthidae), habitats in the dead-tree complex are far more abundantly represented than are attachments to microspores. A transference of diet from mycelium to fungal spores and thence to pollen is not difficult to imagine; Crowson[34] suggests such a transference in *Micrambe* (Cryptophagidae). A close parallel to the situation of these curculionoids exists in the symphytan Hymenoptera; *Xyela* can be compared to *Cimberis*, Siricidae to Brenthidae, and the sawflies to the weevils. Malyshev[7] thought that *Xyela* might betray the diet of the mecopteran-hymenopteran ancestor (i.e. microspores scattered on forest litter). In view of what has been said above a stem through mycetophagy with early branches to spore and pollen feeding seems at least equally likely. Wood, litter, moss, pollen, and even water all cluster near to the faint phantom of the earliest hymenopteran. It is possible that, in the near future, uncovering the lives of the Meropeidae may slightly brighten this dim scene.

Pollen feeding is sometimes far from primitive. In connection with the unexpected pollen-feeding and other evolutionary novelties of heliconiine butterflies[30,88] it is relevant to note that these insects seem unusually capable of creating a structure of small demes through their own social behaviour.[89] In this they show a parallel tendency to polygynous eusocial insects (Chapter 8) and to mammals.[15] Such animals can speed their evolution without need for the kind of *forced* deme structure which, as this essay supposes, dead-tree cavities have provided to so many juvenescent groups of insects.

APPENDIX A

Insect wing polymorphisms from live and dead trees

Table 12.2 lists genera containing species which are both colonial inhabitants of cavities in dead trees and polymorphic for wings in the adults. For this table wing polymorphism is interpreted broadly as any natural co-occurrence among adults of some individuals capable of flight and others incapable: hence sexual flight dimorphisms are potentially included (e.g. *Xyleborus*, with flying wings always vestigial in the male and always functional in the female). The kind of correlation between flight and sex is indicated in the last column; a key to symbols used in this column is given below the table. In some cases a category has been assigned on the basis of pooled information regarding several species.

The last column of Table 12.2 shows that slightly more frequently than not the wing-sex correlation leans towards males being more commonly flightless than females. Flightlessness of males when females can fly is strongly indicative of inbreeding (Chapter 4); likewise an opposite condition, flightless females combined with flying males tends to indicate outbreeding, although this second implication is less definite than the first. Very small winged males of such flightless-female species, as of coccids, *Dusmetia*, some sciarids, etc., probably normally mate in their natal colony even though they might and sometimes do fly to enter other colonies. In insects as a whole such condition, with the female sex the more flightless, is far more common than the reverse. This fact underlines the hint of inbred colonies given by the correlations indicated in Table 12.2. In flightless-female species the sex ratio is usually about normal (1:1) whereas in flightless-male species (including all those known from dead trees) the sex ratio is female-biased. This also has an understandable connection with inbreeding (Chapter 4). If flightless-female species are relative outbreeders and flightless-male species relative inbreeders, the suggestion from Table 12.2 that subcortical habitats of dead trees force on some types of inhabitants a more than usual amount of inbreeding is further reinforced by noting that on the exterior of bark (and on the exterior of the living tree as a whole) we find a majority of all the flightless-female conditions that are known. Flightless-male conditions, on the contrary, are here far more rare and, when they occur, do so along with some kind of special cover that could confine the colony (e.g. *Archipsocus*, with colonies living under self-made webs—it must be admitted,

Table 12.2 Flight-polymorphic insects living inside dead trees

Order	Habitat				
Subgroup*	Under dead bark	In sound wood	In rotted wood	Food	Wing–sex association
ZORAPTERA					
†*Zorotypus*	x		x	Fungi? Mites?	–
PSOCOPTERA					
†*Embidopsocus*	x			Yeasts, etc.	–
Psoquilla	x			Yeasts, etc.	
THYSANOPTERA					
Megathrips	x			Spores	–
Cryptothrips	x			Spores	–
†*Hoplothrips*, etc.	x			Fungi	–
HEMIPTERA					
Aradidae	x			Fungi	–?
Henicocephalidae	x			Arthropods	+
†*Xylocoris*, etc.	x			Arthropods	–
COLEOPTERA					
Micromalthus			x	Wood, fungi?	x
†*Ptinella*	x			Fungi	–
Pteryx	x			?	–
Astatopteryx	x			?	?
†*Xyleborus*		x		Fungi	– –
DIPTERA					
†*Heteropeza*	x		x	Fungi	x
†*Pezomyia*	x		x	Fungi	+
Micropteromyia	?		?	Probably fungi	–
Plastosciara	?		?	Probably fungi	+
Coenosiara	?		?	Probably fungi	+
†*Pnyxia*	x			Probably fungi	+
Sciara semialata	?		?	Probably fungi	– –
HYMENOPTERA					
Sycosoter	x			Beetle larvae	+
Theocolax		x		Beetle larvae	+
†*Cephalonomia*	x		x	Beetle larvae	+
ACARINA					
†*Pygmephorus*‡	x		x	Fungi	–

*Species where either (1) female lays eggs by ovipositor from outside of bark, or (2) male is both large and winged, are considered incapable of continuous colonial life inside dead trees and are excluded from the list. Notable exclusions by this rule are in Hymenoptera, for example *Eupelmella* by (1), and Thynnidae by (2). In Diptera *Chonocephalus* has relatively large males and phoretic copulation like Thynnidae and sometimes breeds under bark; absolute smallness of males and their agility argues for inclusion of this genus, suggesting that continuous colonies would be possible, but the fact that rotting fruit is more typical habitat than bark argues for exclusion.

†Groups personally observed by the author.

‡This genus of mites shows phoretomorphs analogous to the winged morphs of subcortical insects (see Chapter 13 and ref. 90).

– – All males flightless, all females capable of flight; –, one or both sexes polymorphic such that males are more often flightless than females; ?, data on wing–sex correlation is lacking or inconclusive, or, in other columns that the main site of breeding is uncertain; +, one or both sexes polymorphic such that females are more often flightless than males; x, species with thelytokous paedogenesis, outbreeding status unclear.

Table 12.3 Insect groups living externally on trees and showing some species with male winged and female flightless

Order Subgroup*	Habitat			Food
	Bark		Leaves	
	Sometimes inside	Outside		
DICTYOPTERA				
Blattaria	x	x		Omnivorous
Perlamantinae	?	x		Arthropods
EMBIOPTERA		x		Lichens, etc.
PSOCOPTERA		x		
HEMIPTERA				
Coccoidea		x	x	Sap
Microphysidae	?	x		Arthropods
COLEOPTERA				
Ptinidae	?	x		Lichens, etc.
LEPIDOPTERA				
Psychidae		x	x	Leaves
Lymantridae			x	Leaves
Geometridae			x	Leaves

See Table 12.2 for explanation of symbols.

however, that the contrast between polymorphism in *Archipsocus* and in *Reuterella* or bark-dwelling Embioptera, which also live in webs, remains puzzling). Table 12.3 summarizes the well-known flightless-female groups that are known on trees. Trees are so favoured by flightless-female species because trees are particularly exposed to wind and this allows for passive dispersal, usually by young larvae.[91]

To avoid over-encumbering the reference list of this paper, supporting references for the statements of Tables 12.2 and 12.3 are not given. Groups for which the author has personal experience and data are marked with a dagger sign.

Acknowledgements

Dead-tree insects, especially those manifesting wing polymorphism, were studied in Britain with assistance from a grant (GB3/1383) from the Natural Environment Research Council. For help with various facts, ideas, and bibliographic items, thanks are also due to G. Borgia, M. Deyrup, L. A. Mound, U. Nur, V. A. Taylor, and J. Waage.

References and notes

1. A. E. Stubbs, *Wildlife Conservation and Dead Wood* (S. Devon Trust for Nature Conservation, Exeter, Devon, 1972).
2. V. A. T. Taylor, The biology of feather-winged beetles of the genus *Ptinella* with particular reference to coexistence and parthenogenesis (PhD thesis, University of London, 1975).
3. J. D. Hood, The cause and significance of macropterism and brachypterism in certain Thysanoptera, with description of a new Mexican species, *Anuario de Biología Científicas, Mexico* **1**, 497–505 (1940).
4. A. Bournier, Remarques au sujet du brachypterisme chez certaines espèces de thysanoptères, *Bulletin de la Société Entomologique de France* **66**, 188–91 (1961).
5. T. R. New, An introduction to the natural history of the British Psocoptera, *Entomologist* **104**, 59–97 (1971); A new species of *Belaphopsocus* Badonnel from Brazil with notes on its early stages and bionomics (Psocoptera), ibid., 124–33.
6. G. Borgia, Evolution of haplodiploidy: models for inbred and outbred systems. *Theoretical Population Biology* **17**, 103-28 (1980).
7. D. I. Malyshev, *The Genesis of the Hymenoptera and the Phases of Their Evolution* (Methuen, London, 1969; transl. from the Russian).
8. L. A. O. Campos, F. M. Velthuis-Kluppell, and H. H. W. Velthuis, Juvenile hormone and caste determination in a stingless bee, *Naturwissenschaften* **62**, 98–9 (1975).
9. W. D. Hamilton, Evolution sozialer Verhaltensweisen bei sozialen Insekten, in G. H. Schmidt (ed.), *Sozialpolymorphismus bei Insekten*, pp. 60–93 (Wissenschaftliche Verlagsgesellschaft, Stuttgart, 1974).
10. F. A. Turk, A new genus and species of pseudo-scorpion with some notes on its biology, *Proceedings of the Zoological Society of London* **122**, 951–4 (1953).
11. B. M. Mamaev, The significance of dead wood as an environment in insect evolution, *Proceedings of the 13th International Congress of Entomology, Moscow, 1968* **1**, 269 (1971).
12. M. S. Ghilarov, The significance of the soil in the origin and evolution of insects [in Russian], *Entomologicheskoe Obozrenie* **35**, 487–94 (1956).
13. T. R. E. Southwood, The insect/plant relationship—an evolutionary perspective, in *Insect/Plant Relationships, Symposia of the Royal Entomological Society of London* No. **6**, 3–20 (1973).
14. H. E. Hinton, Enabling mechanisms, *Proceedings of the 15th International Congress of Entomology, Washington*, 71–83 (1977).
15. A. C. Wilson, G. L. Bush, S. M. Case, and M. C. King, Social structuring of mammalian populations and rate of chromosomal evolution, *Proceedings of the National Academy of Sciences, USA* **72**, 5061–5 (1975).

16. G. L. Bush, S. M. Case, A. C. Wilson, and J. L. Patton, Rapid speciation and chromosomal evolution in mammals, *Proceedings of the National Academy of Sciences, USA* **74**, 3942–6 (1977).

17. G. L. Bush, Modes of animal speciation, *Annual Review of Ecology and Systematics* **6**, 339–64 (1975).

18. S. Wright, The roles of mutation, inbreeding, cross-breeding and selection in evolution, *Proceedings of the 6th International Congress of Genetics* **1**, 356–66 (1932); and later papers.

19. Chapter 2; and references in E. O. Wilson, *Sociobiology* (Harvard University Press, Cambridge, MA, 1975).

20. S. Wright, Breeding structure of populations in relation to speciation, *American Naturalist* **74**, 232–48 (1940).

21. S. Wright, *Evolution and the Genetics of Populations*, Vol. 2 (University of Chicago Press, Chicago, 1969).

22. R. L. Trivers, The evolution of reciprocal altruism, *Quarterly Review of Biology* **46**, 35–57 (1971).

23. D. S. Wilson, Evolution at the level of communities, *Science* **192**, 1358–60 (1976); *The Natural Selection of Populations and Communities* (Benjamin/Cummings, Menlo Park, CA, 1980).

24. W. D. Hamilton and V. A. T. Taylor, unpublished work.

25. T. R. E. Southwood, The insect/plant relationship—an evolutionary perspective, in *Inspect/Plant Relationships, Symposia of the Royal Entomological Society of London* No. **6**, 3–20 (1973).

26. P. A. Larkin and C. A. Elbourn, Some observations on the fauna of dead wood in live oak trees, *Oikos* **15**, 1–92 (1964).

27. E. W. Fager, The community of invertebrates in decaying oak wood, *Journal of Animal Ecology* **37**, 121–42 (1968).

28. D. A. Levin, The chemical defenses of plants to pathogens and herbivores, *Annual Review of Ecology and Systematics* **7**, 121–59 (1976).

29. T. Swain, The effect of plant secondary products in insect co-evolution, *Proceedings of the 15th International Congress of Entomology, Washington*, 249–56 (1977).

30. L. E. Gilbert and J. T. Smiley, Determinants of local diversity in phytophagous insects: host specialists in tropical environments, in *Diversity of Insect Faunas, Symposia of the Royal Entomological Society of London* No. 9, 89–104 (1978).

31. C. S. Elton, *The Pattern of Animal Communities* (Methuen, London, 1966).

32. E. Mayr, *Animal Species and Evolution* (Harvard University Press, Cambridge, MA, 1963).

33. M. R. McKay, Lepidoptera in Cretaceous amber, *Science* **167**, 379–80 (1970).

34. R. A. Crowson, The evolutionary history of coleoptera as documented by fossil and comparative evidence, *Atti 10° Congresso Nazionale Italiano di Entomologia, Sassari*, 47–90 (1974).

35. L. R. Fox, Cannibalism in natural populations, *Annual Review of Ecology and Systematics* **6**, 87–102 (1975).

36. L. J. Stannard, The thrips or Thysanoptera of Illinois, *Bulletin of Illinois State Natural History Survey* **29**, 1–552 (1968).
37. L. A. Mound and K. O'Neill, Taxonomy of the Merothripidae, with ecological and phylogenetic considerations (Thysanoptera), *Journal of Natural History* **8**, 481–509 (1974).
38. D. J. Brothers, Phylogeny and classification of the aculeate Hymenoptera with special reference to Multillidae, *Kansas University Science Bulletin* **50**, 483–648 (1975).
39. For example, B. M. Mamaev, The significance of dead wood as an environment in insect evolution, *Proceedings of the 13th International Congress of Entomology, Moscow, 1968* **1**, 269 (1971).
40. See also W. Beebe, *Jungle Days*, Ch. 7 (Garden City, New York, 1973).
41. P. D. Ashlock, The uses of cladistics, *Annual Review of Ecology and Systematics* **5**, 81–99 (1974).
42. A. W. Steffan, Morphological and behavioral polymorphism in *Plastosciara perniciosa* (Diptera: Sciaridae), *Proceedings of the Entomological Society of Washington* **77**, 1–14 (1975).
43. N. W. Hussey, W. H. Read, and J. J. Hesling, *The Pests of Protected Cultivation* (Edward Arnold, London, 1969).
44. R. A. Bedding, Parasitic and free-living cycles in entomogenous nematodes of the genus *Deladenus*, *Nature* **214**, 174–5 (1967).
45. R. Tuomikoski, Beobachtungen uber einige Sciariden (Diptera), derem Larven in faulen Holz oder unter der Rinde abgestorbener, *Suomen hyönteistieteellinen aikakauskirja* **23**, 3–35 (1957).
46. H. C. Obreen, Contributions to the biology and taxonomy of *Rhizoglyphus echinopus, Mededelingen Rijksfaculteit Landbouwwetenschappen, Ghent* **32**, 602–6 (1967).
47. R. L. Usinger and R. Matsuda, *Classification of the Aradidae (Hemiptera-Heteroptera)* (British Museum (Natural History), London, 1959).
48. F. J. Turcek, Beitrage zur Okologie der Kiefernrindernwanze *Aradus cinnamoneus,* Panz. *Biológia, Bratislava* **19**, 762–77 (1964).
49. A. D. Imms, O. W. Richards, and R. G. Davies, *A General Textbook of Entomology* (Methuen, London, 1957).
50. R. L. Usinger, Monograph of Cimicidae (Hemiptera-Heteroptera), *Thomas Say Foundation Monograph* **7**, 1–585 (1986).
51. L. G. Saunders, On the life history and the anatomy of the early stages of *Forcipomyia* (Diptera, Nematocera, Ceratopogonidae), *Parasitology* **16**, 164–213 (1924).
52. A. Vandel, Essai sur l'origine, l'évolution et la classification des Oniscoidea (Isopodes terrestres), *Bulletin Biologique de la France et de la Belgique,* Suppl. **30**, 1–136 (1943).
53. H. L. Carson and K. Y. Kaneshiro, *Drosophila* of Hawaii: systematics and ecological genetics, *Annual Review of Ecology and Systematics* **7**, 311–65 (1976).
54. W. Kuhnelt, *Soil Biology*, transl. by N. Walker (Faber, London, 1961).
55. E. C. Zimmerman, Homoptera: Auchenorryncha, *Insects of Hawaii* **4**, 1–268 (1948).

56. A. J. P. Goodchild, Evolution of the alimentary canal in the Hemiptera, *Biological Reviews* **41**, 97–140 (1966).
57. W. H. Thorpe, *Orthezia cataphracta* (Shaw) (Hemiptera, Coccidae) feeding on a basidiomycete fungus, *Collybia* sp., *Entomologists's Monthly Magazine* **103**, 155 (1968).
58. G. F. Bornemissza, Mycetophagous breeding in the Australian dung beetle *Onthophagus dunningi, Pedobiologia* **11**, 133–42 (1971).
59. H. F. Howden, Biology and taxonomy of North American beetles of the subfamily Geotrupinae with revisions of the genera *Bolbocerosoma, Eucanthus, Geotrupes and Peltotrupes* (Scarabaeidae), *Proceedings of the US National Museum* **104**, 151–319 (1955).
60. P. Hurd, Observations on the nesting habits of some New World carpenter bees with remarks on their importance in the problem of species formation, *Annals of the Entomological Society of America* **51**, 365–75 (1958).
61. E. M. Barrows, Soil nesting by wood-inhabiting halictine bees, *Augochlora pura* and *Lasioglossum coeruleum* (Hymenoptera: Halictidae), *Journal of the Kansas Entomological Society* **46**, 496–99 (1973).
62. K. E. Schedl, Breeding habits of arbicole insects in Central Africa, *Proceedings of the 10th International Congress of Entomology, Montreal* **1**, 183–97 (1958).
63. A. C. Scott, Coprolites containing plant material from the Carboniferous of Britain, *Palaeontology* **20**, 59–68 (1977).
64. Z. Massoud, Essai de synthèse sur la phylogenie des Collemboles, *Revue de Ecologie et de Science Biologiques* **13**, 241–52 (1976).
65. W. C. Williamson, On the organisation of the fossil plants of the coal measures. X. Including an examination of the supposed radiolarians of Carboniferous rocks, *Philosophical Transactions of the Royal Society of London* **171**, 493–539 (1880).
66. H. B. Geinitz, *Die Versteinerungen der Steinkohlen-formation sachsen* (Lepizig, 1885).
67. H. E. Hinton, Enabling mechanisms, *Proceedings of the 15th Congress of Entomology, Washington*, 71–83 (1977).
68. J. Smart and N. F. Hughes, The insect and the plant: progressive palaeoecological integration, in *Insect/Plant Relationships, Proceedings of the Royal Entomological Society of London* No. 6, 163–65 (1973).
69. F. M. Carpenter, Geological history and the evolution of the insects, *Proceedings of the 15th International Congress of Entomology, Washington*, 63–70 (1977).
70. For example, E. O. Wilson, *The Insect Societies* (Harvard University Press, Cambridge, 1971).
71. For species suggestive of this see lists in E. A. Hicks, *Check-List and Bibliography on the Occurrence of Insects in Birds Nests* (Iowa State College Press, Ames, Iowa, 1959).
72. J. W. Evans, Concerning the Peloridiidae, *Australian Journal of Science* **4**, 95–7 (1941).
73. W. E. China, South American Peloridiidae, *Transactions of the Royal Entomological Society of London* **114**, 131–61 (1962).

74. A. Balachowsky and L. Mesnil, *Les Insectes Nuisables aux Plantes Cultivées*, 2 vols (Paris, 1935–36).
75. U. Nur, Electrophoretic comparisons of enzymes of sexual and parthenogenetic mealy-bugs (Homoptera: Coccoidea: Pseudococcidae), *Virginia Polytechnic Institute and State University Research Division Bulletin* **127**, 69–84 (1977).
76. W. D. McEnroe, Spreading and inbreeding in the spider mite, *Journal of Heredity* **60**, 343–5 (1969).
77. See, for example, P. J. den Boer, On the significance of dispersal power from populations of carabid beetles (Coleoptera, Carabidae), *Oecologia, Berlin* **4**, 1–28 (1970).
78. See Chapter 15 and H. N. Comins, Evolutionarily stable strategies for localized dispersal in two dimensions, *Journal of Theoretical Biology* **94**, 599–606 (1982).
79. M. J. West-Eberhard, The evolution of social behaviour by kin selection, *Quarterly Review of Biology* **50**, 1–33 (1975).
80. R. L. Trivers and H. Hare, Haplodiploidy and the evolution of the social insects, *Science* **191**, 249–63 (1976).
81. H. H. Ross, The evolution and past dispersal of the Trichoptera, *Annual Review of Entomology* **12**, 169–206 (1967).
82. J. G. Needham and D. Betten, Aquatic insects of the Adirondacks, *Bulletin of the NY State Museum*, **47**, 1–612 (1901).
83. J. C. Bradley, Classification of Coleoptera, *Coleopterists' Bulletin* **1**, 75–85 (1947).
84. C. F. C. Riley, Migratory responses of water-striders during severe droughts, *Bulletin of the Brooklyn Entomological Society* **15**, 1–10 (1920).
85. H. M. Parshley, A note on the migration of certain water-striders, *Bulletin of the Brooklyn Entomological Society* **17**, 136–7 (1922).
86. T. R. E. Southwood, Migration of terrestrial arthropods in relation to habitat, *Biological Reviews* **37**, 171–214 (1962).
87. M. L. Luff, The occurrence of some Coleoptera in grass tussocks, with special reference to microclimatic conditons (PhD thesis, University of London, 1964).
88. L. E. Gilbert, Pollen feeding and reproductive biology of *Heliconius* butterflies, *Proceedings of the National Academy of Sciences, USA* **69**, 1403–7 (1972).
89. J. R. G. Turner, Experiments on the demography of tropical butterflies. II. Longevity and home-range behavior in *Heliconius erato, Biotropica* **3**, 21–31 (1971).
90. J. C. Moser and E. A. Cross, Phoretomorph: a new phoretic phase unique to the Pyemotidae (Acarina: Tarsonemoidea), *Annals of the Entomological Society of America* **68**, 820–2 (1975).
91. M. S. Ghilarov, The evolution of insects during transition to passive dissemination and the principle of reverse links in phylogenetic development [in Russian], *Zoologicheskii Zhurnal* **45**, 3–23 (1966).

CHAPTER 13

Fighting among unrelated parasites

In the cavity of a Brazilian fig (above right) a victor large male *Idarnes* moves on searching for virgin females, leaving just below a paralysed and dying rival he has fought. Two smaller males of the same species (one severed at his 'waist') are already dead (lower centre and right). Unrecognizably different in form from her male, an *Idarnes* female (lower left) is struggling out of a galled fig flower: because already fertilized by a mate who bit into and entered the flower with her, she is now of no interest to the wandering victor. At the left, a healthy male of a different genus also patrols. Although these males are fighters too, the resources they seek—their females—are different; therefore, the males of the two species are unhostile to each other.

CHAPTER 13

DISCORDANT INSECTS

Wingless and Fighting Males in Fig Wasps and Other Insects

Devouring is the impotent ambition to rule, of all Arabians, who are born near to the sheykhly state.

C. M. DOUGHTY[1]

READING about them was not enough, they were too fantastic. Only in 1963 in Brazil did I first thoroughly believe in fig wasps, the tiny black insects that both breed in and pollinate all fig trees. Worldwide this means 600 species, not just the one cultivated species that is most familiar to Northern Europeans for its dried fruits in the brown, oblong packets. The 600 mainly inhabit tropical countries where many are giants of the forest. To this great assemblage the almost shrubby European fig stands an outlier in several ways. For example, besides its geographic placement as the world's most northerly fig, and its small stature, it is dioecious—that is, has male and female trees separate like holly or yew. Again, generally each fig species has one kind of wasp adapted to pollinate it, but the European can sometimes have none; for the most practical and novel pollination feature of the European fig is that although it has a wasp not all its varieties need it, or at least not for their crops. The best varieties still use the wasp to set fruit, but some others produce fruit with no viable seeds and therefore no need for pollination.

If, however, you eat only the best and are now worried by the idea that you swallow a small wasp with each fig, first let me say that the wasps doing the job for the trees are tiny—2 mm or less—and, second, that you will be out of line with much of your successful ancestry if upset at the idea of eating these few small flakes of chitinous bran and protein derived from an embalmed and long dead pollinator. All through the later Tertiary period fig-tree fruiting was certainly a cause for celebration for forest wildlife and our ancestors must have whooped and brachiated towards the fruiting crowns just as our great ape cousins and most monkeys do today. Arriving there, they consumed figs that had many more wasps—hundreds per fruit instead of just one or two—compared with our figs today.[2] This protein addition to a largely vegetarian diet and the perceptual/spatial problems involved in reaching the bonanza must have been among the factors pushing ahead the enlargement of the anthropoid brain, and it is perhaps the extra wasp protein that makes figs such a favourite food not only for primates but for a great variety of birds and bats.[3]

Rumours of the winged and targeted symbiotic pollination of fig trees had intrigued me before my 1963 visit to Brazil and seeing the reality was no disappointment. Yet for more than 10 years I had no time, either in libraries or during my visits to the tropics, to look into the points that remained puzzling. When I did come to take a more serious interest in 1975 and developed the data for the paper in this chapter, it was during my third visit to Brazil in 1975, and even then the study came about more by accident than by plan.

In 1974, I concentrated my courses at Imperial College into a short intensive period so that I could be free from teaching for most of 1975. My idea was unconnected with figs and was rather to pursue the 'rotten-wood' interest of the last chapter into the tropical realm: I would go to Brazil, look at tropical rotting-wood insects, and make comparisons with my observations in temperate Britain. To help with my expenses, especially because my wife Christine and our two small children were coming with me, I had arranged to teach a course in population genetics at a campus in southern Brazil where several people of Warwick Kerr's Rio Claro group (see Chapters 2 and 8) were established and I was therefore known. Two Faculdades, of Medicine and also of Philosophy, Science and Letters, had been built together on a common campus on the outskirts of Ribeirão Prêto. Ribeirão Prêto is a small but wealthy coffee and sugar-cane city and centres a region of the best soils of likewise small and rich São Paulo State. (For the town 'small' means about 100 000 inhabitants; for the state it means about the size of England—this to be contrasted to, say, Amazonas State, which is the size of Europe.) The course I taught was my first wholly at a graduate level and my class was just six. To my surprise I enjoyed the teaching. I had the difficulty of speaking in Portuguese but, against this, how different were these thoughtful, respectful Brazilians from the hundred-headed swarm I faced in my customary nine o'clock lectures at IC. There I was lucky if I could see one or two of the heads caring one Johannsen's polygenic bean for my Hardy–Weinberg equilibrium or my definition of linkage: during the classes most students would be good-humouredly chatting or perhaps deep in their morning newspapers. And towards the end of the course I knew they would have another topic: they would be discussing who was to lead this year's delegation to the professor to complain about the irrelevance and incomprehensibility of the lectures I was giving. In those days apparently 'inapplicable' subjects such as population genetics and evolution were not the zoology that aspiring entomologists and parasitologists had come to Britain's prime technical institute to learn. Instead they wanted insect physiology, practical parasitology, and the like—in short, knowledge to make a living with. It is an immodest comparison but what a pleasure it was to me later to read how Isaac Newton as a lecturer at Cambridge in the seventeenth century seemed to have had similar experiences to mine. Newton sometimes lectured to 'ye bare walls', as his assistant put it—a class of zero rushed in to learn from the world's greatest-ever scientist! At least mine came to the theatre; but then, perhaps, there was some check that they did. Anyway, in contrast, here were Brazilian students really wanting to listen.

My pleasure in teaching was slightly offset by my duties necessarily keeping me close to Ribeirão Prêto, where I was finding little of the wildlife I had hoped to study. The immediate surround of the city was little more than an undulating chequer of large, intensively cultivated fields. Even the once endless *cafezals* were, as the bushes aged beyond their maximum productivity, being rapidly replaced by sugar cane, which if anything presented still more of an opposite to my hoped-for diverse arboreal tropical systems. Almost the only real trees were occasional eucalyptus along field boundaries or in small groves planted for timber and firewood. Unfortunately for me, eucalypts are notoriously healthy trees throughout their now immense range outside their Antipodean home and are correspondingly deficient in insects. This is, of course, why they are so liked by farmers and timbermen and why they grow so crazily fast. I was further disappointed to find a lack of insects even in their dead parts and felled trunks—only a few seemingly uninteresting species were there, although those were usually quite common. In short, almost every scrap of the natural rich forest was gone from the rich basaltic plateau to create its booming agriculture and this had left little for me to look at to compare with my findings in British woods. Another trouble was that the time of my arrival in Brazil was during the onset of the dry season so that logs that might, in the wet season, have been interesting, were usually baked to sterility by the intense winter sun of the plateau. However, raising my sweat-dimmed eyes from such logs towards living trees other than eucalypts, which meant mainly towards the trees planted on the beautiful campus of the university, I soon noticed some that seemed not to be heeding either the winter drought or the sun. These were budding forth, in one crown or another, crops not only of fresh leaves but of green fruits—all while the rest of the landscape was scorched, dusty, and static.

Never planted, indeed weeds of the campus, these small drought-defying trees bearing fruits so strangely unpreceded by any visible flowers, were fig trees. Most of them were, in fact, the '*Ficus* 2' of my paper. In the long term the system of pollination of *Ficus* requires that in every population at least some trees somewhere must be beginning crops of new figs and the gaps between such events must be no more than a matter of a week. Otherwise their short-lived pollinator insects have no living homes to fly to and the local wasp supply dies out, potentially leaving local trees unpollinated and seedless for the rest of their lives. One fig-tree life time after this, populations that have failed in the continuous provision of fig-wasp homes would simply disappear. Facing this necessity, and somehow drawing their water from deep in the rich red soils, figs of the plateau manage to keep active when the rest of southern Brazilian nature is as dormant through winter drought and sunshine as a British winter woodland is dormant through rain and cold. If my *Ficus* 2 were to salute any winter tree of Britain as its peer, I suppose it should be the hazel, the tree that so often shakes out catkins into our equally inhospitable air of February; in fact, this idea seems the more appropriate because the trees are about the same size and both are non-dominant species in their forests. In any case in Ribeirão Prêto in June, if the rest of the landscape was seemingly dead, here among my *Ficus* 2, clinging to or standing under the old terrace walls of the grand coffee estate the universities had taken over, I found greenery; and among the leaves tiny flying creatures were zipping about.

A student, Lucio Campos, during my last brief visit to the Ribeirão Prêto campus in 1968, had shown me the wasp populations of one or two of these local fig trees. I had been amazed then by the variety and strangeness of the crew that ambled or flew forth out of the split fruits. Equally intriguing among those left behind, engaged in restless patrolling activity, were such weird, reddish, square-headed, and tiger-mandibled wingless insects as I had never imagined. By 1975 I had almost forgotten this experience, but when I started to split figs in the *faute de meiux* spirit that I have just mentioned, I was duly stunned again by the contrasts of this emergent and resident crowd. Soon I was an addict, my thumbnail sore all day long from splitting figs and yet ever ready to split more. In spite of pain, how well my thumb, or (in more civilized moments) my scalpel, repaid me. What psychedelic marvels the green or speckled boxes held. It was all the old excitement of *Melittobia*, of the thrips and all those over again. Indeed some forms, habits, and habitats in the trees turned out so similar to features of my former interests that observations I had made in the rotting wood back in Britain could be worked in with the new data I was collecting. Partly because of this, the paper ended quite a rag-bag of species and themes. Loosely connected as some may be I am proud of the work as making up the only real 'data' paper of my life, and it was a pleasure after the trip was ended to see so much of what I now knew or surmised about strategic lives of arthropods being built into a single story.

In the little 'Ecologia' lab on the campus I was often opening the figs and observing and recording their contents all day long or else making the kinds of manipulations where, with a few hairs left on a camel-hair paint brush, I would move wasps from one fig to another in order to see who was choosing to mate with whom, and how hard the big males battled with like dominants drawn from other halved figs. Sometimes Christine worked with me. A great convenience of living in a small convent-cum-primary-school just across the valley from the campus was that we could leave our two small girls, aged three and two, in the care of the infant department of the school for most of the day. Ecologia had no air conditioning but Brazilian buildings are designed to be airy and it was cooler than outside; in any case the heat at Ribeirão Prêto is not too drastic and sometimes in the evening and at night it was very cold. Generally it was not so much a matter of melting over one's microscope as of shaking it, with this referring to the more common kind of addiction than fig splitting one can easily be drawn into in Brazil—addiction to fierce coffee that most labs and businesses have always available. Undeniably the coffee helped to keep up concentration but, alas, if, during the more delicate manipulations of the wasps, you had taken more than two of the tiny syrupy *cafezinhos* in one morning, a caffeine tremor conveyed through a camel hair could so beat up on a male *Idarnes* you were trying to guide gently to a potential confrontation or mating that he would instead dive for an empty gall flower, and there, within the split-fig amphitheatre where he was being staged, he would stay cowed and hidden for the rest of the morning—fearful I suppose that my shaking camel hair was still pursuing him.

Several of the predictions I made in this paper have since been confirmed but more have concerned arthropod sexual fighting in general than anything of specific to the figs. Bernard Crespi showed in due time that gladiatorial-looking thrips we had by then both observed do indeed dagger each other in the course of deadly embraces.[4] Paula Levine showed that the *Acanthocinus* bugs use the bottle-opener femora of their hind legs not, of course, for bottles but instead to embrace rival males, again to their general detriment and repulsion from the thistle heads where such fighting mostly went on.[5] The mentioned *Cardiocondyla* wingless male ants[6,7] also fight for harems as I had suspected. A group of marine crustaceans showing extraordinary convergence to the morphology of fighting fig wasps (which I thought I had mentioned in this paper but seem not to have) fight also[8] and probably guard harems of the extremely different females in small intertidal cavities. As for fighting within figs, by the end of 1975 I had become accustomed to the idea of megadeaths through strife occurring within each quiet leafy dome of a fruiting tree every few months.

In childhood I had once watched two blue tits clinch in a fight and fall into the grass. Only one flew up. When I and my sister had rotted out the skull of the dead loser (for her skull collection) we found its cranium to be dented like a mistruck table-tennis ball. Ever since I have been a doubter of the doctrine that had been so strongly taught in my school and undergraduate years (and endorsed by even such an excellent observer as Konrad Lorenz) that animal fights are seldom injurious and almost never fatal. Since the fighting blue tits, more and more examples of injury and fatality have come to me both through literature and through the evidence of my own eyes. None of this, however, quite prepared me for the mayhem I found going on in the fig trees.

As a result, and with the guidance of further study,[9] I think I now have a better idea when to expect fighting to be damaging and when not. Lethal fights occur when the weaker has nothing to gain either by waiting or by surrender. Towards its end, however, this paper adds that sometimes when instead it is possible to wait things out or to seek other arenas, there is special scope for bluffing one's fighting ability. This has not, so far as I know, yet been followed further either in theory or by study of possible natural cases. Two views on bluffing currently stand starkly contrasted. Amotz Zahavi of Hazeva Field Station in Israel, and famous inventor of The Handicap Principle to be discussed in Volume 2 of this collection, says that it hardly happens: when a male chimpanzee puffs his hair and charges, his aim is not to make himself look bigger but rather the reverse, to give his target a more realistic 'see through' of the muscle and weight that is about to hit him—or of what might hit if the charger does not finally swerve. An honest signal in short. I and some others maintain instead that the more obvious interpretation of the pilo-erection is the true one. The animal is trying to make himself look bigger—terrifying to the point where his rival will defer without risking a fight that could be damaging to both. Such bluff may not always work but it has a chance to when the bluffer guesses that it's not life or death to the other whether, let us say, it keeps the banana that both want. If his combination of size and need are made to seem great enough, the other chimpanzee may yield the fruit even he thinks he could prove himself really the stronger. Fig wasps, much shorter-lived than chimpanzees, may, as I suggest in the paper, also try this bluff a little but for the most part I think they are evolutionarily conditioned to see little point in it. The chaos of the mating activity of the hollow of the fig will be all over within a few hours and a fig wasp who tries for an advantage by waiting is likely to end his life in a bat's stomach with a score of no matings, no wives escaped and carrying his genes, and therefore zero fitness. All the other wasps also end in the bat, of course, but if some of their genes have indeed escaped in the seminal receptacles of the departed females, these—usually successful fighters—do not have zero fitness. In short, Zahavi may be right in this all-or-nothing context and honest signals may be best exchanged. On the other hand, to me the chimpanzee, as well as the puffed mother hen rushing at you as you near her chicks and the fighting *Chiasognathus* beetles discussed in the later part of this paper, are another case; with them, bluff can and reasonably does play a part.

While on the subject of bluffing, does it ever pay to bluff as being *lower* in ability and standing than you actually are? Seldom in the subhuman world, I would guess; but in the human world two personal experiences, looking a few years backwards in my life from 1975, suggest that bluffing low can be advantageous and the stories are tangential to the topic of this paper. Both concern my search for a rare ant, once in a biscuit factory and once in a prestigious London hospital. Neither type of institution likes to admit that it houses ants whether these are common black *Lasius* of the country window sill or the rarest and most exciting ants in the world. Therefore, in the first case of hearing of an infestation of the ant I desired, I found myself having to put on the white uniform of a biscuit-maker and to promise my guide that I would claim to be an assistant ant exterminator, not any sort of collector, in case the disguise should be penetrated. In that case, there was no challenge: I looked a passable biscuit-maker apparently. In the case of the hospital I had similarly to simulate a workman fixing the drains because it happened to be in the old soil and debris of a drain that a few living specimens had been found during a routine inspection by my friend who was the real pest-executive (and I hardly need add a very good naturalist as well). In the hospital, as I climbed back into a bustling corridor through a window that had access to the central court or well of the building, I found myself irately questioned by a nurse. My pocket was bulging with vials containing my first colony of *Hypoponera punctatissima* just culled from a gully of the oldest, dingiest drains of the building (this is the colony referred to in the coming paper). Perhaps because of these prizes I may have looked irritatingly smug and a bit out of character; in any case, with a devious honesty and innocence I told her that I had been 'seeing to a problem in the drains'. An Amazonian *Dinoponera grandis* spotted strolling a hospital pillow could hardly, I think, have produced such hauteur and disgust as appeared on the nurse's face as she heard this explanation. While she was tersely advising me about the nearest exit of the building I suddenly knew how one is addressed and how appreciated when a lowest untouchable of London society—a drain fixer. But like Professor Penrose if he thought me a rabid eugenicist, purely professionally she may have been right. I myself hadn't been delighted by the rotted gauze, old sutures, and rusted syringe needles embedded in the black, rich soil through which I had trowelled towards the tiny and beautiful nest. Clearing drains is not a profession I would choose but it does provide restful and reflective moments. Besides the unusual ants leading me on, I could muse as I dug, like Hamlet with the gravedigger, on the scenes of hospital life evoked one by one by my finds—and I have not mentioned them all.

To end the digression, the various thoughts of this paper on fighting and deception certainly suggest no cosy Beatrix Potter world. Another idea, on which the paper just touches, casts even bleaker reflections into the human world as well as back to a topic of an earlier introduction (see Chapter 8). It is a theme that has been little followed as far as I know. It is the idea of adaptiveness of a threat strategy in mating. It is hinted for the stag beetles discussed near the end but I overlooked at the time many papers relevant to the theme.[10–12] After seeing the events and entering on a parallel speculation about the strange mode of insemination of the *Xylocoris* / bed-bug clade (which, by traumatic insemination through the female's body wall, sends the sperm directly to the female reproductive tract, through the haemocoel). I have encountered so many other cases hinting at a similar strategy of threat, either current or suggested for the evolutionary past, that I have come to think the idea should be taken seriously.[13–15] Unfortunately it is a topic that if found to be valid won't make a heart-warming thesis any more than the parallel topic does for humans. For us, the idea that copulation by violence and threats could ever be a natural and 'best' course under natural selection always evokes—and on one side rightly—an outcry of horror and dismissal; this was discussed before (Chapter 8). The basic question is whether it can ever be advantageous to threaten violence to obtain a resource in circumstances where the threat, if effected, destroys the existence of what was being sought without advantage to the destroyer. It is difficult to avoid a human example to give at least a feel for the situation: sell me your haystack for my low offer or I will come in the night and burn it. But if he still refuses to sell, why should I bother with a cold dark walk to carry out my threat since the burning in itself doesn't benefit me? The person I threaten knows this. The parallel here is to the loss of all possible benefit that follows if the stag beetle kills, blinds, or otherwise seriously damages the female he is trying to persuade to accept his copulation. If he damages her, later she may not be able even to lay the eggs that he hopes to sire. However, may it not be rational to apply a little damage in the course of such threatening, and may there not exist some evolutionarily stable strategy within the range of gradual damage, even possibly entailing a considerable detriment to the females reproduction? In effect this strategy amounts to the oppressive buyer pulling and burning one stook each night from the man's rick in order to show his power to effect his threat. In all such grim imaginings, of course, it is assumed that apart from delay no law can stop or counter threat deter the threatened action. Fortunately, in humans, there usually are laws, and in general I will gladly leave the full resolution of the animal problem to others, more hard-nosed and game-oriented.

Because it is sad to end on such a dark note with Mafiosi and bullies apparently ascendant (ascendant at least until better understood and thereby opposed), let me finally bring you instead other visions from the year when this paper was written. Perhaps I should offer the blue jacaranda flowers snowing the deserted playground of the little school of the convent, where, after school hours, my children played all that year. Or I should take you to those Andes of southern Chile where at sunrise on the frosty lawn beside my hotel in a town as wild-western and as lovely as its name, Coyhaique, my great *Chiasognathus* stags fight on my boots and my trousered shins and have you watch, in the final round, Tongs Two prove himself champion over the monster that I deem the bluffer, Tongs One. Or, a little to the north in Chile where we spent longer, show you the sawyers, the oxen teams, and the great wheels that are taking out the giant rauli logs (a southern beech, *Nothofagus alpina*) from the pristine mountain forests. There again you may see my tiny daughters playing on earthen forest roads with the logs beside them and the immense, doomed, virgin trees arched above. These are all glimpses of places and scenes of the happy year when the paper was gestated. By the time it was born, serious life had come back to me; I had left Silwood and was a professor in another continent and another job.

References and notes

1. C. M. Doughty, *Wanderings in Arabia* (Duckworth, London, 1908).
2. Dioecious figs, which include European *Ficus carica*, 'mislead' one or two pollinator wasps into entering the fruits on a female tree, the young figs there being indistinguishable at this stage from the fruits of male trees. Only in the latter can an entering female successfully rear offspring. Later these male 'fruits' become dryish hollow boxes containing the pollen-producing male flowers and hundreds of wasps. These fruits are normally unattractive to fruit-eaters. In the case of the more primitive monoecious figs (such as banyans and my *Ficus* 1 and 2) seeds, pollen, and wasps to carry the pollen are all produced in the same fruit. The consumer of one of these fruits would eat at the least all the male wasps left after the exodus of the females (which occurs only a matter of an hour before the fig softens and sweetens). If the exodus somehow fails in a monoecious fig the entire fig population of several hundred insects would be eaten.
3. S. B. Hrdy and W. Bennett, The fig connection, *Harvard Magazine* **ix–x**, 25–30 (1979).
4. B. J. Crespi and L. A. Mound, Ecology and evolution of social behavior among Australian gall thrips and their allies; in J. Choe and B. J. Crespi (eds), *Ecology and Evolution of Social Behavior in Insects and Arachnids*, (Princeton University Press, Princeton, 1996).
5. P. L. Mitchell, Combat and territorial defense of *Acanthocephala femorata, Annals of the Entomological Society of America* **73**, 404–8 (1980).
6. R. J. Stuart, A. Francoeur, and R. Loiselle, Fighting males in the ant genus *Cardiocondyla*, in J. Eder and H. Rembold (eds), *Abstracts of the 10th Congress of the International Union for the Study of Social Insects* (Munich, 1986). See also J. Heinze, S. Kühnholz, K. Schilder, and B. Hölldobler, Behavior of ergatoid males in the ant *Cardiocondyla nuda, Insectes Sociaux* **40**, 273–82 (1993).
7. J. Heinze and B. Hölldobler, Fighting for a harem of queens: physiology of reproduction in *Cardiocondyla* ants, *Proceedings of the National Academy of Sciences of the USA* **90**, 8412–14 (1993).
8. M. Hesse, Crustacés rares ou nouveaux des Côtes de la France, *Annales des Sciences Naturelles* **19**, 1–29 (1874).

9. M. G. Murray and R. Gerrard, Conflict in the neighbourhood: models where relatives are in direct competition, *Journal of Theoretical Biology* **111**, 237–46 (1984).
10. J. S. Huxley, A 'disharmony' in the reproductive habits of the wild duck (*Anas boschas*), *Biologisches Zentralblatt* **32**, 621–3 (1912).
11. B. Lutz, Fighting and incipient notion of territory in male tree frogs, *Copeia* **1**, 61–3 (1960).
12. A. Michelson, Observations on the sexual behaviour of some Longicorn beetles, subfamily Lepturinae (Coleoptera, Cerambycidae), *Behaviour* **22**, 152–66 (1963).
13. F. McKinney, S. R. Derrickson, and P. Mineau, Forced copulation and mixed male reproductive strategies, *Behaviour* **86**, 250–94 (1983).
14. R. Thornhill and J. Alcock, *The Evolution of Insect Mating Systems* (Harvard University Press, Cambridge, MA, 1983).
15. B. J. Le Boeuf and S. Mesnick, Sexual behaviour of male northern elephant seals: 1. Lethal injuries to adult females, *Behaviour* **116**, 143–62 (1991).

WINGLESS AND FIGHTING MALES IN FIG WASPS AND OTHER INSECTS†

W. D. HAMILTON

From June 1975 to February 1976, my wife and I gathered life-history and behavioural data on the fig wasps of two species of wild fig trees (*Ficus*) that were common around Ribeirão Prêto, Brazil. This account firstly summarizes these data and outlines the mating and breeding systems that they imply.

The 18 or so species of fig wasps covered by our survey showed (1) many possessing *wingless males*, (2) *lethal combat* among several types of these wingless males, and (3) several cases of profound *male dimorphism*, that is, a normally winged male occurring as alternative to a wingless male. Literature suggests that our findings may be fairly representative throughout the world distribution of *Ficus*: yet only (1) is well known.[1] In insects as a whole (1), (2), and (3) are rare. The second object of this paper is to consider why three rare phenomena should be common in fig wasps, which they are to the extent of occurring not just separately within the aggregate of species that we observed, but in some cases within single species. Other relevant instances of male winglessness, fighting, and dimorphism in insects—which sometimes also concur, although usually no more than two at a time—will be reviewed.

THE FIGS AND FIG WASPS

The species we called *Ficus* 1 provided the largest trees in local natural woodlands and had a few large trees on the campus of Faculdade de Medicina de Ribeirão Prêto. Their figs were apt to be inaccessible for collection and fruitings were infrequent, perhaps not more than once a year. *Ficus* 2 trees were small (not more than about 8 m) and common on terrace walls and in waste places. They fruited at least twice a year. Hence we gathered much more data

†In M. S. Blum and N. A. Blum (eds), *Reproductive Competition, Mate Choice and Sexual Selection in Insects*, pp. 167–220 (Academic Press, New York and London, 1979).

for this species. Both species of *Ficus* belong to the section *Americana* of subgenus *Urostigma*.[2]

A summary of data obtained by rearing fig wasps from ripening figs is presented in Table 13.1. Our makeshift taxonomy of the wasps, which assigns letters for species, is further explained below. Figures 13.1, 13.2, 13.3, and 13.4 give a rough impression of selected species and their morphs. In all the species females were always winged (only three individual exceptions to this were noted and these were obviously runts). Table 13.1 shows that the males in some species were always winged; in some, always wingless; and in some, dimorphic for wings. For each species absence of any wingless male in a fig is made the basis of classifying the fig's females separately from those

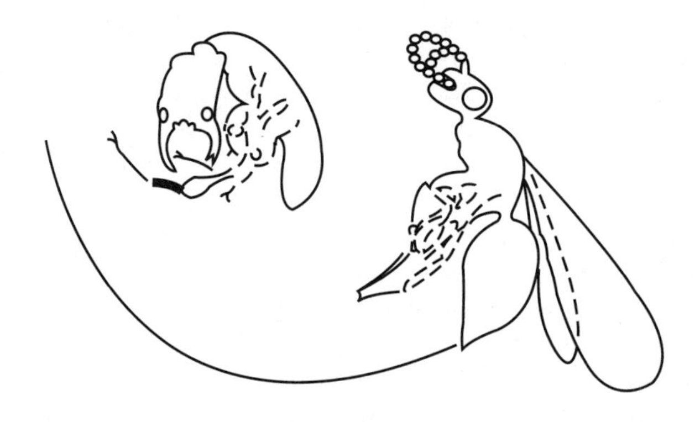

Figure 13.1 Sexual dimorphism in *Idarnes F2*: large male (left) and large female (right).

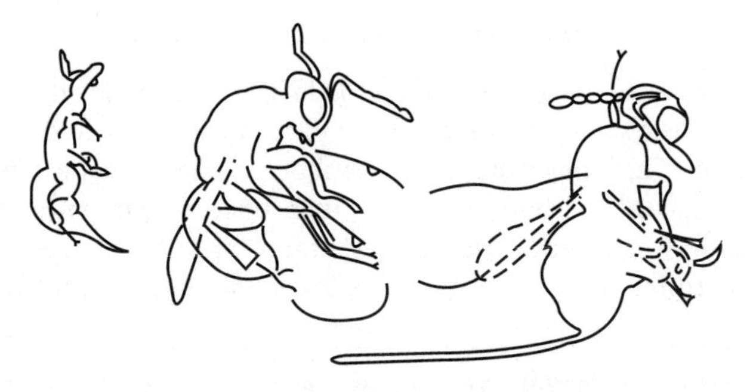

Figure 13.2 Sexual and male dimorphism in parasite species *F*: micropterous male (left), normal male and female (right).

Table 13.1 Polymorphism in fig wasps. Wasps of two *Ficus* of southern Brazil: brood data classified for sex and wings

Species (S)	Frequency of figs with S in random sample	No. of figs with S in total sample	Totals bred					Sex ratio (males ÷ total)	Wing ratio in males (winged males ÷ all males)	Numbers of S in figs having any	
			Males		Females						
			Alate	Flightless	Flightless males absent	Flightless males present	Total			Mean	SD
Blastophaga F1	12/12	11	0	89	0	1177	1266	0.070	0	115.1	33.0
Blastophaga F2	39/39	55	0	303	72	2956	3331	0.090	0	60.6	68.1
Idarnes F1 (Id)	12/12	12	0	169	0	433	602	0.281	0	50.2	33.5
Idarnes F2 (id)	38/39	57	0	1045	0	1900	2945	0.355	0	51.7	27.4
C	5/12	5	12	0	12	0	24	0.500	1	4.8	6.5
c	6/39	15	15	0	11	0	40	0.725	1	2.7	1.8
E	4/12	3	1	2	2	3	8	0.375	0.333	2.7	2.1
e	17/39	36	36	26.6	37	44	143.6	0.436	0.574	4.0	3.8
	1/39	7	7	2.4	17	4	30.4	0.309	0.745	4.3	4.2
F	5/12	5	1	7	1	34	43	0.186	0.125	8.6	6.1
f	16/39	24	40	14	29	16	99*	0.529	0.907	4.1	4.2
θ	15/39	26	11	102	6	17.7	296	0.382	0.097	11.4	13.1
G	2/12	2	0	0	2	0	2	0	1	1.0	0
g	7/39	9	11	0	1	0	12	0.762†	1	1.3	0.5
h	2/39	2	0	0	2	0	2	0.250†	1	1.0	0
i	4/39	17	72	0	137	0	209	0.383‡	1	12.3‡	12.1

* Record omits three very small brachypterous females.
† Sex ratio includes specimens additional to main data.
‡ Common only in one crop of one tree.

Figure 13.3 Sexual and male dimorphism in parasite species *e*: micropterous male (left), normal male and female (right).

Figure 13.4 Parasite species *c*: male (left) and female (right).

occurring in figs with wingless males: the significance of such division will appear later.

Due to the diversity and the often extreme differences between males and females, the task of sorting adult wasps into species was difficult and some uncertainties remain. The genus *Idarnes* was the most problematic. Females easily separated into types (by size, length of ovipositor, and antennae: two types in *Ficus* 1 and four in *Ficus* 2), but males gave a fairly continuous range of variation, and some instances of mating suggested that perhaps any type of male could mate any type of female. Since previously only one *Idarnes* species has been reported for each *Ficus*, it seems possible that our cases also concern only single species which have a very marked polymorphism in the females and more continuous but wide-ranging variation, similar to that described by Grandi[3] and others[4] for males of *Philotrypesis* and other Old World genera, in the males. Polymorphism in ovipositor length could well be concerned with drilling to hit female flowers layered at different depths in the young fig (see Richards[5] and Brues[6] for possible parallels in other Parasitica), but equally striking and correlated variation in the antennae favours separate species. A similar problem with suspicion of conspecificity regarding three or four fig wasps currently put in distinct genera has been raised by Wiebes.[7] Since males could not be ascribed with certainty, the one-species view has been taken, rightly or wrongly, in Table 13.1 and subsequent analyses.

There was no evidence that any one species was breeding on both fig species. Nevertheless most species had an obvious 'sister' in the wasp series of the other *Ficus*, and our use of capital and small letters as species symbols attempts to reflect this. Some similarity groupings among the parasites of each *Ficus* were also apparent. The following arrangement of the symbols illustrates our tentative groupings within and between the two series:

	Agaonidae			Torymidae							
Ficus 1	*Bl*	*Id*		*C*	*D*	*E*	*F*		*G*		
	⋮	⋮		⋮		⋮	⋮		⋮		
Ficus 2	*bl*	*id*	*i*	*c*		$\epsilon \ldots e$	*f*	θ	*g*	*h*	*j*

The dotted lines indicate the most obvious correspondences.

Lists of Hymenoptera do not, of course, exhaust the sycophilous insects found in the figs. Larvae of a beetle, a moth, and a cecidomyiid fly, for example, were fairly common. A tarsonemoid mite was even more common and a nematode was abundant (transmission of these was phoretic and parasitic, respectively). Even with these added, our observed diversity—even for *Ficus*

2—certainly does not exceed that recorded for at least one Old World *Ficus*.[8] Likewise the strange sight when one of our figs at the right stage is opened—the winged wasps in varied shapes and colours that, by the hundreds, run out and fly, and the grotesque, brownish, wingless forms that remain behind, roaming the fig cavity—is also fairly well paralleled by a description for an Old World *Ficus*[9] (see also Fig. 13.9b).

FIGHTING IN FIG WASPS

The family Agaonidae includes all the pollinator fig wasps. For the most part, its species are in one-to-one correspondence with species of *Ficus*. The family Torymidae includes most of the other chalcidoid wasps that breed in figs. The torymid fig wasps have various parasitic relationships (details usually unknown) with agaonids or with the *Ficus*. The great majority of male fig wasps are wingless, and this applies whether we count overall numbers bred (which gives great weight to the always wingless males of the agaonids) or survey taxonomic lists (which tends to emphasize the torymids). Winglessness and precocious mating of male agaonids is classic knowledge. The parallel wingless states that occur in male torymids, however, although often even stranger than those of agaonids, have records and comment confined to specialist journals, and even there the information is esoteric even to most entomologists. This is because the process of pollen transfer and fruiting in *Ficus* is unique, so that the animals dependent on this process form a community which has no close parallel.

Even in specialist literature there has been little attempt to interpret the adaptive functions of the strange modifications of torymid males,[10] although these are often quite different from those of agaonid males. Only one author has reported fighting in Torymidae (*Philotrypesis*: Joseph[11]), and this report did not relate fighting to morphology nor attempt to generalize. Our own information strongly suggests that some of the recurrent modifications of torymid males, notably large head and mandibles, are always concerned with fighting.

We watched mortal fights between wingless males of *Id*, *id*, *E*, *e*, and θ and on grounds of morphology strongly suspect its occurrence in ϵ and *D*. Fighting males were always conspecific. (Except for occasional mistakes, interspecies fighting would be surprising. Milder fighting is known, however, in rather similar circumstances, for two scelionid parasitoids of bug egg masses;[12] its function is unexplained.)

Apart from the large head and mandibles and perhaps shield-like formations of head and pronotum in the neck region, other modifications of wingless male fig wasps are probably attributable to selection not connected with fighting. For example, the common tendency for the head to be most darkened and sclerotized compared to the rest of the body, and for the abdomen to be least sclerotized, can be attributed to the way all wingless male parasites have actively to look for females in galls and to the help (far from altruistic) which they give in opening the hole which lets her out. Some other modifications are probably connected with the exact site of mating; for example, flattened, flexible body (*Id*, *id*, θ; Fig. 13.1) or small size (F, f; Fig. 13.2) for those which enter the gall with the female, or elongated abdomen and enlarged forelegs and hind legs (with reduced mid legs) for those which cling to the gall and insert only the genitalia (*Blastophaga*). (For reasons not yet clear, the wing-reduced F and f males, although entering to mate, show an approach to the same modifications.) As the wingless males of E, e, and ϵ (Fig. 13.3) lack any of these further specializations, predictably they merely mate each female outside the gall after helping to make her exit hole. Perhaps in consequence of this—being mobile fighters and little else—they present a striking, superficial resemblance to soldier ants and may have sometimes been mistaken for ants. The male of D was rather similar, but so rare that we saw nothing of its habits.

Most of our observations on fighting concern *Idarnes*; wasps of this genus were present in almost every fig.

Male behaviour must be at least slightly upset when a fig is cut open. Light alone affects the behaviour to an unknown degree, and manipulation of males even with hairs of a fine brush tends to intimidate them (giving the impression, perhaps, that they are being pushed or lifted by a very powerful adversary). Even so, it was easy to provoke fights by adding a gall with a female about to emerge to a half fig containing two or more live large males. A male's fighting movements could be summarized thus: touch, freeze, approach slowly, strike, and recoil. Their fighting looks at once vicious and cautious—cowardly would be the word except that, on reflection, this seems unfair in a situation that can only be likened in human terms to a darkened room full of jostling people among whom, or else lurking in cupboards and recesses which open on all sides, are a dozen or so maniacal homicides armed with knives. One bite is easily lethal. One large *Idarnes* male is capable of biting another in half, but usually a lethal bite is quite a small puncture in the body. Paralysis follows a small injury so regularly and quickly as to suggest use of venom. The males certainly have mandibular glands and appear not to have any mouth.[13] The glands are present in the male forms of other genera which parallel *Idarnes* (see Wiebes[4,14]). But females have the glands, too.

Once wounded in the body, an *Idarnes* or θ male stops fighting and moves more and more spastically until he dies. Although he does not attempt to hide, he is ignored by the victor. (On the other hand, an *e* male was once seen to remain beside a damaged rival and bite at him repeatedly, eventually severing his body.) If no serious injury results from the first or second reciprocal attempts to bite, one of the males, injured perhaps by loss of a tarsus or in some way sensing himself outmatched, retreats and tries to hide. Usually he

Table 13.2 Live and dead *Idarnes* in one fig, 31 December 1975

Live

Females

15, of which 5 emerged at once and approximately 10 were in galls ready to emerge

Males

1 : large, in cell with female, missing part of one tarsus
1 : large, wandering, missing parts of two tarsi
1 : medium, in empty cell, missing one tarsus
2 : small, both perfect

Moribund

Males

1 : small, missing parts of fore- and midleg on one side, viscera extruding
1 : small, seems perfect

Dead

Large males

1 : seems perfect
1 : one frontleg missing from base of tibia (and almost severed at base of coxa); hindleg on same side missing from base of femur

Medium males

4 : seem perfect (one thorax lopsided congenitally)
4 : seem perfect but dark areas in thorax suggests wounds
1 : one tibia and one tarsus missing
1 : one tarsus missing and darkened midthorax
1 : with head missing
1 : body almost severed in two places; one rear corner of head and three legs missing
1 : almost severed in thorax; one tarsus missing
1 : rear corner of head perforated and darkened; dark spot on thorax
1 : twisted and blackened neck; part of one tarsus missing
1 : midcoxa with black band (cut across?)
1 : one hindleg missing
1 : one front tarsus and one antenna missing
1 : one frontleg from base of femur and one part of a tarsus missing
1 : one antenna missing
1 : abdomen eviscerated, one midtarsus missing
1 : dark spots at back of pronotum, one midtarsus missing
1 : dark spot at neck, antennae missing, one midtarsus almost severed
1 : midleg severed at base of femur and dark wounds on thorax nearby
1 : blackened at neck; part of one tarsus missing

Table 13.2 (*continued*)

Small males

7 : seem perfect
7 : seem perfect but with dark areas on thorax
2 : hindtarsus missing
1 : three tarsi missing on one side
1 : body severed between hind- and midlegs
1 : hindleg missing from base of tibia
1 : seems perfect but dark mark at base of abdomen

SUMMARY

Total number of males		
(1) Apparently perfect	12	54
(2) Obviously or probably injured	42	
Total number of females		15

finds an empty gall into which he plunges, turns, and comes to rest with mandibles agape at the gall's opening. From this position he can bite at the legs of the victor or another passing male with much less danger. Such an inactive male only ventures out again when long undisturbed and then very cautiously. In some figs the males' sharp reaction to contact with other males is muted and they brush past each other unconcernedly. This seems more likely to happen when females are in rapid flush of emergence from their galls (c.f. Browne[15] for a like observation on *Melittobia* males). Perhaps with matings freely available, the gains from fighting come to be less than worth the risk.

A similar lack of reaction holds generally for small *Idarnes* males. These also sometimes fight mortally but seem on the whole less aggressive and do not try to monopolize a fig cavity as large males do. Instead they spend much time forcing their way between the fig flowers and mating females in hidden galls that are not easily reached by large males. I never succeeded in getting them to attack large males, whereas I did sometimes see them killed by large males, which, however, seemed less sensitive to their presence than they are to others of their own size.

For one fig of *Ficus* 2, Table 13.2 lists the *Idarnes* complement emerged from galls but still inside the fig. The extent of injury shown is unusual: it probably follows from the untypically large number of males and small number of females in this fig. Nevertheless, on the basis of a low estimate of 10 males dying in mortal combat per fig, one fruiting of a large tree of *Ficus* 1 (a tree larger than those for which Hill[8] gives a crop of 100 000 figs) probably involves several million deaths due to combat.

The strategy of small *Idarnes* males—of being first to reach hidden galls—is also that adopted by wingless males of *F*. These are small compared both to their females and to conspecific winged males (Fig. 13.2) and, compared to wingless males of other species, have their heads reduced instead of their gasters. I never saw them fight, whereas I did once find two jammed fast, back-to-back, halfway into a gall containing an *F* female. This suggests that they compete for matings by energetic but peaceable means. Small wingless males of *f* were probably similar in habits but were too rare for much observation, and as a morph were so variable that they could be considered runts if their modifications were not so similar to those shown more markedly by the clearly defined wingless morph of *F*.

The explanation of the peaceful competition in these species may lie in an evolutionary difficulty of associating a trend to smallness with a trend to pugnacity. It is very generally true for small taxonomic units throughout the animal kingdom that where pugnacity varies, it is the largest animals of a group that fight most. The present case may illustrate a general reason: the male who is most successful through smallness in reaching hidden females in galls is the male least likely to win a fight, so that the supergene (or switched set of genes) controlling size and morph cannot easily come to include elements promoting aggressiveness.

A more important problem is to explain why the wingless males of *Blastophaga*, which are not particularly small, do not fight. To judge from morphology (small head and mandibles, neck region unshielded), this puzzle may apply to all the pollinating agaonids: the only agaonid males which look much like fighters (*Alfonsiella*) are probably non-pollinator parasites.[16] Yet due to the lower sex ratio, a successfully pugnacious *Blastophaga* male among non-pugnacious rivals could expect to monopolize a much larger harem than could a dominant *Idarnes* male.

At the same time, the low sex ratio may show a path to the answer since it reflects breeding structure (see Chapter 4) and hence also the level of relatedness inside the fig. It is tempting to suggest that, since fighting involves both energy loss and danger for both rivals and since a lone surviving *Blastophaga* male might well be unable to fertilize all of 100 or so females (whereas an *Idarnes* male would have less difficulty with the 50 or so *Idarnes* females), *a difference in mean relatedness between rivals* accounts for the different male behaviours (see Chapter 2, Part II). That is, to anthropomorphize and oversimplify, *Blastophaga* males may be restrained because (1) many of the rival males are brothers, and a male doesn't care so much whether he or a brother does the mating; (2) many of the females are his sisters, and he doesn't wish to risk that some sisters remain unmated; and (3) for those that are his sisters, a male may actually prefer to have them mated by an unrelated male because of the opportunities this will give for useful recombination in the next generation.

A reasonable case for higher relatedness among *Blastophaga* males can be made straight away from the behaviour of their mothers. Once a foundress *Blastophaga* female has struggled through the osteole into a fig, losing her wings and often her antennae, she is committed to laying her whole brood there: she never escapes. *Idarnes* females on the other hand lay a few eggs at most each time they drill with a long ovipositor from the outside. They are free to fly from one fig to another between acts of drilling and normally do so. Other parasites with shorter ovipositors were observed to be even more restless, and these almost certainly lay only one or two eggs per act of drilling. With *Ficus* 2 figs at the right stages, it was common to see several *Idarnes*, *e*, or *f* females drilling at once on single figs. While drilling, females are often chased off or caught by ants. The dense, dark speckling which *Ficus* 2 figs acquire at this stage is a further index to this erratic activity. The speckles are due to pricking by fig wasp ovipositors. Even though most pricks are only trials, a number lead on to deep drilling and, presumably, to egg-laying.

The evidence from sex ratios assessed in the light of recent theory (see Chapters 4 and 8) reinforces these impressions of vagility: it rather strongly indicates that the *Idarnes* are more panmictic than the *Blastophaga* and that the other parasites, taken together, are more panmictic still. Figure 13.5 shows how the sex ratios of the species are distributed and shows that other wingless fighting males have sex ratios similar to *Idarnes* (although this sex ratio alone, as wholly winged species show, is not necessarily connected with fighting). Plots from data (see Table 13.4) of various other non-fig wasp species are also included in Fig. 13.5: the relevance of these will be made clear later.

With *Blastophaga*, the data on which Fig. 13.5 is based offered the possibility of a check on the sex-ratio theory, since the remains of dead foundresses could be counted even when the *Blastophaga* progeny were eclosing. Figure 13.6 shows how foundress number correlated with sex ratio in total progeny in the mature figs that were analysed, and Table 13.3 shows some independent counts of foundresses in young figs. The mean foundress numbers in the two sets of figs were 1.86 and 3.21. On the straightforward theory, these figures demand a much higher level of sex ratio than was observed (0.195 and 0.315 compared with the observed mean sex ratio 0.090). So, according to this test, either the facts are misleading or the theory needs to be modified. However, the trend for sex ratio to rise with increasing foundress number, although slight, is of the right kind.

Two excuses easily suggest themselves for the generally poor success of the test just mentioned. Unfortunately both are beyond the bounds of data that could check them. (Other less-obvious excuses are suggested in Hartl's discussion[17] of low sex ratios in single foundress situations. On the whole, however, the factors discussed by Hartl are as germane to raising evolutionarily stable sex ratio as to depressing them.) The most promising excuse is that earliness of entering or aggression between females[18,19] creates markedly unequal broods so that relatedness is higher in the assemblage of progeny than the equal-brood assumption of the model suggests it should be. Another excuse assumes that

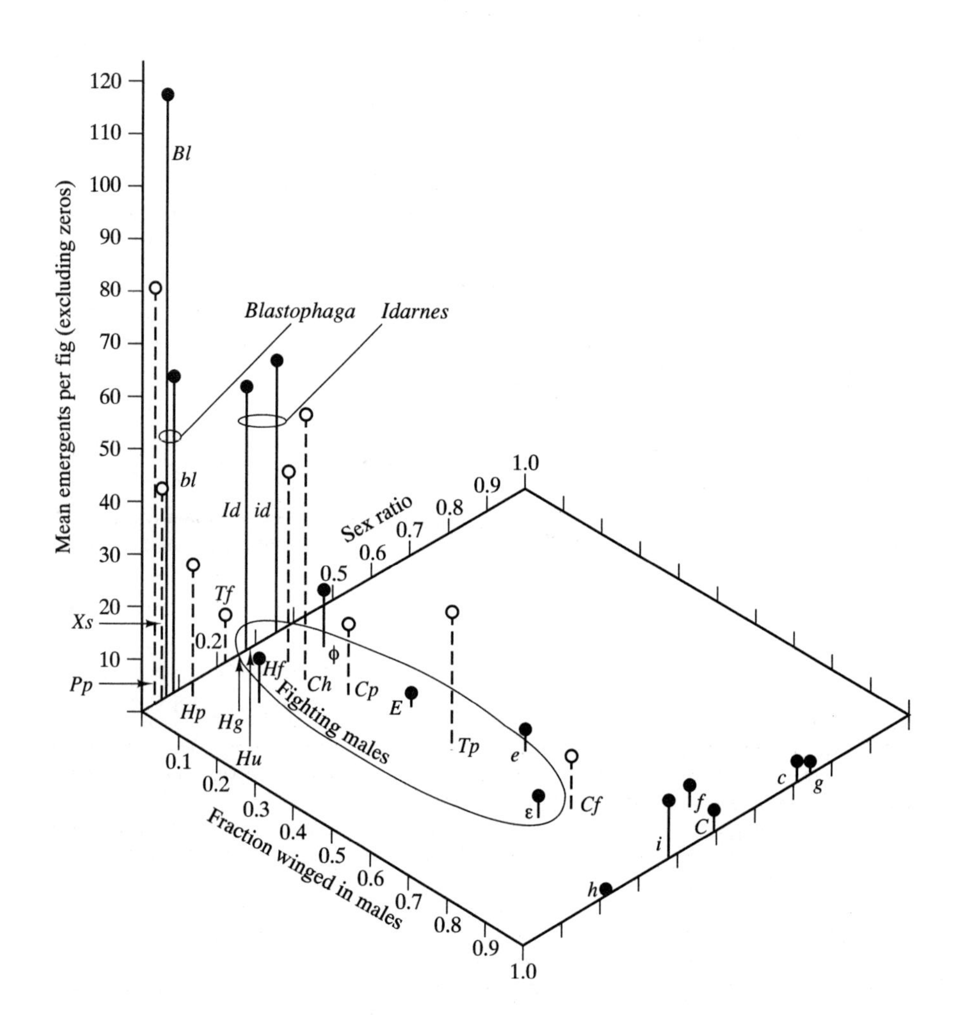

Figure 13.5 Sex ratio, male wings, and group size in some fig wasps and other male haploid insects. For *Ficus* 1, species are indexed as follows: *Bl* = *Blastophaga* 1, *Id* = *Idarnes* 'AB', C, E, F (genera unknown). For *Ficus* 2: *bl* = *Blastophaga* 2, *id* = *Idarnes* '*aαbβ*', *c*, *e*, *f*, *g*, *h*, *i*, ϵ, θ (genera unknown). The remaining species (non-fig wasps) are: *Cf* = *Cephalonomia formiciformis*, *Cp* = *C. perpusilla*, *Ch* = *Chilalictus* sp., *Hf* = *Hoplothrips fungi*, *Hg* = *H.* sp., *Hp* = *H. pedicularius*, *Hu* = *H. ulmi*, *Pp* = *Pygmephorus* sp., *Tf* = *Theocolax formiciformis*, *Tp* = *Telenomus polymorphus*, *Xs* = *Xyleborus saxeseni*.

Heights of columns for fig wasps show the mean numbers of adults emerging per fig when figs with zero counts are excluded. For other insects column height is mean number per brood (*Cf*, P_p, *Tp*) or per site of collection (*Ch*, *Cp*, *Hf*, *Hg*, *Hu*, *Xs*). Black balls at column heads are used for fig wasps, white for all others.

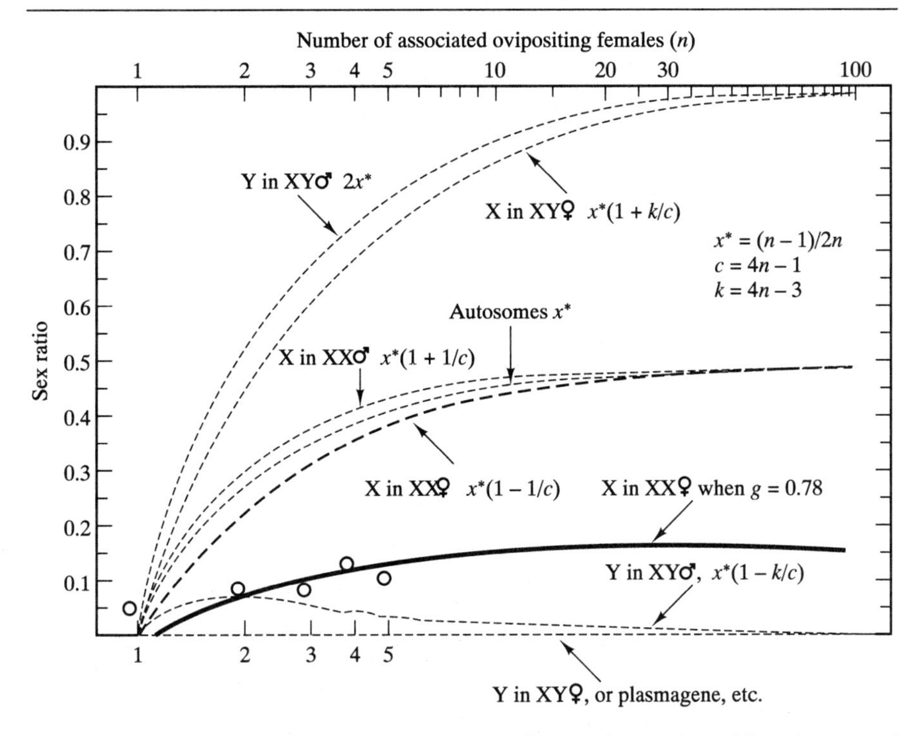

Figure 13.6 Sex ratios of *Blastophaga* in figs according to the number of foundresses and theoretical curves bearing on the struggle for the sex ratio in sperm-storing species with early local mating. Actual mean sex ratios in *Ficus* 2 synconia are shown as open circles and, from left to right, are based on counts in 21, 7, 4, 3, and 2 figs.

In the formulas given $x^* = (n-1)/2n$, $c = 4n - 1$, and $k = 4n - 3$. The heavy dashed line shows evolutionarily stable sex ratio appropriate for male haploids if the recombinational disadvantage of being inbred is not important. The solid line shows the modified curve appropriate if the disadvantage to genes on chromosomes unable to undergo effective crossing over is given by a fitness factor 0.22 (argument and calculation are explained in the text).

The light dashed lines show evolutionarily stable sex ratios for other components of the genome and plasmon under other systems of sex control, including female heterogamety. The two lowest of these curves (corresponding to male-control of sex ratio and control by a plasmagene or symbiont with ovarial transmission) are those most likely to have bearing on the *Blastophaga* sex ratios shown.

the erratic selection pressures on fig wasps give a large extra incentive to having progeny capable of effective recombination. If g is the disadvantage which a chromosome will have through being unable to recombine effectively with genes at other loci, then it always has this disadvantage in the haploid males (fitness factor $1 - g$), but in the females only has it in proportion to the extent that loci have allele pairs that are identical by descent (so that the fitness factor is $1 - Fg$). Combining these factors with consideration of relatedness (see

Table 13.3 Counts of *Blastophaga* females in young figs of individual trees

Date	Foundresses per fig																												Excluding zeros		
	0	**1**	**2**	**3**	**4**	**5**	**6**	**7**	**8**	**9**	**10**	**11**	**12**	**13**	**14**	**15**	**16**	**17**	**18**	**19**	**20**	**21**	**22**	**23**	**26**	**27**	**28**	**31**	**Total**	**$\bar{n}$**	**SD**
	Observed distribution for each tree																														
July 1975	3	18	12	4	3	1		2		2																			42	2.45	2.11
August 1975	0	14	3	2																									19	1.37	0.68
23/9/75	2	55	3																										58	1.05	0.22
25/9/75	103	193	8	1			1			1																			204	1.11	0.69
4/10/75	7	77	32	3	2	2		1																					117	1.50	0.93
21/10/75	1	13	5	13	4	14	9	11	8	9	5	9	14	7	4	3	3	2	2	3	1	2	3	3				1	148	9.03	5.96
26/11/75	11	9	5	7	8	4	4	2	1	1			1		1					1			1						45	4.76	4.46
27/11/75	0	6	6	1	4	3	2																				1		23	4.00	5.50
27/11/75	0	7	8	5	7	4	3																						34	3.06	1.61
1/1/76	25	23	1	1																									25	1.12	0.44
1/1/76	8	31	8	3																									42	1.33	0.61
1/1/76	5	3	1		1																								5	1.80	1.30
21/1/76	11	39	8	4	1	1																							53	1.43	0.86
22/1/76	3	9	8	8	9	8		3	1	2		2																	50	4.20	3.33
3/2/76	4	7	9	7	8	3	2	2	2	1	2	1		2	1	1	1									1			50	5.46	5.08
5/2/76	0	6	4	3	5	6	3	3	1	1	7	1	1	2		1		1	1	1			1	1	1				50	7.76	6.16
																							Unweighted mean of means							3.21	
Totals	183	510	121	62	52	46	24	24	13	17	14	13	16	11	6	5	4	3	3	5	1	2	5	4	1	1	1	1	965	3.484	4.456
																								With zeros included						2.9286	4.280

Chapter 8), we obtain the following genetic fitness valuations of a son and daughter by a mother:

$$\beta_u = \tfrac{1}{2}(1 + F)(1 - g) \quad \text{and} \quad \beta_v = \tfrac{1}{2}(1 + 3F)(1 - Fg).$$

Assuming the model conditions set out in Chapter 4, we have the evolutionarily stable sex ratio ('unbeatable sex ratio' in the earlier terminology; the 'ESS' sex ratio in the sense of Maynard Smith and Price[20]):

$$x^* = \frac{\beta_u}{\beta_u + \beta_v} \cdot \frac{V_{WG}}{V_T}$$

where V_{WG} and V_T are within-group and total genic variances, respectively. The differently arranged and more general result appearing in this formula compared to that of my 1967 account results from (1) the relatedness-weighted fitnesses of offspring now used and (2) reworking of the argument using Price's hierarchical analysis of natural selection (for outline, see Chapter 9). The most reasonable and simple assumption to fix the ratio of the variances is that females settle in groups at random. If the mean size of these groups is n, standard procedures of statistics and of population genetics enable us to find $V_{WG}/V_T = (n - 1)/n$ and $F = 1/(4n - 3)$, respectively, and with these formulas we are in a position to evaluate the unbeatable sex ratio for any given values of n and g. By trial and error it was found that at $g = 0.78$ the model fits the *Blastophaga* data quite well (Fig. 13.6). But due to the uncertainty of the assumptions, little weight can be placed on this fact. In particular, such a high value of g implies that in the hindsight of a distant future generation, the outbred individuals will appear to have been about 40 per cent more fit than their sib-mated sisters. This implies that *Blastophaga* would evolve more outbreeding via winged males or more shared laying in figs if *Ficus* would let it. Perhaps so, but at present this line of thought only evaporates in speculation. It is, of course, very possible that both of our excuses present part of the truth, and with this in mind the theory has no serious difficulty as yet: on present evidence it is reasonable to accept that sex ratios in *Blastophaga* and the other species reflect levels of outbreeding, and fighting adaptations only arise when the average rival has a chance somewhat less than one-third of being a brother.

WINGED MALES AND MALE DIMORPHISM

No less extraordinary than the bizarre forms and pugnacity of some of the males is the fact that these males were sometimes associated with conspecific males, quite possibly their brothers, which had totally different form and behaviour. These other forms were winged and very like the females. When a fig was opened, they were always among the first to run out and attempt to fly. Wingless males, in contrast, hardly ever even wandered onto the outside of their halved fig—if they did so it was usually as fugitives (i.e. in prolonged rapid ambulation following aggression by another male). We never found the winged males mating or even showing interest in the females inside the figs, but after they had been allowed to run or fly outside for a few minutes, they would readily mount females and apparently mate them when put together in a jar. Whether they had to fly before they would do this was not obvious, but it seems probable that normally they are programmed to mate only after a dispersal flight. Obviously they then compete—in a quite different mating arena from the wingless males—for those females which have escaped mating by wingless males in their fig of origin. Possibly they may also compete to remate non-virgin females, but we saw no evidence of this. The only males actually seen at fig trees where females were ovipositing were rare examples of *c*, a wholly winged species: females of *e*, ϵ, *f*, and θ which were searching for sites and drilling were not visited by males.

In addition to the species with male dimorphism (*E*, *e*, ϵ, *F*, *f*, θ), there were an equal number of species (*C*, *c*, *G*, *g*, *h*, *i*) which in all probability are wholly winged in the male (Fig. 13.4 shows *c*). These males were similar in degree of resemblance to their females and in behaviour to the winged males of the other species; they included both the brightest patterned (*c*) and the most sombre uniformly coloured (*h*) males. In no winged males of any kind did we see any fighting.

With the exception of species *i*, the sex ratios of the wholly winged species tend to confirm the outbreeding which their wings and habits imply. In fact, apart from *i*, it was excesses of males that were hard to explain, but, again, apart from *i* in one fig crop of one tree, numbers of these parasites were everywhere very low.

Male dimorphisms and switches from 'homeomorphic' to 'heteromorphic' males within a genus have been noted in Old World fig wasps,[1,14,21] but the startling extent of the divergence in some cases does not seem to have had the emphasis it deserves. So far as I know these are the most extreme dimorphisms of the male sex that are known in the animal kingdom. (Soldier–alate divergences in termites are of similar degree but apply to both sexes.) They must reflect long continued selection of a peculiar, disruptive kind.

Comparing the heights of the columns in Fig. 13.5, the abundance of progeny in figs colonized by a species seems to be important: common parasites tend to have wingless males and rare ones, winged males. This is easily understandable on the grounds that in a common parasite a male is fairly certain to find plenty of females to mate in his fig, and the females are fairly certain to have a male mate them before they fly out. In a rare parasite on the other hand, both sons and daughters would often die without mating if all males were wingless. The wing reduction is probably partly in the interest of redirection of growth into greater sperm production and (sometimes) into fighting adaptations, and partly simply because wings are an encumbrance for the males' activities inside the fig. They are lost here for the same reasons as they are in so many other insects which spend their adult lives in soil or rotting wood and disperse themselves adequately by other means (including flight prior to wing loss in those which shed their wings).

Dimorphisms might be expected to show levels of abundance between those of the species groups with wholly winged and wholly wingless males. Except for species E and i, the levels of abundance in occupied figs (not true means for abundance in all figs because zeros are excluded) are roughly as expected. F and θ are both common in their figs and have high fractions of wingless males, and e, ϵ, and f are less common (while more common than the wholly winged species except C and i) and have higher fractions of winged males. Variability of numbers per fig suggests itself as another factor which might be favourable to polymorphism. It can be seen from the standard deviations of the truncated frequency distributions given in Table 13.1 that the high populations of the *Blastophaga* and the *Idarnes* showed less variation relative to mean level than the rest, but among the rest there was no tendency for the dimorphic-male species to be more variable. Overall the truncated distributions tended to show standard deviations about equal to the means. Thus the variability is indeed great, but this does not seem to be the important factor.

Without necessity to expect greater variability in the dimorphic cases, provided there is some randomness in the way wasps are apportioned to figs, a simple basis for stable dimorphism is easily found. The crucial requirement for this is a reasonable frequency of male-less figs (i.e. figs producing females which have no male capable of mating them prior to their escape). Then, rather on lines that Gadgil[22] has argued for other male dimorphisms, we can see the advantages of wings and winglessness to be frequency-dependent in a way that promotes stability. For a very simple model, the argument can be shown as follows.

Assume females lay one egg per fig and move on; that if male, an egg develops as a wingless male with probability s (this parameter being supposed a genetic trait either of the egg itself or of the mother who lays it); and that a winged male is unable to mate females inside the fig—to do so conflicts too sharply with his adaptations as a flier. Random laying of eggs and random determination of type will give numbers of winged and wingless males (and of females and totals, although this fact will not be needed) in Poisson distributions. If M is the mean of all males reared per fig, then the mean for wingless males is sM. The fraction of all figs having no wingless male is e^{-sM}. Consequently, this is also the fraction of females not mated in the figs and therefore available to the winged males outside. Hence the reproductive values are proportional to $v' = e^{-sM}/(1-s)$ for winged males and to $v = (1 - e^{-sM})/s$ for wingless males. As $s \to 0$, $v' \to 1$ and $v \to M$, so a gene for producing a wingless morph can start to spread if $M > 1$, otherwise not. (If there is a selective advantage, c, to matings by flying males because more certain to be outbred, this condition becomes $M > 1/(1+c)$.) The stable equilibria for dimorphism in this model, possible for all $M > 1$, are shown in Fig. 13.7. The curve is rapidly asymptotic to $s = 1$; already at $M = 5$ less than 1 per cent of males are expected to be winged. Thus although the model shows an advantage to winged males at $s = 1$ for all values of $M(v' \to \infty, v \to 1 - e^{-M})$, the simplicity of the model plus the practical difficulty of maintaining the genetic programme and the switch gear for a complex adaptation that is hardly ever produced make it not at all surprising that the winged morph is completely absent in the more common fig wasps.

Also in Fig. 13.7 are plotted the mean numbers of males per fig (zeros included) for the various species. The graph gives true means (zeros excluded) for the model: unfortunately we cannot easily adjust either the points (downward) or the graph (upward) to remove this discrepancy owing to the involvement of females (and therefore sex ratio) in the truncation implied in the data. Actually any adjustment on these lines would be slight compared to

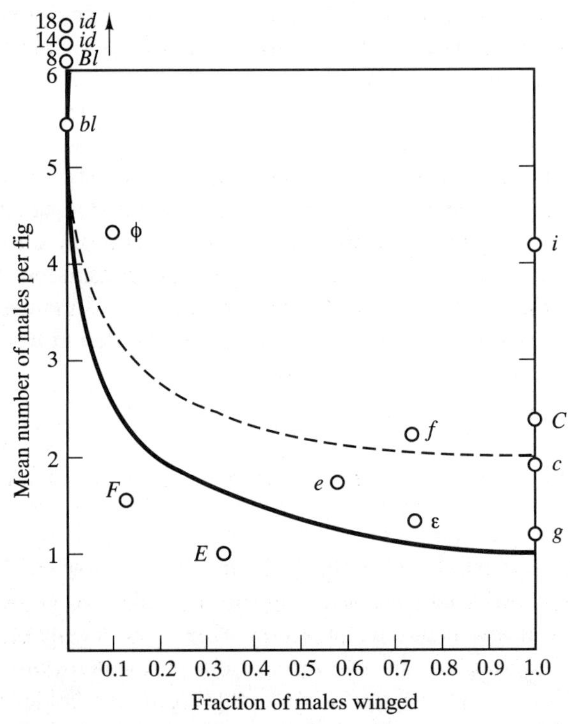

Figure 13.7 Data for fig wasp male dimorphisms compared to the stable states expected in a model which assumes (1) that egg laying and sex and morph determination are at random, and (2) that winged males only mate outside figs and only mate those females which have not been exposed to mating by wingless males inside figs. The solid curve shows evolutionarily stable states if a female has no fitness disadvantage from being mated by a wingless male (i.e. from being more likely to produce inbred offspring); the dashed curve shows the stable states when a wingless-male mating yields half the fitness of a winged-male mating.

the spread of the points, and this, combined with the known crudity of the model, does not encourage an attempt: accepting that the comparison would not be much affected by a slight upward adjustment of the curve towards the right, we may conclude that the fit is far from impressive but perhaps good enough to suggest that the model is of the right general kind.

In the way it has the fraction of wingless males so rapidly asymptotic to zero as population per fig is raised, this model effectively narrows the band of abundances within which dimorphism can occur. To explain the very deep divergence of morphs, we would hope for less rapid convergence and a broader band: this would imply that both morphs could persist over varying fortunes of the species (i.e. varying abundances) and so find time to reach a high degree of adaptive improvement. But the asymptotic curve obtained depends on use of the Poisson distribution, and not only is this not observed in any case, but in some has its underlying assumptions denied by the observed behaviour (e.g. the egg-laying behaviour of *Blastophaga* and *Idarnes*). Various possibilities, including fighting, may help to keep up the fraction of male-less figs in the more common species. Actually the present data do not suggest much in detail about how this 'band of stability' might be broadened. However, the data can provide a comparison which tests the main principle of the model more directly, and so avoids the uncertainties underlying the comparison in Fig. 13.7: we can test for equality of reproductive values of winged and wingless males without assuming random egg laying. The principle of the model was that reproductive values will be equal if the fraction of winged males is the same as the fraction of females that they can expect to mate with (i.e. those not mated by wingless males in the figs). Figure 13.8 compares these fractions for all the species other than those wholly winged. It can be seen that the correlation of the fractions is such that all points lie reasonably close to the 45° line.

OTHER WING-REDUCED, FIGHTING, AND HAPLOID MALES

In a 1967 publication (see Chapter 4), I briefly reviewed situations with flightless males in inbreeding situations. Many more cases could be added now. The correlation with probable incest remains very strong, but some exceptions and evidence—not least the evidence on foundress numbers in *Blastophaga* (refs 23, 24, and this study) and on sex ratio in *Idarnes*—suggest that male winglessness is connected less with incestuous situations *per se* than with having, besides female ability to store sperm, what might be called a 'seraglio' situation (i.e. a predictable abundance of females who are more or less confined at the site where the males mature). Figure 13.5 has various points for species whose males have more or less seraglio situations of this kind. The data from which these points derive are set out in Table 13.4.

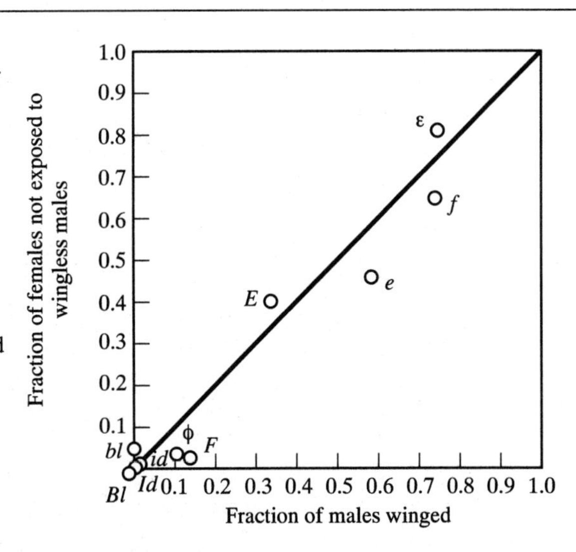

Figure 13.8 The correlation of the fraction of males that are winged with the fraction of females expected to leave figs while still unmated. The equality of these fractions, implying equality of reproductive values of winged and wingless males, tends to support the basis for evolutionary stability explained in the text. Since for wholly winged species the equality of the fractions is trivial (both fractions = 1), these species are omitted.

Rotting logs are a habitat in which inbreeding and male wing reduction are common. Whole logs provide sites for intrademic inbreeding and their smaller inhabited cavities set up potential seraglios that might encourage incest at the family level. To take a case at the extreme, I have observed a species of *Pygmephorus* mite (close to *P. priscus*) whose females suck hyphae of *Stereum* and become extremely physogastric (fig. 17 in Cross and Moser[30] illustrates exactly this state of a *Pygmephorus* female). Their broods, with two to four males among 16–160 females, develop and mate inside the mother; but since the mother usually bursts before all her eggs have hatched and since two mothers are sometimes pressed together (e.g. in crevices of contracting rotted wood), outbreeding can certainly occur. Do these wingless mites really qualify for discussion under the present heading? The females wander away as soon as the mother bursts, but the males are extremely reluctant to do so and continue to patrol the pile of remaining eggs. The females, moreover, are of two forms: with and without strongly developed chelae on the forelimbs. The chelate females are 'phoretomorphs',[31] adapted for dispersion by clinging to the hairs of larger arthropods. In this sense they are analogous to the flying morphs of various wing-polymorphic insects of the habitat and so emphasize the sedentary 'wingless' character of the males. (Variability of morph ratios in various mothers showed that female dimorphism in the *Pygmephorus* is not simply genetic: the same is usually found with wing dimorphism in multivoltine or continuously breeding insects. But, as mentioned below, two genetic cases are claimed). It is of particular interest here to note that the males do not fight at all, and being virtually headless and with all eight legs unmodified and alike, it is difficult to visualize that they could. The scene of peaceably scrambling

Table 13.4 Polymorphism in haplodiploid arthropods: brood data showing classification for sex and 'wings'

	Number of 'broods' included	Totals in broods				Total	Sex ratio	Wing ratio in males	Mean	SD
		Males		Females						
		Alate	Flightless	Alate	Flightless					
Pygmephorus priscus	11	0	36	650.5*	182.5	869	0.041	0	79.0	42.4
Hoplothrips pedicularius	20	4	52	93	354	503	0.111	0.071	25.2	24.0
Hoplothrips fungi	9	13	89	90	136	328	0.311	0.128	36.2	33.8
Hoplothrips ulmi	2	0	38	8	83	129	0.295	0	64.5	10.6
Hoplothrips sp. *g*	2	0	137	6	384	527	0.260	0	263.5	289.2
Cephalonomia formiciformis	96	362	85	229	698	1374	0.325	0.810	14.3	13.0
Cephalonomia perpusilla[25]	18	18	62	42	127	249	0.321	0.225	13.8	19.1
Theocolax formiciformis[26,27]		4	158	11	623	796	0.204	0.025	10	
Telenomus polymorphus[28]	19	78	84	328	0	490	0.331	0.481	25.8	22.6
Chilalictus[29]	1	2	13	35	0	50	0.300	0.133	50	
Xyleborus saxeseni	13	0	35	492	0	527	0.260	0	40.5	23.5

*Phoretomorphs.

and mating males amidst the numerous lethargic females (as seen through the transparent balloon-like hysterosoma of the mother) is very reminiscent of the peaceful mating activity of *Blastophaga* inside a freshly opened fig at a corresponding stage.

At levels of inbreeding which are less extreme, we can find mites with fighting males which correspond in the same rough way to *Idarnes, Tetranychus,*[32,33] *Caloglyphus,*[34] and *Rhinoseius*[35]). In *Caloglyphyus* one of two male morphs is a fighter, and in rotting wood the mite *Pyemotes dimorphus* has a male dimorphism that may be parallel, although there is again a definite dimorphism (phoretic) in females as well.[31] (In *Caloglyphus* induction of the fighting morph is pheromonal, but in the related *Rhizoglyphus echinopus*, common in rotting wood but better known as a pest of bulbs, the induction of a similar male morph may require a gene difference[36] (cf. wing dimorphism in male *Cephalonomia gallicola*[37]). Both *Rhizoglyphus* and *Caloglyphus* can also produce, in both sexes and in response to some environmental switch, a hypopal stage adapted for dispersion by phoresy. Neither *R. echinopus* nor *Caloglyphus* has haploid males.[38])

Prominences and enlargement of the forelimbs in male tubuliferan thrips of the same habitat[39] at least suggest fighting adaptations. Usually there is a gradation from large 'armed' males to smaller ones resembling females, but the changes are complex and not purely allometric,[40] and some females may be better armed than some males. Figure 13.5 includes four points for *Hoplothrips* from field-collected samples. The species with the most pronounced oedymery of forelegs in its males (*H. fungi*) lies well within the region of other fighting males, while a species with much less oedymery (*H. pedicularius*) has fewer males and lies near to *Blastophaga*. Another unknown species is intermediate in oedymery and in sex ratio. Less satisfactory, a sample of *H. ulmi* had almost as high a sex ratio as *H. fungi* yet shows very little oedymery; but this sample was small and from only two collections.

Under bark in Southeast Asia, a larviform male beetle exists which looks like a fighter (see Chapter 8; Fig. 13.9a). It is so strange that two very widely separated families have been proposed for it. Present considerations together with the known forms of parthenogenesis and of male modification in other bark beetles favour its placement in Scolytidae rather than in Histeridae as suggested by Crowson.[41] In other words, I believe the male much more likely to be that of *Ozopemon brownei* as originally suggested by Browne.[42,43] Browne's brief description of the communal galleries of *Ozopemon* under bark outlines a seraglio situation that would make precocity, winglessness, and fighting advantageous in a male; and the fig wasps show how profoundly males can diverge from their females when such conditions hold (Figs 13.1, 13.3, and 13.9b).

Whether or not *Ozopemon* is a parallel to *Idarnes*, another scolytid genus, *Xyleborus*, probably parallels *Blastophaga* fairly well. Adults mature in

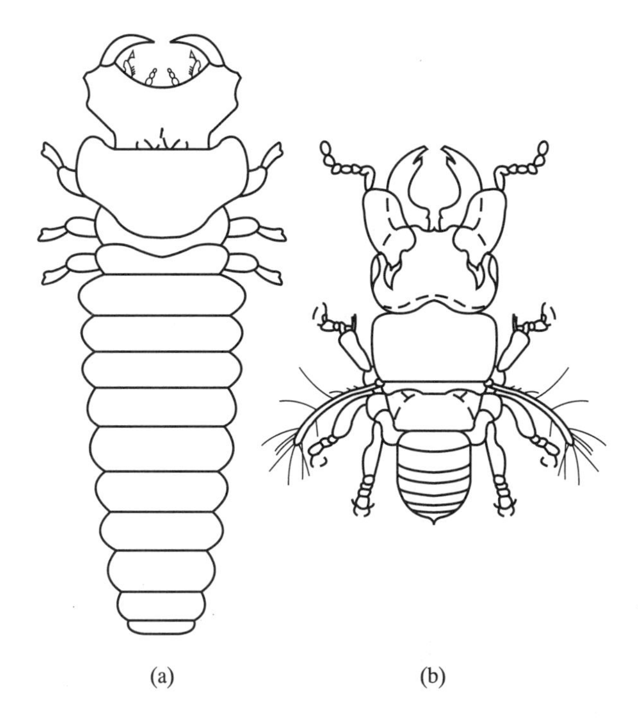

Figure 13.9 Adult male insects predicted to be fighters: (a) presumed male of *Ozopemon brownei*, Scolytidae (from Browne[42,43]); (b) male of *Philosycus monstruosus*, Torymidae (from Grandi[43a]).

crowded ambrosia galleries in live or freshly dead trunks, branches, stalks, etc. Each gallery usually derives from a lone-founding female. Although galleries sometimes coalesce and although cases of associative nest founding (*X. saxeseni* and *X. dispar*) are not rare (Hopkins[44] and personal observation) the mean degree of relatedness seems to be too high for males to fight. No record that I know suggests that the haploid, flightless, small, brown, half-blind, and harmless-looking males ever do fight. Fighting between larvae or killing of pupae by adults (see Matthews[45] and Hopkins[44]) is perhaps a more possible mode of aggression, if any occurs. A difference from *Blastophaga* affecting breeding structure lies in the site's capacity to support more than one generation.

In contrast, in the generally more outbred ambrosia beetles of the Platypodidae, severe fighting has been recorded.[46] But since these males are, as far as I know, winged and diploid (and also horned on the elytra), this case is considered later.

In the tropics rotting logs are a characteristic habitat for ants of the genus *Hypoponera* (Kempf[47] and personal observation). This genus has species with wingless males. I have verified that the ergatoid males of *H. punctatissima* (in a colony from soil in a London hospital) fight each other for occupancy of the chamber where queens are being reared. Fights look rather slow and ineffective; nevertheless males often lose legs and antennae and death soon follows. The squared-off heads of the males (illustration in Wheeler[48]) are larger than those of workers and are reminiscent of *Idarnes* and of our wingless *e*-type males. (In the ant genus *Cardiocondyla*, the wingless males have two types of head form,[49] one of which is suggestive of fighting. This may accentuate a puzzle as to why two free-living genera should have fighting males, while a whole host of parasitic species (see Chapter 8 and Buschinger[50]) have wingless males that look relatively unpugnacious and are not known to fight.) The winning male *H. punctatissima* mates the queens in the nuptial chamber soon after they eclose. If the males are brothers and the females their sisters, why—contrary to the cases of *Blastophaga, Pygmephorus*, and others—should the males fight? Nothing is known of the normal mode of colony foundation; if it could turn out to be associative as in various other ants, this difficulty would be resolved. If association leads on to permanent polygyny (which admittedly would be unusual—see Chapter 8 and West Eberhard[51]), a crop of sexuals may not be a sibling group. Transferences which I tried between colonies suggested no strong hostility to strangers, so adventive queens might be accepted by a colony. Since males are wingless either such acceptance, or associative founding, or fusion of colonies, would be needed to secure any useful outbreeding. Similar speculations have to be applied to the similarly unexpected pugnacity of *Melittobia* (see Chapter 8 and Dahms[52]). In *Melittobia* it is known that males do sometimes fight with actual brothers,[53] but perhaps in nature multiple colonization of a host is more common than we know. (In a recent report on fighting in two species of *Melittobia*, Matthews[45] mentioned another possible factor, that a male may need to feed on rivals before he becomes equal to his task of mating a hundred or so females. The same report raises the suspicion that the sex ratio of *Melittobia* may be underestimated because the first one or two males to mature may destroy others as pupae. For these reasons the genus is not included in Fig. 13.5, where, on known data, it would come close to the *Pygmephorus*.)

Panmixia of females combined with associative founding could be the normal background to all the cases of lethal fighting discussed so far. Although mites can't fly, tetranychid females can balloon on silk threads, and *Caloglyphus* is a genus that is commonly phoretic.[54,55] I know of no information about phoretic tendency for either of the two morphs in *Caloglyphus*, but already it seems that the unarmed males are probably more nearly analogous to the small males of *Idarnes* than to the winged males of fig wasps *E*, *F*, etc.

In a different superfamily (Proctotrupoidea) in the Hymenoptera, a complex polymorphism, which sounds rather similar to those in the chalcidoid fig wasps, has been described. It concerns a situation that is surprisingly 'open air' for a wingless male (egg masses of a bug, apparently attached to leaves and twigs), but one not wholly without parallel (e.g. *Trichogramma semblidis* on *Sialis* egg masses). The bug is the reduviid *Heza insignis* and the parasite is

Telenomus polymorphus.[28] The females are all winged and the males are polymorphic in melanization, wing reduction, and head size. Wingless males are always yellow rather than brown and may or may not have specially enlarged heads. Lima[28] gives neither statistics for large-headedness in his rearings nor any behavioural notes that might reveal fighting, but the position his data take on Fig. 13.1 suggests that his large-headed male is a fighter like the wingless *e*-type males that it most resembles. It is very possible that the winged males also fight in this species, although probably less fiercely. Such likelihood arises from the known pugnacity of other male scelionids on bug egg masses (see Chapter 4). Eberhard[12] described the circumstances of fighting for two species that attack the eggs of a pentatomid. Fighting was never fatal, but occasionally a leg or antenna was lost. The eggs, and consequently the females at stake in the fighting, were in much smaller batches than those of *Heza*. Eberhard's data do not quite match with kinship expectations since male behaviour seemed slightly *less* fierce and exclusive in the species where associative parasitism seemed more likely to occur. On the other hand, the sex ratio in this species did deviate from the other slightly in the direction expected (mean broods, male:female, were 3.1:13.3 as against 1.5:4.8).

A large-headed, short-winged male bee which looks like a fighter, coexisting with a normal winged male, has been described in the genus *Chilalictus*.[29] A very similar situation exists in a species of *Perdita* (Andrenidae) in Arizona.[56] For this species, as for Houston's *Chilalictus*, there is reason to suspect communal nest founding. In British andrenids no short-winged males are known, but large-headedness is marked in exactly those two species which have communal nests.[57] Data from the one described colony are included in Fig. 13.5. It falls among the fighting fig wasps, but since this is only one instance and there is uncertainty about presences and absences in the nest, this carries little weight. Similar male head variations reported in *Evylaeus ohei* seem not to be associated with wing reduction and not to show bimodality. Similar head enlargements are known in various female bees.[58] Wherever they occur, I suspect that, like the similar head modifications in the cuckoo aculeates in *Sulcopolistes* and *Psithyrus*, they are for threatened or actual fighting.

Sakagami's list of cases includes several of bees that are communal or semisocial and none that are known to be solitary. In the recently studied *Centris pallida*,[59] nests are merely aggregated. The males show a wide range of size in which enlargement of the head appears to be no more than isometric, but perhaps they are, nevertheless, on the road to the more extreme kind of polymorphism known in at least one other *Centris*.[60] In *C. pallida* the large males, accompanied by some small males, tend to specialize in patrolling the ground surface and in digging for females that are about to emerge from the soil, while small males specialize in high aerial mating with females that have somehow escaped the attentions of the ground males. On the ground, males fight to usurp excavations and to defend them, and large males almost always win. Gregarious mating, the markedly different mating strategies, and this fighting combine to suggest a distant but most interesting parallel to the dimorphic fig wasps. Other observations[61] suggest that a divided arena of mating—near to the nests and also some place distant from them—is quite widespread in bees. It may well occur in sphecids and eumenids as well.

In males of the eumenid wasp *Synagris cornuta*, we find another grotesquely enlarged and modified head. Here the size and showiness of the male wasp, with his unreduced wings and hypertrophied horned mandibles, seem a link from *Chilalictus* to showy and grotesquely horned beetles (see below). In the present context it is noteworthy (1) that a larger-headed male has been observed driving off others with lesser heads from a clay nest on a leaf where a female was emerging,[62] and (2) that although this nest was solitary with only four cells, large aggregations of nests occur,[63] giving the male a situation which may be rather similar to situations in *Chilalictus* and *E. ohei*. In *Synagris* low relatedness of fighting males would tend to be assured by the nature of the aggregation (which is obviously 'communal' in the sense of Michener[64]) and the free flight of both sexes, whereas in the two bee species (probably in the former, more certainly in the latter), it is assured by nest sharing by several independent free-flying females, and, in *E. ohei*, potential flight by the fighting males.

In sperm-storing species, wing reduction in females obviously has quite different implications for the breeding system from wing reduction in males.

If females are totally wingless, this usually means that either they or the young stages have alternative means of dispersion (generally wind or phoresy, the latter including transport as a parasite—as in stylops—and by their own winged males—as in thynnids, some mutillids, phorids of *Chonocephalus*, etc.). Although in some ectoparasitic cases the males are wingless too, such species are outside the present purview: constantly hitchhiking around on their large hosts the parasitic species, for example, have a mobility almost as if both sexes were winged. Winglessness in phoretic and parasitic insects gives little guidance as to relatedness and inbreeding, although naturally species do exist among them with incestuous habits and sex ratios biased accordingly (most are mites—e.g. *Dichrocheles, Syringophilus, Demodex, Pyemotes*, and *Iponemus*—but there is also one possible case in a flea[65]).

If merely some females are wingless, the situation is quite different. It suggests that a habitat, when found, can support a colony for at least a few generations before exhaustion of food, discovery by enemies, or other terminating events prevail. The wingless female is adapted to increase the colony while the going is good; the winged female, to be prepared to seek a new habitat when conditions worsen—or when reliable signs show that they are about to. Aphids provide the most studied example. We would generally expect that in the growth phase the colony would be largely isolated: the sex ratio would therefore be low and, if this phase is sexual at all, the few males worked hard to accomplish all the matings available to them. However, if colony founding is commonly associative, we would expect ideally that the first generation sex ratio would be high (about $(n - 1)/2n)$) and that its males would fight, while in subsequent generations the frequency and pugnacity of males would fall off, perhaps rising again whenever colonies coalesce.

It seems possible that such an outline could apply to the underbark thrips already mentioned and account for the graded variability of their males. But published data on another hymenopteran parasitoid in dead wood (*Theocolax formicformis*: see Fig. 13.5) show sex and wing ratios similar to those of the *Hoplothrips*, and yet no unusual variation or behaviour in the male is reported.[26,27] This ant-like parasitoid searches for larvae of wood-boring beetles (Taylor mentions that *females* sometimes fight for larvae). This is a claustral habitat very like that of *Hoplothrips* or *Melittobia*. However, only small broods are possible in particular larvae (not more than nine), and this can create only a rather unattractive seraglio compared to that encountered by *Melittobia* or even *Idarnes*: to mate many females the *Theocolax* male probably needs to be adapted for tunnelling and searching rather than for fighting. (In discussing fighting generally, Ghiselin[66] rightly emphasizes the factor of high density—to evolve fighting for females the males must at least frequently meet. On the other hand, defendability of resource is hardly mentioned in his discussion and the factor of relatedness is ignored.) The same applies to the wingless males of other ant-like dead-wood parasitoids in the Bethylidae. In Fig. 13.1, data for *Cephalonomia perpusilla*[25] give a point in the midst of the fighting fig wasps, but my own data for *C. formiciformis*—in which the winged forms were much more common, especially in the males—give a point outside. In many parasitoids wingless varieties seem to have the status of occasional underfed runts;[67] this probably applies to the wingless females we found in fig wasp *f* and perhaps to some extent to the uncommon wingless males of this species). In *Cephalonomia* wingless males are indeed smaller and nutrition may be a factor, but at the same time they show clear morphological differences that parallel those in the wingless females, which are quite as large as their winged sisters and clearly not produced by underfeeding. In *Cephalonomia gallicola* evidence has been recorded that the wing polymorphism in males is simply genetic;[37] but in *Sclerodermus domesticum* the occasional wingless males are described definitely as runts,[68] although the species seems to have much the same conditions of life as *Theocolax*. Wing development and dispersal in Bethylidae need further study to resolve these contrasts.

WINGLESS DIPLOID MALES

All the cases of polymorphism, male winglessness, and male fighting discussed so far are in groups for which male haploidy is either known or can be suspected. Others where males are more wingless, or more often wingless than their females, occur scattered in various groups which are unlikely to be, or known not to be, male haploid. Not all these have mating situations especially favourable to incest and not all have female-biased sex ratios, although I know of no exception to the rule that a marked bias exists if the male is always wingless and the female usually winged. Even in male haploid groups, always-wingless males can occur in non-incestuous situations. For example, while wingless male thrips in galls[69] and in the similarly flask-like flowers of *Erica*[70] extend the range of cases of ready-made seraglios, some other wingless male thrips seem to be fairly normally phytophagous although still tending to specialized diets and to secluded windless environments.[71,72] To these external phytophages can be added others that have come to notice in halticid flea beetles.[73,74]Presumably in these cases there is less brother–sister incest, but the 'perenniality' and patchy distribution of the plant hosts are such that there may be fairly intense intrademe inbreeding. The same applies to wingless male Psocoptera feeding on lichens under dense webs on bark and sometimes on leaves and twigs (New[75] and personal observation) and to the flightless male Plecoptera that mate their females among stones by the mountain streams where they breed. But just why some species and not others should become inbreeders with female-biased sex ratios and reduced wings in the male is not clear. For example, another psocid making webs on bark, *Reuterella helvimacula*, has the female wingless and the male winged.

Mention may also be made in passing of some other phytophagous insects, including more flea beetles (e.g. *Longitarsus rubiginosus*: personal observation) and sundry leafhoppers, which show wing reduction in most specimens but have an uncommon macropterous form. In such species the macroptera are more common among females than among males, although sex ratios are fairly normal.

Under dead bark a situation similar to this is shown by the minute beetle *Ptinella aptera*[76] and the psocid *Psoquilla marginepunctata*.[77] In the latter, at least, the fraction of winged individuals can increase very markedly in some circumstances, winged females remaining more common. In *Ptinella* normality of the sex ratio probably reflects the females' inability to store many of the relatively enormous sperm, which are as long as the animals.[76] Zoraptera present a similar case, but in some species the winged morph is much more rare in the male (e.g. *Zorotypus hubbardi*[78]). Also under bark, Hemiptera (Anthocoridae; e.g. *Xylocoris cursitans*: personal observation) and Psocoptera[77] show parallel features, although *Embidopsocus*, with males always precocious and wingless, can have the sex ratio under one-third (personal observation).

In *Ectopsocus richardsi*, in the related habitat of human-stored products, males are always brachypterous and the only sex ratio recorded was one-sixth.[79] Similar cases in rotting wood are known in sciarids and there is one case (sex ratio not recorded) in Cecidomyiidae.[80]

The absence of male structural modifications which could be for fighting (and absence of records of fighting) in all these non-male haploid cases is in striking contrast to their frequent presence in the haploid males already discussed. Perhaps this difference will prove less real than it seems. For example, the excuse given for the non-fighting *F*-type wingless males of the fig wasps can serve as well for *Embidopsocus* males, which are small through being sexualized one instar earlier than the females (Broadhead[81]; they are thus 'protothetelic' in the sense of Southwood[82]). And on the other side, it is conceivable that the enlarged and sometimes spined hind femora of *Zorotypus* males are used for fighting. Further, in *Xylocoris*, males do show a kind of sexual assault on other males and may succeed in certain species in making them effective vehicles of their sperm.[83] However, on the whole such points are weak when set against the proven pugnacity of so many haploid males. In so far as it is real, this contrast is not easy to explain.

When inbreeding occurs, but is less than total, haploid males are expected to be less related to male rivals in their colony than diploid males would be. But the slight difference seems unlikely to be important. It is maximal when the fixation index is 1/2, that is, when the relatednesses of brothers are 3/4 and 5/6, respectively. With sex ratio under the control of mothers and strong selection incentives for mothers to gamble with the sexes of progeny in a sex ratio game (see Chapter 4), the haplodiploid species are much more apt to create situations where a non-sibling male can greatly increase his inclusive fitness, through size of his expected harem, if he is a fighter. Non-haplodiploid species have more difficulty in changing the sex ratio, and when they find a means (as *Embidopsocus*, *Ectopsocus*, *Xylocoris* have apparently found one), the female bias they achieve is steady, moderate, and probably closely connected to the mean male relatedness. It follows that where harems are potentially largest, a male would have to be fighting closely related rivals in order to win them.

FIGHTING DIPLOID MALES; FIGHTING IN GENERAL

Examples of fighting in normally diploid males are also found in rotting wood. Hubbard[46] recorded lethal fighting in polygynous males of a *Platypus*. These males have small horns on the elytra. Platypodids are ambrosia feeders like *Xyleborus* but usually have a very different breeding system, with mating *after* flight to the new host tree, and a normal sex ratio. The situation is not known for certain for Hubbard's species, but it is likely that the males are much less related than are *Xyleborus* males in one gallery. I have observed fighting in

Gnatocerus cornutus[84] and have seen signs of it in *Siagonium quadricorne*. Unfortunately the latter is too photophobic to be observed easily. It is found under bark in groups usually consisting of one male and several females (and often larvae as well). Both *S. quadricorne* and *G. cornutus* have striking horns on their mandibles. *Gnatocerus*, too, probably comes originally from dead phloem,[85,86] and for horned beetles in general, rotting wood is the most favoured habitat. In a related habitat, mining in tough bracket fungi, the horned males of *Cis bilamellatus* fight at entrances of burrows where females are feeding.[87] Almost all these horned beetles, including those just mentioned, are winged in both sexes, but apart from the implication of similar dispersive abilities this may not be very important. Wingless primitive crickets with huge-jawed males (Stenopelmatidae) live in and under dead trunks,[88] and one is recorded in polygynous situations[89] like that mentioned for *Siagonium*.

Similar habitat and habits are probable for cockroaches of the genus *Gromphadorhina*.[90] In captivity males of this genus are seen to fight by butting. They possess (most conspicuously in the large males of *G. portentosa*) short, forward-directed horns bordering a scooped declivity of the pronotum. *Cryptocercus* males have a faint trace of a similar armature and these, although apparently monogamous, fight over females in burrows.[91] Stick insects normally evoke a greener, more open world than that of rotting wood; but in its last oceanic refuge (an islet off Lord Howe Island), a large primitive wingless phasmid, *Dryocelus australe*, with grotesquely enlarged nipper-like hindlegs in the male, lives in large colonies in beetle holes in dead wood.[92] And a few other related eurycanthine phasmids that are similarly sexually dimorphic may also hide away in rotting wood, emerging at night (like *Dryocelus*) to feed.[93]

It seems open to conjecture whether the enlarged, spined hind femora of these primitive phasmids (which in *Eurycanthus horridus* are so strong that they are used as fish hooks) are functionally convergent with similar developments of the hind femora of males in certain coreid bugs and in certain gonyleptids in the Arachnida. For the last two groups, it is at least recorded that the structures can give a sharp nip to a human finger.[94]

So far as I am aware, the swollen-legged coreids are unexceptional plant feeders and are all winged. Yet the occurrence of a 'high–low' range of variation in the legs of males, in, for example, *Acanthocephala femorata*[95] so like the variation of stag beetle mandibles, strongly suggests that males fight. If they do, it would be of interest to know what special circumstances make it worthwhile. Gonyleptids live in cavities; and in general, outside of rotting wood, the most characteristic habitats for horned beetles are burrows around dung and carrion.[96] Thus horned beetles along with various pugnacious crickets,[97] the bees already mentioned, stomatopods,[98] and other marine crustaceans and worms[99,100] suggest an important common factor: defence of burrows.

Fighting is expected to evolve when an object valuable for fitness is not obtainable (or retainable) by other means and when not too much inclusive fitness is lost through any damage a fighter has to inflict (i.e. the antagonist is not too closely related: Chapter 2, Part II; see also Eberhard[12] and Alexander[101]). Any object that has been laboriously made (for mating, egg protection, larval feeding, etc.) is worth the risks and the energy costs of violence to usurp or defend. A burrow, a dung ball,[102,103] for example, can be such an object. Its value, of course, lies in what can be done with it and is in no way sentimental. Similarly any compact resource which can simply be *found* and then counted on to yield richly in fitness—what Wilson[104] called a 'bonanza' situation—is also worth fighting for. Here might be cited a carrion,[105,106] a nest aggregation with emerging females,[107] and a position near or in special flowers or food sources where mateable females come, as in some bees, some Hawaiian *Drosophila*,[108] some lucanid beetles, and perhaps some nitidulids.[96] Since, obviously it is best to avoid the risks of an outright fight if there are strong signs that the opponent will win, we would expect the amount of overt fighting, but not of weaponry, to be reduced according to how well an individual is able to form a true appreciation of the fighting ability of rivals and also according to the probability that alternative opportunities can be found elsewhere if a fight is declined. Probably part of the reason that fighting in the fig is so damaging (as also between parasitoid larvae: Chapter 2, Part II) is that there is no possibility of going off to another fig. Nevertheless, as already pointed out, the males in the fig do seem to assess the strength of rivals and if weaker, often employ a 'waiting' strategy which occasionally, no doubt, brings mating opportunities later. (Such waiting by defeated males is also observed in a scelionid, *Phanuropsis semiflaviventris*.[109])

When an owner of an object has an initial positional advantage, the assessment that an intending usurper has to make is perhaps more difficult and a trial of strength correspondingly more likely. The fighting which W. G. Eberhard has observed in the dynastine beetle *Podischnus agenor* rather suggests this. Here it seems probable that chances of mating with more than one female are at stake in the fighting; in other words, that the ownership of the object of fighting—a feeding gallery—sometimes enables a male to attract and mate several females (Eberhard[110] and personal communication).

When I watched a scolytid *Pityogenes bidentatus* with a situation apparently similar to that hypothesized for the dynastine just mentioned (except that the polygyny is simultaneous and continued), I did not see fighting or usurping, but I did see males blocking entrances to their nuptial chambers against other males but admitting females (these observations were very brief). Among scolytids it is particularly the males of polygynous species like *P. bidentatus* that often have small horns on the ends of the elytra.[111] In *P. bidentatus*, males are on the average slightly larger than females and use the flattened and horned elytral declivity to block their entrances. Table 13.5 shows a census of polygamy in gallery systems very soon after their establishment and some counts made a little later. Obviously some males are getting many more females than others. Yet the distributions seem to end up almost as if females had arrived at random ($\mu \simeq \sigma^2$), which seems to imply little favouritism for particular males. Although as in other outbreeding polygamous scolytids (see Chapter 4), there is an unaccounted deficiency of males (since the sex ratio at the pupal stage is about 1:1), and although the change in distribution from 'May' to 'June' shows that some females are leaving the most overcrowded systems, the superficially random distribution that results suggests that sexual selection of males either through combat or female preference is not very strong.

A case of fighting in tunnels by a polygynous male with small horns on the elytra has already been mentioned. This was in Platypodidae. Yet another family, Bostrychidae, has examples converging strongly to the morphological facies, at least, of the polygynous scolytids and platypodids, but here there is sometimes a complication in the existence of a sharply defined class of variant males.[112] (Some marine wood-boring Crustacea converge to the same facies,[113] but social organization and behaviour are hardly known. Hole-dwelling rather than wood-boring, however, is probably the important factor for the convergence of male adaptations.[114]) In *Bostrychopsis uncinata*, variant males have sharp points around the elytral declivity exactly like ordinary males, but completely lack a pair of barbed prongs at the front of the pronotum. As with most bostrychids, the gallery system is in wood (not phloem) and is not ambrosial. Based on a few observations which I made in Brazil, the gallery system is not unlike that of *Pityogenes* or *Ips*. Its shape is somewhat glove-like with an entrance near the base of the 'palm'. Close inside a male is normally waiting with head inwards. Some 'fingers' of the glove are close below the surface of the wood and admitting to these is sometimes another window or entrance. The barbed prongs appear designed to give a purchase in the wood when a male is initiating a burrow. Slight as they are, my data suggest that an initiating male is usually, or perhaps always, pronged; and that prongless males may enter galleries later.

In three systems in October 1975, I found one such prongless male alone with six females, two together with one pronged male and two females, and one pronged male alone with three females. A month earlier I had found in the same log just one pronged male together with 11 females. If, even using prongs, initiating a system is a difficult and dangerous task, and if, through sexual dissatisfaction of females or simply their clumsy mining, a fraction of the systems acquire windows through which prongless males can enter, we see a possible basis for the stable dimorphism akin to that suggested for the fig wasp dimorphism.

Other equally sharp and even more impressive dimorphisms are known in horned beetles.[115] This is quite distinct from the usually graded series from 'low' to 'high' males which is often found in the same species, but is connected with it in the sense that only one morph shows the more graded 'high–low' variation, while the other occurs only among large ('high') males. The function

Table 13.5 Distribution of polygyny in *Pityogenes bidentatus*

Date	Number of females per system												Total	$\bar{n}$	S^2
	0	**1**	**2**	**3**	**4**	**5**	**6**	**7**	**8**	**9**	**10**	**11**			
	Systems with 0 males present														
May 1967		5	1	2	2								10	2.10	1.66
June 1967	4	16	1										21	0.86	0.23
	Systems with 1 male present														
May 1967	13	4	9	6	5	7	5	5	5	4	1	1	65	3.95	9.73
June 1967	6	8	13	11	13	8	5	3					67	3.13	3.57
	Systems with 2 males present														
May 1967	1												1	0	
June 1967	1		1										2	1	
	Systems in 1970 (all one-male)*														
June 1970	4	6	16	50	79	87	51	18	14	5	1		331	4.65	2.82
	All one-male systems surveyed in June														
	10	14	29	61	92	95	56	21	14	5	1		398	4.39	3.26
	***P. hopkinsi* (mature systems)[118]**														
		6	17	22	12	2	1						60	2.83	1.16

*Data gathered by R. Kowalski.

of this variation is unknown, but one might suggest that if, as in *Idarnes*, large males do more real fighting, an uncommon variant may get an advantage similar to that of a left-handed boxer, while perhaps being slightly disadvantaged in other respects because the genes for his armament are not so well coadapted with the rest of the genome or with the environment (as may also be the case with left-handedness). The rare occurrence, noted by Arrow,[96] or bilateral mosaics for the two male forms in the beetles slightly suggests that, besides adequate feeding (to give large size), a different genotype is needed for the variant to appear.

Although male horns seem to be most extravagant in the species that are known to be polygamous (or reasonably supposed to be so), many of the apparently monogamous hole-dwelling males have horns (e.g. *Lethrus apterus*, *Typhoeus typhoeus*), and even where lacking (as with *Cryptocercus punctatulus*;[91] *Dendroctonus ponderosae*[116]), the male may fight to defend the burrow. Paralleling male adversity to other males, we would expect females in these monogamous situations to resent the presence of other females if they ever got in. Within-sex resentment is actually illustrated in beetles of *Necrophorus*, where, after the phase of co-operation during burial of a carcass, fighting begins and continues until only one male–female pair remains. Similar intrasexual conflict over dung balls by both males and females is known in a ball-rolling scarab.[103] In some scarabaeines, however, there are hints that pairs may fly as pairs from one site to another[117] and, if they can, they might be expected to fight for their burrow or food object as a bonded pair too. This might help to explain the very occasional but almost equal horn development in the two sexes (e.g. *Megaphanaeus*). In rotting wood, again, there are some strange cases of horns developed equally in both sexes. Of the habits in these also almost nothing is known, even though one such genus (*Priochirus*, Staphylinidae) has been the subject of a phylogenetic study based on morphology and biogeography.[119] That horns of some bisexually armed tenebrionids are used in fighting is shown by Eberhard.[110]

Since Darwin, appreciation of fighting in the selection of epigamic characters has a rather chequered history. Such appreciation has been overdaring and vitalistic,[120] or overcautious[121,122] or even almost denied.[96] In view of the freqent difficulty of demonstrating fighting in horned beetles and the apparent ineffectiveness of fighting when seen, varied alternative hypotheses have been tried. The most reasonable of these, due to Darwin, emphasizes selection through female preference. The bright colours which some of the most grotesque males have in addition to their horns suggest that this factor does play a part. But various objections have been urged against the factor of female preference. One of the most weighty is that the female seems to accept any male. Nevertheless it is generally not excluded on present evidence that she may have some means to control sperm precedence in favour of the male she likes best.

Another point de-emphasizing female preference is that biologists seem to have too hastily assumed that large horns cannot be efficient weapons of combat. This point is strongly made by Eberhard.[110] The huge mandibles of *Chiasognathus granti* obviously have a poor mechanical advantage: Charles Darwin in his *Voyage of the Beagle* commented that he found them 'not strong enough to pinch my finger so as to cause actual pain'. I observed these beetles in Chile in March 1975. With regard to a finger, I found much the same as Darwin, but I was not sure about lack of strength: I thought they were not biting as hard or holding as tenaciously as they do when fighting each other and suspect that they restrain their strength in deterrent biting of a large animal to lessen the risk of a lethal reaction. In any case, the literature shows that in nature fighting males do at times pierce each other and even tear each other apart.[123,124] Fights are not normally sanguinary but are hardly less decisive: one male secures a grip that enables him to lift his opponent until all legs are free (the long mandibles and long legs play their part here, although, of course, the lifted beetle can use his for hanging on) and then drops him. Fighting and mating usually occur on tree trunks around sap flows (Claude-Joseph[123] and informants in Chile) or flowers of a tree vine so that dropping disposes of the loser, at least for a while. Probably he does not normally return, for I found a loser reluctant to engage the victor again, and if he did so at my prompting, the result was always the same. A dropped male walks hurriedly away, climbing if possible, and tries to fly. Males with small mandibles fought each other much more readily than they fought large males, and they always lost to large males if they did fight them. I have found the same with *Lucanus cervus* in Britain, and Mathieu[125] the same for other lucanids in the US; Mathieu also found that large males (in *Lucanus* and *Platycerus*) sometimes pierced and killed small males. Very similar methods of male fighting on tree trunks have recently been recorded for a grotesquely elongate brenthid, *Lasiorhynchus barbicornis*, which also showed (1) the use of a long appendage (the rostrum) for holding out and dropping an opponent and (2) reluctance of small males to engage large ones.[126]

I suspect that horned-beetle fights have too often been staged in the bottoms of boxes, and that here the lifting-and-dropping theme looks unduly like a fight to rules, resembling human wrestling.[127,128] Had the fights been watched on the more natural elevated surfaces (like the branches of the fallen tree where Beebe[127] found his *Megasoma* males), the effectiveness of the methods would have been obvious. I suspect that similar methods will be found for many male insects that are clumsy fliers and fight on elevated objects. This may be suggested, for example, for the strange horned males of plataspids (Heteroptera) in those cases where the females live on tree trunks, as some do, and perhaps for some membracid males as well.

COMMUNICATION IN SEXUAL SELECTION

It is possible to imagine selection processes by which horns can evolve, at least in some variants, beyond the point of their maximum effectiveness in fighting, or can evolve when not used in fighting at all. These possible processes ascribe a more subtle function than pinching or throwing an opponent: they require that, perhaps prior even to contact, the horns can be used to communicate something (see Fisher[129] and O'Donald[130–132]). What they can communicate is either 'true' information—regarding principally (1) *ability to fight* (to males) or (2) *desirability as a mate* (to females)—or else, as their owner may at least hope, 'false' information—such as (3) *bluff*, mainly about (1) (to males) but also about (2) as implied by (1) (to females). Perhaps because of the intriguing suggestion of a positive evolutionary feedback,[129,130] it is, as already mentioned, desirability as a mate (2) that has received most attention from the time of Darwin. The other factors cannot cause such feedback, but it is questionable how often data really call for it or show sexual preference to be effective. Ability to fight (1) has the recommendation of simplicity and seems unduly neglected. On behalf of desirability as a mate (2), the positive allometry so often observed in the features in question is just what is needed to make characters so conspicuous that they will impress even half-blind females—as females of the horned species are sometimes claimed to be in the course of a somewhat illogical objection to the role of female preference.[96] But the same allometric exaggeration of certain parts will make them conspicuous and sensitized advertisements of size, with which goes, in most cases, real superiority in fighting. Thus again, whether eyesight is good or bad, the allometry could usefully impress other males as well. (Regarding insects with no special weapons, the advantage of size in fighting is mentioned by Alexander[133] for crickets and documented statistically by Alcock *et al.*[59] for the bee *Centris pallida*, and, even more thoroughly, by Marshall[103] for the ball-rolling scarab *Canthon imitator*. Some other cases of small horned males being reluctant to engage large ones, and of real fighting advantages to the latter, have already been mentioned.)

Horns that communicate fighting ability to other males would be expected to be, in most cases, the actual weapons used in any fighting that occurred, but they would not have to be. Aspects of size and large-headedness in *Scatophaga*, as discussed by Borgia,[134] may be relevant here as non-weapon characters that communicate strength. McAlpine[135,136] discusses this matter for the eight or so dipteran families in which strikingly megacephalic and 'stalk-eyed' males are known and indicates a connection with head-to-head fighting. Bristowe[94] described the modestly 'stalk-eyed' males of the drosophilid *Zygothricha dispar* as fighting by butting amid a crowd of females on a fungus. He likened them to fighting male bovids, but it is unlikely that the points of the 'horns' are used to jab opponents since these bear the compound eyes. (Horns of rather similar aspect occur on the thorax in some male pentatomids: I suspect these males fight and that the horns display size.) Among the uniquely numerous and diverse *Drosophila* of Hawaii are species whose males form 'leks' on surfaces close to sites where females come to feed and oviposit. These species include the largest known *Drosophila* and also the most sexually dimorphic. Among other differences, the males are larger than their females. In some cases males wrestle head-to-head and, in a preliminary display, the elongate forelegs are used in a special upright posture. This, combined with bobbing movements, shows off both the legs and size generally. Small males retreat from obviously larger males after such display. However, the long legs and the head (also especially enlarged) are used in the fighting if this ensues.[108,137] In one case, that of *Drosophila heteroneura*, the male head has a distinct stalk-eyed tendency. Taking advantage of interfertility with another species, an ingenious genetic analysis has already strongly implicated the X chromosome as the site for the switch gene for this male-limited character.[138,139]

Acanthocinus aedilis, a cerambycid beetle whose larvae feed in dead phloem of conifers, has males which have long been known to fight,[140] although of how and for what they fight there seems to be no detailed description. Even for this 'long horn' family (and compared to their own females), males of *A. aedilis* have extremely long antennae. Since it is very difficult to imagine they can be weapons, their most probable function is perhaps long-range detection-assessment of rivals and females. But parallel to our hypothesized function of eye-stalks in flies, these antennae could also be serving as advertisements of size and strength.

Such advertisement might be expected to accrete an element of bluff. Returning to beetle horns used as weapons, my own scanty observations tend to favour a largely honest communicative function towards other males, but as regards a residuum—a small margin of excess that may not correlate with real fighting ability—they tend to favour bluff rather than Darwin–Fisher exaggeration through female preference. The subject of animal 'deceit' is beginning to attract attention,[141] but as regards bluffed displays fixed by genotype, I know of no theory beyond brief comments by Fisher.[129] In three 'tournaments' that I have been able to arrange (two of *Lucanus cervus* and one of *Chiasognathus*), I found that the winner was the second largest male. In *Chiasognathus* the males fought vigorously although no female was present. In *Lucanus* the presence of a receptive female was necessary to stimulate more than brief posturing by the males (see also Mathieu[125]), but the female herself showed no interest and passively allowed mounting by any male. However, her co-operation—by slight raising of the elytra—may have been needed before intromission could take place. Sometimes a mounted male bit quite hard at a female's head so that an element of bullying to gain co-operation also seems possible.

In leaving little necessity to invoke restraint or ritual, the view which these observations (and points of theory) tend to support departs rather widely from a long-held general interpretation of animal fighting (e.g. Arrow,[96] Harrison Matthews,[142] and critique in Ghiselin[66]). Yet it accords well with various scattered eyewitness accounts and, as regards horned animals generally, is by no means original.[143]

Most combinations of the selection processes able to affect horns are not mutually exclusive. Female preference, however, will hardly work for horns if the interactions of males are effectively eliminating all basis for choice. But even in this aspect, if females are freely mobile, the fighting of males will affect choice only within one mating site: it is open to mobile females to go off and be available at another. For example, *Chiasognathus* females could go off to another sap flow, perhaps to that with the most 'splendid' male in charge. The changing distribution of females in *Pityogenes* single-male systems (Table 13.5) has given a hint of movements of this kind, although in that case the way the females were opting did not seem strongly selective and, correspondingly, the special structures of their males are modest. But, as also with *Chiasognathus* and all the other cases I have discussed, the work of apportioning functions to these structures—tools, weapons, or ornaments, lies or truth—is hardly begun.

CONCLUSIONS

The coincidence of winglessness, fighting, and dimorphism in male fig wasps is not accidental. Correlations of the same kind are widely present in other insects.

Aggregation of females is evidently a factor underlying all three phenomena, and it seems essential for the extremes of male fighting.

The evolution of *fighting* requires the distribution of a resource in masses compact enough to be guarded and valuable enough to be worth risks to obtain—as reckoned by inclusive fitness. In fig wasps, moderate aggregation of colonizing mothers often leads to this situation through the intense aggregation of their daughters and the formation of groups of males that are not too closely related.

To judge from cases surveyed, structures that look like weapons in males often are weapons. Theory remains open, however, to the existence of positively allometric structures which, while not used as weapons, serve as valid (or slightly bluffed) signals of fighting strength for direction at weaker (and slightly stronger) males. Contrary to Darwin, appreciation by the females appears less important than such male–male display and fighting for most weapon-like and size-related characters.

Male fighting correlates with size both within and between species: small size and pugnacity do not easily coevolve. Enlargement of the head and of mandibles are particularly indicative of fighting; but alternative massive enlargements elsewhere (e.g. particular legs, thoracic horns) are frequent.

Provided (1) that males are with females initially and (2) that the females can carry stored sperm, *male winglessness* tends to evolve in situations similar to those favourable to male fighting (as in parasite fig wasps). But situations with relatedness too high for fighting can still easily be favourable to winglessness (as with pollinator fig wasps). The economy of winglessness provides for virility and sometimes for the adaptations for fighting. Wings have no value inside figs and are probably an encumbrance there, as suggested by their convergent tendency to disappear in such similarly claustral habitats as soil and rotting wood. Outside of figs, and other claustral habitats wings have value at least sometimes. Total male aptery is expected only if the average value of having wings is very low, which tends to be the case if average fig populations are high. If populations are low, wholly winged males are expected; and, given that a substantial proportion of females are maturing in male-less figs, intermediate average densities in figs are expected to lead to *male dimorphism*. These expectations are reasonably fulfilled. It is not clear yet whether the kind of frequency dependence here postulated is especially likely to maintain both morphs over long periods and varying abundance of the species. Disruptive selection that is either long continued or very intense seems needed to explain the wide divergence of male morphs in some fig wasps.

Acknowledgements

My visit to Brazil was arranged by the Royal Society of London and financed by awards from the Leverhulme Trust and the Conselho Nacional de Pesquisas of Brazil. I thank R. Zucchi and L. A. de O. Campos for showing me the fig wasps, and R. Zucchi, F. A. M. Duarte, I. Ferrari, L. S. Goncalves, and W. E. Kerr for the facilities kindly provided at various stages during my stay at the two Faculdades of Ribeirão Prêto. Opportunity to observe *Chiasognathus* arose during an expedition assisted by the Royal Society, the Forestry Commission, the Manchester University Museum, and the Corporacion Nacional Forestal of Chile. Wing-polymorphic insects of rotting wood in Britain were studied with assistance of a grant (GR3/1383) from the Natural Environment Research Council. Work of J. Auckland, R. Wilkinson, and J. Fethney contributed much to my knowledge of these insects. R. Edwards and J. Newton helped me to obtain colonies of *Hypoponera punctatissima*. For various facts, ideas, and bibliographic items, I also thank R. W. Barth, A. Camonsseign, W. G. Eberhard, G. Gordh, R. Kowalski, P. Levin, D. K. McAlpine, D. Otte, T. J. Palmer, V. A. Taylor, and J. T. Wiebes.

ADDENDUM

After this manuscript had been completed, I found that the existence of 'trimorphisms' (i.e. dimorphisms of males) in fig wasps of southern Brazil had been previously noted by Fritz Müller in 1886.[144] Müller found the phenomenon, then known through only one previous example in the Old World, in three or perhaps four of the fig wasp genera of his area and pointed out that correspondingly several genera previously erected to contain the wingless types must be dropped. Some of my species are almost certainly congeneric with his. He seems to have said little of the wingless males' known or possible biology beyond remarking that it was unreasonable to suppose that wingless males would consistently be collected from figs where their females—treated till then as wholly unknown—did not occur. These observations further attest to the breadth of interest of a great naturalist. Müller, an esteemed correspondent of Darwin, is perhaps best known for his concept of 'Müllerian mimicry.'

References and notes

1. G. Grandi, The hymenopterous insects of the super family Chalcidoidea developing within the receptacles of figs: their life-history, symbioses and morphological adaptations, *Bollettino dell'Istituto Entomologia Università degli Studi, Bologna* **26**, 1–13 (1961).
2. E. J. H. Corner, An introduction to the distribution of *Ficus*, *Reinwardtia* **4**, 15–45 (1958).
3. G. Grandi, Monografia del generi *Philotrypesis* Forst., *Bollettino del Laboratorio di Entomologia Istituto Superiore Agrario, Bologna* **3**, 1–181 (1930).
4. For example, J. T. Wiebes, Bornean fig wasps from *Ficus stupenda* Micquel (Hymenoptera, Chalcidoidea), *Tijdschrift voor Entomologie* **109**, 163–92 (1966).
5. O. W. Richards, An introduction to the study of polymorphism in insects, in J. S. Kennedy (ed.), *Insect Polymorphism*, Symposia of the Royal Entomological Society No. 1, 298–364 (1961).
6. C. T. Brues, Some hymenopterous parasites of lignicolous Itonididae, *Proceedings of the American Academy of Arts and Science* **57**, 263–88 (1922).
7. J. T. Wiebes, Fig wasps from Israeli *Ficus sycomorus* and related East African species (Hymenoptera, Chalcidoidea). 2. Agaonidae (concluded) and Sycophagini, *Zoologische Mededelingen, Leiden* **42**, 307–20 (1968).
8. D. S. Hill, *Figs (*Ficus *spp.) of Hong Kong* (Hong Kong University Press, Hong Kong, 1967).
9. C. F. Baker, A study of caprification in *Ficus nota*, *Philippine Journal of Science* D **8**, 63–83 (1913).
10. G. Grandi, The problems of 'morphological adaptation' in insects, *Smithsonian Miscellaneous Collections* **137**, 203–30 (1959).

11. K. J. Joseph, Recherches sur les chalcidiens *Blastophaga psenes* (L.) et *Philotrypesis caricae* (L.) du figuier *Ficus carica* (L.), *Annales des Sciences Naturelles* **20**, 197–260 (1958).

12. W. G. Eberhard, The ecology and behaviour of a subsocial pentatomid bug and two scelionid wasps, *Smithsonian Contributions to Zoology* **205**, 1–39 (1975).

13. G. Gordh, The comparative external morphology and systematics of the neotropical parasitic fig wasp genus *Idarnes* (Hymenoptera: Torymidae), *University of Kansas Science Bulletin* **50**, 389–455 (1975).

14. J. T. Wiebes, Redescription of Sycophaginae from Ceylon and India, with a description of betotypes, and a world catalogue of Otitesellini (Hymenoptera, Chalcidoidea, Torymidae), *Tijdschrift voor Entomologie* **110**, 399–442 (1967).

15. F. B. Browne, On the life-history of *Melittobia acasta*, Walker; a chalcid parasite of bees and wasps, *Parasitology* **14**, 349–70 (1922).

16. J. T. Wiebes, The genus *Alfonsiella* Waterston (Hymenoptera, Chalcidoidea, Agaonidae), *Zoologisches Mededeelingen* **47**, 321–30 (1971).

17. D. L. Hartl, Some aspects of natural selection in arrhenotokous populations, *American Zoologist* **11**, 309–25 (1971).

18. B. W. Ramirez, Taxonomic and biological studies of neotropical fig wasps (Hymenoptera: Agaonidae), *University of Kansas Science Bulletin* **49**, 1–44 (1970).

19. Female aggression is itself a problem for kinship theory. Taking other cases, there may be a negative correlation with fighting in the males. Thus ovipositing females of *Melittobia* and of two scelionids (see ref. 12) are unaggressive but produce males that fight.

20. J. Maynard Smith and G. R. Price, The logic of animal conflict, *Nature* **246**, 15–18 (1973).

21. K. J. Joseph, A proposed revision of the classification of the fig insects of the families Agaonidae and Torymidae (Hymenoptera), *Proceedings of the Royal Entomological Society* B **33**, 63–6 (1964).

22. M. Gadgil, Male dimorphism as a consequence of sexual selection, *American Naturalist* **106**, 574–80 (1972).

23. G. Grandi, Studio morfologico e biologico della *Blastophaga psenes* (L.), *Bolletino del Laboratorio di Entomologia Istituto Superiore Agraria, Bologna* **2**, 1–147 (1929).

24. J. Galil and D. Eisikovitch, On the pollination biology of *Ficus sycamorus* in East Africa, *Ecology* **49**, 259–69 (1968).

25. H. E. Evans, A new species of *Cephalonomia* exhibiting an unusually complex polymorphism (Hymenoptera, Bethylidae), *Psyche* **70**, 151–63 (1963).

26. G. Becker and W. Weber, *Theocolax formiciformis* Westwood (Hymenoptera: Chalcididae) ein Anobien-parasit, *Zeitschrift für Parasitologie* **15**, 339–56 (1952).

27. J. M. Taylor, Studies on *Theocolax formiciformis* Westw. (Hymenoptera, Pteromalidae), a parasite of *Anobium punctatum* (Deg.) (Coleoptera, Anobiidae), *Bulletin of Entomological Research* **54**, 797–803 (1964).

28. A. da Costa Lima, Quarta contribuicão ao conhecimento da biologia do *Telenomus polymorphus* n.sp. (Hymenoptera: Scelionidae), *Anais da Academia Brasiliera de Ciências* **15**, 211–27 (1944).

29. T. F. Houston, Discovery of an apparent male soldier caste in a nest of a halictine bee (Hymenoptera: Halictidae) with notes on the nest, *Australian Journal of Zoology* **18**, 345–51 (1970).

30. E. A. Cross and J. C. Moser, Taxonomy and biology of some Pyemotidae (Acarina: Tarsonemoidea) inhabiting bark beetle galleries in North American conifers, *Acarologia* **13**, 47–64 (1971).

31. E. A. Cross and J. C. Moser, A new dimorphic species of *Pyemotes* and a key to previously described forms (Acarina: Tarsonemoidea), *Annals of the Entomological Society of America* **68**, 723–32 (1975).

32. B. Lee, Cannibalism and predation by adult males of the two spotted mite *Tetranychus urticae* (Koch) (Acarina: Tetranychidae), *Journal of the Australian Entomological Society* **8**, 210 (1969).

33. D. A. Potter, D. L. Wrensch, and D. E. Johnston, Aggression and mating success in male spider mites, *Science* **193**, 160–1 (1976).

34. J. P. Woodring, Environmental regulation of andropolymorphism in tyroglyphids (Acari), *Proceedings of the International Congress of Acarology* **2**, 433–40 (1969).

35. R. K. Colwell, Competition and coexistence in a simple tropical community, *American Naturalist* **107**, 737–60 (1973).

36. A. Foa, Studio sul polimorfismo unisessuale del *Rhizoglyphus echinopus* (Fum. and Rob.) Murray, corredato da osservazioni biologiche, anatomiche e citologiche, *Memorie Accademia Pontificia dei Nuovi Lincei, Roma, Ser.* V **12**, 3–109 (1919).

37. C. W. Kearns, Method of wing inheritance in *Cephalonomia gallicola*, Ashmead (Bethylidae), *Annals of the Entomological Society of America* **27**, 533–41 (1934).

38. J. H. Oliver, Parthenogenesis of mites and ticks (Arachnida: Acari), *Annual Review of Entomology* **22**, 407–29 (1977).

39. T. N. Ananthakrishnan, *Indian Thysanoptera*, CSIR Zoological Monographs, Vol. 1 (New Delhi, 1969).

40. T. N. Ananthakrishnan, Patterns of structural diversity in the males of some phlocophilous Tubulifera (Thysanoptera), *Annales de la Société Entomologique de France NS*, **4**, 413–18 (1968).

41. R. A. Crowson, Observations on Histeroidea with descriptions of an apteous larviform male and of the internal anatomy of *Sphaerites, Journal of Entomology* B, **42**, 133–40 (1974).

42. F. G. Browne, Notes on two Malayan scolytid bark beetles, *Malayan Forester* **22**, 292–300 (1959).

43. F. G. Browne, The biology of Malayan Scolytidae and Platypodidae, *Malayan Forest Records* **22**, 1–255 (1961).

43a. G. Grandi, Gli insette a regime speciallizzato ed i loro 'adattimenti morfologici', *Atti dell'Accademia Nazionale dei Lincei, Memorie. Clessa di Scienze Fisiche e Matematiche,* Series 8, **5**, Sezione 3a, 1–48 (1955), Plate X.

44. A. D. Hopkins, On the life history and habits of the 'wood-engraver' ambrosia beetle—*Xyleborus xylographus* (Say.), *saxeseni* (Ratz.), *Canadian Entomologist* **30**, 21–9 (1898).

45. R. M. Matthews, Courtship in parasitic wasps, in P. W. Price (ed.), *Evolutionary Strategies of Parasitic Insects and Wasps*, pp. 66–86 (Plenum Press, New York, 1975).

46. H. G. Hubbard, The ambrosia beetles of the United States, *US Department of Agriculture, Division of Entomology Bulletin NS* **7**, 9–30 (1897).

47. W. W. Kempf, Miscellaneous studies on Neotropical ants, II. *Studia Entomologia* **5**, 1–38 (1962).

48. W. M. Wheeler, *Mosaics and other Anomalies among Ants* (Harvard University Press, Cambridge, MA, 1937).

49. M. R. Smith, Ants of the genus *Cardiocondyla* Emery in the United States, *Proceedings of the Entomological Society of Washington* **46**, 30–41 (1944).

50. A. Buschinger, Polymorphismus und Polyethismus Sozialparasitischer Hymenopteren, in G. H. Schmidt (ed.), *Sozialpolymorphismus bei Insekten*, pp. 604–23 (Wissenschaftliche Verlagsgesellschaft MBH, Stuttgart, 1974).

51. M. J. West Eberhard, The evolution of social behaviour by kin selection, *Quarterly Review of Biology* **50**, 1–34 (1975).

52. E. Dahms, The courtship behaviour of *Melittobia australica* Girault 1912 (Hymenoptera: Eulophidae), *Memoirs of the Queensland Museum* **16**, 411–14 (1973).

53. A. M. Alston, The life-histories and habits of two parasites of blowflies, *Proceedings of the Zoological Society of London*, 195–243 (1920).

54. A. M. Hughes, *The Mites Associated with Stored Food Products* (His Majesty's Stationery office, London, 1948).

55. J. Théodorides, Contribution à l'étude des parasites et phoretiques de coléoptères terrestres, *Actualités Scientifique Industrielles* No. 1217, 1–310 (1955).

56. J. G. Rozen, personal communication.

57. I. H. Yarrow and W. M. Guichard, Some rare Hymenoptera Aculeata, with two species new to Britain, *Entomologist's Monthly* **77**, 2–13 (1941).

58. S. K. Sakagami, Sozialstruktur und Polymorphismus bei Furchen- oder Schmalbienen (Halictinae), in G. H. Schmidt (ed.), *Sozialpolymorphismus bei Insekten*, pp. 257–93 (Wissenschaftliche Verlagsgesellschaft MBH, Stuttgart, 1974).

59. J. E. Alcock, E. Jones, and S. L. Buchmann, Male mating strategies in the bee *Centris pallida* Fox (Anthophoridae: Hymenoptera), *American Naturalist* **111**, 145–55 (1977).

60. J. S. Moure, Sôbre a origim do Meloponinae parasitas (Hymenoptera: Apoidea), *Ciência e Cultura, São Paulo* **15**, 183–4 (1963).

61. M. J. Orlove, A contribution to the biology of the great carpenter bee *Xylocopa virginica*: geographical variation in courtship, mimicry and rearing methods (Thesis (part) for the Diploma of Imperial College London, 1973).

62. E. B. Poulton, Mr Lambourn's observations on marriage by capture in a West African wasp, *Report of the British Association for the Advancement of Science* **1913**, 511 (1913).

63. See photograph in W. M. Wheeler, *The Social Insects* (Kegan Paul, London, 1928).

64. C. D. Michener, Comparative social behavior of bees, *Annual Review of Entomology* **14**, 299–342 (1969).

65. F. J. Radovsky, Fixed parasitism in the Siphonaptera, *Journal of Medical Entomology* **9**, 487–94 (1972).

66. M. T. Ghiselin, *The Economy of Nature and the Evolution of Sex* (University of California Press, Berkeley, 1974).

67. G. Salt, Trimorphism in the ichneumonid parasite *Gelis corruptor, Quarterly Journal of Microscopical Science* **93**, 453–74 (1952).

68. H. Kühne and G. Becker, Zur Biologie und Ökologie von *Scleroderma domesticum* Latreille (Bethylidae, Hymenoptera), einem Parasiten holzzerstorender Insektenlarven, *Zeitschrift für Angewandte Entomologie* **76**, 278–303 (1974).

69. L. A. Mound, Gall-forming thrips and allied species (Thysanoptera: Phlaeothripinae) from *Acacia* trees in Australia, *Bulletin of the British Museum (Natural History) B (Entomology)* **25**, 387–466 (1971).

70. E. Hagerup and O. Hagerup, Thrips pollination of *Erica tetralix, New Phytologist* **52**, 1–7 (1953).

71. G. D. Morison, Observations and records for British Thysanoptera. XV. Thripidae, *Thrips discolor* (Haliday), *The Entomologist* **104**, 276–81 (1971).

72. G. D. Morison, Observations and records for British Thysanoptera. XVII. Thripidae, *Baliothrips dispar* (Haliday), *The Entomologist* **106**, 157–64 (1973).

73. G. C. Champion, Notes on the sexual characters of *Longitarsus agilis* Rye, *Entomologist's Monthly Magazine* **46**, 261 (1910).

74. L. G. Gentner, The systematic status of the mint flea beetle (Chrysomelidae, Coleoptera), with additional notes, *Canadian Entomologist* **60**, 264–6 (1928).

75. T. R. New, South American species of *Nepionorpha* Pearman and *Notiopsocus* Banks (Psocoptera), *The Entomologist* **106**, 121–32 (1973); The Archipsocidae of South America (Psocoptera), *Transactions of the Royal Entomological Society of London* **125**, 57–105 (1973).

76. V. A. Taylor, The biology of feather-winged beetles of the genus *Ptinella* with particular reference to coexistence and parthenogenesis (PhD thesis, University of London, 1975).

77. E. Broadhead, The biology of *Psoquilla marginepunctata* (Hagen) (Corodentia, Trogiidae), *Transactions of the Society of British Entomology* **14**, 223–36 (1961).

78. G. T. Riegel, The distribution of *Zorotypus hubbardi, Annals of the Entomological Society of America* **56**, 744–7 (1963).

79. J. V. Pearman, Third note on Psocoptera from warehouses, *Entomologists' Monthly Magazine* **78**, 289–92 (1942).

80. W. Mohrig and B. M. Mamaev, Neue flügelreduzierte Dipteren der Familien Sciaridae und Cecidomyiidae, *Deutsche Entomologische Zeitschrift* **7**, 315–36 (1970).

81. E. Broadhead, The life-history of *Embidopsocus enderleini* (Ribaga) (Corrodenta, Liposcelidae), *Entomologist's Monthly Magazine* **83**, 200–3 (1947).

82. T. R. E. Southwood, A hormonal theory of the mechanism of wing polymorphism in Heteroptera, *Proceedings of the Royal Entomological Society* A, **36**, 49–88 (1961).
83. J. Carayon, Insemination traumatique hétérosexuelle et homosexuelle chez *Xylocoris maculipennis* (Hemiptera: Anthocoridae), *Comptes Rendus de l'Académie des Sciences, Paris* **278**, 2803–6 (1974).
84. G. D. Morison, Notes on the broad-horned flour beetle *Gnathocerus cornutus* Fab., *Proceedings of the Royal Physical Society of Edinburgh* **21**, 14–18 (1925).
85. A. Sokoloff, *The Biology of Tribolium with Special Emphasis on Genetic Aspects*, Vol. 2 (Clarendon Press, Oxford, 1972).
86. M. J. D. Brendell, *Coleoptera; Tenebrionidae*, Handbooks for the Identification of British Insects, Vol. V, Part 10 (Royal Entomological Society of London, 1975).
87. C. T. David, personal communication.
88. For example, C. Coquerel, Observations entomologiques sur divers insectes recueillis à Madagascar (2° partie), *Annales de la Société Entomologique de France*, 275–84 (1848).
89. G. H. Hudson, On some examples of New Zealand insects illustrating the Darwinian principle of sexual selection, *Transactions of the New Zealand Institute* **52**, 431–8 (1920).
90. L. Chopard, Sur l'anatomie et le développement d'une blatte vivipare, *Proceedings of the 8th International Congress of Entomology, Stockholm*, 218–22 (1950).
91. H. Ritter, Jr, Defence of mate and mating chamber in a wood roach, *Science* **143**, 1459–60 (1964).
92. A. N. Lea, Notes on the Lord Howe Island phasma and an associated longicorn beetle, *Transactions and Proceedings of the Royal Society of South Australia* **40**, 145–7 (1916).
93. A. B. Gurney, Notes on some remarkable Australasian walking sticks, including synopsis of the genus *Extatosoma* (Orthoptera: Phasmatidae), *Annals of the Entomological Society of America* **40**, 373–96 (1947).
94. W. S. Bristowe, Notes on the habits of insects and spiders in Brazil, *Transactions of the Entomological Society of London* **1924**, 475–504 (1924).
95. E. H. Gibson and A. Holdridge, Notes on the North and Central American species of *Acanthocephala* Lap., *Canadian Entomologist* **50**, 237–40 (1918).
96. G. J. Arrow, *Horned Beetles* (W. Junk, The Hague, 1951).
97. R. D. Alexander and D. Otte, The evolution of genitalia and mating behaviour in crickets (Gryllidae) and other Orthoptera, *Miscellaneous Publications of the Museum of Zoology, University of Michigan* **133**, 1–62 (1967).
98. R. L. Caldwell and H. Dingle, Stomatopods, *Scientific American* **234** (1), 80–9 (1976).
99. J. L. Brown and G. H. Orians, Spacing patterns in mobile animals, *Annual Review of Ecology and Systematics* **1**, 239–62 (1970).
100. S. M. Evans, A study of fighting reactions of some nereid polychaetes, *Animal Behaviour* **21**, 138–46 (1973).

101. R. D. Alexander, The evolution of social behaviour, *Annual Review of Ecology and Systematics* **5**, 325–83 (1974).

102. E. G. Matthews, Observations on the ball-rolling behavior of *Canthon pilularis* (L.), *Psyche* **70**, 75–93 (1963).

103. W. E. Marshall, Optimization of parental investment: pre-oviposition care of brood balls in the dung beetle *Canthon imitator* (MA thesis, University of Texas, Austin, 1976).

104. E. O. Wilson, *The Insect Societies* (Belknap/Harvard University Press, Cambridge, MA, 1971).

105. E. Pukowski, Ökologische Untersuchungen an *Necrophorus* F., *Zeitschrift für Morphologie und Okologie der Tiere* **27**, 518–86 (1933).

106. L. J. Milne and M. Milne, The social behavior of burying beetles, *Scientific American* **235** (2), 84–9 (1976).

107. J. Alcock, Territorial behaviour by males of *Philanthus multimaculatus* (Hymenoptera: Sphecidae) with a review of territoriality in male sphecids, *Animal Behaviour* **23**, 889–95 (1975).

108. H. T. Spieth, Mating behaviour and evolution of the Hawaiian *Drosophila*, in M. J. D. White (ed.), *Genetic Mechanisms of Speciation in Insects*, pp. 94–101 (Reidel, Dordrecht, 1974).

109. W. G. Eberhard 1975 (ref. 12) and personal communication.

110. W. G. Eberhard, The function of horns in *Podischnus agenor* (Dynastinae), in M. S. Blum and N. A. Blum (eds), *Reproductive Competition, Mate Choice and Sexual Selection in Insects*, pp. 231–58 (Academic Press, New York and London, 1979).

111. O. W. Richards, Sexual selection and allied problems in the insects, *Biological Reviews* **2**, 298–364 (1927).

112. P. Lesne, Revision des coléoptères de la famille des Bostrychides. 3e Mémoire. Bostrychinae, *Annales de la Société Entomologique de France* **67**, 438–621 (1898).

113. H. Kühne, Über Beziehungen zwischen *Teredo, Limnoria* und *Chelura*, in G. Becker and W. Liese (eds), *Holz und Organismen*, pp. 447–56 (Duncker und Humblot, Berlin, 1965).

114. See Plate VIII in J. H. Barrett and C. M. Yonge, *Collins Pocket Guide to the Sea Shore* (Collins, London, 1958).

115. G. J. Arrow, Polymorphism in giant beetles, *Proceedings of the Zoological Society of London* **113**, 113–16 (1944). The contrasting morphs in *Enema pan* are well shown in a photograph in A. Smith, *Mato Grosso* (Michael Joseph, 1971).

116. J. H. McGhehey, Territorial behaviour of bark-beetle males, *Canadian Entomologist* **100**, 1153 (1968).

117. G. Halffter and E. G. Matthews, The natural history of dung beetles of the subfamily Scarabaeinae (Coleoptera: Scarabaeidae), *Folia Entomologica Mexicana* **12–14**, 1–312 (1966).

118. N. W. Blackman, Observations on the life history and habits of *Pityogenes hopkinsi* Swaine, *New York College of Forestry Technical Publication* No. 2, 11–66 (1915).

119. P. J. M. Greenslade, Distribution patterns of *Priochirus* species (Coleoptera: Staphylinidae) in the Solomon Islands, *Evolution* **26**, 130–42 (1972); Evolution in the staphylinid genus *Priochirus* (Coleoptera), ibid., 203–20.
120. R. W. G. Hingston, *The Meaning of Animal Colour and Adornment* (Arnold, London, 1933).
121. A. Lameere, L'évolution des ornaments sexuels, *Bulletin de Classe Scientifique de Académie Royale de Belgique*, 1327–64 (1904).
122. J. S. Huxley, Darwin's theory of sexual selection and the data subsumed by it, in the light of recent research, *American Naturalist* **72**, 416–37 (1938).
123. F. Claude-Joseph, El *Chiasognathus grantii* Steph., *Revista Universitaria, Santiago* **13**, 529–35 (1928).
124. R. E. Ureta, Sobre algunas costumbres del *Chiasognathus grantii*, Steph., *Revista Chilena de Historia Natural* **38**, 287–92 (1934).
125. J. Mathieu, Mating behaviour of five species of Lucanidae (Coleoptera: Insecta), *Canadian Entomologist* **101**, 1054–62 (1969).
126. M. J. Meads, Some observations on *Lasiorhynchus barbicornis* (Brenthidae: Coleoptera), *New Zealand Entomologist* **6**, 171–6 (1976).
127. W. Beebe, The function of secondary sexual characters in two species of Dynastidae (Coleoptera), *Zoologica* **29**, 53–8 (1944).
128. W. Beebe, Notes on the hercules beetle, *Dynastes hercules* (Linn.) at Rancho Grande, Venezuela, with special reference to combat behavior, *Zoologica* **32**, 109–16 (1947).
129. R. A. Fisher, *The Genetical Theory of Natural Selection* (Clarendon Press, Oxford, 1930).
130. P. O'Donald, Sexual selection and territorial behaviour, *Heredity* **18**, 361–4 (1963).
131. P. O'Donald, A general model of sexual and natural selection. *Heredity* **22**, 499–518 (1967).
132. P. O'Donald, Models of sexual and natural selection in polygamous species, *Heredity* **31**, 145–56 (1973).
133. R. D. Alexander, Aggressiveness, territoriality, and sexual behaviour in field crickets (Orthoptera: Gryllidae), *Behaviour* **17**, 130–223 (1961).
134. G. Borgia, Sexual selection and the evolution of mating systems, in M. S. Blum and N. A. Blum (eds), *Reproductive Competition, Mate Choice and Sexual Selection in Insects*, pp. 19–80 (Academic Press, New York and London, 1979).
135. D. K. McAlpine, Combat between males of *Pogornortalis doclea* (Diptera, Platystomatidae) and its relation to structural modification, *Australian Entomological Magazine* **2**, 104–7 (1975).
136. D. K. McAlpine, Agonistic behavior in *Achias australis* (Diptera, Platystomatidae) and the significance of eyestalks, in M. S. Blum and N. A. Blum, *Reproductive Competition, Mate Choice and Sexual Selection in Insects*, pp. 221–30 (Academic Press, New York and London, 1979).
137. D. E. Hardy, *Insects of Hawaii*, Vol. 12 (University of Hawaii Press, Honolulu, 1965).

138. F. C. Val, Genetic analysis of the morphological differences between two interfertile species of Hawaiian *Drosophila*, *Evolution* **31**, 611–29 (1977).

139. A. R. Templeton, Analysis of head shape differences between two interfertile species of Hawaiian *Drosophila*, *Evolution* **31**, 630–41 (1977).

140. E. C. Rye, *British Beetles* (Reeve, London, 1866).

141. See, for example, B. Wallace, Misinformation, fitness, and selection, *American Naturalist* **107**, 1–7 (1972); D. Otte, On the role of intraspecific deception, *American Naturalist* **109**, 239–42 (1975).

142. L. Harrison Matthews, Overt fighting in mammals, in J. D. Carthy and F. J. Ebling (eds), *The Natural History of Aggression*, pp. 23–32 (Academic Press, London, 1964).

143. V. Geist, The evolution of horn-like organs, *Behaviour* **27**, 175–213 (1966); On fighting strategies in animal combat, *Nature* **250**, 354 (1974).

144. F. Müller, *Transactions of the Entomological Society of London*, x–xii (1886); Feigenwespen, *Kosmos, Leipzig*, **18**, 55–62; **19**, 54–6 (1886).

CHAPTER 14

Optimal indigestion?

A caterpillar of a 'winter moth', *Operophtera brumata*, rests during the tattering of an elm leaf.

C H A P T E R 14

ASTRINGENT LEAVES

Low Nutritive Quality as Defence Against Herbivores

Consideration of these egoistic drives brings us back to the hen-coops from which we started.

W. C. ALLEE[1]

THIS paper was the first to be generated entirely from within my new job in America. Why had I moved? Partly it was the Brazilian experience of teaching attentive and interested graduate students that set me thinking that there might be better places to be a lecturer than at Imperial College. More, however, it was the return from the same Brazilian trip to find myself still in the same teaching rank that I had held 12 years before when I joined the college, and with various later comers promoted past me, which indeed had happened for some time. I knew that my teaching was not considered a success, but my experience in Ribeirão Prêto had suggested there could be some place where it might go better. Some time in 1976 I circulated a letter to several people in the USA who on one occasion or another had invited me to give seminars, asking if they knew of any job I could be fitted to. To my pleasure several wrote indicating a possibility at their own institution. I almost immediately resigned at IC and set out to be interviewed at the three universities that seemed most enthusiastic and to offer the best prospects. At the University of Michigan, Richard Alexander, long a prime mover in the study of social evolution study in the US and also an early and generous supporter of my work, had reacted with characteristic energy on receiving my letter. He had persuaded his colleague Donald Tinkle, likewise an exceptional evolutionist and currently the Director of the University Museum, to consider that I might be a useful colleague in the museum and in the graduate teaching programme of the university. Hence, after a 6-month stay at Harvard as a visiting professor, I took a permanent position in Biological Sciences and the Museum at the University of Michigan. I occupied an office within the museum, an environment that I soon came to love.

I found myself in the unusual position of having status equivalent to a curator but not having any particular duty towards the collections. Perhaps, however, a 5-year editorship of the museum publications could be counted as a curatorial duty in terms of managing commas, italics, and silverpoint plate proofs of what became permanent matters of record, and of making sure that any newly described species did not lapse either as non-descripts or 'into synonymy' through such slips as any of our journals not going out on the right day. (Did you know that if you name a new species in a paper you mustn't mention the new name—in the abstract or introduction, for example—until immediately before the new formal description? If you do your name may easily become invalid. I am glad to say, however, that most work in the museum and most issues in the papers were over much more substantial matters than this, most of them without connection to any kind of naming; nevertheless, it mattered for the editor to get the procedures right, and for a lover of nature's vast diversity it was a pleasure to participate, even at such far fringes as these, in the primary descriptive and systematic activity of biology.) Sometimes I also told puzzled listeners at other universities that I curated the museum's mathematical models; no one else was modelling at the time when I arrived, so this was just a way of saying that I took care of my own.

I was required only to teach graduate students. My classes were usually held in the room that terminates a long, echoing museum corridor on the first floor of the old building, a corridor (or 'hall' by the American notation—which notation also, of course, had my first floor as the second) redolent of sweet museum alcohol with its touch of formalin and besides this (just here) the special scent added by long-pickled frogs, lizards, and snakes. Sometimes I taught in my home. As much as in Brazil, all this was a stunning change from IC. First, all my students were graduates (as also in my short period as a visiting professor at Harvard, in 1978). Second (and I suppose partly in consequence of their level), they were neither drifters nor narrowly profession-oriented. Instead I found them boldly and almost aggressively interested in whatever I could teach. The degree of interest in evolution may have been a special Michigan feature. As a result of, I think, the influence of Richard Alexander and later other faculty members he had encouraged to be there, the university had acquired a reputation throughout the whole USA as one of the very best campus destinations to which you might 'U-haul' your possessions, including your bachelor's or master's degree, but, most important of all, your research-oriented curiosity about fish, ferns, birds, lakes, lizards, insects—or simply about evolution *per se*.

On arrival in Ann Arbor such students formed for themselves, or were quickly guided to form, a coherent social group within our Division of Biological Sciences. The graduate group seemed particularly enthused and held together by a remarkable and dedicated elderly secretary, Ursula Freimarck. This woman not only helped each new student to know what to do but organized periodic social gatherings and discussions for them. Graduate students were also expected to help entertain visiting speakers and, even more, the candidates coming for interview for teaching positions. Their opinions about the latter were carefully sought and heeded in the making of appointments. So was generated a surprising self-confidence and group identity. I think the graduate students felt themselves a kind of academy within the university, a young blood already perfusing an old body and keeping it active. Certainly the biology graduate students gained a nationwide reputation, if for nothing else, as being the most argumentative in the land. Was this all, perhaps, a meme complex emanating from one source—my patron Alexander? In any case, such young men and women came to the classes expecting the best in teaching, also expecting to be heard when they disagreed. They could be fierce if they thought instruction was substandard or ideas were being skated by them without proper explanation or support. One slightly negative side to the independence and unity that was encouraged was that sectors of students came to foster strong political commitments. These may have been worthy in themselves but on occasions interfered seriously with the transfer of knowledge. Many students, however, the best as I thought, stood apart from all the pressure groups and yet still participated in the positive spirit of the school.

Compared to the IC undergraduates, who seemed to me to have little that was intermediate between silent shyness and insolence, Michigan graduates were always a challenge but also a great pleasure. If it had not been for the first storm signs on the horizon of the early 1980s of political correctness appearing to replace Marxism as the top university social issue (and also as an escape route, I have always thought, for the less creative minds), combined with an exceptional offer of a Royal Society Professorship for a rather unsociable and research-oriented scientist like myself, I suspect I would be contentedly teaching those graduates still.

Nancy Moran, my co-author on this paper, was a student in the first graduate class I taught at Michigan, or perhaps it was the second. While attending a graduate seminar by another professor, Michael Martin, which Nancy took for credits towards her doctorate and which I attended by invitation and out of interest, we came to notice that we both had the same objection to a plausible and accepted line of reasoning about why tannins might be adaptively present in leaves. Nancy and I exchanged brief formal models noting our criticisms in more detail. We worked the best from each into the single model of the paper that follows. It is no shattering new discovery but I like to think it may still provide a useful warning about claimed adaptations and it is also a good example of the spirit of the Michigan Graduate School I have outlined above.

It is so easy to accept that something might be a 'good thing' if the whole population or species somehow has it: but how does the 'good thing' come to be there? Contrary even to some able theorists whom one might expect to see the trouble better,[2] it is no answer to suggest that populations or species of trees with the character would expand faster than ones without it. The more proper question to ask is whether mutations causing the character will be selected into the population against competition from genes determining the existing condition. The case of the tannins, which certainly aren't good for caterpillars but might not be good for the leaves either if they oblige caterpillars to bite more holes during a lengthened development, had a teasing, uncertain status that it seemed one couldn't decide without some careful specification of assumptions and a little analysis.

Always independent in her ideas and taking nothing on authority, Nancy Moran, now on the faculty of the University of Arizona, was just such a student as made teaching at Michigan a pleasant change. Long after our association at Michigan we continued to discuss and collaborate on more than would appear from this one joint publication.

References and notes

1. W. C. Allee, Where angels fear to tread: a contribution from general sociology to human ethics, *Science* **97**, 517–25 (1943).
2. A. J. Lotka, Population analysis as a chapter in the methematical theory of evolution; in W. E. Le Gros Clark and P. B. Medawar (eds), *Essays on Growth and Form, a Testimonial Volume presented to D'Arcy Wentworth Thompson*, pp. 355–84 (Clarendon, Oxford, 1945).

LOW NUTRITIVE QUALITY AS DEFENCE AGAINST HERBIVORES†

NANCY MORAN and W. D. HAMILTON

Contrary to a widespread assumption in the literature on plant–herbivore interactions, individual plants do not necessarily benefit by possessing traits which lower herbivore fitness. In particular, genes conferring lowered nutritive quality could even increase herbivore damage under certain circumstances. Three special sets of conditions are outlined in which low nutritive quality would lower herbivore-induced damage to an individual plant. These sets are far from exhaustive. It is concluded that the adaptiveness of lowered nutritive quality in herbivore defence is widely possible but in no case demonstrated

Discussions of plant–herbivore interactions often assume that plant traits are favoured by natural selection to the extent that they lower herbivore fitness.[1] Thus, Feeny[2] claims that:

> plants . . . may use for protection substances of relatively subtle effect; these may not even be immediately toxic to an invading insect individual but nevertheless can reduce its fitness. This serves the function both of minimizing a population buildup by an attacker and of selecting against those invaders which attempt to colonize the plant. Thus a plant chemical which does not prove immediately toxic ... may nevertheless represent a significant defense against insects.

Feeny suggests that poor nutritional quality of foliage (resulting from low nutrient concentrations or from presence of compounds such as tannins which lower nutrient digestibility) is a widespread anti-herbivore adaptation in forest trees. Clearly, such a plant trait will tend to decrease the fitness of insect herbivores by lowering growth rate—and thus survivorship—of larvae or by lowering adult reproductive output. However, adaptiveness of the trait does not follow automatically from lowered herbivore fitness, at least not when

†*Journal of Theoretical Biology* **86**, 247–54 (1980).

selection is supposed to be effective primarily at and below the level of the individual. Some plant characteristics which lower herbivore fitness could have the effect of lowering fitness of the individual plant as well. Consider the example of low nutritive quality of forest tree foliage, and assume that larvae remain upon the same tree until they reach a certain size, whereupon they pupate, emerge as adults, and fly off. A larva on an 'adapted' plant may consume many more leaves than would be the case if nutritive quality were high. If the interests of the plant are best served by losing as few leaves as possible and if nutrition-rich leaves are not much different in cost to the plant, then one might expect the immediate selection to be towards foliage of high nutritive quality.

There are, however, circumstances where poor nutritive quality of foliage could evolve as an adaptation to insect herbivory. Three possible pathways are presented below. Each requires restrictive assumptions concerning properties of the herbivore and, as will be seen, cannot cover all herbivore–plant interactions.

1. If herbivores are able to detect differences in the nutritive quality of individual plants, and if they preferentially feed upon more nutritious host individuals, then low nutritive quality of leaves will be advantageous. Selectivity may be exercised by the stage which produces the feeding damage, provided that the feeding stage has the necessary mobility and sensory capacity. Even in the case of leaf-feeding larvae lacking these requirements, the argument will work if females selectively oviposit on plants of higher nutritive quality. Ovipositing female Lepidoptera often appear choosy about the individual plants and the parts of them where they place eggs but beyond attributes confirming host species little seems known of what they are able to be choosy about.[3] On current information this first model seems most likely to be valid for vertebrate herbivores.
2. If successive herbivore generations tend to feed upon the same host individual then low nutritive quality will act to prevent future buildup of herbivore numbers, thus increasing the plant's fitness. To the extent that neighbouring plants are relatives, the requirement for low between-generation mobility in the herbivore is eased. Insects with wingless females and very limited larval dispersal, such as coccids, may be sufficiently sedentary for this model to apply even where neighbouring plants are unrelated. It may apply more generally to plants which clone extensively: even with some movement between plants, successive herbivore generations are likely to feed upon the same genetic individual.

   This explanation, assuming very non-mobile herbivores, is at an opposite extreme from the first. However, for many important herbivores both models seem too extreme. Many insects both fly freely as adults and do not apparently sample food quality on behalf of their prospective offspring. Even for the relatively immobile coccids, winter moths, psychids, and others—albeit these are often important antagonists of plants—it is far from certain that rates of immigration are low enough for this explanation to work.

3. Consider again the larva which feeds upon an individual host plant from hatching until it reaches a given pupation size. Assume that the amount of leaf eaten per day is proportional to larval size. If leaves are low in nutritive quality, both more leaves and more time are required to reach pupation. The bulk of the feeding damage occurs during the latter part of the larva's growth, when its greatly increased size enables it to ingest leaves faster. On plants of low nutritional quality, early larval stages are prolonged, increasing the likelihood of mortality—by predation or some other factor—before reaching the most voracious stages. As a result, plants of lower nutritive quality might lose fewer leaves overall to herbivory. But late larval stages are prolonged too, so that those which do survive are eating more. Some simple formal model is evidently needed to clarify the conditions where low nutritive quality can evolve through this kind of pathway.

The approach based on (3) will be to construct a function which describes the expected amount of herbivore damage incurred by a plant from the feeding of a cohort of larvae. We assume that a larva's rate of consumption is proportional to its size so that the following integral should be proportional to the total damage to the plant:

$$D = \int_0^T l_t S_t \, \mathrm{d}t,$$

where l_t is survivorship, s_t is size, and T is the time at which the larva reaches pupation size, S. Obviously, T is dependent on the growth function, s_t, for which we will take the simple and fairly realistic assumption of exponential growth throughout the larval period:

$$s_t = \mathrm{e}^{gt}, \quad g \geq 0.$$

For simplicity of analysis, the unit of size is taken to be size at hatching, and g, growth rate, incorporates the variable we wish to study, the effect of nutritive quality. Note that when g is increased, nutritive quality is higher, and $g = 0$ means nutritive quality is so low that growth is prevented completely.

By definition of pupation size, $S = \mathrm{e}^{gT}$, so that $T = (\ln S)/g$.

So now, changing focus to damage as a function of g, we have

$$D(g) = \int_0^{(\ln S)/g} l_t \mathrm{e}^{gt} \, \mathrm{d}t.$$

The form of l_t is the crucial factor determining whether an impediment to nutrition can benefit the plant. We will first take the case of constant mortality and later consider deviations from this pattern.

Let there be a constant mortality μ, so that

$$l_t = e^{-\mu t}.$$

Hence

$$D(g) = \int_0^{(\ln S)/g} e^{(g-\mu)t}\,dt$$

$$= \frac{1}{g-\mu}[e^{(g-\mu)(\ln S)/g} - 1].$$

Now we want to see how D is affected by changes in g so we find the derivative

$$\frac{dD(g)}{dg} = \frac{1}{(g-\mu)^2}\left[\frac{\mu(\ln S)(g-\mu)}{g^2}e^{(\ln S)[1-(\mu/g)]} - (e^{(\ln S)[1-(\mu/g)]} - 1)\right].$$

Lowered nutritional quality will benefit the plant if $dD(g) > 0$; that is, if

$$\frac{\mu}{g}\left(1-\frac{\mu}{g}\right)\ln S > 1 - e^{-(\ln S)[1-(\mu/g)]}.$$

Figure 14.1 shows the values of (μ/g) for which this inequality is true. In Lepidoptera, ratios of final to initial size (our S) usually fall between 500 and 10 000.[4] For ratios greater than 500 ($\ln S = 6.2$), decreased nutritional quality is certainly favoured if $(\mu/g) > 0.2$. To obtain a rough expectation for (μ/g), note that, for insects feeding primarily as larvae and assumed to be in ecological equilibrium with their host plants,

$$\frac{S}{c}l_T = 1,$$

where $1/c$ represents the proportion of biomass loss (for the population) between pupae of one generation and newly hatched larvae of the next. In other words, $1/c$ takes into account factors such as mortality of non-larval stages, the 'wastage' of male production (in so far as males do not contribute biomass to the eggs and so do not directly support population growth), and the

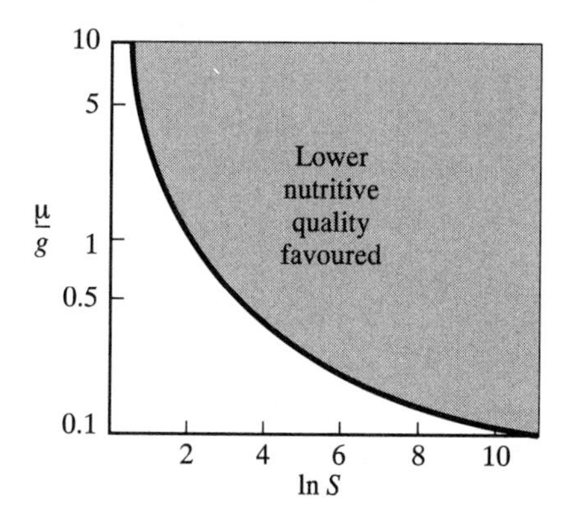

Figure 14.1 Values of μ/g and $\ln S$ for which lower nutritive quality is favoured under the constant mortality submodel.

proportion of adult female biomass not converted into progeny. In our model we can substitute

$$l_T = \mathrm{e}^{-\mu T} = \mathrm{e}^{-(\ln S)/g}$$

and the above equilibrium equation then gives

$$\frac{\mu}{g} = 1 - \frac{\ln c}{\ln S}.$$

The model thus certainly indicates adaptiveness of no-cost changes towards low nutritive value if

$$1 - \frac{\ln c}{\ln S} > 0.2;\ \text{that is, if } 0.8 \ln S > \ln c.$$

The lowest S for the above lepidopteran examples is about 500. Even at this ratio the condition holds if c is less than approximately 150; that is, if more than 1/150th of female pupal weight is converted to newly hatched female larvae of the next generation. As this level of efficiency is probably greatly exceeded in most insects, this constant-mortality submodel tends to confirm the possibility of the evolution of defence through low nutritive quality.

Constant mortality, however, is a weak assumption. What are the effects of disproportionate mortality of either large or small larvae? We consider two extreme cases to see whether benefits to the plant are increased or decreased by a change in either direction.

Suppose all mortality occurs in a certain narrow 'size window' which occurs either just after the eggs hatch (I: early instar mortality) or just before the larvae pupate (II: late instar mortality). We suppose that this much higher localized mortality is just enough to bring the population survivorship curve down to l_T; so that, at sizes outside the window, mortality is zero (Fig. 14.2).

The effect of lowering nutritive quality is to widen the window in time so that the constant mortality goes on longer and lowers l_T. At the same time, the function s_t is 'stretched' over its entire length (in both cases). Thus, on the side of the advantage—the lowering of l_T—case I gains more than case II, because its survivorship function falls much earlier; the same proportion of larvae die in either case, but in case I they die before eating as much. As regards the disadvantage—the increase in T and the need to consume more—both cases are affected the same.

Thus, starting from a constant mortality case which is just neutral for the adaptive advantage of lowering nutritive quality (i.e. a case where $(\mu/g)[1-(\mu/g)] \ln S = 1 - \mathrm{e}^{-(\ln S)[1-(\mu/g)]}$), it is clear that any tendency to form a type I window will make the strategy become adaptive while any tendency to form a type II window will make it maladaptive. In short, if late instars die relatively more, the strategy is less likely to be adaptive.

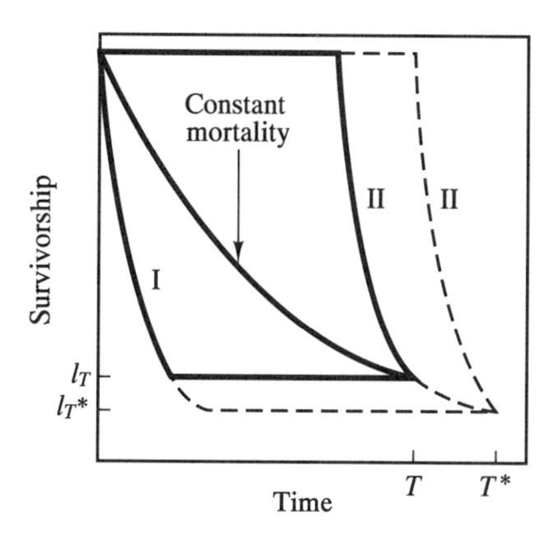

Figure 14.2 Effects of low nutritive quality on survivorship in the cases of early instar mortality (I), constant mortality throughout larval stages, and late larval mortality (II). Starred values and dashed lines indicate effects of low nutritive quality.

One further simple submodel can be used to illustrate a case where late instar mortality is greater. Consider the survivorship pattern $l_t = 1 - pt$. Such linear survivorship is admittedly artificial but approximations occur in nature. If we replace the constant mortality expression with this one and treat p as constant, then following the same reasoning as before our model yields the result that lower nutritive quality is favoured whenever

$$c > S - \frac{1}{2}\frac{S \ln S(S-1)}{S \ln S - (S-1)}.$$

For the region of values of c and S already treated as reasonable this is easily fulfilled. However, constant p means that the mortality pattern is wholly dependent on age and not at all on size. Thus benefit to the plant occurs because larvae are being forced to defer their fast-feeding stages until times when fewer are left to feed. In nature, decreased survivorship of late instars, implicit in this survivorship pattern, is likely to result from their larger size rather than from age, so we might expect the mortality rate to ease off on slower growing larvae: this would be the case if, for example, the higher late mortalities occur because birds see large larvae more easily. So another assumption, erring at a different extreme with respect to the likely real situation, would have p dependent on g in such a relationship that l_t maintains its descending linear form but aims, whatever the growth rate, to hit the same survivorship (c/S) at time of pupation. This gives the condition

$$c - 1 + \frac{(S-c)(S-1)}{S \ln S} < 0,$$

which is not true for any feasible c and S. In this case, then, lowered nutritive quality is never favoured. It is, no doubt, artificial in that the mutant plant, besides failing to prevent damage, fails even to lower fitness of the herbivore. It is, however, no more unrealistic than the case which has mortality independent of size.

Price[5] divides survivorship curves for herbivorous insects into two categories. One set approximates our original case of constant mortality, with perhaps slightly higher mortality of late stages. The other group shows much greater late stage mortality, approximating the linear case just considered. Our analysis has indicated that more data on the effect of an unusually low nutritive quality on pattern of survivorship in nature is needed before we can say whether insects of this group could be effective selective agents.

We have not considered costs of being unnutritious. Even where one or more of the three pathways is applicable, the evolution of low nutritive quality depends upon costs to plants of producing digestibility-reducing compounds such as tannins or of maintaining low nutrient concentrations in tissues.

In summary, the point of this note is to caution that what hurts herbivores is not necessarily a help to that plant which steps out of line to inflict hurt. Some herbivore–plant interactions probably do not fit any of our three models, suggesting that sometimes both plant and animal may suffer through mutation to lowered nutritive quality. The common view that low nutritive quality of plant tissue is an anti-herbivore adaptation must be allowed as plausible but it remains uncertain. It will be hard to refute. If life history and mobility parameters of winter moth eliminate it as a force favouring tannin production in oaks (for example), tannin might still be conjectured to be adaptive against some other herbivore. Many other insects besides winter moths ravage oaks and the proposal to seek life history and behavioural data for them all[6] is daunting.

References and notes

1. For example, D. F. Rhoades, in G. A. Rosenthal and D. H. Janzen (eds), *Herbivores: Their Interaction with Plant Secondary Metabolites*, pp. 3–54 (Academic Press, New York, 1979); P. Feeny, *Recent Advances in Phytochemistry* **10**, 1–40 (1976); D. F. Rhoades and R. G. Cates. *Recent Advances in Phytochemistry* **10**, 168–213 (1976).
2. P. Feeny, in L. E. Gilbert and P. H. Raven (eds), *Coevolution of Animals and Plants*, pp. 3–19, here p. 4 (University of Texas Press, Austin, 1975).
3. M. C. Singer, personal communication.
4. Examples in O. W. Richards and R. G. Davies, *Imms' General Textbook of Entomology*, 10th edn, Vol. 1: *Structure, Physiology and Development*, p. 364 (Halsted Press, New York, 1978); I. M. Campbell, *Canadian Journal of Genetics and Cytology* **4**, 272–88 (1962); G. C. Rock, in J. G. Rodriguez (ed.), *Insect and Mite Nutrition* (North-Holland, Amsterdam, 1972).
5. P. W. Price, *Insect Ecology*, pp. 139, 140 (Wiley, New York, 1975).
6. T. R. E. Southwood, *Journal of Animal Ecology* **30**, 1–8 (1961).

CHAPTER 15

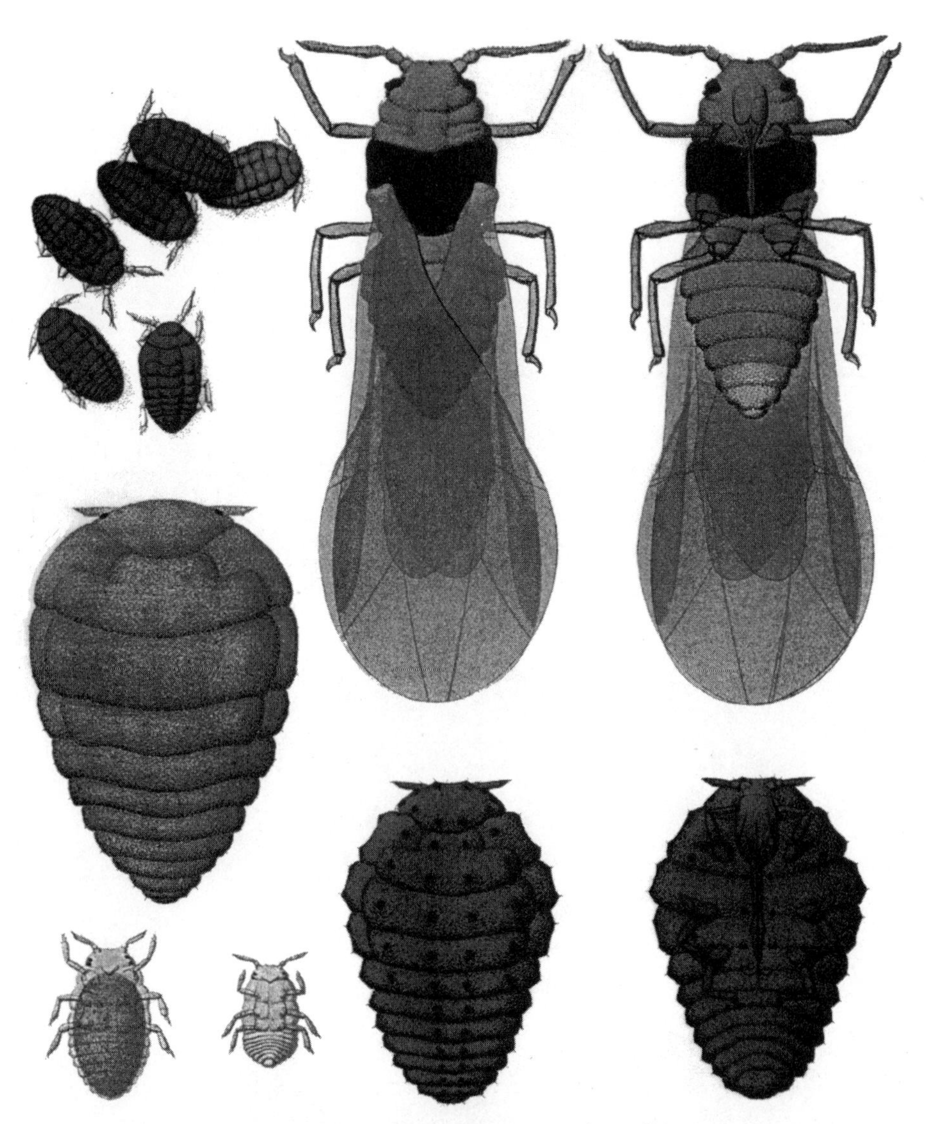

Troglodytes, fatties and fliers — and a walking egg

The many lives of a dreaded pest of European vines, *Phylloxera vastatrix*. Most individuals reproduce viviparously and parthenogenetically on roots (bottom right: back and underside of a laying mother), and also hibernate there as juveniles (top left). In late summer, however, some root dwellers (what proportion being the question of the paper that follows) produce winged dispersing females (top right) who fly each to lay a few eggs on the bark of twigs of vines. Hatching usually only a single sex from each mother, such eggs give female and male non-feeding forms (bottom left), which then mate. Deep in the bark crevice each mated female matures a single egg from which, in spring, a female crawler goes to form a gall on a young leaf. Later, becoming large and obese (mid-left) in the gall, she reproduces the first of a series of parthenogenetic gall broods. From each such brood some crawlers return to the soil. All drawings are to the same scale: winged forms are about 1.6 mm long.

CHAPTER 15

ADVANCED ARTS OF EXIT

Evolutionarily Stable Dispersal Strategies

'Papakena and Vanavana are off there to the westward, or west-nor'westward a hundred miles and a bit more', he said. 'One is uninhabited, and I heard that the people on the other one had gone off to Cadmus Island. Anyway, neither lagoon has an entrance. Ahunui is another hundred miles on to the nor'west. No entrance, no people.'

'Well, forty miles beyond them are two islands?' Captain Davenport queried, raising his head from the chart.

McCoy shook his head.

'Paros and Manuhungi—no entrances and no people. Nengo-Nengo is forty miles beyond them, in turn, and it has no people and no entrance. But then there is Hao Island. It is just the place. The lagoon is forty miles long and five miles wide. There are plenty of people.'

JACK LONDON[1]

HUGH COMINS came to Imperial College Field Station in May 1974. Like Robert May he was Australian and had a doctorate in theoretical physics. Like May he came a viking into the quiet hall of Silwood, brandishing an analytical skill against which we entomologists and ecologists had no defence. Vikings, extraterrestrials, Ned Kellys blazing against Authority and the Old Country—our only hope was to get them on our side. Our pacification efforts seem to have worked reasonably well as, I think, Chapter 11, the paper discussed here, and many others coming out of Silwood around that time attest.

When I had exhausted May's patience with my requests for generalizations of the evolutionarily stable strategies (ESSs) concerned with dispersal, or perhaps when he had decided that other fields of ecology or evolution now needed his battleaxe more than mine did, I found myself passing my challenges to Hugh Comins instead. At first (as with Bob May initially) it typically took only a few days for each new generalization to be resolved: as an analyst Hugh was proving inexhaustible. Working in this manner we added to the published M&H prototype generalizations to 'island' population structure with multiple adults per site, finite brood sizes, site extinctions, recurrent mutation by the strategy-determining gene, etc. Towards the end my demands began to slow Hugh down a bit, the interval to each successful new outline lengthening from days to weeks. However, he was still analysing and writing almost faster than I could read. At last I came up with a request that I hoped might really slow him: I asked for the whole model to be recast into the framework of a 'stepping-stone' meta-population—I will explain this in a moment. It was a big request and took him a month or so but eventually even this was done. By this point I think it had become fairly clear to both of us that I was not understanding much about the equations he was bringing back to me, whether they came the next day or in a month: I was losing even a feel for what his solutions meant. In contrast, May was following the new windings and was still able to make pertinent comments. I think it was while Hugh was still working on the stepping-stone version that we had a meeting and decided that he should write up all he had done short of stepping-stones and that from then on Bob and I would drop out, treating what followed as Hugh's project alone. Hence came to be defined the scope of the paper in this chapter and quite a while later, when I had already been 2 years in America, it was published.

Even with Hugh's labour divided and rounded off to be the present joint paper, I remain somewhat doubtful about my right to a place on its author list. I feel like Lestrade the detective putting my name to a case that I began but which ended being solved wholly by Sherlock Holmes—solved, moreover, contrary to a lot of Lestrade's ideas about how it was going. All the same I am proud that my simple idea ignited such a long-continuing chain of analytical fireworks by the two Australians. As noted, the results in the paper of this chapter were only the central third of the display.

'Stepping-stone structure' implies considering all sites where organisms occur as arrayed in a square lattice and then specifying that dispersal occurs wholly or more commonly between neighbouring sites. A classic stepping-stone excludes moving to neighbour sites along diagonals—these, of course, are not strictly nearest neighbours. Yet, obviously, any spatial restriction of this kind gives the situation a geographic and particularly a two-dimensional character that is completely lacking in the model of the paper that follows, and this character leads to much extra complication. While Hugh was coping with my first stepping-stone demand, I thought up an additional twist, which, in spite of the optimistic sounds I was already hearing about the simple stepping-stone picture, I thought still had a chance to stump him.

When a fat envelope with the ESS solution for the square-lattice stepping-stone array of sites had, via my mail pigeon hole in the corridor, found its way to my somewhat listless hands (perhaps particularly listless just then because I was busy with rotting wood again) I started seriously to procrastinate the effort of understanding the solutions. I complained to Hugh about defects of the classic square lattice, how it was not really a natural model of biological grouping. This in itself is a quite valid objection: in nature the placement of organisms or groups on the ground generally approximates a lattice of 'packing' much more nearly based on triangles than on squares. It can be, actually, a matter of packing in a most literal sense, usually of packing against mutual territorial pressures. Pour dried peas, even peas of fairly uneven sizes, to form a layer in the bottom of a box and you will see a rough triangular lattice of centres surrounded by hexagonal 'territories'—in short, a 'honey-comb' arrangement for the peas. It is thus much more typical for a biological entity to have six neighbours; four is unnatural. Again, look at national or state boundaries on a map: most have three-way meeting points (within the USA there is only one that is of four).

To try to get Hugh involved in my new request and to take me seriously, I pointed out to him something that it needs no analytical skill to see. This is that if the migration rate rises above 50 per cent, and in particular if it ever approaches 100 per cent, then in the square-lattice stepping-stone a situation arises where the original unitary population begins to divide into two separate populations that, while mixed geographically, are no longer in contact, having undergone a kind of sympatric speciation. To see this at 100 per cent migration, imagine a chess board completely populated with bishops but (un-chess-like) having (1) many bishops to a square and also (2) bishops that make a very unorthodox move: just one step up, down, left, or right (therefore never in the diagonal line of a true bishop's move). Half the bishops are black and half white; all initially are on their own colour as in chess. Then think how each round of 100 per cent migration with classic stepping-stone nearest-sites migration (the moves just described) will carry all the black bishops on to white squares and at the same time all the white on to black. In the next generation all are back on their own colours, and so on. In this scheme, as with true bishops in chess, white bishop never meets black bishop, therefore in the analogy to demes they can never interbreed and remain two separate populations—for this extreme case and the purpose of my analogy instead of bishops I might perhaps better have called them monks and nuns. In the case of the 'more natural' triangular lattice it is easy to see that such separation does not happen because individuals from two corners of each lattice triangle always can meet at the third corner even when migration is 100 per cent (while they also, of course, meet at the third corner of another triangle that is on the other side). Thus, except at zero migration, the population is always united and becomes steadily more so as the migration rate is raised. Rates near to 100 per cent, with as many animals rushing on to site as are simultaneously rushing off, are absurd biologically, of course, whichever type of lattice one thinks of; yet it seemed to me very possible that something significantly different might hold for six-way as against four-way migration even for moderate migration rates. At any rate it was a good puzzle, and good enough to hook Hugh's attention.

This one really slowed Hugh down. For stepping-stone models in general the maths is much tougher than most in population genetics. Already it gets to be hairy stuff in which Green, elliptic, and Bessel functions bristle freely. It may have been 6 months or a year before

Hugh told me (now by post from Australia) that he had the full answer for six-way stepping-stone migration. The new lattice structure, he said, had turned out to affect things surprisingly little—disappointing, but he now understood why. I was disappointed too. I had hoped that when the Comins extension to six-way stepping-stone models was found it might revolutionize meta-population theory, and show that henceforth for realism we must always model six-way spatial structures. This was not a particularly constructive thought, but it is nevertheless sometimes quite pleasant to tell experts that they need to rethink something. Results of analysis of the square lattice structure were generally known already and, as might be guessed, even these are generators of pretty bad nightmares both for innocent readers and for typesetters—about as bad, in fact, as population biology ever generates. Nevertheless it is good to see alphabet soup of such quality being poured out on page after page of manuscript especially when (1) it is poured at your request, (2) you are very certain of its grade, much more than if you had somehow screwed such soup out of yourself (Hugh is careful and gets things right), and (3) it tells something true about dispersal—tells it at least truly for somewhere, for some conceivable conditions, may be those of some distant planet where nunmonks may indeed march and countermarch for ever on triangular or on square lattices (reproducing themselves, perhaps by pure thought). What is far from satisfying for me, however, is that I have only a moderate inkling of the whole, or rather I grasp results with effort and the aid of a computer, for particular sub-sub-cases, but can't hold more than very small parts of the whole outcome in my mind at any time—the overall shape remains for me a mountain in fog.

As I said, I had been hoping for the six-way version to outbid the four in the richness and naturalism of its implications and that there would emerge some dramatic and easily understood points. Hugh, however, pointed out to me that the 'pathologies' of population mixing that would attend 100 per cent migration (the case of the mutually passing bishops) were never even nearly attained in the ESS results. Consequently it turned out more simply the two-dimensionality of the meta-population array that was important at the low to moderate migration rates that were realistic, and this two-dimensionality had the same effect as for the classic model. In short, it seemed that ESS dispersal stepping-stone results under other assumptions about two dimensions were never going to be much different from the results Hugh had already worked out for us, first, in the paper of this chapter, by then published and, second, under the 'classical' stepping-stone conditions. I wrote thanking Hugh again for his splendid efforts, and said I would skim through his latest sketch with great interest as soon as I had time (Napoleon wrote in a similar suave and evasive style, I recalled later, to Laplace, the great French mathematician, when the latter had sent him a copy of his prodigiously analytical *Méchanique Céleste*). Meanwhile I told Hugh I hoped he would write it all up for publication. Possibly my letter had a weary tone; it contained no further challenges. Wearily himself, as I judge from the time taken, he did publish. His result stands in the *Journal of Theoretical Biology*,[2] a monument so perfect and so elaborate it seems to have slowed further theoretical research into evolutionarily stable dispersal strategies pretty much to a crawl since it came out. It has created, it almost seems, a playing field where Hugh Comins plays alone.[3]

References and notes

1. J. London, The seed of McCoy, in E. C. Parnwell (ed.), *Stories of the South Seas*, pp. 266–301 (Oxford University Press, 1928).
2. H. N. Comins, Evolutionary stable strategies for localised dispersal in two dimensions, *Journal of Theoretical Biology* **94**, 579–606 (1982).
3. H. N. Comins and I. R. Noble, Dispersal, variability, and transient niches: species coexistence in a uniformly variable environment, *American Naturalist* **126**, 706–23 (1985).
4. M. Cornu, *Mémoires de l'Académie des Sciences de l'Institut de France* **26** 1–357 (1878).

EVOLUTIONARILY STABLE DISPERSAL STRATEGIES†

HUGH N. COMINS, WILLIAM D. HAMILTON, and ROBERT M. MAY

Using the idea that life-history parameters are subject to natural selection and should approach values that are stable optima, with the population immune to invasion by mutant individuals, we derive an analytic expression for the evolutionarily stable dispersal rate in a stochastic island model with random site extinction. The results provide interesting contrasts between three different optimization criteria: species survival, individual fitness, and gene fitness. We also consider the effects of sexual reproduction, and of localized migration (stepping-stone structure).

1. INTRODUCTION

Movement leading to dispersal is an essential property of living things. Dispersal enables a species to extend its range, and reduces the chance of extinction due to local fluctuations (including such fluctuations as might result from excessive local population pressure and exploitation of resources). However, notwithstanding the undoubted benefits to the species (and indeed to the stability of the ecosystem), it is by no means obvious what advantage an individual organism gains by undertaking a perilous dispersal movement instead of staying back to compete more safely in the locality where it was reared.

In a simulation study using a 'stepping-stone' model[1] of population structure, Roff[2] showed that the dispersal strategy which maximizes site occupancy, and is thus to be regarded as optimal from the species point of view,[3] differs significantly from that produced by genetic selection. Thus evolution will lead to the species exploiting less of its habitat than it conceivably might, a result of considerable ecological interest.

The determination of the evolutionarily stable dispersal strategy (ESS) by simulation is by nature a very lengthy process, particularly when stochastic

†*Journal of Theoretical Biology* **82**, 205–30 (1980).

elements are present. Thus the investigation of how the ESS varies with different chances of surviving migration and differing degrees of environmental variability would be greatly facilitated if the ESS could be obtained analytically. Hamilton and May (see Chapter 11) made a start in this direction and obtained results for a model which had a simple stochastic pattern. The present paper extends their results to transient environments and general probability distributions for the number of progeny, and allows more than one organism to occupy each site.

In addition to the main result, which is in qualitative agreement with Roff's model, the effects of two other features of the simulation model are considered in less detail. These are sexual reproduction and stepping-stone (as opposed to island-model) migration. Both these additions reduce the ESS migration rate, although the effect of sexual reproduction is usually small.

The full mathematical derivation of the ESS migration rate is given in Appendix A. However, the important points are repeated in the text, so that the appendices need only be used for reference.

2. THE MODEL

The basic model (see Fig. 15.1 and Table 15.1) is essentially a discrete generation island model,[1] in which the processes of reproduction, migration, and competition are stochasticized. Because of the stochastic effects it is possible for a site to become empty if all offspring in the site happen to emigrate and there are no immigrants. In addition a degree of environmental variability will be introduced in the form of a fixed probability X that a site is destroyed by exogenous forces during a one-generation time interval. As in Hamilton and May's (1977) model reproduction is assumed to be asexual, and individuals of each genotype have a certain probability of being born with a predisposition to migrate.

Table 15.1 Parameters and variables in the model

k	number of adults per site
τ	average number of offspring per adult
n	average number of offspring per site $= k\tau$
v	probability of migration
X	exogenous extinction rate
p	chance of surviving migration
f	proportion of sites occupied
ρ	mutant gene frequency
F	equilibrium variance parameter
C	ratio of destroyed sites to remaining sites $= X/(1-X)$
θ	product of p and f

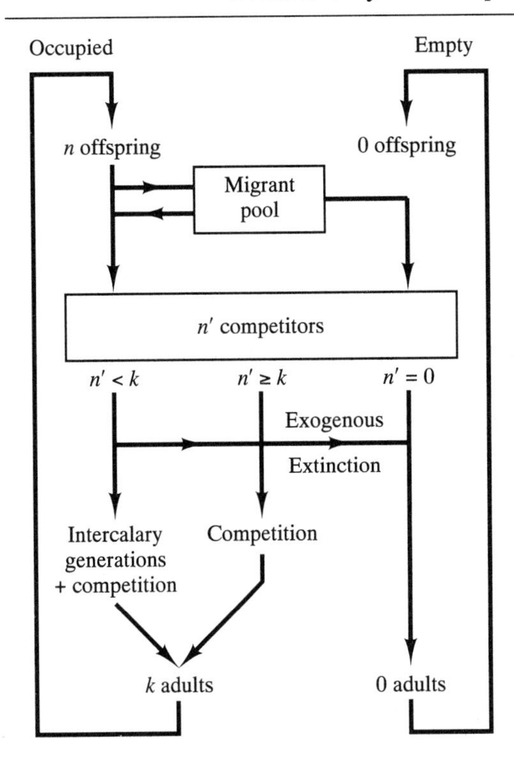

Figure 15.1 Cycle of events in each generation.

Suppose that there are a very large number of sites each capable of supporting k reproductive adults. Each adult independently produces a number of offspring according to a probability distribution which is independent of genotype. The mean number of offspring per adult is τ, giving an average total of $n = k\tau$ offspring per site. Each of these offspring is independently determined to be migratory with probability v or sedentary with probability $1 - v$, where v depends on the genotype. Migrants enter a pool from all sites, where a proportion $1 - p$ is removed (hence p is the survival rate for migration). The survivors are then redistributed among the sites; the number of a given genotype arriving at a given site is assumed to be Poisson-distributed, with the mean for each type equal to the number of that type in the pool divided by the total number of sites.

Sedentary local offspring and immigrants compete for each site in such a manner that by the time of the next generation there are exactly k reproducing adults. If there are more than k competitors the excess must die or fail to reproduce. If there are fewer than k competitors then we require intercalary non-migratory generations in order to fill the site. This assumption of fixed site size may seem somewhat artificial but it is necessary to make the analysis tractable; at the same time, bearing in mind possibilities of growth and vegetative reproduction, it is not very unreasonable. In either case we assume fair competition, so that each of the k adults has the same chance of being descended from each of the original competitors. This represents an approximation unless the number of competitors is much greater than k. The case where immigrants have a systematically lower chance of surviving competition can be incorporated as an additional effective mortality, decreasing p.

In summary, this model extends that of Hamilton and May in having the number of adults k at each site allowed to be more than one, and in having a degree of environmental variability due to the exogenous extinction rate X. The extinction of sites can be regarded either as a random process, or as an ageing process with type II 'survivorship' (i.e. probability of death independent of age). This is reminiscent of a feature of the earlier model which gave to the adult organism occupying a site a certain probability of dying; however, in the current model it is as if the site itself disappears or, if the site is considered to persist while occupants die within it, then all are assumed to die together. Disappearance of the old sites is assumed to be balanced by the creation of an equal number of new empty sites; these can only be colonized by migrants, so it is at once evident that large values of X make migration more attractive.

At this stage of its development our model differs from Roff's simulation model most significantly in the assumptions of asexual reproduction and island-model migration (as opposed to Roff's 6×6 stepping-stone model).

3. EVOLUTIONARILY STABLE STRATEGY (ESS)

It is assumed that migration rate is a continuous genetic character. An evolutionarily stable state in terms of such a character can consist of either a single phenotype or a mix of phenotypes at specific frequencies with the characteristic that all genes coding for small changes in the migration rate of a phenotype are disadvantageous. In addition to this genetic stability criterion, which must be satisfied for every phenotype, the stability of a polymorphism requires that there be frequency-dependent selection to stabilize the relative frequency of each phenotype.

The possibility of polymorphism in the ESS will not be pursued in this paper; in the remainder of the paper it will be assumed that the ESS is a single phenotype. The observational difference between a polymorphism and a single phenotype having the same average migration rate is an increased correlation between migration tendencies of offspring of the same parent. Such a correlation could conceivably contribute to stability in a way that would exclude invasion by a single 'average' type. However, a polymorphism cannot exist unless all the component phenotypes are stable with respect to small changes in migration rate. In the case of a single phenotype only one ESS is ever found, representing a compromise between colonizing ability and risk of loss of the home site. It is therefore hard to see how two or more phenotypes could simultaneously be stable, so that a polymorphism seems rather unlikely, although the possibility has not been entirely ruled out.

In order to derive the ESS dispersal rate we thus assume that there are two types of individuals present; the 'wild type' use the putative ESS dispersal rate v while those bearing the mutant gene have a slightly different dispersal rate $v + \epsilon$. We must then find v such that the mutants are at a disadvantage for any small ϵ, whether positive or negative. That is, the frequency of mutant genes ρ' in the next generation must be less than the current frequency ρ (see Table 15.1). Since $\rho' - \rho$ has a term linear in ϵ, a necessary condition for an ESS is stationarity; that is, the derivative of $\rho' - \rho$ with respect of v must be zero.

The complete description of the state of the system requires $k + 2$ parameters; the proportion of sites which are occupied f, and $k + 1$ fractions $\phi_0, \phi_1, \ldots, \phi_k$, where ϕ_i is the fraction of occupied sites in which i out of the k adults are mutants. Since $\phi_0 + \ldots + \phi_k = 1$, the ϕ's can be regarded as the probability distribution of i/k in occupied sites. Fortunately the ESS can be determined without knowing this full distribution, since the ESS condition only depends on f, and the mean ρ and equilibrium variance S of i/k. It can be shown, as in Wright's island model,[1] that (ignoring terms in ϵ) the site occupancy fraction f', the mean mutant frequency ρ' and the mutant frequency variance S' for the next generation can all be obtained from the current f, ρ, and S without reference to the rest of the distribution. For conformity with Wright's (1969) notation we will express the variance in terms of the gene-frequency independent parameter $F = S/(\rho(1 - \rho))$; it will turn out that the state of the system is described adequately by the three parameters f, ρ, and F.

Both f and F rapidly approach equilibria which are independent of ρ except for terms of order ϵ. The equilibrium value of f is given by the implicit equation (see Table 15.1 and equation (A3)):

$$f = \left[\frac{C + \mathrm{e}^{-n(1-v)}\mathrm{e}^{-nvpf}}{1 - \mathrm{e}^{-nvpf}} + 1\right]^{-1}, \tag{1}$$

where $C = X/(1 - X)$. Although Appendix A derives the ESS for a general distribution of family size, we will consider here only the case where the number of offspring is Poisson-distributed. The equilibrium value of the variance parameter is then given by

$$F = \frac{1 - (1 - X)(1 - k^{-1})[H(nv\theta/\alpha) + (f^{-1} - 1)H(nv\theta)]}{1 - (1 - X)(1 - k^{-1})(1 - \alpha)^2 H(nv\theta/\alpha)}, \tag{2}$$

where $\alpha = v\theta/;[(1 - v) + v\theta], \theta = pf$, and the sigmoid-shaped function $H(a)$ is defined in equation (A44). This function has the series expansion $H(a) = \mathrm{e}^{-a}\sum_{i>1}[(i - 1)a^i/ii!]$, but can be evaluated more efficiently using standard techniques for exponential integrals. Note that if $k = 1$ we obtain the simple result $F = 1$ as expected. The ESS condition (for Poisson-distributed offspring) is found to be

$$C + \{1 - nv(1 - \theta)\}Z = [(\alpha/\theta)(1 - Z) - nvZ][(1 - \theta) - (1 - \alpha)F], \quad (3)$$

where $Z = \exp[-n\{1 - v + v\theta\}]$. These equations can be solved as follows for given p, X, and n. Select a value of v; calculate f by solving (1) (the result is unique). Provided the results are feasible, determine F from (3) and then k from (2). Invert the graph of k as a function of v to obtain the ESS migration rate as a function of number of adults per site k. Since the graphs of k against v are invariably monotonic decreasing there is only one ESS for any p, X, n, and k.

4. ESS RESULTS

Figures 15.2–15.5 show some typical results for the ESS dispersal rate given by equations (1)–(3). Note that when the number of adults per site k varies in these graphs, it is the total expected offspring n which is kept constant, rather than the average offspring per adult τ, since this is much more convenient for calculation. If, however, we wish to take the limit of very large site size $k \to \infty$ then it is only reasonable to let the expected total offspring n become large as well and in referring to the limit $k \to \infty$ we shall assume that this happens.

All the graphs have the property that the ESS migration rate decreases as the number of adults per site is increased. It also decreases whenever the total number of offspring increases. At the same time the ESS migration rate always remains above a certain threshold v at which non-migration ceases to offer the individual offspring the best option for becoming an established adult. This threshold v, which we symbolize v_{IND}, merits further explanation. If a pre-dispersal offspring has a view only to maximizing its chance of becoming a reproductive adult, and if both in its group and in the population at large the migration rate is v, then its chance is maximized by migrating if $v < v_{\text{IND}}$, by

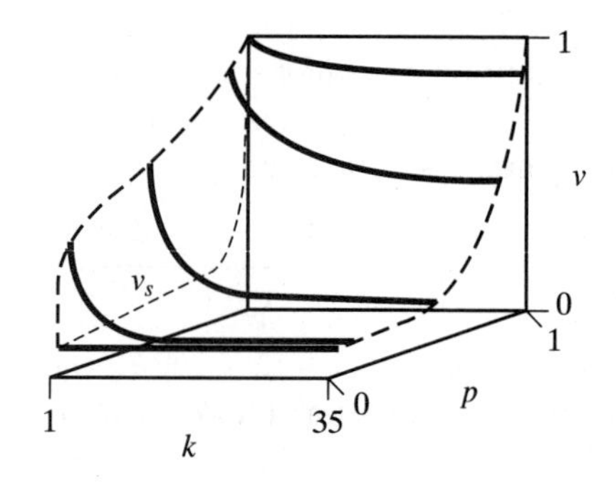

Figure 15.2 Three-dimensional plot of the ESS migration probability v as a function of site size k and probability p of surviving migration. The average number of offspring per site is $n = 10$ and there is no exogeneous extinction. In the $k = 1$ plane we show the species optimal migration probability v_s as a function of migration survival p.

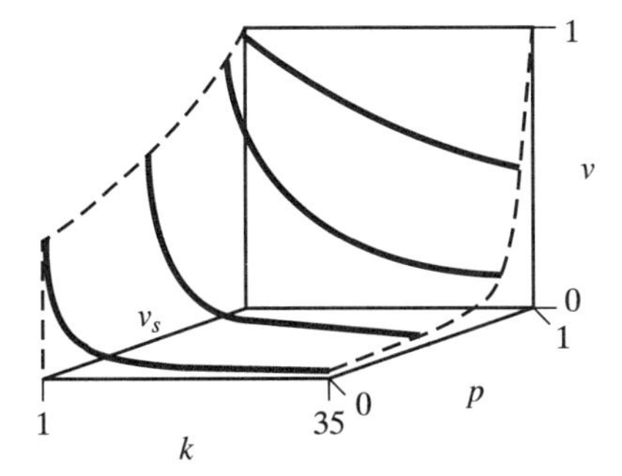

Figure 15.3 As Fig. 15.2, but with the average number of offspring per site $n = 500$.

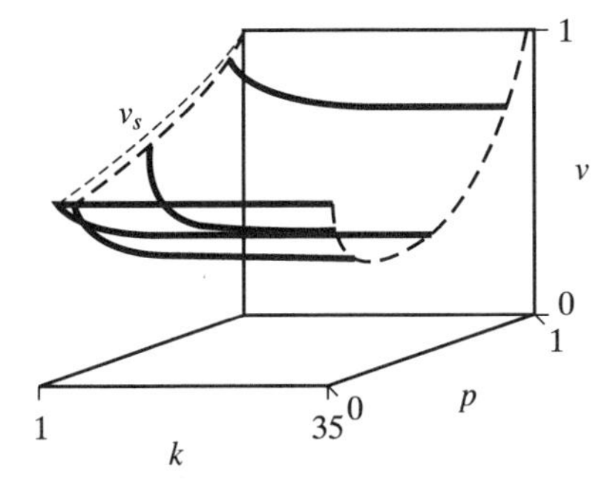

Figure 15.4 As Fig. 15.2, but with 10 per cent exogenous extinction ($X = 0.1$).

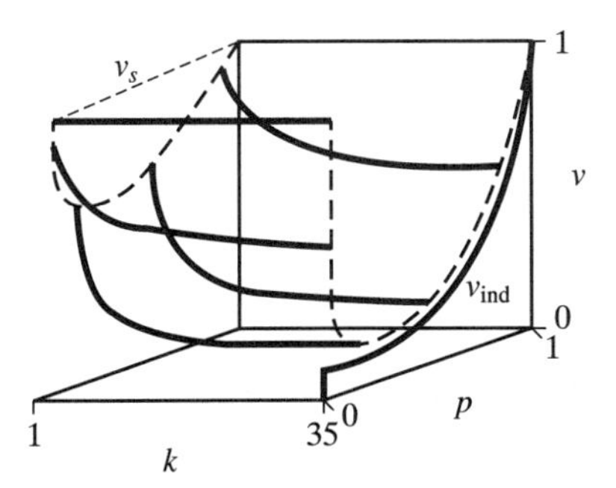

Figure 15.5 As Fig. 15.2, but with 10 per cent exogenous extinction ($X = 0.1$) and the average number of offspring per site $n = 500$. An additional graph in the $k = 35$ plane shows the individual optimum migration probability v_{IND} as a function of migration survival p.

not migrating if $v > v_{\text{IND}}$, while if $v = v_{\text{IND}}$ chances either way are equal. Thus if the individual's short-term selfish view were representative of natural selection, v_{IND} would establish as the ESS. It hardly need be said that the suggested view is too limited; to look no further ahead than immediate adulthood is a very inadequate evolutionary viewpoint. This is particularly obvious in the present asexual model where long-term individual and genic viewpoints actually coincide. Apart from demonstrating the inadequacy of this apparently plausible but simplistic view, the evaluation of v_{IND} provides a true lower limit to which the ESS rate must converge as $k \to \infty$, as explained below. In this limit it is easily known that

$$\nu_{\mathrm{IND}} = \frac{X}{1-(1-X)p}. \tag{4}$$

This equation represents a balance between the chance of establishing in the home site and the overall value of migrating. The latter depends on the frequencies of the three possible outcomes: dying *en route*, arriving at another occupied site, and finding an empty site. Since the competition at another occupied site is exactly the same as at the home site, only the third outcome can be considered favourable. It may indeed be verified that if $X = 0$ (giving no empty sites when $k \to \infty$) then $\nu = 0$ unless migrants incur no additional mortality ($p = 1$).

An example of (4) with $X = 0.1$ is plotted in Fig. 15.5 and placed so as to indicate the limit towards which the ESS rate (heavy dashed curve) is falling as k increases.

There are two ways of understanding the decline in the ESS with increasing numbers of adults or number of offspring, and also the convergence just mentioned. From the point of view of a gene, migration has the advantage of allowing competition for sites at which it is under-represented. Clearly this will only be worthwhile if there is likely to be a significant difference in gene frequencies between the starting and end points of the migration. This is the case when k and n are small, but as $k \to \infty$, $n \to \infty$ the genetic variance between sites declines to zero, so that there is no longer a difference between occupied sites from the gene's point of view. This now coincides with the selfish individual point of view expressed by (4), in which the only advantage of migration results from the possibility of colonizing an extinct site.

The second way of regarding this phenomenon is in terms of inclusive fitness. A particular offspring is genetically identical to about $1/k$ of the other offspring at its site, and is unrelated to the others. Thus its tendency to migrate when this is not individually favourable should be smaller for large k. The equivalence of these approaches to problems of selection in grouped populations is explained for a more elementary example in Chapter 9.

The ESS can also be compared with the optimum migration rate for the species (i.e. the value ν_s which maximizes the proportion of sites occupied). A qualitative difference between Hamilton and May's (1977) results (see Chapter 11) and Roff's[2] is that in the former the ESS migration rates are always found to be greater than the species optimum, whereas Roff always finds the ESS to be smaller.

A comparison of the migration rates for maximum site occupation plotted in the $k = 1$ planes of Figs 15.2–15.5 shows that for a large total number of offspring (as in Roff's model) the maximum occupation ν is extremely sensitive to Hamilton and May's (1977) assumption of no exogenous extinction ($X = 0$). Roff's model, on the other hand, includes environmental variations, which have a similar effect to $X > 0$ in making migration more rewarding. Thus the differences in the two earlier models can be understood from the abrupt change found in the present model as X departs from a value of zero.

Another unexpected feature of the model for $X > 0$ is related to this effect. With $X > 0$ the ESS surfaces curve sharply upwards for low migration survival p. The idea of increasing migration rate when the chance of surviving decreases is at first sight paradoxical. It suggests, however, that the first priority for low p is to maintain a certain level of colonization of empty sites, and that competition for the home site is being left to any surplus offspring. This effect only occurs when the species is hard-pressed to maintain itself; in the extreme case only one migration rate can ensure continuance—applying both for individual lines and for species. It is clear that the ESS and species optimium v must coincide in this case. Figures 15.4 and 15.5 show that the 'width' of the high ESS 'rim' (i.e. roughly the distance of this upturning portion of the surface from the plane $p = 0$) decreases with n, tending to zero as $n \to \infty$. This is understandable if we note that for a non-stochastic version of the model the species just maintains itself if $np > 1/(1 - X)$; hence it is to be expected that the width will be inversely proportional to n.

5. ADDITIONAL EFFECTS

It is possible to infer the effect on the ESS of two additional considerations which cannot be included in the full model; sexual reproduction and stepping-stone migration.

Appendix C derives the ESS for hermaphroditic sexually reproducing organisms in the limit of large total number of offspring n. The analysis proceeds in much the same way as before except that the k adult organisms are replaced by $2k$ genes. The result is surprisingly simple: the ESS for sexual reproduction with site-size k is obtained by substituting $k + \frac{1}{2}$ for k in the ESS formula for asexual reproduction with site-size k. Thus sexual reproduction effectively increases the site size by $\frac{1}{2}$. This is only important if k is very small. This simple result is not expected to extend to the stochastic case; however, the relative importance of sexual reproduction is presumably no greater.

Introducing stepping-stone migration[4] has the effect of producing local correlations in gene frequencies. Thus the endpoint of migration is more likely to have a similar gene frequency to the starting point. As seen in the last section, this makes migration less worthwhile from the gene point of view, so the ESS v should be closer to the individual optimum. The change to a stepping-stone model seems also to be of minor importance. A derivation of this result (for the case of no exogenous extinction, $X = 0$) is published elsewhere.[5]

6. DISCUSSION

From the gene or 'inclusive fitness' point of view, there are two advantages in having a high dispersal rate. The first advantage is the colonization of empty sites, which also aids individual emigrants by enhancing their reproductive potential. The second advantage is the opportunity to compete for other sites in which the migrator allele is under-represented. Merely changing one arena of competition for another where numbers are the same is clearly of no advantage to an individual emigrant, and making the change may entail considerable risk. Nevertheless this 'recolonization' effect is important on the genetic level, as can be seen from the extreme case of a completely sedentary phenotype in an environment where no sites are ever empty (see Chapter 11). In such a case there is never any advantage to individual emigrants; yet still at the ESS much migration may occur.

The chance that a migrant entering a non-empty site finds a significantly different gene frequently clearly depends on the equilibrium variance in gene frequency. If the number of adults per site and the number of offspring per site are very large, all sites tend to approach the average gene frequency and there is little advantage in moving from one occupied site to another. If, on the other hand, either number is small the resulting sampling variance maintains a degree of variability between the sites. In this case it becomes advantageous to a genotype to distribute its competition more widely through migration.

With this in mind we return now to the more general case having site extinctions, and summarize the major features of our model. First, it is understandable that the ESS migration rate is expected to be large when either there is a high probability of sites becoming extinct (giving immediate individual advantage) or the variance in gene frequency is high due to small numbers of adults or offspring (giving inclusive fitness advantage). Second, allowing for exceptions near $p = 0$ (where occurs the 'rim' effect already discussed—probably of little biological significance), it is also understandable that ESS migration rate increases as migration is made less perilous (see Figs 15.2–15.5).

The effect of introducing sexual reproduction is a decrease (usually slight) in the ESS rate. Introducing stepping-stone migration would also decrease the ESS rate because it reduces the effective variance in gene frequency (i.e. that variance which is perceived by migrants).

Comparison of the ESS and the species optimum (maximum occupancy) migration rate reconciles the results of Roff [2] and Hamilton and May (Chapter 11). Unless the environmental variance is very small the ESS is constrained to be close to the species optimum. This accounts for the 'rims' apparent in Figs 15.4 and 15.5 for low migration survival rate, and also for the 'trough' apparent in the corresponding position in Fig. 15.2 (which is present also in principle in Fig. 15.3 but is too narrow to be seen).

Acknowledgements

This research was carried out at Imperial College Field Station, Silwood Park, Ascot, Berkshire, UK, and the numerical results were obtained on Imperial College's CDC computer. H.N.C. was financed by a Natural Environment Research Council grant to the Environmental Management Unit at Silwood Park; R.M.M. was supported in part by NSF grant DEB 79–03290.

References and notes

1. S. Wright, *Evolution and the Genetics of Populations* (Chicago University Press, Chicago, 1975).
2. D. A. Roff, *Oecologia, Berlin* **19**, 217 (1975).
3. R. M. May, *Stability and Complexity in Model Ecosystems*, 2nd edn (Princeton University Press, Princeton, NJ, 1974).
4. M. Kimura and G. H. Weiss, *Genetics* **49**, 561 (1964).
5. H. N. Comins, ESS dispersal in stepping-stone models, *Journal of Theoretical Biology* **94**, 579–606 (1982).
6. W. Feller, *An Introduction to Probability Theory and its Applications*, Vol. 1 (Wiley, New York, 1950).
7. R. A. Fisher, *The Genetical Theory of Natural Selection* (Oxford University Press, Oxford, 1930).
8. J. F. Crow and M. Kimura, *An Introduction to Population Genetics Theory* (Harper and Row, New York, 1970).

APPENDIX A

This appendix outlines the derivation of the evolutionary stable strategy for dispersal probability in the stochastic island model with random extinction. Generating functions are used extensively; all the theorems required are grouped together in Appendix B.

In the classical island model the mean ρ and the variance S of the distribution of numbers of mutants per site suffice to determine the mean and the variance in the next generation. This is almost still true for the present model; however, it is now possible for some sites to become empty. Thus a third parameter is required to describe the state of the system, namely the fraction of sites occupied f. The three parameters ρ, S, and f suffice to determine the new values ρ', S', and f' for the next generation.

The expression for the ESS migration probability is found to depend solely on f and S, and not on ρ or any other details of the probability distribution. Thus in calculating the ESS we need only be concerned with ρ, S, and f (ρ is still required, as it affects S).

The ESS migration probability v has the property that it can coexist stably with a phenotype having slightly different migration probability $v + \epsilon$. This is a consequence of the fact that v represents an extremum of 'gene fitness', which should therefore have zero derivative with respect to v at this point. We therefore wish to discover what values of v will lead to an equilibrium of ρ, S, and f when the two genotypes have migration probabilities v and $v' = v + \epsilon$.

This appendix derives the conditions for no change between generations in each of the parameters ρ, S, and f, given that the other two have fixed values. Taking the limit $\epsilon \to 0$, the second and third conditions specify the equilibrium values of S and f, while the first gives a condition on migration rate v which must be satisfied for an equilibrium to exist. Since this condition is independent of gene frequency ρ, it gives the value of v which is an ESS with respect to small variations.

Note that although the S and f equilibria are individually stable if the other two state variables are held constant, it has not been shown that the overall equilibrium is stable. Thus it is conceivable, although unlikely, that ρ, S, and f execute a complicated limit cycle, with different values of v being optimal at each point. The mathematics required to dismiss this possibility is rather formidable; however, some qualitative stability considerations are discussed in the main text.

1. EQUILIBRIUM OCCUPATION PROBABILITY

The equilibrium of f represents a balance between the colonization of empty sites, and the depopulation (by emigration) and extinction (by exogenous forces) of occupied sites. The new value is

$$f' = (1 - f)(1 - p_{ee}) + f(1 - p_{fe}), \tag{A1}$$

where p_{fe} = probability that an occupied site becomes empty, p_{ee} = probability that an empty site remains empty. For $v' = v$ and Poisson arrival of immigrants, and arbitrary distribution of offspring with generating function $g(x)$, it can be shown, using equations (B1) and (B2), that:

$$\begin{aligned} p_{fe} &= X + (1 - X)g(v)^k \, \mathrm{e}^{-nv\theta} \\ p_{ee} &= X + (1 - X)\, \mathrm{e}^{-nv\theta}, \end{aligned} \tag{A2}$$

where the quantities X, k, n, v, and θ are as defined in Table 15.1. If both immigrants and offspring are Poisson-distributed then $g(x) = \exp[\tau(x - 1)]$, where $\tau = n/k$ is the average reproductive rate. Substituting in (A1) and taking the equilibrium $f' = f$, we can rearrange to get

$$p = \theta \left[\frac{C + \mathrm{e}^{-n(1-v)} \mathrm{e}^{-nv\theta}}{1 - \mathrm{e}^{-nv\theta}} + 1 \right], \tag{A3}$$

where $C = X/(1 - X)$.

It can be shown that the right-hand side of (A3) is monotonic increasing with θ. Thus (A3) can be inverted to give θ as a function of v, n, and p. We then obtain f using $f = \theta/p$. The equilibrium of f exists provided

$$(1 - X)[nvp + (1 - e^{-n(1-v)})] > 1. \tag{A4}$$

If it exists the equilibrium is necessarily stable. This can be shown by graphical arguments, given that f' is a monotonic increasing function of f.

Note that if $v' = v + \epsilon$ the equilibrium value of f contains terms linear in ϵ; these have been ignored by setting $v' = v$ in the above derivation.

2. CONDITION FOR NO CHANGE IN GENE FREQUENCY

We first derive the probabilities that a certain number of mutant or wildtype juveniles compete for a site which previously contained i mutant adults and $k - i$ wildtype adults. The probability of j mutants is given by

$$M_j^{(i)} = \sum_r J_r' I_{j-r}' \tag{A5}$$

where I_{j-r}' is the probability of $j - r$ mutant immigrants, given by a Poisson distribution with mean $n\theta v'\rho$. J_r' is the probability that r mutant offspring remain behind:

$$J_r' = \sum_l G_l^{(i)} B_r^{(l)}(1 - v'), \tag{A6}$$

where $B_r^{(l)}(1 - v')$ is the probability that r out of l events occur if each has independent probability $1 - v'$ (i.e. binomial distribution). $G_l^{(i)}$ is the probability (same for both genotypes) that i parents have l offspring.

If $g(x)$ is the generating function (g.f.) of the number of offspring per adult, then the g.f. of $G^{(i)}$ is $[g(x)]^i$ (cf. B2), and the g.f. of J' is $[g(u')]^i$ (cf. B1), where u' is the g.f. of the binomial distribution with one trial, that is

$$u'(x) = v' + (1 - v')x. \tag{A7}$$

Substituting (A7) and the Poisson generating function (B3) into (A5), and using (B4), we obtain the g.f. of $M^{(i)}$:

$$m^{(i)}(x) = [g(u')]^i \, e^{n\theta v'\rho(x-1)}. \tag{A8}$$

A similar derivation shows that the g.f. of the probability $W_j^{(k-i)}$ of j wildtype competitors is:

$$w^{(k-i)}(x) = [g(u)]^{k-i} \, e^{n\theta v(1-\rho)(x-1)}. \tag{A9}$$

Given that the site does not become depopulated, the mean number of mutant adults resulting from fair competition is given by

$$\langle i' \rangle = \left[k \sum_{s>0} \sum_{j} (j/s) M_j^{(i)} W_{s-j}^{(k-i)} \right] \Big/ (1 - M_0^{(i)} W_0^{(k-i)}) \\ \equiv k\eta(i)/(1 - M_0^{(i)} W_0^{(k-i)}). \tag{A10}$$

Note that it was necessary to correct for sites which died out, and thus had no definite value of i'.

A similar expression is obtained for the mean number of mutant adults in newly colonized sites:

$$\langle i' \rangle = \left[k \sum_{s>0} \sum_{j} (j/s) I_j' I_{s-j} \right] \Big/ (1 - I_0' I_0) \\ \equiv k\zeta/(1 - I_0' I_0), \tag{A11}$$

where I_j' and I_{s-j} are, respectively, the probabilities that there are j mutant and $s - j$ wild-type immigrants. These are Poisson-distributed with means $n\theta v'\rho$ and $n\theta v(1 - \rho)$.

Suppose that a fraction ϕ_i of the occupied sites contained i mutants. Then the previous mutant gene frequency was

$$\rho = \sum_i \phi_i (i/k). \tag{A12}$$

Including the appropriate weighting factors the new gene frequency is

$$\rho' = (f')^{-1} \left[\frac{(1 - f)(1 - p_{ee})\zeta}{1 - I_0' I} + f \sum_i \phi_i \frac{(1 - p_{ie})\eta(i)}{1 - M_0^{(i)} W_0^{(k-i)}} \right], \tag{A13}$$

where p_{ee} = the probability that an empty site remains empty, p_{ie} = the probability that a site with i mutants becomes empty, f' = new value of f.

Taking account of random (exogenous) extinction we have

$$p_{ee} = X + (1 - X) I_0' I_0 \\ p_{ie} = X + (1 - X) M_0^{(i)} W_0^{(k-i)}. \tag{A14}$$

Using (B5) and (B3) it can be shown that

$$\zeta = \sum_{s>0} \sum_j (j/s) I_j' I_{s-j} \\ = \frac{v'\rho}{v'\rho + v(1 - \rho)} [1 - I_0' I_0]. \tag{A15}$$

Substituting (A14) and (A15) into (A13), and using the equilibrium condition for f

$$\rho' = (1-X)\sum_i \phi_i \eta(i) + \frac{v'\rho}{v'\rho + v(1-\rho)}\left[X + (1-X)\sum_i \phi_i M_0^{(i)} W_0^{(k-i)}\right]. \tag{A16}$$

The square bracket is just the average probability that an occupied site will become empty. Thus (A16) has the interpretation that the change in gene frequency consists of one term resulting from competition for already occupied sites, and a second term resulting from the divison (in the ratio of representation in the migrant pool) of sites which have been allowed for the present to become empty, but which will be recolonized some time in the future.

Using (B5) we have that

$$\begin{aligned}\eta(i) &= \sum_{s>0}\sum_j (j/s) M_j^{(i)} W_{s-j}^{(k-i)} \\ &= \int_0^1 \dot{m}^{(i)}(x) w^{(k-i)}(x)\,\mathrm{d}x\end{aligned} \tag{A17}$$

where the dot denotes differentiation. Integrating by parts and substituting into (A16) we get after rearrangement (using B6):

$$\begin{aligned}\frac{\rho'}{1-X} &= \left[1 - \frac{v(1-\rho)}{v'\rho + v(1-\rho)}\right] C + 1 - \sum_i \phi_i \gamma(i) \\ &\quad - \frac{v(1-\rho)}{v'\rho + v(1-\rho)}\left[\sum_i \phi_i m^{(i)}(0) w^{(k-i)}(0)\right],\end{aligned} \tag{A18}$$

where

$$\gamma(i) = \int_0^1 m^{(i)}(x)\dot{w}^{(k-i)}(x)\,\mathrm{d}x. \tag{A19}$$

When $\epsilon = 0$ (i.e. $v' = v$) it must be that $\rho' = \rho$ since the two genes are indistinguishable. If ϵ is small but non-zero, a necessary condition for an ESS is that the first-order term in ϵ must be zero, otherwise ϵ could be chosen (either positive or negative) such that $\rho' > \rho$. Thus we have the condition

$$\left[\frac{\partial \rho}{\partial v'}\right]_{v'=v} = 0. \tag{A20}$$

Substituting (A8) and (A9) in (A18) and differentiating gives the result:

$$0 = \rho(1-\rho)\left[\frac{C}{v} + \sum_i \phi_i \left\{\frac{1}{v} + k\left(\tau\theta - \frac{i}{k\rho}\frac{\dot{g}(v)}{g(v)}\right)\right\} g(v)^k \, e^{-nv\theta}\right]$$
$$- \sum_i \phi_i[\partial\gamma(i)/\partial v'] - (1-\rho)\sum_i \frac{\partial\phi_i}{\partial v'} g(v)^k \, e^{-nv\theta}$$
$$- \sum_i \frac{\delta\phi_i}{\delta v'}\int_0^1 [n\theta v(1-\rho) + (k-i)(1-v)\dot{g}(u)/g(u)]g(u)^k \, e^{n\theta v(x-1)} \, dx. \tag{A21}$$

The $\partial\phi_i/\partial v'$ terms are easily disposed of, since we have (using A20):

$$\sum_i \frac{\partial\phi_i}{\partial v'} = \frac{\partial}{\partial v'}\sum_i \phi_i = \frac{\partial}{\partial v'}(1) = 0$$

$$\sum_i \frac{\partial\phi_i}{\partial v'}(k-i) = \frac{\partial}{\partial v'}(k) - k\frac{\partial}{\partial v'}\sum_i\left(\frac{i}{k}\right)$$
$$= -k\frac{\partial\rho}{\partial v'} = 0.$$

The first sum over i in (A21) is easily performed using the definition $\rho = \sum_i \phi_i(i/k)$:

$$0 = \rho(1-\rho)\left[\frac{C}{v} + \left\{\frac{1}{v} + k\left(\tau\theta - \frac{\dot{g}(v)}{g(v)}\right)\right\} g(v)^k \, e^{-nv\theta}\right]$$
$$- \sum_i \phi_i \partial\gamma(i)/\partial v'. \tag{A22}$$

From (A19) we have

$$\frac{\partial\gamma(i)}{\partial v'} = \int_0^1 \left[\frac{\partial m^{(i)}}{\partial v'}\right]\left[\frac{\partial w^{(k-i)}}{\partial x}\right] dx. \tag{A23}$$

Calculating the derivatives from (A8) and (A9) and substituting gives

$$\left[\frac{\partial\gamma(i)}{\partial v'}\right]_{v'=v} = \int_0^1 K(y)\, dy[-n\theta\rho + i\dot{g}/g][n\theta v(1-\rho) + (k-i)(1-v)\dot{g}/g], \tag{A24}$$

where

$$y = 1 - x$$
$$K(y) = yg(u)^k \acute{e}^{-n\theta v y}$$
$$u = 1 - (1-v)y$$

and g and $\dot{g}$ are evaluated at u.

Using the definitions of the mean and variance of i/k:

$$\rho = \sum_i \phi_i(i/k)$$
$$S = \sum_i \phi_i(i/k - \rho)^2 \qquad \text{(A25)}$$

we can perform the second sum over i in (A22), giving

$$-\sum_i \phi_i[\partial\gamma(i)/\partial v']_{v'=v} = \rho(1-\rho)T_1 + ST_2, \qquad \text{(A26)}$$

where

$$T_1 = n^2 \int_0^1 K(y)\,\mathrm{d}y[\theta - \tau^{-1}\dot{g}/g][v\theta + (1-v)\tau^{-1}|\dot{g}/g] \qquad \text{(A27)}$$

$$T_2 = n^2(1-v)\int_0^1 K(y)\,\mathrm{d}y[\tau^{-1}\dot{g}/g]^2. \qquad \text{(A28)}$$

Thus the ESS condition is that the sum of (A26) and the first part of (A22) should be zero.

For the case of Poisson-distributed offspring $g(x) = \exp[\tau(x-1)]$. Thus $\tau^{-1}\dot{g}/g = 1$ and the integrals are easily performed to give

$$C + \{1 - nv(1-\theta)\}Z = [(\alpha/\theta)(1-Z) - nvZ][(1-\theta) - (1-\alpha)F], \qquad \text{(A29)}$$

where

$$\alpha = v\theta/[(1-v) + v\theta]$$

and

$$Z = \exp[-n\{(1-v) + \theta v\}]$$

is the probability that an occupied site will become depopulated, and

$$F = S/[\rho(1-\rho)]. \qquad \text{(A30)}$$

It will be shown below that the equilibrium value of F is independent of the mutant gene frequency ρ. Thus in general $\rho(1-\rho)$ factors out of the ESS condition, so that the ESS is never frequency dependent. Also $S \leq \rho(1-\rho)$ at equilibrium, so F lies between 0 and 1.

3. EQUILIBRIUM VARIANCE IN GENE FREQUENCY (USING $v' = v$)

In the model there are benefits of migration; colonization of empty sites, and competition for occupied sites in which the gene frequency is significantly different from the present site. Therefore the variance in gene frequency is an important factor in determining the ESS migration rate, which is higher if the variance is high. Note that in a stepping-stone model the variance is effectively smaller, due to correlations between neighbouring sites, so the ESS migration rate would be somewhat smaller.

Let P_r be the probability that a previously occupied site is still occupied and has r mutants:

$$P_r = (1 - X)\sum_i \phi_i \sum_{\substack{s>0 \\ j}} B_r^{(k)}(j/s) M_j^{(i)} W_{s-j}^{(k-i)} \tag{A31}$$

and let P_{Nr} be the probability that a previously empty site is occupied and has r mutants:

$$P_{Nr} = (1 - X)\sum_{\substack{s>0 \\ j}} B_r^{(k)}(j/s) I_j' I_{s-j}. \tag{A32}$$

Then the overall probability that a site is both occupied and has r mutants is $fP_r + (1-f)P_{Nr}$; at equilibrium the probability that an occupied site has r mutants is

$$\phi_r' = P_r + [(1-f)/f]P_{Nr}. \tag{A33}$$

In order to derive the new variance S' we calculate the moment generating function of this distribution:

$$\phi'(t) = (1 - X)\sum_{\substack{s>0 \\ j}} \sum_r B_r^{(k)}(j/s)\, e^{rt} R_{sj}, \tag{A34}$$

where

$$R_{sj} = \sum_i \phi_i M_j^{(i)} W_{s-j}^{(k-i)} + \{(1-f)/f\} I_j' I_{s-j}.$$

Using the binomial theorem

$$\sum_r B_r^{(k)}(j/s)\, e^{rt} = \left(\frac{s-j}{j} + \frac{je^t}{s}\right)^k. \tag{A35}$$

Substituting this, and expanding in powers of t, we get (using B7):

$$\begin{aligned} \rho' &= (1 - X)\sum_{\substack{s>0 \\ j}} (j/s) R_{sj} \\ S' + (\rho')^2 &= (1 - X)\sum_{\substack{s>0 \\ j}} \left\{ (j/s) - \frac{k-1}{k}\frac{(s-j)j}{s^2} \right\} R_{sj}. \end{aligned} \tag{A36}$$

The first equation is equivalent to (A13) and at equilibrium we should have $\rho' = \rho$. Substituting this in the second equation gives

$$S' = \rho(1-\rho) - (1 - X)\frac{k-1}{k}\sum_{\substack{s>0 \\ j}} \frac{(s-j)j}{s^2} R_{sj}. \tag{A37}$$

Thus

$$S' = \rho(1-\rho) - (1-X)\frac{k-1}{k}\left[\sum_i \phi_i Q_i + \{(1-f)/f\}Q_N\right], \qquad \text{(A38)}$$

where

$$\begin{aligned} Q_i &= \sum_{s>1}\sum_j \frac{(s-j)j}{s^2} M_j^{(i)} W_{s-j}^{(k-i)} \\ Q_N &= \sum_{s>1}\sum_j \frac{(s-j)j}{s^2} I_j' I_{s-j}. \end{aligned} \qquad \text{(A39)}$$

Using (B8) and the generating functions (A8) and (A9) gives

$$Q_i = \int_0^1 \frac{dy}{y}\int_0^y dx L(x)[nv\theta\rho + i(\dot{g}/g)(1-v)][nv\theta(1-\rho) + (k-i)(\dot{g}/g)(1-v)], \qquad \text{(A40)}$$

where

$$\begin{aligned} u &= v + (1-v)x \\ L(x) &= g(u)^k \, e^{nv\theta(x-1)} x. \end{aligned}$$

As with (A22) this simplifies when summed over i:

$$\sum_i \phi_i Q_i = \rho(1-\rho)A_1 - SA_2, \qquad \text{(A41)}$$

where

$$\begin{aligned} A_1 &= n^2 \int_0^1 \frac{dy}{y}\int_0^y [v\theta + (1-v)\tau^{-1}\dot{g}/g]^2 L(x)\, dx \\ A_2 &= n^2(1-v)^2 \int_0^1 \frac{dy}{y}\int_0^y [\tau^{-1}\dot{g}/g]^2 L(x)\, dx. \end{aligned} \qquad \text{(A42)}$$

A similar process leads to

$$Q_N = \rho(1-\rho)A_3,$$

where

$$A_3 = n^2(v\theta)^2 \int_0^1 \frac{dy}{y}\int_0^y e^{nv\theta(x-1)} x\, dx. \qquad \text{(A43)}$$

We define the function

$$\begin{aligned} H(a) &= a^2 \int_0^1 \frac{dy}{y}\int_0^y x\, dx\, e^{a(x-1)} \\ &= 1 - e^{-a}[1 + Ei(a) - \ln a - \gamma], \end{aligned} \qquad \text{(A44)}$$

where γ is Euler's constant (0.57721 ...) and Ei is a standard exponential integral. $H(a)$ is sigmoid in form, and ranges from 0 to 1 for $0 \leq a < \infty$. Then (A38) becomes

$$F' = 1 - (1 - X)\frac{k-1}{k}[A_1 - FA_2 + \{(1-f)/f\}H(nv\theta)], \tag{A45}$$

where $F = S/[\rho(1-\rho)], F' = S'/[\rho(1-\rho)]$. At equilibrium we have $F' = F$, so

$$F = \frac{1 - (1-X)\dfrac{k-1}{k}[A_1 + \{(1-f/f\}H(nv\theta)]}{1 - (1-X)\dfrac{k-1}{k}A_2}. \tag{A46}$$

Since $\dot{g}/g > 0$, we have $A_1 > A_2$, so $F \leq 1$.

For Poisson-distributed offspring $\dot{g}/g = \tau$ and the integrals can be evaluated to give

$$F = \frac{1 - (1-X)\dfrac{k-1}{k}[H(nv\theta/\alpha) + \{(1-f)/f\}H(nv\theta)]}{1 - (1-X)\dfrac{k-1}{k}(1-\alpha)^2 H(nv\theta/\alpha)}, \tag{A47}$$

where as before $\alpha = v\theta/[(1-v) + v\theta]$. Thus the equilibrium conditions for an ESS consist of (A3), (A29), and (A47).

4. A SPECIAL CASE

We will now demonstrate that the ESS condition reduces to Hamilton and May's (1977) equation (3) in the special case $k = 1, X = 0$. For this case (A47) gives $F = 1$, and hence (A29) becomes

$$\{1 - nv(1-\theta)\}Z = [(\alpha/\theta)(1-Z) - nvZ](\alpha - \theta). \tag{A48}$$

Solving this as a linear equation in Z and simplifying the denominator we get

$$Z = (\alpha - \theta)/[\alpha + \theta(\alpha^{-1} - 1)(1 - nv)]. \tag{A49}$$

Substituting $\alpha = v\theta/[(1-v) + v\theta]$ and multiplying both top and bottom by $v(1 - v + v\theta)$:

$$Z = \frac{v^2\theta - v\theta(1 - v + v\theta)}{v^2\theta + (1-v)(1 - v + v\theta)(1 - nv)}. \tag{A50}$$

Substituting the definitions $Z = \exp[-n(1 - v + v\theta)]$ and $\theta = pf$ we obtain

$$\exp[-n(1 - v + vpf)] = \frac{v^2pf - vpf(1 - v + vpf)}{v^2pf + (1-v)(1 - v + vpf)(1 - nv)}. \tag{A51}$$

This is equivalent to Hamilton and May's equation, disregarding several misprints in their first published version (corrected in Chapter 11 where (A51) appears as equation (3)).

APPENDIX B

The following theorems for generating functions are used in Appendix A. These are standard theorems[6] except for (B5) and (B8).

(B1) If N is the sum of n variates chosen independently from the distribution $\{A_i\}$ then the generating function $c(x)$ of the distribution of N is[7]

$$c(x) = f(a(x)),$$

where $f(x)$ is the g.f. of the distribution of the number of samples n, and $a(x)$ is the g.f. of $\{A_i\}$.

(B2) A particular case of (B1). If the number of samples n is constant then

$$c(x) = [a(x)]^n.$$

(B3) The Poisson distribution with mean τ has generating function

$$f(x) = e^{\tau(x-1)}.$$

(B4) If $a(x)$ is the g.f. of $\{A_i\}$ and $b(x)$ is the g.f. of $\{B_i\}$, then $c(x) = a(x)b(x)$ is the g.f. of the distribution $\{C_i\}$, where

$$C_i = \sum_j A_i B_{j-i}.$$

(B5) If $a(x)$ is the g.f. of $\{A_i\}$ and $b(x)$ is the g.f. of $\{B_i\}$ and

$$P = \sum_{\substack{s,j \\ (s>0)}} A_j B_{s-j} j/s,$$

then

$$P = \int_0^1 (\mathrm{d}a/\mathrm{d}x) b(x)\, \mathrm{d}x.$$

This can be proved using projection operators, which are defined as follows:

$$\hat{T}_i\left(\sum_i a_i x^i\right) = a_j x^j.$$

(B6) If $a(x)$ is the g.f. of $\{A_i\}$ then $a(0) = A_0$ and $a(1) = 1$.

(B7) If $a(x)$ is the g.f. of $\{A_i\}$ then

$$\begin{aligned}\phi(t) &= a(e^t) \\ &= \mu_1 + \frac{\mu_2 t^2}{2!} + \frac{\mu_3 t^3}{3!} + \dots,\end{aligned}$$

where μ_i is the ith moment of $\{A_i\}$ about zero.

(B8) If $a(x)$ is the g.f. of $\{A_i\}$ and $b(x)$ is the g.f. of $\{B_i\}$ and

$$Q = \sum_{\substack{s,j \\ (s>1)}} A_j B_{s-j} j(s-j)/s^2,$$

then

$$Q = \int_0^1 \frac{\mathrm{d}y}{y} \int_0^y (\mathrm{d}a/\mathrm{d}x)(\mathrm{d}b/\mathrm{d}x) x \mathrm{d}x.$$

This can be proved using projection operators in a similar manner to (B5).

APPENDIX C

The analogue of Appendix A for sexual reproduction is most readily approached by considering the $2k$ genes in the gene pool at each site, rather than the k adult organisms. In the full model a number of serious complications arise; for example, all migration involves pairs of genes, and there are correlations between the numbers of immigrant genes of different types. We will therefore only consider the limiting case where the number of offspring per site is extremely large. For asexual reproduction (and provided $v \neq 1$, $p \neq 0$) (A3), (A29), and (A47) reduce in this limit to:

$$\begin{aligned} f &= 1 - X \\ C &= (\alpha/\theta)[(1-\theta) - (1-\alpha)F] \\ F &= [k - (k-1)(1-X)(1-\alpha)^2]^{-1}. \end{aligned} \tag{C1}$$

Suppose that reproduction is sexual, and that there are two genotypes: wild-type with migration probability v and mutant with migration probability $v + \epsilon$. Heterozygotes are supposed to have migration probability $v + \epsilon/2$.

The expected numbers of wild type and mutant immigrant genes at a site are respectively:[8]

$$\begin{aligned} W: \quad & kfp \sum_i \phi_i \{2(1 - i/2k)^2 v + 2(i/2k)(1 - i/2k)(v + \epsilon/2)\} \\ & = k\theta\{2v(1-\rho) + \epsilon(\rho(1-\rho) - S)\} \end{aligned}$$

$$\begin{aligned} M: \quad & kpf \sum_i \phi_i \{2(i/2k)^2 (v + \epsilon) + 2(i/2k)(1 - i/2k)(v + \epsilon/2)\} \\ & = k\theta\{2v\rho + \epsilon(\rho(1+\rho) + S)\}, \end{aligned} \tag{C2}$$

where

$$\rho = \sum_i (i/2k)\phi_i$$
$$S = \sum_i (i/2k - \rho)^2 \phi_i.$$

The expected numbers of sedentary competitors (genes) at a site which previously had i out of $2k$ mutant genes are

$$W: \quad k[2(1 - i/2k)^2(1 - v) + 2(i/2k)(1 - i/2k)(1 - v - \epsilon/2)]$$
$$M: \quad k[2(i/2k)^2(1 - v - \epsilon) + 2(i/2k)(1 - i/2k)(1 - v - \epsilon/2)] \qquad (C3)$$

Since the number of offspring is extremely large the probability of any gene in the next generation being a mutant is given by the ratio of the expected total number of mutant genes competing to the expected total of both types. Letting $\eta = i/2k$ be the previous fraction of mutants and η' the expected new fraction, we obtain

$$\eta' = (A + B\epsilon)/(E + D\epsilon), \qquad (C4)$$

where

$$A = (1 - v)\eta + \theta v \rho$$
$$B = \tfrac{1}{2}[-\eta(1 + \eta) + \theta\{S + \rho(1 + \rho)\}]$$
$$E = (1 - v) + v\theta$$
$$D = -\eta + \theta\rho.$$

To first order in ϵ this is

$$\eta' = A/E + \epsilon[BE - AD]/E^2. \qquad (C5)$$

Note that the denominator E is independent of η. Thus to obtain the average value of η' for all previously occupied sites we sum (C5) over i to get

$$\langle\eta'\rangle_{\mathrm{OCC}} = \rho + \epsilon G/[(1 - v) + \theta v], \qquad (C6)$$

where

$$G = S\{(1 - \alpha) - \tfrac{1}{2}(1 - \theta)\} - \tfrac{1}{2}(1 - \theta)\rho(1 - \rho).$$

For previously empty sites we obtain the analogous result

$$\langle\eta'\rangle_{\mathrm{EMPTY}} = \rho + \epsilon[S + \rho(1 - \rho)]/2v. \qquad (C7)$$

The ESS condition is that $(1 - X)$ times the ϵ term in (C6) plus X times the ϵ term in (C7) should be zero. This can be rewritten as

$$F = \frac{\{1 - \theta(1 + C/\alpha)\}}{2(1 - \alpha) - \{1 - \theta(1 + C/\alpha)\}}, \qquad (C8)$$

where

$$C = X/(1 - X)$$
$$F = S/\{\rho(1 - \rho)\}.$$

As in the asexual case we require the equilibrium genetic variance S (calculated for $\epsilon = 0$). We observe that the first term in (C5) is

$$\begin{aligned}\eta' &= A/E = (1 - \alpha)\eta + \alpha\rho \\ &= \eta + \alpha(\rho - \eta).\end{aligned} \tag{C9}$$

This formula, together with the number of genes selected ($2k$) and $\langle\eta'\rangle_{\text{EMPTY}} = \rho$ from (C7), determines the binomial choice of the next generation gene pool, and thus the genetic variance of the next generation. However it is found that the identical formulae are obtained for the asexual case with $2k$ adults per site in the limit $n \to \infty$, so the sexual genetic variance is simply given by using $2k$ instead of k in the third of equations (C1):

$$F = [2k - (2k - 1)(1 - X)(1 - \alpha)^2]^{-1}. \tag{C10}$$

Thus the ESS is given by (C8) and (C10).

If we define $z = k + \frac{1}{2}$, then the ESS condition can be written

$$1 - \alpha = \{1 - \theta(1 + C/\alpha)\}[z - (z - 1)(1 - X)(1 - \alpha)^2]. \tag{C11}$$

Except for the substitution $k \to z$ this equation is equivalent to (C1). Thus, in the $n \to \infty$ limit, the effect of sexual reproduction is formally equivalent to an increase of $\frac{1}{2}$ in the number of adults in each site.

Epilogue

THE last paper, being so little mine, perhaps should be considered more an Appendix to this Volume 1 of my papers; in any case it came appropriately at the end. It is obviously no 'narrow road' of my own. My involvement with two (!) co-authors is untypical likewise of the earlier behaviour and writings of that solitary bear-like animal who began the 'dispersal act' from his tribe that is represented in the volume that now ends. For some time, it seems, this animal had been feeling a wind of change, a degree of acceptance, less need to shrink from shadows either of himself or possible future companions. Who can refuse to call Science a line of work, which, even if only at the hands of a superior analyst, grasps Euler's constant and the equations of the last chapter—to say nothing of Bessels and elliptics moulded in the same friendly hands and to appear later? Who can deny prediction in the face of the tests of theory through sex ratio that have now been shown (pages 132 and 141), or call pointless the simple, three-term equation of an ailing man as it is now unfurled and used (see notes to Chapter 8)?

In the early days of evolution applied to humans a rival notion was floated, and a book written I think, claiming that the family Ursidae, or bears—not any of the great apes—was the group closest to *Homo*. I wonder if this weird idea reflected not so much the logic of bones as the style and the sociability of the theorist. I could well sympathize. And I wonder whether, as his theory became more and more rejected, this man stood out, continually more isolated and angry, as its best example. If I had found myself the crank that I feared, how easily I might have joined him. Instead, in new valleys after this 'dispersal', my Furies asleep, and my 'confession to murder' seemingly accepted and forgotten, I drew in a little my savage muzzle, sloughed some winter fur; shortly as I took another path I began even to imagine myself human.

Acknowledgements

We thank the publishers for permission to reproduce the copyright material listed below:

CHAPTER 1

'The evolution of altruistic behaviour', *The American Naturalist*, Vol. 97, pp. 354–6 (1963). © 1963 University of Chicago Press

CHAPTER 2

'The genetical evolution of social behaviour I and II', *Journal of Theoretical Biology*, Vol. 7, pp. 1–16 and 17–52 (1964). © 1964 Academic Press Ltd. 'Addendum' in G. C. Williams, *Group Selection*, pp. 87–9 (Aldine-Atherton, Chicago 1971).

CHAPTER 3

'The moulding of senescence by natural selection', *Journal of Theoretical Biology*, Vol. 12, pp. 12–45 (1966). © 1966 Academic Press Ltd.

CHAPTER 4

'Extraordinary sex ratios'. *Science*, Vol. 156, pp. 477–88 (1967). © 1967 American Association for the Advancement of Science.

CHAPTER 5

'Selfish and spiteful behavior in an evolutionary model', *Nature*, Vol. 228, pp. 1218–20 (1970). © 1970 Macmillan Magazines Ltd.

CHAPTER 6

'Selection of selfish and altruistic behavior in some extreme models', Ch. 2 in J. F. Eisenberg and W. S. Dillon (eds), *Man and Beast: Comparative Social Behavior*, pp. 57–91 (Smithsonian Press, Washington, DC, 1971). 'Population control' [letter], *New Scientist*, 30 October (1969).

CHAPTER 7

'Geometry for the selfish herd', *Journal of Theoretical Biology*, Vol. 31, pp. 295–311 (1971). © 1971 Academic Press Ltd.

CHAPTER 8

'Altruism and related phenomena, mainly in social insects', *Annual Review of Ecology and Systematics*, Vol. 3, pp. 193–232 (1972). © 1972 Annual Reviews Inc.

CHAPTER 9

'Innate social aptitudes of man: an approach from evolutionary genetics', in R. Fox (ed.). *ASA Studies 4: Biosocial Anthropology*, pp. 133–53 (Malaby Press, London, 1975).

CHAPTER 10

'Gamblers since life began: barnacles, aphids, elms' [review], *The Quarterly Review of Biology*, Vol. 50, pp. 175–80 (1975). © 1975 University of Chicago Press.

CHAPTER 11

'Dispersal in stable habitats', *Nature*, Vol. 269, pp. 578–81 (1977). © 1977 Macmillan Magazines Ltd.

CHAPTER 12

'Evolution and diversity under bark', Ch. 10 in L. A. Mound and N. Waloff (eds), *Diversity of Insect Faunas*, Symposia of the Royal Entomological Society of London No. 9, pp. 154–75 (Blackwell Scientific, Oxford, 1978).

CHAPTER 13

'Wingless and fighting males in fig wasps and other insects', in M. S. Blum and N. A. Blum (eds), *Reproductive Competition, Mate Choice, and Sexual Selection in Insects*, pp. 167–220 (Academic Press, New York, 1979),
© 1979 Academic Press Inc.

CHAPTER 14

'Low nutritive quality as defence against herbivores', *Journal of Theoretical Biology*, Vol. 86, pp. 247–54 (1980). © 1980 Academic Press Ltd.

CHAPTER 15

'Evolutionarily stable dispersal strategies', *Journal of Theoretical Biology*, Vol. 82, pp. 205–30 (1980). © 1980 Academic Press Ltd.

Chapter frontispieces

Photographs by the author except as follows:

Chapter 1: Provided by Dr. D. W. Macdonald, Department of Zoology, Oxford University. For further details see D. W. Macdonald, *The Velvet Claw* (BBC Books, London, 1992).
Chapter 2: Location thanks to Cristina Lorenzi.
Chapter 3: Provided by Frank Kierman, Yerkes Regional Primate Center.
Chapter 5: Taken from *Fairy Tales of the Brothers Grimm* (illus by Arthur Rackham), Freemantle & Co, 1900.
Chapter 6: Hephaestus and Thetis. Hephaestus making new armour for Achilles at the bidding of his mother, Thetis. From a red-figure amphora, *c* 480 BC. Boston Museum of Fine Arts. This illustration courtesy The Mansell Collection, London.
Chapter 7: Location and animals thanks to Jeremy Railton.
Chapter 10: Recomposed from W. Giesbrecht, Systematik und Faunistik der Pelagischen Copepoden des Golfes von Neapel und der angrenzenden Meeresabschnitte. *Fauna und Flora des Golfes von Neapel* **19** 1–831 (1892).
Chapter 15: Recomposed from M. Cornu, *Mémoires de l'Academie des Sciences de l'Institut de France* **26** 1–357 (1878).

Name Index

Names are indexed if present in the text or notes; they are not indexed if mentioned only in the tables, figures, captions to these, or references

Subject Index

Taxon names are indexed if present in the text or notes; they are not indexed if mentioned only in the tables, figures, captions to these, or references

A

D